AF595326

# Recycled Materials in Civil and Environmental Engineering

# Recycled Materials in Civil and Environmental Engineering

Editors

**Andrea Petrella**
**Michele Notarnicola**

MDPI • Basel • Beijing • Wuhan • Barcelona • Belgrade • Manchester • Tokyo • Cluj • Tianjin

*Editors*
Andrea Petrella
Department of Civil, Environmental, Land, Building Engineering and Chemistry
Polytechnic University of Bari
Bari
Italy

Michele Notarnicola
Department of Civil, Environmental, Land, Building Engineering and Chemistry
Polytechnic University of Bari
Bari
Italy

*Editorial Office*
MDPI
St. Alban-Anlage 66
4052 Basel, Switzerland

This is a reprint of articles from the Special Issue published online in the open access journal *Materials* (ISSN 1996-1944) (available at: www.mdpi.com/journal/materials/special_issues/Recyc_Material_Envir_Eng).

For citation purposes, cite each article independently as indicated on the article page online and as indicated below:

LastName, A.A.; LastName, B.B.; LastName, C.C. Article Title. *Journal Name* **Year**, *Volume Number*, Page Range.

**ISBN 978-3-0365-5234-7 (Hbk)**
**ISBN 978-3-0365-5233-0 (PDF)**

# Contents

# About the Editors

**Andrea Petrella**

Andrea Petrella (Associate Professor)

Andrea Petrella was born in Bari in 1976. He graduated with an M.Sc. in Chemistry at the University of Bari in July 2001 and received his Ph.D in Chemical Sciences in 2005 at the University of Bari. In 2007, he became the Assistant Professor in Materials Science and Technology at the Polytechnic University of Bari. Since March 2022, he has been an Associate Professor in Materials Science and Technology at the Polytechnic University of Bari.

He is an author of more than 100 papers in refereed scientific journals and conference proceedings (SCOPUS: h-index 23, citations 1463). His research fields are summarized as follows: use of recycling organic and inorganic materials in the building trade and/or in the removal of heavy metals present in wastewater; photocatalytic materials for the degradation of bio-persistent pollutants in water and wastewater; and nanocomposites for energy conversion and for novel optical devices.

**Michele Notarnicola**

Michele Notarnicola (Full Professor)

Michele Notarnicola was born in Foggia (Italy) in 1965, graduated in Mechanical Engineering at the Technical University of Bari in 1991. From 1992 to 1995, he was project engineer of Public Consult SpA and Waste Management Italia SpA, Italian private firms involved in design, construction, and the management of waste and wastewater treatment plants. Since 1995, he has been with the Technical University of Bari, Dept. of Environmental Engineering and Sustainable Development (DIASS), now Department of Civil, Environmental, Building Engineering and Chemistry (DICATECh), with the present position of Full Professor of Material Science and Technology.

He is an author of more than 200 papers in refereed scientific journals and conference proceedings (SCOPUS: h-index 25, citations 2815), with major research interests in the fields of solid waste management (inertization treatment of hazardous waste and reuse of not-hazardous waste in construction materials), characterization, and remediation of contaminated sites, as well as wastewater technology (advanced treatment and disinfection of municipal wastewater for agricultural or industrial reuse).

He is scientific coordinator of local research units in international and national R&D projects and consultant for public administrations, authorities and major corporations in the environmental field.

# Preface to "Recycled Materials in Civil and Environmental Engineering"

Waste represents a huge reserve of resources that, after appropriate management, can guarantee a sustainable and continuous supply of materials and energy over the years.

Waste management includes the activities aimed at managing the entire waste process, which involves the collection, transport, treatment (recovery or disposal), reuse, and recycling of waste materials in order to reduce the effects on human health and the impact on the environment.

The proper management of hazardous and not-hazardous waste deriving from civil and industrial operations is the basis of the principles of circular economy that the European Union has indicated in specific regulations and directives devoted to the control of the entire waste cycle, from production to disposal, with specific attention to recovery and recycling.

The aim of this book was to report recent innovative studies based on the use of secondary raw materials for applications in civil and environmental engineering. To this purpose, papers were related to the preparation of innovative construction materials and to the treatment of wastes for environmental applications. The investigations were characterized by a common purpose, i.e., finding a way to reduce the amount of waste generated, thus reducing the need for landfilling and optimizing the values of these novel materials, which are an abundant resource that can be easily reused for different applications.

**Andrea Petrella and Michele Notarnicola**
*Editors*

Editorial

# Recycled Materials in Civil and Environmental Engineering

**Andrea Petrella *** **and Michele Notarnicola**

Department of Civil, Environmental, Land, Building Engineering and Chemistry, Polytechnic University of Bari, Via E. Orabona, 4, 70125 Bari, Italy; michele.notarnicola@poliba.it
* Correspondence: andrea.petrella@poliba.it; Tel.: +39-0805963275

**Citation:** Petrella, A.; Notarnicola, M. Recycled Materials in Civil and Environmental Engineering. *Materials* **2022**, *15*, 3955. https://doi.org/10.3390/ma15113955

Received: 22 May 2022
Accepted: 30 May 2022
Published: 2 June 2022



## 1. Introduction

Waste represents a huge reserve of resources that, after appropriate management, can guarantee a sustainable and continuous supply of materials and energy over the years [1–4].

Waste management includes the activities aimed at managing the entire waste process which involves the collection, transport, treatment (recovery or disposal), reuse, and recycling of waste materials in order to reduce the effects on human health and the impact on the environment [5–7].

The proper management of hazardous and non-hazardous waste deriving from civil and industrial operations is the basis of the principles of Circular Economy that the European Union has indicated in specific Regulations and Directives devoted to the control of the entire waste cycle, from production to disposal, with specific attention to recovery and recycling.

In this respect, the aim of this collection of papers was to report recent innovative studies based on the use of secondary raw materials for applications in Civil and Environmental Engineering.

For the purpose, eleven papers were related to the preparation of innovative construction materials, whereas six papers were related to the treatment of wastes for environmental applications.

The investigations were characterized by a common purpose, i.e., to find a way to reduce the amount of waste generated, thus reducing the need for landfilling and optimising the values of these novel materials which become an abundant resource that can be easily reused for different applications.

## 2. Recycled Waste for Construction Materials

In the papers by Petrella and coworkers [8,9], industrial and agricultural by-products such as end-of-life tire rubber and wheat straw were used as aggregates for the production of unconventional cement mortars prepared by a cheap and environmentally safe process. The artifacts resulted in thermal insulating with respect to the sand-based references, while the presence of another aggregate as perlite was used to improve the mechanical strengths with no detrimental effects on the thermal conductivities due to the porous nature of the inorganic material [10,11]. The conglomerates with end-of-life tire rubber also showed hydrophobic behavior due to the low water absorption, while the conglomerates with wheat straw showed high impact resistance and acoustic absorption. Non-structural insulating products, specifically for indoor applications, may be a possible use of these composites.

Tsai and coworkers [12] presented a trend analysis on the generation and treatment of industrial waste in the 2010–2020 period, with specific reference on promotion policies and regulatory measures for mandatory renewable resources from industrial sources in Taiwan. To the purpose, the renewable resources considered in the present study were reclaimed asphalt pavement (RAP) material, water-quenched blast furnace slag, and ilmenite chlorination furnace slag for the production of recycled aggregate materials in concrete and construction applications.

The paper by Stevulova and coworkers [13] was focused to study the properties of sustainable cement mortars with cellulose fibers derived from waste paper recycling operations. This represents an economic, technically feasible, and ecological approach to waste management for indoor applications. Moreover, the study showed the benefits of these innovative composites for providing healthy living solutions, due to the cellulose aggregate property to regulate humidity inside buildings.

The paper by Dilbas and coworkers [14] showed the influence of treatment and mixing methods on recycled aggregate concretes (RAC) obtained after various techniques. Specifically, a statistical study was carried-out after evaluation of the compressive strength test results of concretes designed with Absolute Volume Method (AVM) and Equivalent Mortar Volume Method (EMV) and containing natural aggregate (NA), recycled aggregate (RA), and silica fume (SF).

The paper by Karimipour and coworkers [15] studied steel-concrete-steel (SCS) sandwich panels based on two thin high-strength steel plates separated by a core made of low-density and low-strength thick concrete. The panels were characterized by a new stud-bolt connector able to regulate its shear behaviour, and the bolts' diameter, concrete core's thickness, and bolts' spacing were analysed. Furthermore, the core was realised with normal-strength concrete and steel fibre concrete (SFC) and with recycled coarse aggregate. It was observed that increasing the stud-bolts diameter or using SFC improved the shear strength and ductility ratio of the sandwich panels.

The paper by Pederneiras and coworkers [16] studied the physical and mechanical properties of environmentally friendly and low-cost products based on coir fibers in rendering mortars with two different binders. It was observed that the addition of coir fibers led to improvements in the cracking behavior and that the volume fraction of the fibers and the binder used were the main factors which improved the brittleness of the conglomerate. A cement-lime binder together with higher fiber content determined the highest fracture toughness, thus showing the potential feasibility of these fibers for reinforcement of cement and cement-air-lime mortars.

In the work by Poulose et al. [17], HDPE/strontium aluminate-based auto glowing composites ($SrAl_2O_4$: Eu, Dy (AG1) and $Sr_4Al_{14}O_{25}$: Eu, Dy (AG2)) were prepared and the phosphorescence, thermal, mechanical and rheological characteristics were analysed. Due to the better mechanical characteristics and long afterglow time of these HDPE composites with the addition of the AG1 and AG2 fillers, applications in roadway nighttime displays, fluorescent lamps, etc. can be considered.

In the paper by Rocha et al. [18], waste tire steel fibers (WTSF) were used for the production of soil-cement blocks. Specifically, threefold mixtures with 10% by weight of Portland cement and with 0%, 0.75%, and 1.5% volumetric additions of WTSF were prepared and characterized. The soil-cement blocks with 1.5% WTSF showed an average compressive strength increase of 20%. It was observed that all the results of the investigated samples were in agreement with the minimum requirements of the different standards considered.

The paper by Deb et al. [19] was devoted to the use of recycled brick aggregates in the preparation of pervious concrete (PC) which were characterized by mechanical tests together with an analysis of pore structure distribution. Replacement ratio and particle size were investigated on structural performance and pore feature through laboratory testing and image processing techniques.

In the paper by Khan et al. [20], the compressive strength of recycled aggregate concrete was predicted by the use of machine learning techniques like decision tree and two approaches as gradient boosting and bagging regressor. Correlation coefficients and statistical tests were carried out to evaluate the performance of each method and the validity was confirmed by k-fold evaluation and error dispersals.

## 3. Recycled Materials for Environmental Applications

In the paper by Valentukeviciene and coworker [21], sorbents from agricultural and cement manufacturing waste were used to remove oil residues in wastewater with a turbidity removal efficiency of 64%, and color removal efficiency of 56%, and total iron concentration removal ~68%. Moreover, straw sorbent oil adsorption capacity was up to 33 mg/g, peat sorbent 37 mg/g, and mineral sorbent 1.83 mg/g, demonstrating the effectiveness of these recycled sorbents as environmentally sustainable materials for wastewater treatments.

In the review by Borjan and coworkers [22] different existing recycling technologies for carbon fiber-reinforced composites were presented with advantages and drawbacks. This is because the production of these materials has dramatically increased in the last years together with the quantity of waste. Specifically, the study was focused on chemical recycling using sub- and supercritical fluids which were demonstrated to be more efficient systems to recover clean fibers characterized by good mechanical properties.

In the paper by Medynska-Juraszek and coworkers [23], wheat straw, sunflower husk, and pine-chip biochars produced by torrefaction/pyrolysis conditions were chemically modified with ethanol and HCl for applications as $Na^+$ sorbents from aqueous solution. Wheat straw biochar, obtained by pyrolysis (550 °C), showed the best $Na^+$ sorption capacity. The pre-treatments with EtOH and HCl showed an improvement in sorption capacity in biochars produced through torrefaction (<300 °C). Moreover, the biochars produced by torrefaction may be used for Na+ removal with lower energy demand and carbon footprint by ethanol treatment, resulting in a very promising method for applications in Environmental Engineering.

In the review by Amari and coworkers [24] different types of clay-polymers nanocomposites (CPNs) were studied together with their efficiency in the removal of different pollutants from water/wastewater by flocculation and adsorption. Specifically, the ability of these novel and cost-effective adsorbents to remove bacteria, metals, phenol, tannic acid, pesticides, dyes, etc., was reported. The adsorption capacity was higher than bare clay and polymer adsorbents together with a better regeneration.

In the paper by Irtiseva et al. [25], three granular sorbents from secondary raw materials were produced and studied for the sorption of diesel fuel and motor oil. Specifically, homogenised peat, waste tyre rubber, and fly ash cenospheres were used. The granules' morphology, porosity, mechanical properties, and sorption kinetics were analyzed. Moreover, the hydrophobicity and the sorption capacity of the pellets were determined according to the type of oil product.

In the review by Loffredo [26] new technologies of biomass conversion were developed with the primary goal of recycling biowaste and producing bioenergy. The co-products and by-products of these processes such as biochar, hydrochar, and digestate showed physicochemical properties that suggested their use as biosorbents of phenolic and steroidal environmental estrogens in wastewater treatment and soil remediation. The review presented recent advances in the characterization of these materials and qualitative and quantitative aspects of the adsorption/desorption of these pollutants, including data modeling.

**Funding:** This research received no external funding.

**Acknowledgments:** We would like to thank all the authors and the reviewers. Special acknowledgments to all the staff of the Materials Editorial Office for their great support during the preparation of this Special Issue.

**Conflicts of Interest:** The authors declare no conflict of interest.

## References

1. Nanda, S.; Berruti, F. Municipal solid waste management and landfilling technologies: A review. *Environ. Chem. Lett.* **2021**, *19*, 1433–1456. [CrossRef]
2. Zhang, C.; Hu, M.; Di Maio, F.; Sprecher, B.; Yang, X.; Tukker, A. An overview of the waste hierarchy framework for analyzing the circularity in construction and demolition waste management in Europe. *Sci. Total Environ.* **2022**, *803*, 149892. [CrossRef] [PubMed]
3. Salmenperä, H.; Pitkänen, K.; Kautto, P.; Saikku, L. Critical factors for enhancing the circular economy in waste management. *J. Clean. Prod.* **2021**, *280*, 124339. [CrossRef]

4. Prajapati, P.; Varjani, S.; Singhania, R.R.; Patel, A.K.; Awasthi, M.K.; Sindhu, R.; Zhang, Z.; Binod, P.; Awasthy, S.K.; Chaturvedi, P. Critical review on technological advancements for effective waste management of municipal solid waste—Updates and way forward. *Environ. Technol. Innov.* **2021**, *23*, 101749. [CrossRef]
5. Rizzi, V.; D'Agostino, F.; Gubitosa, J.; Fini, P.; Petrella, A.; Agostiano, A.; Semeraro, P.; Cosma, P. An alternative use of olive pomace as a wide-ranging bioremediation strategy to adsorb and recover disperse orange and disperse red industrial dyes from wastewater. *Separations* **2017**, *4*, 29. [CrossRef]
6. Petrella, A.; Petrella, M.; Boghetich, G.; Basile, T.; Petruzzelli, V.; Petruzzelli, D. Heavy metals retention on recycled waste glass from solid wastes sorting operations: A comparative study among different metal species. *Ind. Eng. Chem. Res.* **2012**, *51*, 119–125. [CrossRef]
7. Ranieri, E.; Tursi, A.; Giuliano, S.; Spagnolo, V.; Ranieri, A.C.; Petrella, A. Phytoextraction from chromium-contaminated soil using Moso Bamboo in mediterranean conditions. *Water Air Soil Poll.* **2020**, *231*, 408. [CrossRef]
8. Petrella, A.; Notarnicola, M. Lightweight cement conglomerates based on end-of-life tire rubber: Effect of the grain size, dosage and addition of perlite on the physical and mechanical properties. *Materials* **2021**, *14*, 225. [CrossRef]
9. Petrella, A.; De Gisi, S.; Di Clemente, M.E.; Todaro, F.; Ayr, U.; Liuzzi, S.; Dobiszewska, M.; Notarnicola, M. Experimental investigation on environmentally sustainable cement composites based on wheat straw and perlite. *Materials* **2022**, *15*, 453. [CrossRef]
10. Petrella, A.; Petruzzelli, V.; Basile, T.; Petrella, M.; Boghetich, G.; Petruzzelli, D. Recycled porous glass from municipal/industrial solid wastes sorting operations as a lead ion sorbent from wastewaters. *React. Funct. Polym.* **2010**, *70*, 203–209. [CrossRef]
11. Petrella, A.; Petrella, M.; Boghetich, G.; Petruzzelli, D.; Calabrese, D.; Stefanizzi, P.; De Napoli, D.; Guastamacchia, M. Recycled waste glass as aggregate for lightweight concrete. *Proc. Inst. Civ. Eng. Constr. Mater.* **2007**, *160*, 165–170. [CrossRef]
12. Tsai, C.H.; Shen, Y.H.; Tsai, W.T. Reuse of the materials recycled from renewable resources in the civil engineering: Status, achievements and government's initiatives in Taiwan. *Materials* **2021**, *14*, 3730. [CrossRef] [PubMed]
13. Stevulova, N.; Vaclavik, V.; Hospodarova, V.; Dvorský, T. Recycled cellulose fiber reinforced plaster. *Materials* **2021**, *14*, 2986. [CrossRef] [PubMed]
14. Dilbas, H.; Güneş, M.Ş. Mineral addition and mixing methods effect on recycled aggregate concrete. *Materials* **2021**, *14*, 907. [CrossRef]
15. Karimipour, A.; Ghalehnovi, M.; Golmohammadi, M.; De Brito, J. Experimental investigation on the shear behaviour of stud-bolt connectors of steel-concrete-steel fibre-reinforced recycled aggregates sandwich panels. *Materials* **2021**, *14*, 5185. [CrossRef]
16. Pederneiras, C.M.; Veiga, R.; de Brito, J. Physical and mechanical performance of coir fiber-reinforced rendering mortars. *Materials* **2021**, *14*, 823. [CrossRef]
17. Poulose, A.M.; Shaikh, H.; Anis, A.; Alhamidi, A.; Kumar, N.S.; Elnour, A.Y.; Al-Zahrani, S.M. Long persistent luminescent hdpe composites with strontium aluminate and their phosphorescence, thermal, mechanical, and rheological characteristics. *Materials* **2022**, *15*, 1142. [CrossRef]
18. Rocha, J.H.A.; Galarza, F.P.; Chileno, N.G.C.; Rosas, M.H.; Peñaranda, S.P.; Diaz, L.L.; Abasto, R.P. Compressive strength assessment of soil–cement blocks incorporated with waste tire steel fiber. *Materials* **2022**, *15*, 1777. [CrossRef]
19. Deb, P.; Debnath, B.; Hasan, M.; Alqarni, A.S.; Alaskar, A.; Alsabhan, A.H.; Khan, M.A.; Alam, S.; Hashim, K.S. Development of eco-friendly concrete mix using recycled aggregates: Structural performance and pore feature study using image analysis. *Materials* **2022**, *15*, 2953. [CrossRef]
20. Khan, K.; Ahmad, W.; Amin, M.N.; Aslam, F.; Ahmad, A.; Al-Faiad, M.A. Comparison of Prediction Models Based on Machine Learning for the Compressive Strength Estimation of Recycled Aggregate Concrete. *Materials* **2022**, *15*, 3430. [CrossRef]
21. Valentukeviciene, M.; Zurauskiene, R. Investigating the effectiveness of recycled agricultural and cement manufacturing waste materials used in oil sorption. *Materials* **2022**, *15*, 218. [CrossRef] [PubMed]
22. Borjan, D.; Knez, Ž.; Knez, M. Recycling of carbon fiber-reinforced composites—Difficulties and future perspectives. *Materials* **2021**, *14*, 4191. [CrossRef] [PubMed]
23. Medyńska-Juraszek, A.; Álvarez, M.L.; Białowiec, A.; Jerzykiewicz, M. Characterization and sodium cations sorption capacity of chemically modified biochars produced from agricultural and forestry wastes. *Materials* **2021**, *14*, 4714. [CrossRef]
24. Amari, A.; Alzahrani, F.M.; Mohammedsaleh Katubi, K.; Alsaiari, N.S.; Tahoon, M.A.; Rebah, F.B. Clay-polymer nanocomposites: Preparations and utilization for pollutants removal. *Materials* **2021**, *14*, 1365. [CrossRef]
25. Irtiseva, K.; Mosina, M.; Tumilovica, A.; Lapkovskis, V.; Mironovs, V.; Ozolins, J.; Stepanova, V.; Shishkin, A. Application of granular biocomposites based on homogenised peat for absorption of oil products. *Materials* **2022**, *15*, 1306. [CrossRef]
26. Loffredo, E. Recent advances on innovative materials from biowaste recycling for the removal of environmental estrogens from water and soil. *Materials* **2022**, *15*, 1894. [CrossRef]

 materials MDPI

Article

# Lightweight Cement Conglomerates Based on End-of-Life Tire Rubber: Effect of the Grain Size, Dosage and Addition of Perlite on the Physical and Mechanical Properties

**Andrea Petrella * and Michele Notarnicola**

Department of Civil, Environmental, Land, Building Engineering and Chemistry, Polytechnic University of Bari, Via E. Orabona, 4, 70125 Bari, Italy; michele.notarnicola@poliba.it
* Correspondence: andrea.petrella@poliba.it; Tel.: +39(0)8-0596-3275; Fax: +39(0)8-0596-3635

**Abstract:** Lightweight cement mortars containing end-of-life tire rubber (TR) as aggregate were prepared and characterized by rheological, thermal, mechanical, microstructural, and wetting tests. The mixtures were obtained after total replacement of the conventional sand aggregate with untreated TR with different grain sizes (0–2 mm and 2–4 mm) and distributions (25%, 32%, and 40% by weight). The mortars showed lower thermal conductivities (≈90%) with respect to the sand reference due to the differences in the conductivities of the two phases associated with the low density of the aggregates and, to a minor extent, to the lack of adhesion of tire to the cement paste (evidenced by microstructural detection). In this respect, a decrease of the thermal conductivities was observed with the increase of the TR weight percentage together with a decrease of fluidity of the fresh mixture and a decrease of the mechanical strengths. The addition of expanded perlite (P, 0–1 mm grain size) to the mixture allowed us to obtain mortars with an improvement of the mechanical strengths and negligible modification of the thermal properties. Moreover, in this case, a decrease of the thermal conductivities was observed with the increase of the P/TR dosage together with a decrease of fluidity and of the mechanical strengths. TR mortars showed discrete cracks after failure without separation of the two parts of the specimens, and similar results were observed in the case of the perlite/TR samples thanks to the rubber particles bridging the crack faces. The super-elastic properties of the specimens were also observed in the impact compression tests in which the best performances of the tire and P/TR composites were evidenced by a deep groove before complete failure. Moreover, these mortars showed very low water penetration through the surface and also through the bulk of the samples thanks to the hydrophobic nature of the end-of-life aggregate, which makes these environmentally sustainable materials suitable for indoor and outdoor elements.

**Keywords:** cement conglomerates; end-of-life tire rubber; perlite; thermal insulation; mechanical strength

**Citation:** Petrella, A.; Notarnicola, M. Lightweight Cement Conglomerates Based on End-of-Life Tire Rubber: Effect of the Grain Size, Dosage and Addition of Perlite on the Physical and Mechanical Properties. *Materials* **2021**, *14*, 225. https://doi.org/10.3390/ma14010225

Received: 7 December 2020
Accepted: 31 December 2020
Published: 5 January 2021



## 1. Introduction

Industrial waste recycling and reuse are considered important issues to face the need for a more sustainable and environmentally friendly building trade in order to obtain an appropriate management of a large quantity of by-products such as agro-food waste [1], plastics [2], batteries [3], municipal solid waste [4], and glass [5–7]. Indeed, the construction industry has an extensive impact on raw materials consumption and waste production; accordingly, the reuse and the conversion of a waste into a new resource (recycling operation) is fundamental to increasing the sustainability of a product, the so-called secondary raw material that can be re-used in the construction industry [8–11] or in environmental applications [12–15].

In this respect, the large amount produced and the disposal of abandoned waste tires represent fundamental issues to be addressed from an environmental point of view [16–19].

Tire rubber shows biopersistence, chemical inertia, resistance to organic agents (mold and bacteria), temperature changes, and atmospheric agents; accordingly, the performances of this material do not decrease over time. For this reason, tire stockpiles can generate health and safety risks through air, water, and soil pollution, and thus tire burning can represent an easy and cheap solution for the management of the accumulated rubber, although substantial pollution in the air, ground, and surface water can occur [20,21].

For this reason, the conversion of this waste into a new resource through recycling operations is an alternative to incineration and landfilling in the sustainable tire rubber management.

Recycled tire rubber can be considered a valuable resource because many properties tend to be unchanged after its transformation, which can take place through different processes. The most common operation is the mechanical grinding, carried out at room temperature, the result of which is represented by the rubber granulate and powder [22,23]. The granulate recovery is characterized by tire shredding and chipping, by which small pieces of different sizes (crumb rubber: 0.1–5 mm, chips: 15–75 mm, and shreds: 25–450 mm) are obtained.

This material shows elastic properties [24,25], acoustic absorption [26], and has a good thermal resistance [27]. Applications of recycled tire rubber include asphalt conglomerates [9,18], thermo-acoustic insulators [26,27], geotechnical systems and structures [28], waterproofing and absorbing vibrations materials [23,29], energy production [30], and playground equipment, all of which have been proven to be effective in protecting the environment and preserving natural resources.

Several studies have been carried out on the incorporation of waste tire rubber as aggregate in cement conglomerates, which were characterized by investigations on the physico-mechanical [31–34], thermo-acoustic [26,35], and durability [32,36] properties. These lightweight materials can be considered a resource for an appropriate management of the large quantity of industrial by-products, and the production of these cheap and environmentally friendly composites is considered one of the most effective alternatives to tire incineration and landfilling [37,38].

Rubber–cement composites show lower compression resistances as compared to the conventional conglomerates based on sand and gravel because these organic aggregates are softer than the surrounding media, acting like "holes" inside the cement mixture. Nonetheless, although nonstructural applications can be considered, an enhancement of toughness and ability to absorb impact energy can be attributed to these unconventional materials.

Moreover, the low density of tire rubber is beneficial because the cement conglomerates show lightweight properties with respect to the conventional building materials due to the decrease of the specific mass, which enhances sound and thermal insulation. This is in perfect agreement with the current policies of environmental sustainability based on the use of recycled industrial products as secondary raw by-products together with the advantage of the production of composites with low costs and that generate large energy savings as building materials.

The aim of this research was to prepare an environmentally beneficial cement composite based on end-of-life tire rubber (TR) as aggregate, realized as a cheap process without any treatment of the organic material. Specifically, the rheological, mechanical, thermal, microstructural, and wetting properties were studied in order to characterize the conglomerate for indoor and outdoor nonstructural applications. The mixtures were obtained after full substitution of the conventional sand aggregate with TR, and we evaluated the effect of the different grain sizes (0–2 mm ($TR_F$) and 2–4 mm ($TR_L$)) and dosages (25%, 32%, and 40% by weight).

We also discuss the contribution of tire rubber to macroscopic deformation—in particular, the super-elastic properties of this soft material to deform more under loading and the interface affecting the composite strength due to the lower stiffness of the rubber particles with respect to the cement paste [39,40]. Moreover, the thermal insulating properties of these lightweight composites were studied by analyzing the effects of the grain size and of the composition of the mixtures. Moreover, the wetting properties are discussed in consideration of the low surface energy of the rubber particles that tend to inhibit the absorption of water in the artifacts. These properties were also studied after the addition of expanded perlite (P, 0–1 mm grain size), with the aim of improving the mechanical strengths of the resulting composites with negligible modification of the thermal and wetting properties.

## 2. Experimental Part

### 2.1. Materials and Mortar Specimens Preparation

A limestone Portland cement, CEM II A-LL 42.5 R with resistance to compression $Rc_{(2\ days)} > 25.0$ MPa, $Rc_{(28\ days)} > 47.0$ MPa, and 3100–4400 $cm^2/g$ Blaine specific surface area, was provided by Buzzi Unicem, Barletta, Italy, and used for the preparation of the cement mortars [41]. In a first set of tests, an investigation on the mechanical, thermal, and microscopical properties of composites based on bare end-of-life tire rubber as aggregate was carried out. End-of-life tire rubber (0–2 mm ($TR_F$) and 2–4 mm ($TR_L$) size range, 500 $kg/m^3$ and 550 $kg/m^3$, respectively) was provided by Maltek Industrie S.r.l. (Terlizzi, Bari, Italy). TR grains were added as total replacement of the conventional aggregate, which was made on a volume basis rather than on a weight basis due to the low specific weight of the waste material. In the present case, the total volumes of aggregate were set at 450, 600, and 750 $cm^3$, corresponding to weight percentages in the range of 25%, 32%, and 40%, respectively. Tables 1 and 2 report, respectively, the aggregate and mortar composition. The samples were prepared with a water/cement ratio equal to $0.5 \pm 0.01$, specifically with 225 g of water and 450 g of cement [42]. A normalized (sand) mortar was prepared as a control with the same water to cement ratio and with 1350 g of sand. Dry siliceous natural sand in the range of 0.08–2 mm from Societè Nouvelle du Littoral (Leucate, France) was used as aggregate with clean, isometric, and rounded-shaped particles. The rheology of the fresh mixtures was evaluated by the flow-test [43], which allowed us to determine the consistency of the mixtures after evaluation of the percentage increase of the diameter of the non-consolidated sample over the base diameter. Successively, all the specimens were prepared as prisms (40 × 40 × 160 mm) for the flexural and compressive tests (28 days curing) [42]. Moreover, the specimens were prepared as cylinders for the thermal ($\varphi$ = 100 mm; H = 50 mm) tests (28 days curing).

**Table 1.** Type, aggregate composition, density ρ, and porosity of the tire rubber (TR) mortar specimens. Samples prepared with 225 g of water and 450 g of cement. $TR_F$ = fine (0–2 mm) tire rubber, $TR_L$ = large (2–4 mm) tire rubber.

| Sample | Type | Aggregate Composition | ρ Kg/m$^3$ | Porosity % |
|---|---|---|---|---|
| REF | Control | Normalized mortar | 1950 | 21 |
| 1 | $TR_F$ (25%) | 100% TR (0–2 mm) | 1160 | 43 |
| 2 | $TR_L$ (25%) | 100% TR (2–4 mm) | 1215 | 42 |
| 3 | $TR_F/TR_L$ (25%) | 50% TR (0–2 mm)/50% TR (2–4 mm) | 1180 | 43 |
| 4 | $TR_F$ (32%) | 100% TR (0–2 mm) | 1080 | 45 |
| 5 | $TR_L$ (32%) | 100% TR (2–4 mm) | 1130 | 44 |
| 6 | $TR_F/TR_L$ (32%) | 50% TR (0–2 mm)/50% TR (2–4 mm) | 1100 | 45 |
| 7 | $TR_F$ (40%) | 100% TR (0–2 mm) | 970 | 47 |
| 8 | $TR_L$ (40%) | 100% TR (2–4 mm) | 1005 | 45 |
| 9 | $TR_F/TR_L$ (40%) | 50% TR (0–2 mm)/50% TR (2–4 mm) | 990 | 46 |

**Table 2.** TR mortar composition.

| Sample | Type | Cement (g) | Water (cm$^3$) | TR$_F$ Volume (cm$^3$) | TR$_F$ Weight (g) | TR$_L$ Volume (cm$^3$) | TR$_L$ Weight (g) |
|---|---|---|---|---|---|---|---|
| REF | Control | 450 | 225 | 0 | 0 | 0 | 0 |
| 1 | TR$_F$ (25%) | 450 | 225 | 450 | 230 | 0 | 0 |
| 2 | TR$_L$ (25%) | 450 | 225 | 0 | 0 | 450 | 250 |
| 3 | TR$_F$/TR$_L$ (25%) | 450 | 225 | 225 | 115 | 225 | 125 |
| 4 | TR$_F$ (32%) | 450 | 225 | 600 | 300 | 0 | 0 |
| 5 | TR$_L$ (32%) | 450 | 225 | 0 | 0 | 600 | 330 |
| 6 | TR$_F$/TR$_L$ (32%) | 450 | 225 | 300 | 150 | 300 | 165 |
| 7 | TR$_F$ (40%) | 450 | 225 | 750 | 380 | 0 | 0 |
| 8 | TR$_L$ (40%) | 450 | 225 | 0 | 0 | 750 | 420 |
| 9 | TR$_F$/TR$_L$ (40%) | 450 | 225 | 375 | 190 | 375 | 210 |

In a second set of tests, an investigation on the mechanical, thermal, microscopical, and wetting properties of cement mortars based on end-of-life tire rubber (TR$_F$) and dry expanded perlite as aggregate was carried out. Expanded perlite (P) (0–1 mm size range, 100 kg/m$^3$) is a chemically inert, sterile, and incombustible silica material obtained after heat treatment at 760–1100 °C of a vulcanic rock, which determines expansion in granular form and consequent high thermo-insulating properties [44]. It showed the following chemical composition: $SiO_2$ 74.5%, $Al_2O_3$ 12.3%, $K_2O$ 4.2%, $Na_2O$ 4%, $Fe_2O_3$ 1%, CaO 1.4%, and was provided by Maltek Industrie S.r.l. (Terlizzi, Bari, Italy). Moreover, in this case, the samples were prepared with 225 g of water and 450 g of cement [42], and the total volume of aggregate was set at 450, 600, and 750 cm$^3$. Tables 3 and 4 report, respectively, the aggregate and mortar composition. The rheological, thermal, and mechanical [42,45] properties of the mortars were evaluated as reported previously. These specimens were also molded in the form of cylinders for impact resistance ($\varphi$ = 150 mm; H = 60 mm) tests and cured in water for 28 days after demolding [46]. Moreover, the TR$_F$ and perlite samples were characterized by wettability tests.

**Table 3.** Type, aggregate composition, density ρ, and porosity of the expanded perlite and TR mortar specimens. Samples prepared with 225 g of water and 450 g of cement. P = perlite (0–1 mm), TR$_F$ = fine (0–2 mm) tire rubber.

| Sample | Type | Aggregate Composition | ρ Kg/m$^3$ | Porosity % |
|---|---|---|---|---|
| 10 | P (450 cm$^3$) | 100% perlite (0–1 mm) | 1250 | 37 |
| 11 | P/TR$_F$ (450 cm$^3$) | 50% perlite (0–1 mm)/50% TR (0–2 mm) | 1210 | 39 |
| 12 | P (600 cm$^3$) | 100% perlite (0–1 mm) | 1180 | 37 |
| 13 | P/TR$_F$ (600 cm$^3$) | 50% perlite (0–1 mm)/50% TR (0–2 mm) | 1130 | 41 |
| 14 | P (750 cm$^3$) | 100% perlite (0–1 mm) | 1100 | 38 |
| 15 | P/TR$_F$ (750 cm$^3$) | 50% perlite (0–1 mm)/50% TR (0–2 mm) | 1060 | 42 |

**Table 4.** Expanded perlite and tire rubber mortars composition.

| Sample | Type | Cement (g) | Water (cm$^3$) | TR$_F$ Volume (cm$^3$) | Perlite Volume (cm$^3$) |
|---|---|---|---|---|---|
| 10 | P (450 cm$^3$) | 450 | 225 | 450 | 0 |
| 11 | P/TRF (450 cm$^3$) | 450 | 225 | 275 | 275 |
| 12 | P (600 cm$^3$) | 450 | 225 | 600 | 0 |
| 13 | P/TRF (600 cm$^3$) | 450 | 225 | 300 | 300 |
| 14 | P (750 cm$^3$) | 450 | 225 | 750 | 0 |
| 15 | P/TRF (750 cm$^3$) | 450 | 225 | 375 | 375 |

### 2.2. *Mechanical and Thermal Tests*

Compression strengths were obtained as the average of the measurements carried out on 12 semi-prisms (2400 $\pm$ 200 N/s loading rate) obtained by flexural tests carried out on 6 prisms (40 $\times$ 40 $\times$ 160 mm) (50 $\pm$ 10 N/s loading rate) [42]. The mechanical tests were carried out by the use of a MATEST device (Matest S.p.A., Milan, Italy).

The impact resistance test [46] was carried out by dropping a 4.50 kg weight on a steel ball (63 mm diameter) from a height of 45 cm. The steel ball was placed centrally on the upper surface of the specimen and it was evaluated the number of weight drops before the fracture. Thermal conductivity ($\lambda$) measurements were carried out by inducing a constant thermal flow through a heating probe, which was applied on the sample surface, and the temperature was recorded during the time period. The results were obtained by the comparison between the experimental temperature values with the analytical solution of the heat conduction equation [45]. Measurements were carried out by ISOMET 2104 device (Applied Precision Ltd., Bratislava, Slovakia).

### 2.3. *Microscopical, Wetting, and Porosimetric Characterization*

A FESEM-EDX Carl Zeiss Sigma 300 VP (Carl Zeiss Microscopy GmbH, Jena, Germany) electron microscope was used for the microstructural characterization of the aggregates and of the composites. The samples were applied onto aluminum stubs and sputtered with graphite (Sputter Quorum Q150 Quorum Technologies Ltd., East Sussex, UK) before detection.

A Premier series dyno-lyte portable microscope and background cold lighting allowed the aggregate distribution of the mortars to be shown, allowing us to evaluate the wettability of the specimens after deposition of a drop of water onto the side and fracture surface of each sample.

Porosimetric measurements were carried out by an Ultrapyc 1200e Automatic Gas Pycnometer (Quantachrome Instruments, Boynton Beach, FL, USA). The tests were carried-out on three specimens of the same type in which an inert gas (helium) penetrated the finest pores. The results were the average of three measurements on each sample.

## 3. Results and Discussion

Figure 1A shows the end-of-life TR grains used for the mortars preparation, whereas Figure 1B shows the scanning electron micrograph (SEM) of a tire rubber grain.

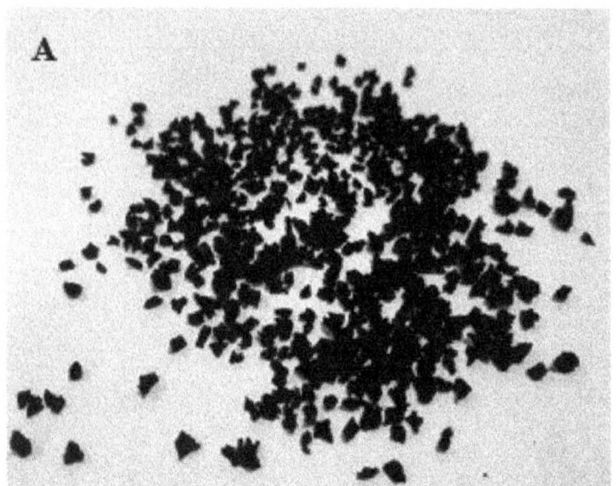

**Figure 1.** (**A**) Tire rubber (TR) and (**B**) tire rubber grain (≈1 mm diameter) from electron microscope detection.

Table 1 shows that the TR samples were lighter and with a much higher porosity than the reference (control) because of the density of the lightweight aggregates, whereas, among the lightweight composites, the $TR_F$ samples (samples 1, 4, and 7) were lighter than the $TR_L$ samples (samples 2, 5, and 8) because fine particles tend to cause more air entrapment during mixing. Specimens with mixture of aggregates ($TR_F/TR_L$) showed intermediate values of specific mass with respect to samples with only one type of aggregate. Finally, TR (40%) mortars resulted in the lightest and the most porous due to the highest dosage of the organic aggregate.

The consistency of the fresh specimens was determined by the flow test measurements (Figure 2). In the case of the lowest aggregate dosage (25% of TR), the TR sample with finer grains ($TR_F$, sample 1) showed similar workability of the sand reference (control), while the TR sample with larger grains ($TR_L$, sample 2) resulted as more fluid (+24%). This result can be ascribed to the lower specific surface area of the $TR_L$ aggregates with respect to the higher specific surface area of the $TR_F$ aggregates, which contributed to the decrease of cohesiveness of the specimen. A huge decrease of fluidity (−30%, −45%) and increase of cohesiveness was observed with the increase of the tire dosage in the TR samples with finer grains ($TR_F$, samples 4 and 7). $TR_L$ samples also showed a decrease of fluidity (+6%, −10%) with the increase of the tire dosage (samples 5 and 8) and, in the case of the intermediate dosage (32% of $TR_L$, sample 5), the flow percentages resulted in being similar to the control (plastic behavior). Specimens with a mixture of aggregates ($TR_F/TR_L$) showed intermediate values with respect to samples with only one type of aggregate; similar workability to the control was found in the case of the sample 6 (32% of $TR_F/TR_L$), although less fluid than sample 5 (32% of $TR_L$).

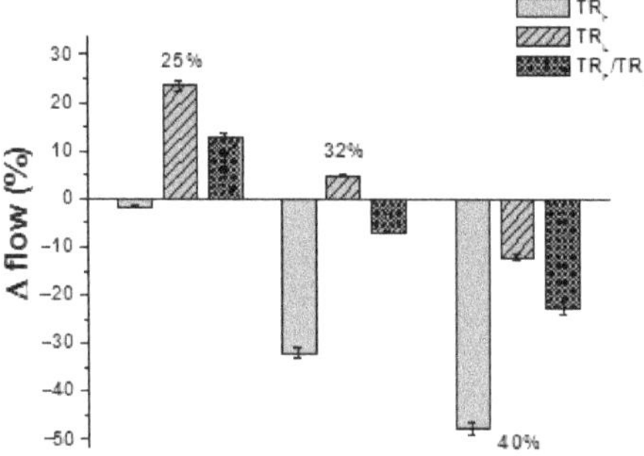


**Figure 2.** Flow test results of the TR samples with respect to the normalized mortar (control).

Figure 3 reports the flexural (Figure 3A) and compressive (Figure 3B) strengths of the TR samples with relative correlation and the evaluation of the percentage increase of the diameter of the non-consolidated sample over the base diameter. Successively, all the specimens with larger grains ($TR_L$, samples 2, 5, and 8) resulted in being more resistant than the composites based on finer grains ($TR_F$, samples 1, 4, and 7) due to the higher specific mass that induced an increase of the flexural resistance and of the compressive resistance in the range of +15% and +20%, respectively. Specimens with a mixture of aggregates ($TR_F/TR_L$) showed intermediate values with respect to samples with only one type of aggregate. A decrease of the mechanical strengths was observed with the increase of the TR weight percentage because of the decrease of the specific mass of the composites. Moreover, TR mortars did not show the typical flexural and compressive brittle rupture that can be observed in the reference mortars, but we observed the presence of discrete cracks after failure without collapse, an effect that can be ascribed to the rubber residual strength contribution, with particles bridging the crack faces (Figure 3C) [47,48].

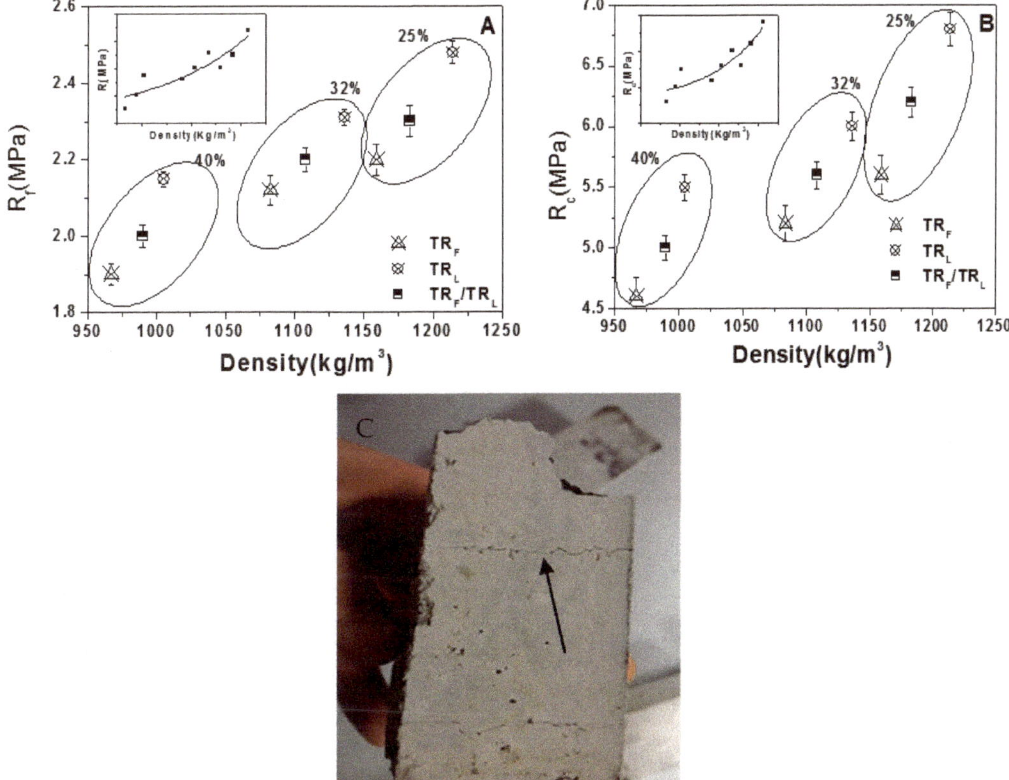


**Figure 3.** (**A**) Flexural and (**B**) compressive strengths of the TR mortars. The Rf and Rc of the control sample were, respectively, 8 ± 1 MPa and 45 ± 2 MPa. (**C**) Discrete cracks after rupture in the TR specimens (evidenced by the arrow), with the two parts of the sample still connected by tire rubber.

The sections of the specimens after the flexural tests are shown in Figure 4, where the different dimensions of the TR grains and the different and homogeneous distribution of the aggregates can be observed.

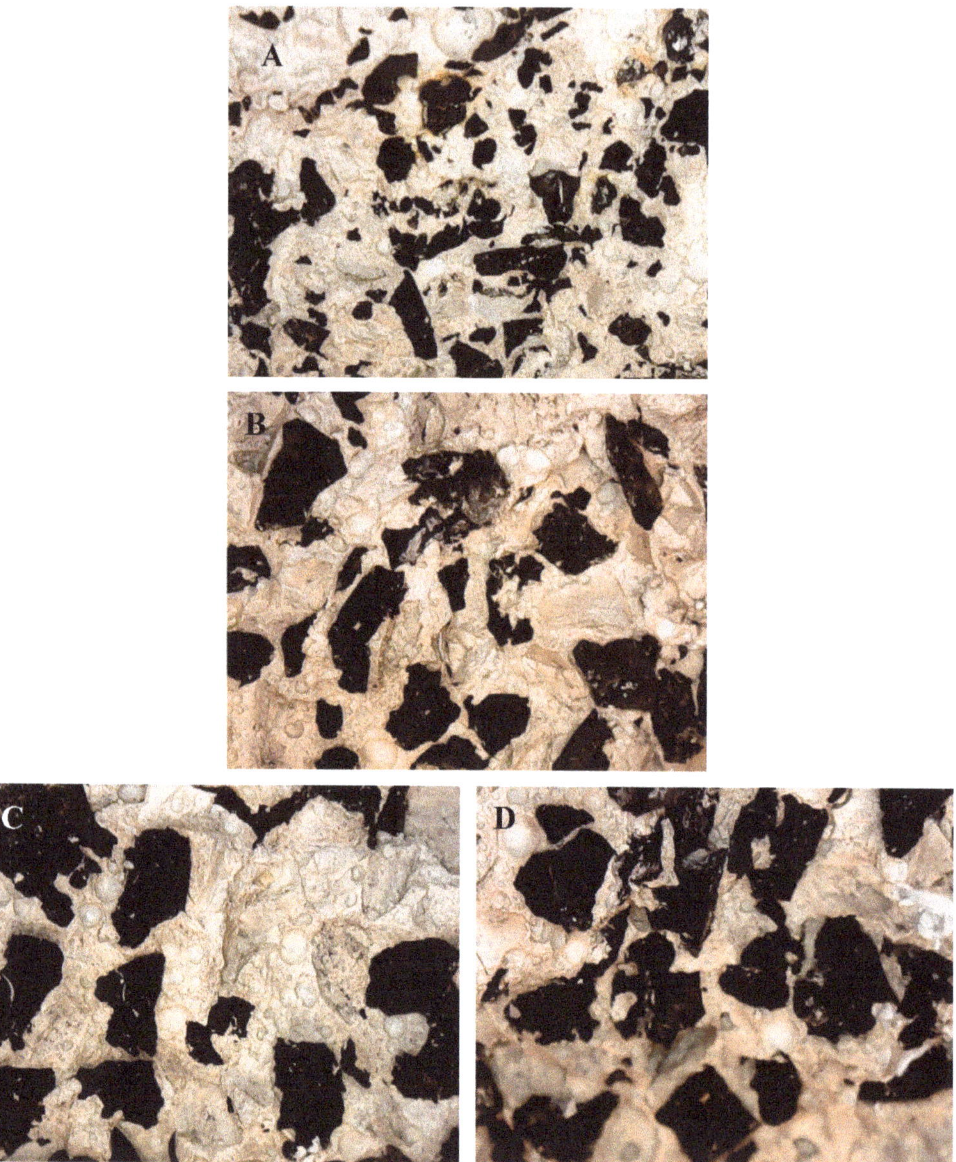


**Figure 4.** Sections of the (**A**) $TR_F$ (25%) sample (sample 1), (**B**) $TR_F/TR_L$ (25%) sample (sample 3), (**C**) $TR_L$ (25%) sample (sample 2), (**D**) $TR_L$ (40%) sample (sample 8). The cross-section of these samples was analyzed by optical microscope after flexural rupture of the specimens.

The mortars also showed lower thermal conductivities (≈85–90%) with respect to the sand reference (≈2 W/mK) due to the lower density of bare TR samples (Figure 5A). The specimens with finer grains ($TR_F$) showed lower thermal conductivities than the composites based on larger grains ($TR_L$) due to the lower specific mass that induced an increase of thermal insulation in the range of 25–30%. Specimens with a mixture of aggregates ($TR_F/TR_L$) showed intermediate values with respect to samples with only one type of aggregate, whereas a decrease of the thermal conductivities was observed with the increase of the TR weight percentage because of the lower specific mass of the composites. An exponential decrease of the conductivity data was observed with the decrease of the density of the composites (Figure 5B).

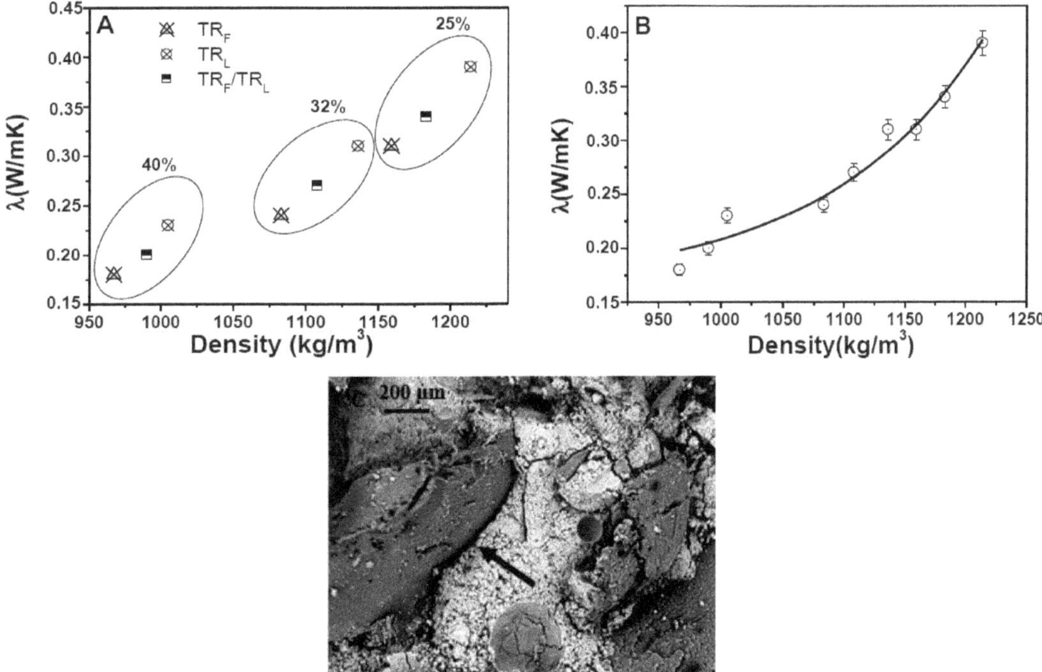


**Figure 5.** (**A**) Thermal conductivity of the TR specimens. The thermal conductivity of the sand reference was ≈2 W/mK. (**B**) Exponential increase of the thermal conductivity with the density increase. (**C**) SEM image of the cement/TR interface (the arrow represents the void at the interface).

An explanation of the decrease of the mechanical performances and of the remarkable increase of the thermal insulation of these composites is associated with the low density of the tire rubber compared to the cement paste [36] and, to a minor extent, to the lack of adhesion of tire to the cement paste, as observed in microscopical observations. The sand replacement with tire rubber determined the formation of voids in the composite (Figure 5C) at the organic/inorganic interface due to the unfavorable adhesion of the aggregate to the cement paste [49–54], with a decrease of the specific mass and increase of the porosity of the samples with respect to the references (Table 1). The hydrophobic tire rubber with its organic compounds can explain the poor adhesion to the inorganic and hydrophilic cement matrix, on the contrary to what was observed with the sand in the reference sample with a good adhesion of this aggregate to the cement paste [55].

Thus, these conglomerates can be considered very low thermal insulating composites with low mechanical strengths. For this reason, a second set of experiments was carried-out with the aim of improving the resistances of the more lightweight mortars (based on finer tire rubber) without significant modification of the thermal properties [44,56–58]. Thus, an investigation on the rheological, mechanical, thermal, and microscopical properties of cement mortars based on end-of-life tire rubber ($TR_F$) with the addition of a lightweight but more stiff aggregate such as expanded perlite was carried out.

The scanning electron micrographs (SEMs) of a perlite bead can be observed in Figure 6, which shows large open (Figure 6A) and closed (Figure 6B) porosity.

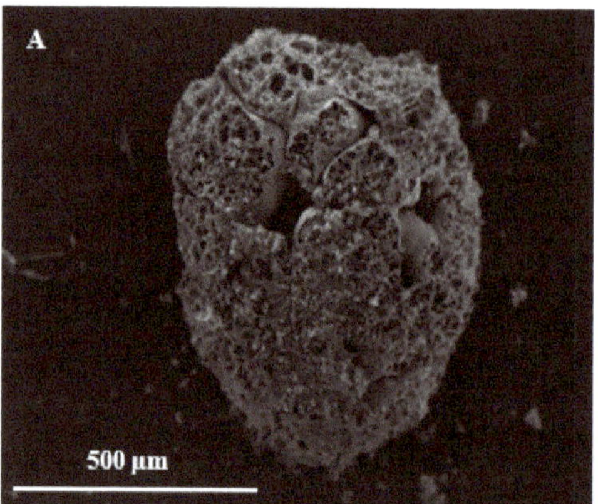


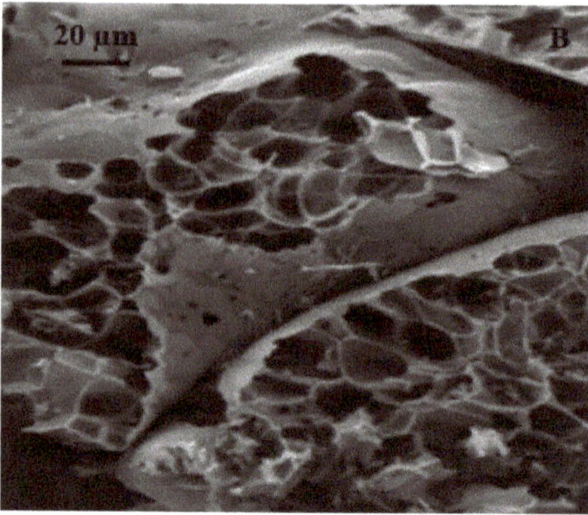


**Figure 6.** SEM image of a (**A**) perlite bead and (**B**) bead magnification (inner porosity).

Table 3 shows that the perlite and TR/perlite specimens were lighter and had a higher porosity than the reference (control), while bare perlite samples (samples 10, 12, and 14) showed higher specific mass (>100 Kg/m$^3$) and lower porosity (<6–9%) than bare $TR_F$ samples (samples 11, 13, and 15). The mortars were lighter and more porous with the increase of the aggregate dosage.

The addition of perlite to the mixture determined a large decrease of fluidity in the fresh specimens (Figure 7) because of the low grain size (high surface area) and large porosity of the silico-aluminate aggregates (Figure 6B). This effect was remarkable in terms of the increase of dosage, which determined an increase of cohesiveness of the specimens.

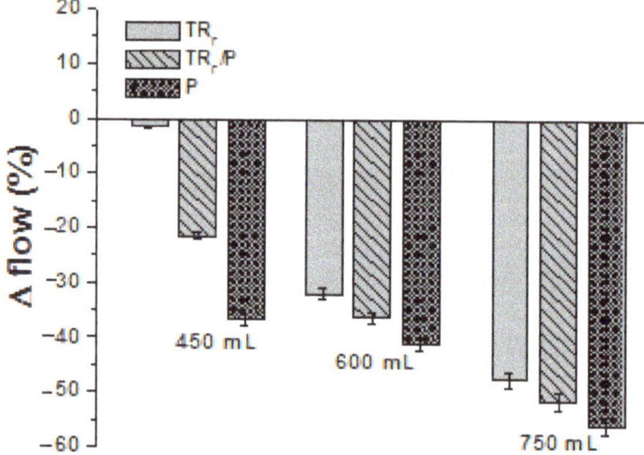


**Figure 7.** Flow test results of the $TR_F$ and perlite samples with respect to the normalized mortar (control).

Figure 8 reports the flexural (Figure 8A) and compressive (Figure 8B) strengths of the $TR_F$, perlite (P), and $TR_F$/P samples. It was observed that bare perlite samples (samples 10, 12, and 14) showed flexural resistances almost double that of bare $TR_F$ composites (samples 1, 4, and 7), and compressive resistances three times higher. This was ascribed to the presence of the stiffer material perlite [44,56–58]. Intermediate values were in the case of mixtures of aggregates (samples 11, 13, and 15). A decrease of the mechanical strengths was observed with the increase of the $TR_F$ and perlite volume because of the decrease of the specific mass of the composites.

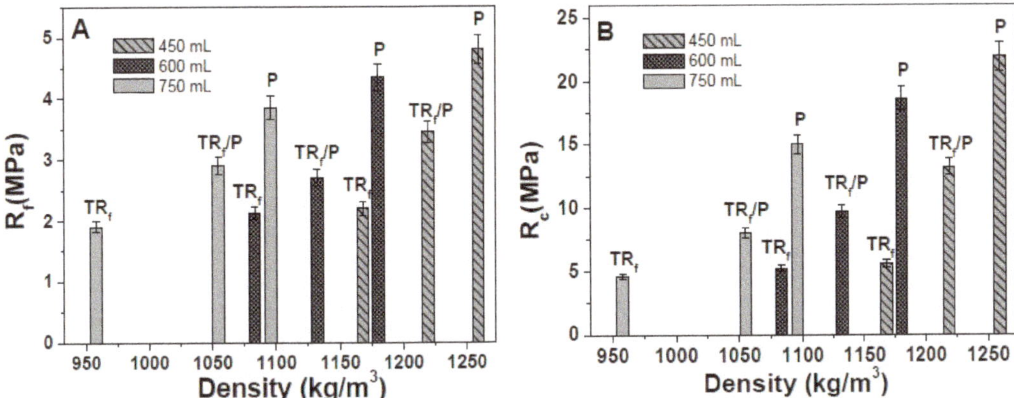


**Figure 8.** (**A**) Flexural and (**B**) compressive strengths of the $TR_F$ and perlite mortars. The Rf and Rc of the control sample were, respectively, 8 ± 1 MPa and 45 ± 2 MPa.

A brittle flexural and compressive failure mode of the perlite mortars was observed, which was very different respect to the bare TR behavior, while a semi-brittle rupture was observed in the TR/perlite samples, where an evident separation of the two parts of the specimens was not observed, with this being ascribed to the tire contribution [47].

The sections of the specimens after the flexural tests are shown in Figure 9, where the homogeneous distribution of the organic and inorganic aggregates can be observed.

**Figure 9.** Sections of the (**A**) $TR_F$ sample (sample 1), (**B**) P sample (sample 10), (**C**) P/$TR_F$ sample (sample 11). The cross-section of these samples was analyzed by optical microscope after flexural rupture of the specimens.

The mortars with bare perlite (samples 10, 12, and 14) were less insulating than the mortars with bare tire rubber (<35–40%), while P/TR mortars (samples 11, 13, and 15) showed similar thermal conductivities to bare TR samples (samples 1, 4, and 7), with λ values in the range of 80–85% lower than the sand reference (Figure 10A). Moreover, in this case, an exponential decrease of the conductivity data was observed with the decrease of the density of the composites (Figure 10B). Thus, these composites showed lightweight features due to the low density but also better mechanical properties ascribed to the stiffness of perlite.

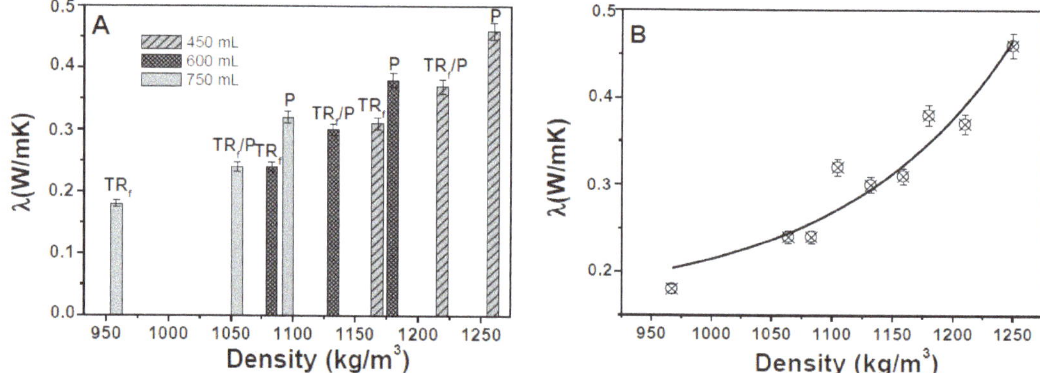


**Figure 10.** (**A**) Thermal conductivity of the $TR_F$ and perlite specimens. The thermal conductivity of the sand reference was ≈2 W/mK. (**B**) Exponential increase of the thermal conductivity with the density increase.

The low thermal and mechanical strengths of these samples with respect to the conventional sand-based mortar is ascribed to the low density of the aggregates. In this specific case, an explanation of the mechanical and thermal results of the lightweight mortars can be obtained after microscopical observations. In the case of the aggregate mixtures, it we confirmed the unfavorable adhesion of the TR aggregate but a good adhesion of the perlite to the cement paste (Figure 11), a result ascribed to the similar chemical composition of the ligand and of the inorganic aggregates based on silicates and aluminates together with the beads' roughness [23]. The microscopical observations can explain the low mechanical strengths in the presence of tire rubber and the higher values with perlite characterized by stiffness and adhesion to the cement paste, which determined an increase of the specific mass and a decrease of the porosity of the samples. Accordingly, better mechanical performances were obtained in the case of perlite/TR mixtures with respect to bare TR mixtures, together with good thermal properties associated with the low specific mass of these composites characterized by both lightweight aggregates.

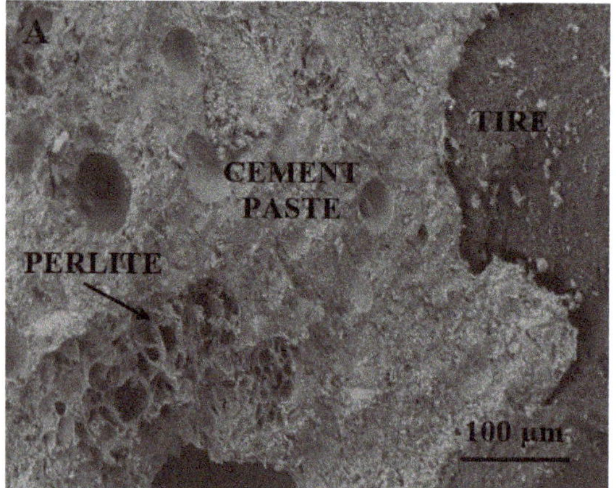


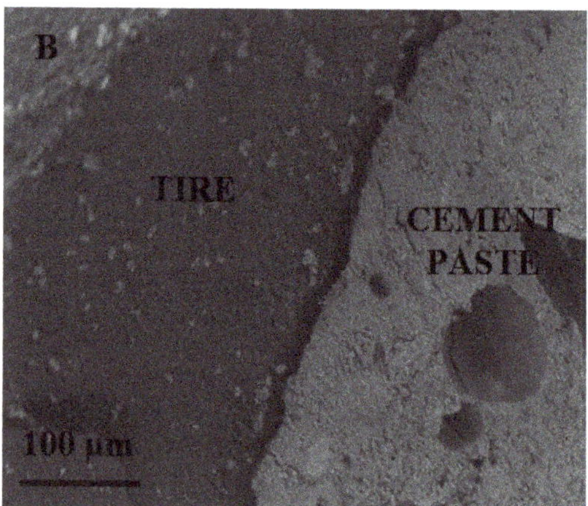


**Figure 11.** *Cont.*

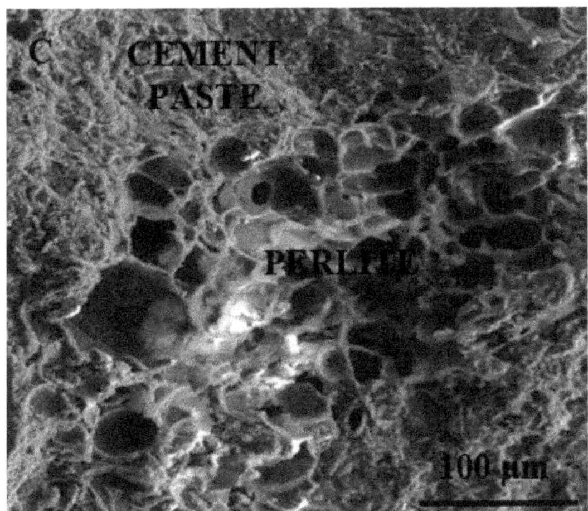


**Figure 11.** SEM images of (**A**) P/$TR_F$ bead, (**B**) cement/$TR_F$ interface, (**C**) cement/perlite interface.

The impact compression tests (Figure 12A) showed that the perlite samples were extremely fragile and the breakage occurred after few blows due to the presence of the brittle aggregate (Figure 12B), while the specimens characterized by the presence of tire rubber (Figure 12C,D) showed the best results due to the super-elastic properties of the elastomeric material. Indeed, the highest energy absorption capacity and a deep groove before complete failure [49,59] were specifically observed in the case of the $TR_F$ samples while average values were observed in samples with 50% of TR and 50% of perlite. As reported previously, a decrease of the mechanical strengths was observed with the increase of the $TR_F$ and perlite volume because of the decrease of the specific mass of the composites; nonetheless, the decrease of the specific mass of the $TR_F$ and P/$TR_F$ samples determined an increase of the impact resistance because of the higher tire dosage. In the case of the bare perlite samples, the variation of the composition did not influence the performances because of the extremely brittle behavior of these conglomerates.

The wetting properties of these specimens was also carried out. The wettability is defined as the attraction of a liquid to a solid surface with an interaction determined by a balance between adhesive and cohesive forces. A poor wettability is associated with a hydrophobic behavior of the surface of a material while a high wettability is associated with a hydrophilic behavior [60]. Tire rubber specimens showed poor wettability (negligible water penetration) on the surface and on the bulk [23] (Figures 13A and 14A). These results were totally ascribed to the hydrophobic nature of the organic aggregate in spite of the higher porosity with respect to the reference and the bare perlite samples. The surface and the bulk of the bare perlite samples showed a fast water absorption and a hydrophilic behavior (Figures 13B and 14B) due to the hydrophilic porous domains of the inorganic aggregate and of the cement paste, a result also observed in the case of the reference sand mortars [23,55]. In the case of the rubber/perlite sample (P/$TR_F$), the water absorption was significantly lower than the perlite and reference samples but higher than the TR mortars ($\approx$20% on the side surface, $\approx$20% on the fracture surface), thanks to the opposite contribute of the hydrophilic and porous perlite and of the hydrophobic tire rubber (Figures 13C and 14C).

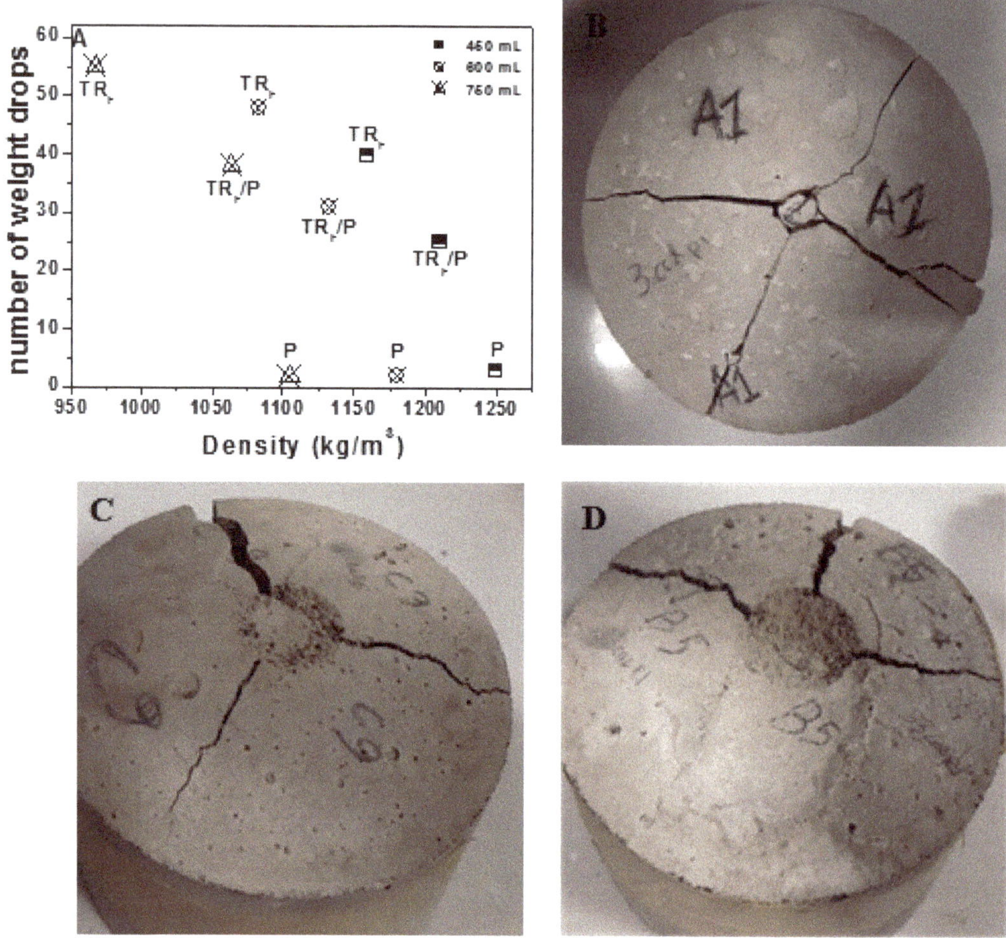


**Figure 12.** (**A**) Impact resistance of the $TR_F$ and perlite mortars; (**B**) P sample (sample 10); (**C**) P/$TR_F$ sample (sample 11); (**D**) $TR_F$ sample (sample 1).

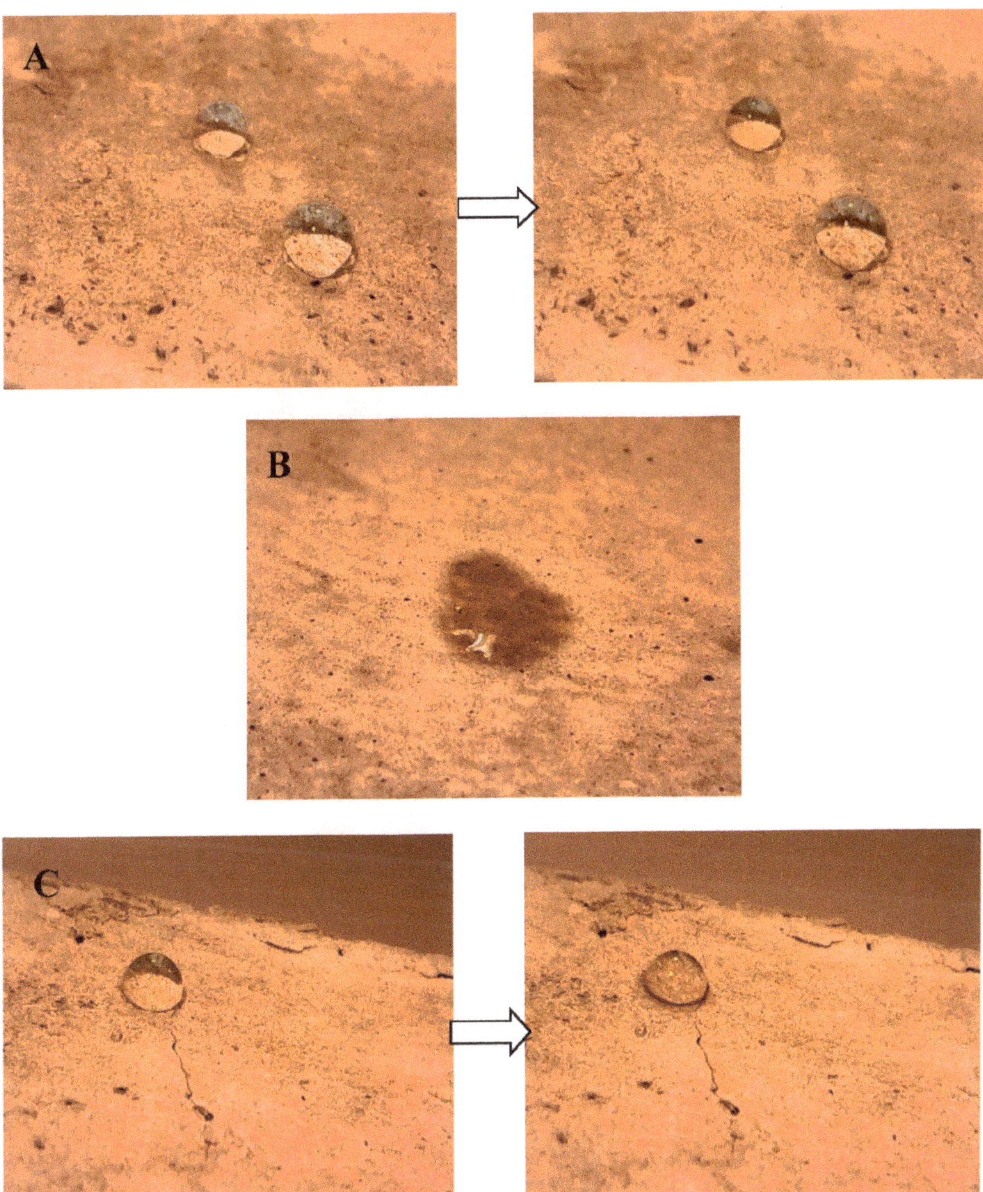


**Figure 13.** Wettability tests on the side surface of (**A**) $TR_F$ sample (sample 1) at t = 0 s (**left**) and at t = 150 s (**right**), (**B**) P sample (sample 10) at t = 1 s, (**C**) P/$TR_F$ sample (sample 11) at t = 0 s (**left**) and at t = 150 s (**right**).

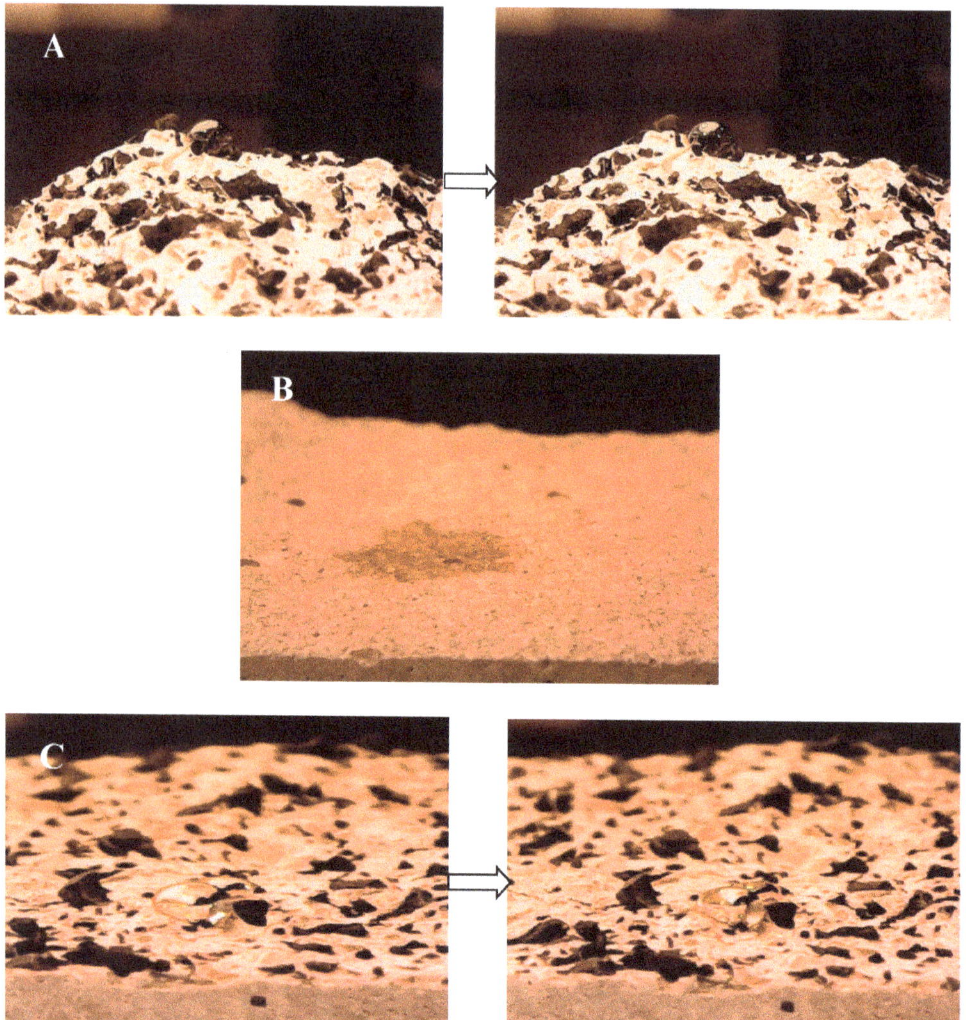


**Figure 14.** Wettability tests on the fracture surface of (**A**) $TR_F$ sample (sample 1) at t = 0 s (**left**) and at t = 150 s (**right**), (**B**) P sample (sample 10) at t = 1 s, (**C**) P/$TR_F$ sample (sample 11) at t = 0 s (**left**) and at t = 150 s (**right**).

## 4. Summary and Conclusions

The rheological, thermo-mechanical, wetting, and microstructural properties of lightweight cement mortars containing end-of-life tire rubber (TR) as aggregate were evaluated. The mixtures were obtained after total replacement of the conventional sand aggregate with untreated TR having different grain sizes (0–2 and 2–4 mm) and distributions (25%, 32%, and 40% by weight).

The main results showed that

1. The fresh mortars showed a decrease of fluidity with the increase of dosage. The $TR_L$ mixtures resulted in having more fluid than the $TR_f$ mixtures. This result can be ascribed to the lower specific surface area of the $TR_L$ aggregates with respect to the higher specific surface area of the $TR_F$ aggregates, which contributes to the decrease

of cohesiveness of the $TR_L$ specimens. The $TR_F$ (25%) specimen showed a plastic behavior as in the case of the sand reference, and similar results were found in the case of the $TR_L$ (32%) fresh mixture.

2. The mortars showed lower thermal conductivities (≈85–90%) and lower mechanical strengths (Rf and Rc) with respect to the sand reference due to the decrease of specific mass of the conglomerates associated with the low density of the aggregates and, to a minor extent, to the voids at the TR/cement interface, which were microstructurally detected.
3. The specimens with larger grains (TRL) showed higher mechanical strengths (Rf and Rc) but higher thermal conductivities than the composites based on finer grains (TRF) due to the higher specific mass of the conglomerates associated with the different density of the aggregates.
4. A decrease of the thermal conductivities and of the mechanical strengths were observed with the increase of the TR weight percentage, which determined a decrease of the specific mass of the conglomerates.
5. TR mortars showed discrete cracks after failure without separation of the two parts of the specimens due to the rubber residual strength contribution, with particles bridging the crack faces.
6. The addition of expanded perlite (P, 0–1 mm grain size) to the mixture allowed us to obtain less fluid mortars because of the low grain size (high surface area) and large porosity of the silico-aluminate aggregates.
7. An improvement of the mechanical strengths was obtained with the addition of perlite. Indeed, the flexural resistances were almost double with respect to bare $TR_F$ composites and the compressive resistances three times higher due to the stiffness of the inorganic aggregate.
8. Negligible modification of the thermal insulating properties (≈80–85% lower than the sand reference) was obtained due to the high porosity of perlite.
9. P/TR mortars also showed discrete cracks after failure without separation of the two parts of the specimens, and this behavior, although less evident than bare TR samples, was exclusively ascribed to the contribute of the elastomeric particles, as opposed to the brittle failure obtained by bare perlite samples.
10. From the impact compression tests, we found the best performances of the tire and, to lesser extent, of the P/TR composites were evidenced by a deep groove before complete failure. Moreover, in this case, this result was associated to the super-elastic properties of the end-of-life tire rubber.
11. TR mortars showed very low water penetration through the surface and also through the bulk of the samples, thanks to the hydrophobic nature of the end-of-life aggregate. Interesting results were obtained in the case of the P/TR samples.
12. The present composites can be considered environmentally sustainable materials because they are prepared with recycled materials and without any treatment of the aggregates. Moreover, the lightweight properties can be effective for thermal insulating elements (vertical elements, screeds, panels), which can be applied for indoor and outdoor structures.

**Author Contributions:** Conceptualization, A.P.; methodology, A.P.; validation, M.N.; formal analysis, A.P.; investigation, A.P.; resources, M.N.; data curation, M.N.; writing—original draft preparation, A.P.; writing—review and editing, M.N.; visualization, M.N.; supervision, M.N. All authors have read and agreed to the published version of the manuscript.

**Funding:** This research received no external funding.

**Institutional Review Board Statement:** Not applicable.

**Informed Consent Statement:** Not applicable.

**Data Availability Statement:** Data sharing not applicable.

**Acknowledgments:** Special thanks to Pietro Stefanizzi and Stefania Liuzzi for thermal analysis. Adriano Boghetich is acknowlwdged for SEM-EDX analysis and also Regione Puglia (Micro X-Ray Lab Project–Reti di Laboratori Pubblici di Ricerca, cod. n. 45 and 56). Acknowledgments to the DICATECh of the Polytechnic of Bari for SEM analyses.

**Conflicts of Interest:** The authors declare no conflict of interest.

## References

1. Garcia, D.; You, F. Systems engineering opportunities for agricultural and organic waste management in the food–water–energy nexus. *Curr. Opin. Chem. Eng.* **2017**, *18*, 23–31. [CrossRef]
2. Rahman, M.T.; Mohajerani, A.; Giustozzi, F. Recycling of waste materials for asphalt concrete and bitumen: A review. *Materials* **2020**, *13*, 1495. [CrossRef] [PubMed]
3. Larouche, F.; Tedjar, F.; Amouzegar, K.; Houlachi, G.; Bouchard, P.; Demopoulos, G.P.; Zaghib, K. Progress and status of hydrometallurgical and direct recycling of Li-ion batteries and beyond. *Materials* **2020**, *13*, 801. [CrossRef] [PubMed]
4. Asefi, H.; Lim, S. A novel multi-dimensional modeling approach to integrated municipal solid waste management. *J. Clean. Prod.* **2017**, *166*, 1131–1143. [CrossRef]
5. Petrella, A.; Petrella, M.; Boghetich, G.; Petruzzelli, D.; Ayr, U.; Stefanizzi, P.; Calabrese, D.; Pace, L.; Guastamacchia, M. Thermo-acoustic properties of cement-waste-glass mortars. *Proc. Inst. Civ. Eng. Constr. Mater.* **2009**, *162*, 67–72. [CrossRef]
6. Spasiano, D.; Luongo, V.; Petrella, A.; Alfè, M.; Pirozzi, F.; Fratino, U.; Piccinni, A.F. Preliminary study on the adoption of dark fermentation as pretreatment for a sustainable hydrothermal denaturation of cement-asbestos composites. *J. Clean. Prod.* **2017**, *166*, 172–180. [CrossRef]
7. Petrella, A.; Petrella, M.; Boghetich, G.; Petruzzelli, D.; Calabrese, D.; Stefanizzi, P.; de Napoli, D.; Guastamacchia, M. Recycled waste glass as aggregate for lightweight concrete. *Proc. Inst. Civ. Eng. Constr. Mater.* **2007**, *160*, 165–170. [CrossRef]
8. Ossa, A.; García, J.L.; Botero, E. Use of recycled construction and demolition waste (CDW) aggregates: A sustainable alternative for the pavement construction industry. *J. Clean. Prod.* **2016**, *135*, 379–386. [CrossRef]
9. Gómez-Meijide, B.; Pérez, I.; Pasandín, A.R. Recycled construction and demolition waste in cold asphalt mixtures: Evolutionary properties. *J. Clean. Prod.* **2016**, *112*, 588–598. [CrossRef]
10. Tavira, J.; Jiménez, J.R.; Ledesma, E.F.; López-Uceda, A.; Ayuso, J. Real-scale study of a heavy traffic road built with in situ recycled demolition waste. *J. Clean. Prod.* **2020**, *248*, 119219. [CrossRef]
11. Xuan, D.X.; Molenaar, A.A.A.; Houben, L.J.M. Evaluation of cement treatment of reclaimed construction and demolition waste as road bases. *J. Clean. Prod.* **2015**, *100*, 77–83. [CrossRef]
12. Rizzi, V.; D'Agostino, F.; Gubitosa, J.; Fini, P.; Petrella, A.; Agostiano, A.; Semeraro, P.; Cosma, P. An alternative use of olive pomace as a wide-ranging bioremediation strategy to adsorb and recover disperse orange and disperse red industrial dyes from wastewater. *Separations* **2017**, *4*, 29. [CrossRef]
13. Petrella, A.; Petrella, M.; Boghetich, G.; Basile, T.; Petruzzelli, V.; Petruzzelli, D. Heavy metals retention on recycled waste glass from solid wastes sorting operations: A comparative study among different metal species. *Ind. Eng. Chem. Res.* **2012**, *51*, 119–125. [CrossRef]
14. Petrella, A.; Petruzzelli, V.; Basile, T.; Petrella, M.; Boghetich, G.; Petruzzelli, D. Recycled porous glass from municipal/industrial solid wastes sorting operations as a lead ion sorbent from wastewaters. *React. Funct. Polym.* **2010**, *70*, 203–209. [CrossRef]
15. Petrella, A.; Spasiano, D.; Race, M.; Rizzi, V.; Cosma, P.; Liuzzi, S.; de Vietro, N. Porous waste glass for lead removal in packed bed columns and reuse in cement conglomerates. *Materials* **2019**, *12*, 94. [CrossRef] [PubMed]
16. Kroll, L.; Hoyer, S.; Klaerner, M. Production technology of cores for hybrid laminates containing rubber powder from scrap tyres. *Procedia Manuf.* **2018**, *21*, 591–598. [CrossRef]
17. Thomas, B.S.; Gupta, R.C. A comprehensive review on the applications of waste tire rubber in cement concrete. Renew. *Sustain. Energy Rev.* **2016**, *54*, 1323–1333. [CrossRef]
18. Presti, D.L.; Izquierdo, M.A.; del Barco Carrión, A.J. Towards storage-stable high-content recycled tyre rubber modified bitumen. *Environments* **2018**, *172*, 106–111. [CrossRef]
19. Ramirez-Canon, A.; Muñoz-Camelo, Y.; Singh, P. Decomposition of used tyre rubber by pyrolysis: Enhancement of the physical properties of the liquid fraction using a hydrogen stream. *Environments* **2018**, *5*, 72. [CrossRef]
20. Escobar-Arnanz, J.; Mekni, S.; Blanco, G.; Eljarrat, E.; Barceló, D.; Ramos, L. Characterization of organic aromatic compounds in soils affected by an uncontrolled tire landfill fire through the use of comprehensive two-dimensional gas chromatography–time-of-flight mass spectrometry. *J. Chromatogr. A* **2018**, *1536*, 163–175. [CrossRef]
21. Artíñano, B.; Gómez-Moreno, F.J.; Díaz, E.; Amato, F.; Pandolfi, M.; Alonso-Blanco, E.; Coz, E.; Garcia-Alonso, S.; Becerril-Valle, M.; Querol, X.; et al. Outdoor and indoor particle characterization from a large and uncontrolled combustion of a tire landfill. *Sci. Total Environ.* **2017**, *593–594*, 543–551. [CrossRef] [PubMed]
22. Roychand, R.; Gravina, R.J.; Zhuge, Y.; Ma, X.; Youssf, O.; Mills, J.E. A comprehensive review on the mechanical properties of waste tire rubber concrete. *Constr. Build. Mater.* **2020**, *237*, 117651. [CrossRef]
23. Petrella, A.; di Mundo, R.; de Gisi, S.; Todaro, F.; Labianca, C.; Notarnicola, M. Environmentally sustainable cement composites based on end-of-life tyre rubber and recycled waste porous glass. *Materials* **2019**, *12*, 3289. [CrossRef]

24. Radheshkumar, C.; Karger-Kocsis, J. Thermoplastic dynamic vulcanisates containing LDPE, rubber, and thermochemically reclaimed ground tyre rubber. *Plast. Rubber Compos.* **2002**, *31*, 99–105. [CrossRef]
25. Karger-Kocsis, J.; Mészáros, L.; Bárány, T. Ground tyre rubber (GTR) in thermoplastics, thermosets, and rubbers. *J. Mater. Sci.* **2013**, *48*, 1–38. [CrossRef]
26. Angelin, A.F.; Miranda, E.J.P., Jr.; Santos, J.M.C.D.; Lintz, R.C.C.; Gachet-Barbosa, L.A. Rubberized mortar: The influence of aggregate granulometry in mechanical resistances and acoustic behavior. *Constr. Build. Mater.* **2019**, *200*, 248–254. [CrossRef]
27. Thai, Q.B.; Chong, R.O.; Nguyen, P.T.; Le, D.K.; Le, P.K.; Phan-Thien, N.; Duong, H.M. Recycling of waste tire fibers into advanced aerogels for thermal insulation and sound absorption applications. *J. Environ. Chem. Eng.* **2020**, *8*, 104279. [CrossRef]
28. Tasalloti, A.; Chiaro, G.; Banasiak, L.; Palermo, A. Experimental investigation of the mechanical behaviour of gravel-granulated tyre rubber mixtures. *Constr. Build. Mater.* **2020**, 121749, in press. [CrossRef]
29. Mazzotta, F.; Lantieri, C.; Vignali, V.; Simone, A.; Dondi, G.; Sangiorgi, C. Performance evaluation of recycled rubber waterproofing bituminous membranes for concrete bridge decks and other surfaces. *Constr. Build. Mater.* **2017**, *136*, 524–532. [CrossRef]
30. Czajczyńska, D.; Krzyżyńska, R.; Jouhara, H.; Spencer, N. Use of pyrolytic gas from waste tire as a fuel: A review. *Energy* **2017**, *134*, 1121–1131. [CrossRef]
31. Najim, K.B.; Hall, M.R. Mechanical and dynamic properties of self-compacting crumb rubber modified concrete. *Constr. Build. Mater.* **2012**, *27*, 521–530. [CrossRef]
32. Bisht, K.; Ramana, P.V. Evaluation of mechanical and durability properties of crumb rubber concrete. *Constr. Build. Mater.* **2017**, *155*, 811–817. [CrossRef]
33. Eldin, N.N.; Senouci, A.B. Rubber-tire particles as concrete aggregate. *J. Mater. Civ. Eng.* **1993**, *5*, 478–496. [CrossRef]
34. Benazzouk, A.; Mezreb, K.; Doyen, G.; Goullieux, A.; Quèneudec, M. Effect of rubber aggregates on the physico-mechanical behaviour of cement–rubber composites-influence of the alveolar texture of rubber aggregates. *Cem. Concr. Compos.* **2003**, *25*, 711–720. [CrossRef]
35. Benazzouk, A.; Douzane, O.; Mezreb, K.; Laidoudi, B.; Quéneudec, M. Thermal conductivity of cement composites containing rubber waste particles: Experimental study and modelling. *Constr. Build. Mater.* **2008**, *22*, 573–579. [CrossRef]
36. Azevedo, F.; Pacheco-Torga, F.; Jesus, C.; de Aguiar, J.B.; Camões, A.F. Properties and durability of HPC with tyre rubber wastes. *Constr. Build. Mater.* **2012**, *34*, 186–191. [CrossRef]
37. Jie, X.U.; Yao, Z.; Yang, G.; Han, Q. Research on crumb rubber concrete: From a multi-scale review. *Constr. Build. Mater.* **2020**, *232*, 117282.
38. Li, X.; Ling, T.C.; Mo, K.H. Functions and impacts of plastic/rubber wastes as eco-friendly aggregate in concrete–A review. *Constr. Build. Mater.* **2020**, *240*, 117869. [CrossRef]
39. Çakmak, U.D.; Major, Z. Experimental thermomechanical analysis of elastomers under uni-and biaxial tensile stress state. *Exp. Mech.* **2014**, *54*, 653–663. [CrossRef]
40. Çakmak, U.D.; Hiptmair, F.; Major, Z. Applicability of elastomer time-dependent behavior in dynamic mechanical damping systems. *Mech. Time-Depend. Mat.* **2014**, *18*, 139–151. [CrossRef]
41. Italian Organization for Standardization (UNI). Cement Composition, Specifications and Conformity Criteria for Common Cements. EN 197-1. Available online: http://store.uni.com/magento-1.4.0.1/index.php/en-197-1-2011.html (accessed on 30 December 2020).
42. Italian Organization for Standardization (UNI). Methods of Testing Cement-Part 1: Determination of Strength. EN 196-1. Available online: http://store.uni.com/magento-1.4.0.1/index.php/en-196-1-2016.html (accessed on 30 December 2020).
43. Italian Organization for Standardization (UNI). Determination of Consistency of Cement Mortars Using a Flow Table. UNI 7044:1972. Available online: http://store.uni.com/magento-1.4.0.1/index.php/uni-7044-1972.html (accessed on 30 December 2020).
44. Sengul, O.; Azizi, S.; Karaosmanoglu, F.; Tasdemir, M.A. Effect of expanded perlite on the mechanical properties and thermal conductivity of lightweight concrete. *Energy Build.* **2011**, *43*, 671–676. [CrossRef]
45. Gustafsson, S.E. Transient plane source techniques for thermal conductivity and thermal diffusivity measurements of solid materials. *Rev. Sci. Instrum.* **1991**, *62*, 797–804. [CrossRef]
46. ACI Committee 544. ACI 544.2R-89. Measurement of properties of fibre reinforced concrete. In *ACI Manual of Concrete Practice, Part 5: Masonry, Precast Concrete and Special Processes*; American Concrete Institute: Farmington Hills, MI, USA, 1996.
47. Aiello, M.A.; Leuzzi, F. Waste tyre rubberized concrete: Properties at fresh and hardened state. *Waste Manag.* **2010**, *30*, 1696–1704. [CrossRef] [PubMed]
48. Bompa, D.V.; Elghazouli, A.Y. Stress–strain response and practical design expressions for FRP-confined recycled tyre rubber concrete. *Constr. Build. Mater.* **2020**, *237*, 117633. [CrossRef]
49. Khalil, E.; Abd-Elmohsen, M.; Anwar, A.M. Impact resistance of rubberized self-compacting concrete. *Water Sci.* **2015**, *29*, 45–53. [CrossRef]
50. Li, G.; Wang, Z.; Leung, C.K.; Tang, S.; Pan, J.; Huang, W.; Chen, E. Properties of rubberized concrete modified by using silane coupling agent and carboxylated SBR. *J. Clean. Prod.* **2016**, *112*, 797–807. [CrossRef]
51. Huang, B.S.; Li, G.Q.; Pang, S.S.; Eggers, J. Investigation into waste tire rubber filled concrete. *J. Mater. Civ. Eng.* **2004**, *16*, 187–194. [CrossRef]

52. Khaloo, A.R.; Dehestani, M.; Rahmatabadi, P. Mechanical properties of concrete containing a high level of tire-rubber particles. *Waste Manag.* **2008**, *28*, 2472–2482. [CrossRef]
53. Marie, I. Thermal conductivity of hybrid recycled aggregate–Rubberized concrete. *Constr. Build. Mater.* **2017**, *133*, 516–524. [CrossRef]
54. Karakurt, C. Microstructure properties of waste tire rubber composites: An overview. *J. Mater. Cycles Waste* **2015**, *17*, 422–433. [CrossRef]
55. Petrella, A.; di Mundo, R.; Notarnicola, M. Recycled expanded polystyrene as lightweight aggregate for environmentally sustainable cement conglomerates. *Materials* **2020**, *13*, 988. [CrossRef] [PubMed]
56. Petrella, A.; Spasiano, D.; Rizzi, V.; Cosma, P.; Race, M.; de Vietro, N. Lead ion sorption by perlite and reuse of the exhausted material in the construction field. *Appl. Sci.* **2018**, *8*, 1882. [CrossRef]
57. Yu, L.H.; Ou, H.; Lee, L.L. Investigation on pozzolanic effect of perlite powder in concrete. *Cem. Concr. Res.* **2003**, *33*, 73–76. [CrossRef]
58. Erdem, T.K.; Meral, Ç.; Tokyay, M.; Erdoğan, T.Y. Use of perlite as a pozzolanic addition in producing blended cements. *Cem. Concr. Compos.* **2007**, *29*, 13–21. [CrossRef]
59. Mastali, M.; Dalvand, A.; Sattarifard, A. The impact resistance and mechanical properties of the reinforced self-compacting concrete incorporating recycled CFRP fiber with different lengths and dosages. *Compos. Part B Eng.* **2017**, *112*, 74–92. [CrossRef]
60. Palumbo, F.; di Mundo, R. Wettability: Significance and measurement. In *Polymer Surface Characterization*; Sabbatini, L., Ed.; De Gruyter: Berlin, Germany, 2014; pp. 207–241.

Article

# Experimental Investigation on Environmentally Sustainable Cement Composites Based on Wheat Straw and Perlite

**Andrea Petrella [1,*], Sabino De Gisi [1], Milvia Elena Di Clemente [1], Francesco Todaro [1], Ubaldo Ayr [2], Stefania Liuzzi [2], Magdalena Dobiszewska [3] and Michele Notarnicola [1]**

1 Dipartimento di Ingegneria Civile, Ambientale, del Territorio, Edile e Chimica, Politecnico di Bari, 4, Via E. Orabona, 70125 Bari, Italy; sabino.degisi@poliba.it (S.D.G.); milviaelena.diclemente@poliba.it (M.E.D.C.); francesco.todaro@poliba.it (F.T.); michele.notarnicola@poliba.it (M.N.)

2 Dipartimento di Scienze dell'Ingegneria Civile e dell'Architettura, Politecnico di Bari, 4, Via E. Orabona, 70125 Bari, Italy; ubaldo.ayr@poliba.it (U.A.); stefania.liuzzi@poliba.it (S.L.)

3 Faculty of Civil and Environmental Engineering and Architecture, Bydgoszcz University of Science and Technology, Al. Kaliskiego 7, 85-796 Bydgoszcz, Poland; magdalena.dobiszewska@pbs.edu.pl

* Correspondence: andrea.petrella@poliba.it; Tel.: +39-(0)8-0596-3275

**Citation:** Petrella, A.; De Gisi, S.; Di Clemente, M.E.; Todaro, F.; Ayr, U.; Liuzzi, S.; Dobiszewska, M.; Notarnicola, M. Experimental Investigation on Environmentally Sustainable Cement Composites Based on Wheat Straw and Perlite. *Materials* **2022**, *15*, 453. https://doi.org/10.3390/ma15020453

Academic Editor: Konstantinos G. Dassios

Received: 21 December 2021
Accepted: 5 January 2022
Published: 7 January 2022

**Publisher's Note:** MDPI stays neutral with regard to jurisdictional claims in published maps and institutional affiliations.



**Abstract:** Environmentally sustainable cement mortars containing wheat straw (Southern Italy, Apulia region) of different length and dosage and perlite beads as aggregates were prepared and characterised by rheological, thermal, acoustic, mechanical, optical and microstructural tests. A complete replacement of the conventional sand was carried out. Composites with bare straw (S), perlite (P), and with a mixture of inorganic and organic aggregates (P/S), were characterised and compared with the properties of conventional sand mortar. It was observed that the straw fresh composites showed a decrease in workability with fibre length decrease and with increase in straw volume, while the conglomerates with bare perlite, and with the aggregate mixture, showed similar consistency to the control. The thermal insulation of the straw mortars was extremely high compared to the sand reference (85–90%), as was the acoustic absorption, especially in the 500–1000 Hz range. These results were attributed to the high porosity of these composites and showed enhancement of these properties with decrease in straw length and increase in straw volume. The bare perlite sample showed the lowest thermal insulation and acoustic absorption, being less porous than the former composites, while intermediate values were obtained with the P/S samples. The mechanical performance of the straw composites increased with length of the fibres and decreased with fibre dosage. The addition of expanded perlite to the mixture produced mortars with an improvement in mechanical strength and negligible modification of thermal properties. Straw mortars showed discrete cracks after failure, without separation of the two parts of the specimens, due to the aggregate tensile strength which influenced the impact compression tests. Preliminary observations of the stability of the mortars showed that, more than one year from preparation, the conglomerates did not show detectable signs of degradation.

**Keywords:** cement mortar; wheat straw; perlite; thermal insulation; acoustic absorption; secondary raw materials

## 1. Introduction

Today, concern for environmental protection is growing, especially in the agro-food industry which generates wastes from direct consumption of primary products. Most of these by-products are non-hazardous and are currently underutilised or simply wasted. For this reason, the concept of bioeconomy is spreading as a new approach to production that gives new life to materials which would otherwise be destined for destruction [1–8]. Accordingly, the recycling of agro-food wastes as biofuels [9–11], fertilisers [12,13], energy [14], chemicals [15] and sorbents [16–20] is considered a valuable alternative to landfilling.

The main agricultural products in Italy include sugar beet, wheat, corn, tomatoes, oranges, potatoes, apples, barley, and rice. Therefore, a large amount of waste needs to be disposed of.

In the context of an environmental and sustainable approach, considerable effort is being invested in the exploitation of renewable cheap agricultural residues for the development of eco-building materials to limit greenhouse gas emissions, save natural resources and develop more energy efficient buildings [21]. The aim of bio-architecture is to construct healthy buildings with little ecological impact based on the use of sustainable, eco-friendly and cheap materials [7,22–28], such as cellulose fibres, which are among the most suitable secondary raw materials for this purpose. Specifically, wheat straw has been used as an aggregate in many low density and environmental safe, construction materials which have shown valuable mechanical, thermal and acoustic properties [22–24,29–37].

In this paper, eco-sustainable cement conglomerates, containing wheat straw from the Apulia region, Italy (33–38% cellulose, 26–32% hemicelluloses, 17–19% lignin [38]) and perlite beads, were prepared and characterised by rheological, thermal, acoustic, mechanical and microstructural measurements. A complete replacement of the conventional sand aggregate was carried out with straw cuttings of different length and dosage and with perlite beads. Composites with bare straw (S), perlite (P), and with a mixture of organic and inorganic aggregates (P/S), were characterised and compared with the properties of conventional sand mortar [39–41]. The addition of the perlite beads to the straw mixture was carried out to improve the mechanical properties of the conglomerates with little modification of the thermal insulation.

Many studies have considered the possibility of using specific treatments to prevent the degradation of this type of construction material resulting from the dissolution of the main constituents of the straw, ascribed to water absorption and alkaline pore solution [42–45]. Modification of the matrix composition, by addition of pozzolanic compound addition or by applying a carbonation process, was carried out to overcome the problems associated with the presence of alkaline compounds [46–51]. The stability of the natural fibres was improved by applying specific procedures, such as interface coatings, chemical structure modification, chemical products additions or combined processes [29,52–61].

The main purpose of the present research was to obtain eco-friendly thermo-insulating cement composites with natural and local by-products as aggregates for indoor applications [62–65]. The idea was to prepare and characterise these materials without the addition of fillers, additives, matrix modifiers or specific straw treatments, which, in some cases, are based on the use of reagents, such as silanes, gasoil, varnish, sodium hydroxide and sulfuric acid. In terms of the circular economy, these conglomerates were in accordance with current policies for environmental sustainability. Worthwhile processes were being carried out since these artifacts were prepared by a cheap process in which pre-treatment of renewable aggregates and complex or expensive procedures were not required [37,66–70].

## 2. Materials and Methods

Cement mortars were prepared with CEM II A-LL 42.5 R [39] from Buzzi Unicem ($Rc_{(2\ days)}$ > 25.0 MPa, $Rc_{(28\ days)}$ > 47.0 MPa) which is characterised by 80–94% clinker, 6–20% limestone LL (<0.2% organic carbon), gypsum (0–5%), minor additional constituents, and which shows 3100–4400 $cm^2/g$ Blaine specific surface area. Natural wheat straw was used as total replacement of sand with cuttings (1.5–2.5 mm diameter) of variable length (0.4–0.6 cm, 1.3–1.7 cm, 3.4–3.7 cm, 5.8–6.2 cm).

Expanded perlite (P) (3–4 mm size range) is an inorganic material derived from volcanic rock with the following chemical composition: $SiO_2$ 74.5%, $Al_2O_3$ 12.3%, $K_2O$ 4.2%, $Na_2O$ 4%, $Fe_2O_3$ 1%, CaO 1.4%. It was provided by Maltek Industrie S.r.l., Terlizzi, Bari, Italy. It is chemically inert, sterile and incombustible and with a granular form obtained after heat treatment at 760–1100 °C of the silica material which induces expansion [71,72].

Figure 1A shows the straw before use and after cutting (see inset), while Figure 1B shows a picture of the perlite beads. Figure 2A shows an SEM image of a perlite bead surface, while Figure 2B shows its closed porosity.

**Figure 1.** (**A**) Straw before use and, in the inset, after cutting. (**B**) Perlite beads.

In the case of the mortar preparation, a volume replacement of the conventional aggregate was carried out due to the lower density of the cellulose fibres and of the perlite beads with respect to sand [41]. The samples S1, S2, S3 and S4 were prepared with cuttings of different length (0.4–0.6 cm, 1.3–1.7 cm, 3.4–3.7 cm, 5.8–6.2 cm); the P sample was prepared with bare perlite, while the P/S1, P/S2, P/S3 and P/S4 samples were prepared with a mixture of perlite and straw. For all these samples, the volume of the aggregate was equal to 400 mL. The choice of this volume was a compromise to obtain comparable values for straw and perlite. Finally, S3A, S3B and S3C samples were prepared with the same straw length (3.4–3.7 cm) but with different volumes of aggregate (340 mL, 470 mL and 550 mL).

The conglomerates were prepared with tap water and with a water/cement ratio equal to 0.5 as in the case of conventional mortar preparation characterised by 225 g of water, 450 g of cement and 1350 g of normalised sand [40,73]. All the rheological, thermo-acoustic and mechanical measurements were compared with the results obtained with this normalised sand control. Table 1 reports the composition of all the mortars that were prepared for the present investigation.

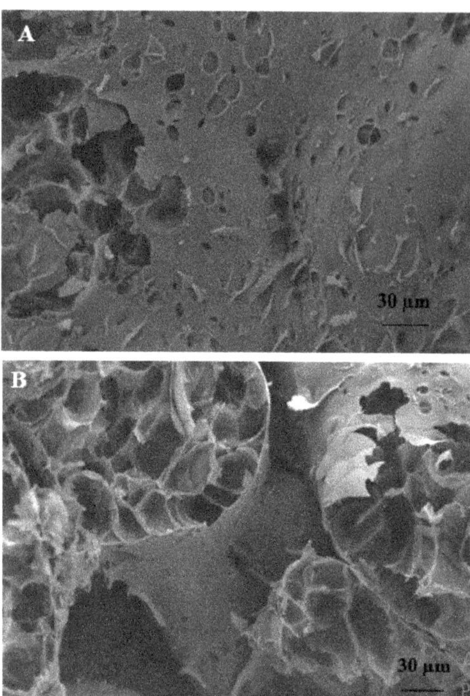

**Figure 2.** SEM image of (**A**) a perlite bead surface and (**B**) a perlite bead inner structure.

**Table 1.** Mortars composition.

| Sample | Type | Cement (g) | Water ($cm^3$) | Perlite Weight (g) | Perlite Volume ($cm^3$) | Straw Volume ($cm^3$) | Straw Weight (g) |
|---|---|---|---|---|---|---|---|
| S1 | straw 0.5 ± 0.3 cm | 450 | 225 | 0 | 0 | 400 | 55 |
| S2 | straw 1.5 ± 0.3 cm | 450 | 225 | 0 | 0 | 400 | 44 |
| S3 | straw 3.5 ± 0.3 cm | 450 | 225 | 0 | 0 | 400 | 34 |
| S3A | straw 3.5 ± 0.3 cm | 450 | 225 | 0 | 0 | 340 | 29 |
| S3B | straw 3.5 ± 0.3 cm | 450 | 225 | 0 | 0 | 470 | 40 |
| S3C | straw 3.5 ± 0.3 cm | 450 | 225 | 0 | 0 | 550 | 47 |
| S4 | straw 6.0 ± 0.4 cm | 450 | 225 | 0 | 0 | 400 | 25 |
| P | perlite 3–4 cm | 450 | 225 | 42 | 400 | 0 | 0 |
| P/S1 | perlite 3–4 cm/straw 0.5 cm | 450 | 225 | 21 | 200 | 200 | 27.5 |
| P/S2 | perlite 3–4 cm/straw 1.5 cm | 450 | 225 | 21 | 200 | 200 | 22 |
| P/S3 | perlite 3–4 cm/straw 3.5 cm | 450 | 225 | 21 | 200 | 200 | 17 |
| P/S4 | perlite 3–4 cm/straw 6.0 cm | 450 | 225 | 21 | 200 | 200 | 12.5 |

The rheology of the fresh mixtures was evaluated using a flow-test [74]. Thermal and acoustic tests were carried out with cylindrical specimens ($\varphi$ = 100 mm; H = 50 mm) after 28 days curing. Thermal measurements were based on the analysis of temperature response of dried specimens to heat flow impulses through a heating probe applied onto the surface of the sample [75]. For this purpose, an ISOMET 2104 device, from Applied Precision

Ltd. (Bratislava, Slovakia), was used for the tests. An estimation of the thermal diffusivity ($\alpha$) and thermal conductivity ($\lambda$) was obtained by comparison between the experimental temperature values and the analytical solution of the heat conduction equation. Acoustic absorption data were obtained after emission of a pure tone of known frequency at 250, 500, 1000 and 1600 Hz through a Kundt tube [76] characterised by a diameter sufficiently small with respect to the wavelength of the sound emission for stationary conditions measurements. A loudspeaker was positioned at one end of the tube and the sample placed at the other end.

Compression tests were carried out with a loading rate of 2400 $\pm$ 200 N/s on twelve semi-prisms which were obtained from flexural tests on six prisms (40 $\times$ 40 $\times$ 160 mm) at 50 $\pm$ 10 N/s loading rate [40]. For this purpose, a MATEST device, Milan, Italy, was used. In the case of the impact resistance tests, a steel ball (63 mm diameter) was placed on the upper surface of a specimen and a 4.50 kg weight was dropped from a height of 45 cm; after evaluation of the number of blows before fracture, the energy absorbed by the sample was obtained [77].

The aggregates and the conglomerates were also characterised by scanning electron microscope (SEM) and energy dispersive X-ray (EDX) analysis. For this purpose, an electron microscope FESEM-EDX Carl Zeiss Sigma 300 VP (Carl Zeiss Microscopy GmbH, Jena, Germany) was used after sputtering the samples with graphite (Sputter Quorum Q150 from Quorum Technologies Ltd., East Sussex, UK). The specimens were also characterised by an optical microscope (Dyno-lyte Digital Microscope, New Taipei City, Taiwan), while porosimetric measurements were carried out by Ultrapyc 1200e Automatic Gas Pycnometer (Quantachrome Instruments, Boynton Beach, FL, USA). For this, helium gas was used.

## 3. Results and Discussion

### 3.1. Rheological Tests

The rheological tests carried out by the flow-test method enabled understanding of the flow and deformation of the fresh conglomerates, by modification of the aggregate composition and distribution, maintaining the same water and cement dosage. Figure 3 shows the workability of the non-consolidated specimens obtained with the flow-test in comparison with the conventional normalised sand mortar [74]. Fresh conglomerates with the least length of straw (S1 and S2) were drier than the normalised mortar ($-90\%$ and $-50\%$ respectively), with specific reference to the S1 sample. The mortars named S3 and S4 contained more fluid, with the S3 mixture showing the same workability as the reference. These results were attributed to the size of the cuttings—in particular, the shortest fibres showed the highest specific surface which determined an increase in water absorption with consequent reduction in workability. The mixture with perlite (P) showed similar consistency to the control, while the straw/perlite mortars showed intermediate values between bare straw and bare perlite composites. In this respect, the presence of perlite improved the fluidity of the S1 and S2 mortars; the P/S1 and P/S2 flows were, respectively, in the range $-30\%$ to $-25\%$ with respect to the control. The P/S3 and P/S4 mixtures showed similar consistency to the normalised mortar. It was also observed that the fresh composites showed a sensible decrease in the flow with increase in straw volume due to the increasing water absorption which determined the manufacture of the dry materials, as in the case of the S3B and S3C samples ($-60\%$ and $-100\%$, respectively).

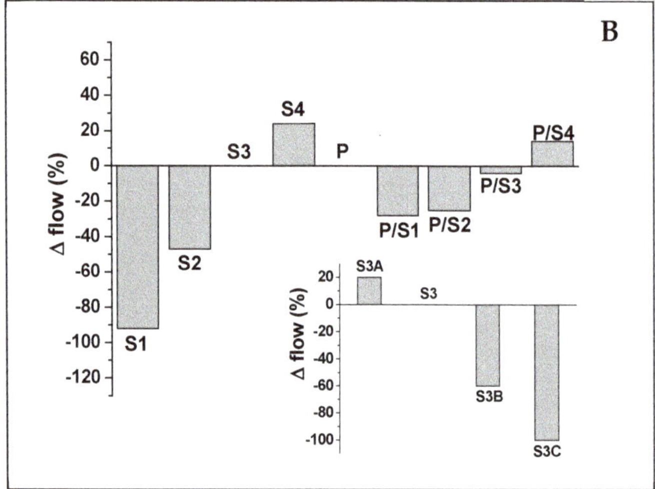


**Figure 3.** (**A**) Flow-test apparatus. (**B**) Flow-test results of the S, P and P/S samples with respect to the normalised mortar (Control).

## *3.2. Thermal and Acoustic Measurements*

The thermal conductivities and diffusivities of the straw mortars were very much lower compared to the sand reference which showed a thermal conductivity in the range of 1.8–2.0 W/mK and a thermal diffusivity in the range of 1.2–1.4 × $10^{-6}$ $m^2/s$. Specifically, the thermal conductivity decrease (%) for the entire set of straw based composites (S1, S2, S3, S4, S3A, S3B and S3C) was in the range of 86–91%, while the thermal diffusivity decrease (%) was in the range of 85–89% (Figures 4 and 5).

The reduction in thermal conductivity and diffusivity of the straw-containing mortars can be attributed to the hollow lumen structure of the organic aggregate, as observed in the analysis of the cross-section of the straw and can be ascribed to the action of the aggregate in modifying the structure of the mortars (Figure 6A) [29–31]. From the SEM detections it can be seen that poor adhesion of the organic straw fibres to the inorganic cement matrix was responsible for the formation of voids at the organic/inorganic interface. For all these reasons, a reduction in the specific mass (in the range of 39–52%) and an increase in the porosity (in the range of 45–54%) of the composites was observed (Table 2) [41].

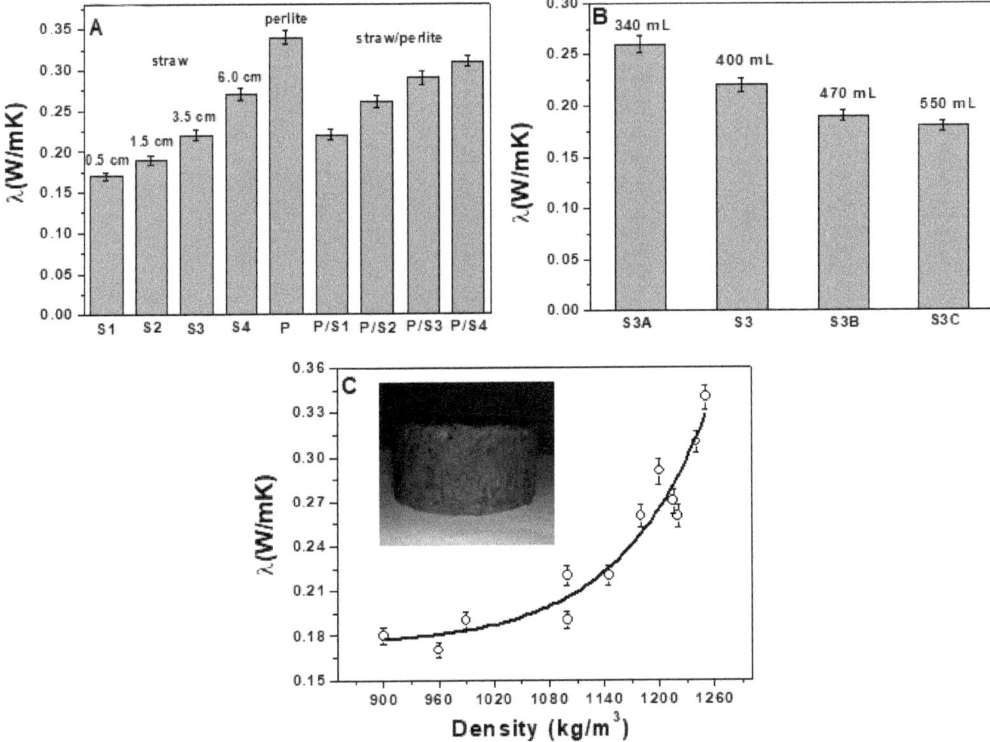


**Figure 4.** (**A**) Thermal conductivity of the specimens with the same volume of aggregates (400 mL). (**B**) Thermal conductivity of the S3 specimens with different volume of aggregates. (**C**) Exponential increase in the thermal conductivity with density increase. In the inset: image of a sample for thermal detections.

The generation of voids in the cement matrix contributes to limiting heat transport with increase in thermal insulation [29–31]. The composite S1, with the lowest size of fibres, showed the highest thermal insulation ($\lambda$ = 0.17 W/mK, $\alpha = 0.13 \times 10^{-6}$ m$^2$/s) at the same volume of aggregate in the mixture, which tended to decrease with increase in straw length (Figures 4A and 5A). This result can be explained by the highest specific surface of this type of fibre which is responsible of the generation of the highest percentage of voids at the interface, together with the increased number of voids attributed to the porous structure of the bare aggregate. In fact, the S1 specimen showed the highest porosity (48%) and the lowest density (960 Kg/m$^3$) among the straw composites (S2, S3 and S4, Table 2). This demonstrates the effect of different length fibres on the conglomerates

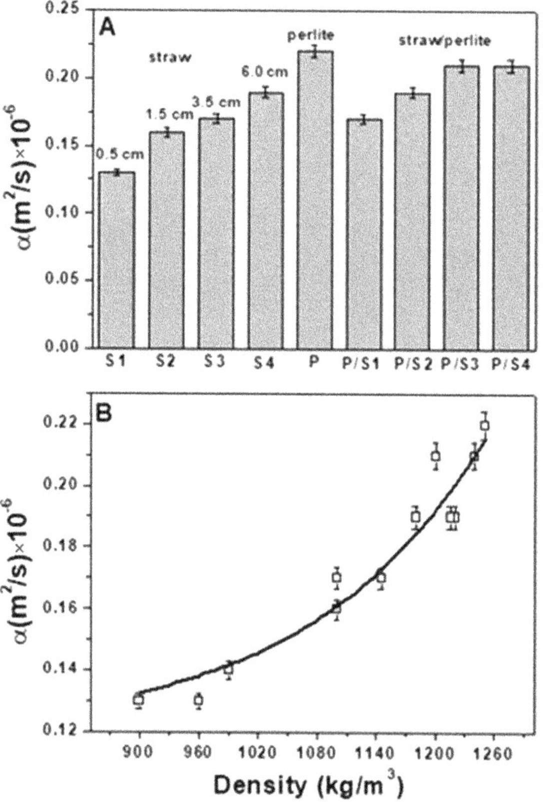


**Figure 5.** (**A**) Thermal diffusivity of the specimens with the same volume of aggregates (400 mL). (**B**) Exponential increase in the thermal diffusivity with density increase.

**Table 2.** Density and porosity of the mortars.

| Sample | Density (kg/m$^3$) | Porosity (%) |
|---|---|---|
| S1 | 960 | 48 |
| S2 | 1100 | 46 |
| S3 | 1145 | 44 |
| S3A | 1220 | 40 |
| S3B | 990 | 46 |
| S3C | 900 | 48 |
| S4 | 1215 | 41 |
| P | 1250 | 37 |
| P/S1 | 1100 | 44 |
| P/S2 | 1180 | 42 |
| P/S3 | 1200 | 40 |
| P/S4 | 1240 | 40 |

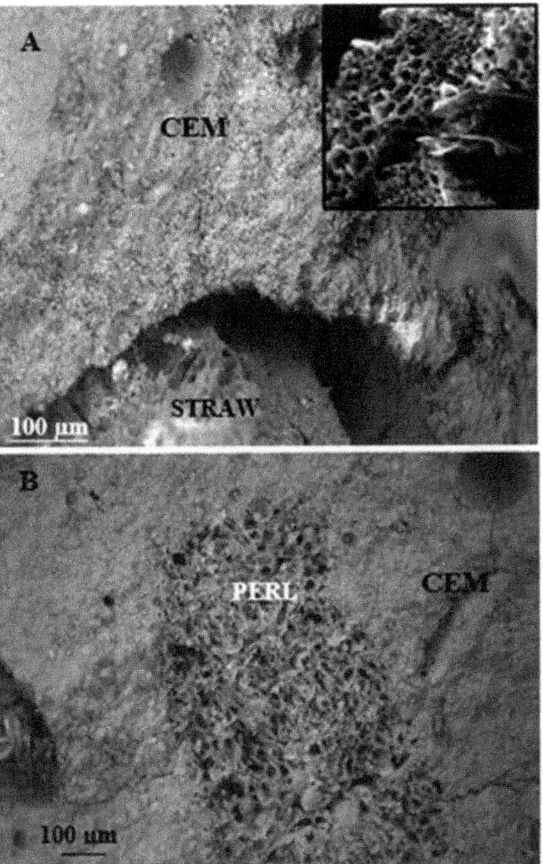


**Figure 6.** (**A**) SEM image of the cement/straw interface, in the inset: porous structure of the straw. (**B**) SEM image of the cement/perlite interface.

The results show that an increase in straw content decreased the thermal conductivity and diffusivity of the composites (Figures 4 and 5). In fact, the S3C sample, with the highest dosage and volume of fibres (550 $cm^3$), was characterised by the lowest density (900 $kg/m^3$), the highest porosity (48%) and the lowest thermal conductivity (0.18 W/mK) and diffusivity ($\alpha = 0.13 \times 10^{-6}$ $m^2/s$), with respect to S3B, S3 and S3A composites which were characterised by an increase in the specific mass (990 $kg/m^3$, 1145 $kg/m^3$, 1220 $kg/m^3$, respectively), decrease in porosity (46%, 44%, 40%, respectively) and increase in thermal conductivity (0.19 W/mK, 0.22 W/mK and 0.26 W/mK, respectively) and diffusivity ($0.14 \times 10^{-6}$ $m^2/s$, $0.17 \times 10^{-6}$ $m^2/s$ and $0.19 \times 10^{-6}$ $m^2/s$, respectively). This result could be due to the increasingly lower encapsulation of the fibres in the cement matrix which increased the voids at the organic/inorganic interface. From the rheological tests, it was also observed that a sensible decrease in the flow of the fresh conglomerates with rise in the straw volume occurred which caused the production of increasingly dry specimens with consequent increase in the porosity of the hardened artifacts. The sample with bare perlite (P) showed the highest value of density (1250 $Kg/m^3$) and the lowest porosity (37%) among all the other lightweight samples; the thermal conductivity was in the range of 0.34 W/mK, accordingly, this specimen resulted in the lowest thermal insulating properties. The perlite thermal conductivity was ~50% higher than the value obtained with the S1 sample with a density increase and porosity decrease in the range of ~23%. Figure 6B demonstrates these results; good adhesion of the perlite to the cement paste can be observed due to the beads roughness and the similar chemical compounds in both the mixture components (silicates and aluminates) [70]. Thus, the porosity of this type of mortar was exclusively associated with the closed porosity of the perlite (Figure 2B) and not the presence of empty spaces at the ligand/aggregate interface, as in the case of the straw-based samples. The straw/perlite samples (P/S1, P/S2, P/S3 and P/S4) showed intermediate values of thermal conductivity and diffusivity as a result of the intermediate values of density and porosity. An exponential increase in thermal conductivity and diffusivity was observed with increase in conglomerate density (Figures 4C and 5B).

To evaluate the acoustic characteristics of the cellulose-cement composites, the normal incident absorption coefficient (α) was determined. When a sound wave strikes a material, a portion of the sound energy is reflected while a portion is absorbed. This coefficient is the ratio of the absorbed energy to the total incident energy and is determined by the Kundt impedance tube [78]. It is calculated as:

$$\alpha = 1 - \rho \tag{1}$$

where ρ is the reflection coefficient of the acoustic energy, expressed as the ratio between the reflected and the incident energy. Figure 7 shows the acoustic absorption data carried-out at 250, 500, 1.000 and 1.600 Hz. Specifically, the percentage increase for the entire set of straw-based composites (S1, S2, S3, S4, S3A, S3B and S3C) was in the range of 10–54% at 250 Hz, 77–89% at 500 Hz, 27–54% at 1000 Hz and 54–70% at 1600 Hz with respect to the control which showed the following results: 9% at 250 Hz, 5% at 500 Hz, 11% al 1000 Hz and 6% at 1600 Hz. As previously reported, this result can be ascribed to the intrinsic porosity of the natural aggregate (inset Figure 6A) and to the action of straw in modifying the structure of the mortars by creating pores in the cement matrix (Figure 6A) with consequent reduction in the specific mass [29–31]. Accordingly, in these specimens, straw-induced formation of voids occurred where acoustic energy was likely to be attenuated, in particular, at 500 Hz where, after resonance phenomena, the closed cavities might play a major role [34,62,79,80]. The S1 sample, with the lowest size of fibres, at the same volume of aggregate in the mixture, showed the highest acoustic absorption at all the frequencies, with specific reference to the 500–1.000 Hz range. This value decreased with increase in straw length (S2, S3 and S4 specimens) because of the decrease in composite porosity (Figure 7A and Table 2).

The increase in the straw content and volume determined an increase in acoustic absorption (Figure 7B). The S3C sample, with the highest dosage and volume of fibres, showed at 500 and 1000 Hz values in the range of 33% and 24%, respectively, while the S3B (31% and 21%, respectively), the S3 (27% and 20%, respectively) and the S3A (22% and 15%, respectively) were characterised by increasingly lower values of α at lower straw content. This result can be explained by the highest porosity of S3C with respect to S3B, S3 and S3A composites which were characterised by a decrease in porosity with increase in the straw volume in the matrix. The sample with bare perlite (P) showed the lowest acoustic absorption because of the highest value of density (1250 $Kg/m^3$) and the lowest porosity (37%) among all the other lightweight samples. The straw/perlite samples (P/S1, P/S2, P/S3 and P/S4) showed intermediate values of α as a result of the intermediate values of density and porosity.

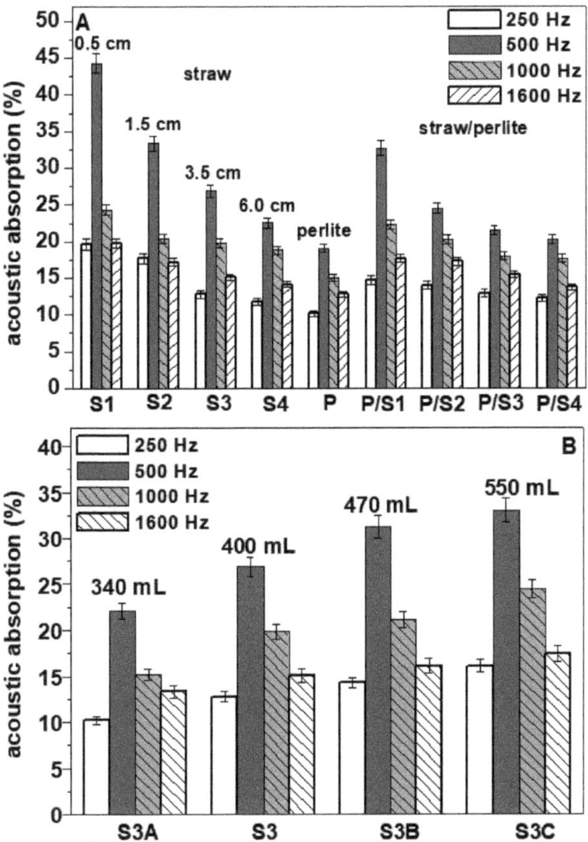


**Figure 7.** (**A**) Acoustic absorption of the specimens with the same volume of aggregates (400 mL). (**B**) Acoustic absorption of the S3 specimens with different volume of aggregates.

### 3.3. Mechanical Tests

Flexural and compressive strengths (at 28, 60, and 90-days ageing) of the samples are shown in Table 3 and Figure 8A,B. A general increase between 24 and 60 days and a final stabilisation between 60 and 90 days was observed. Moreover, Figure 8A,B show an exponential increase in mechanical resistance with the specific mass of the conglomerates.

From a general overview of data referring to straw-based composites (S1, S2, S3, S4, S3A, S3B and S3C), a sensible decrease in the mechanical performances with respect to the sand reference was observed which showed flexural and compressive resistances at 28-days ageing in the range of 8.5–9.0 MPa and 48–50 MPa, respectively.

The conglomerates with bare straw aggregates showed flexural resistances in the range of 1.3–2.5 MPa and compressive resistances in the range of 1.6–6.2 MPa. The decrease in mechanical strength can be attributed to the already mentioned low density of the straw fibres compared to the cement paste and to lack of adhesion of the organic aggregate to the cement paste [22,30,31]. Straw particles showed lower stiffness than the surrounding cement paste; accordingly, under loading, cracks initiated around the straw accelerated the failure in the matrix. The increase in porosity associated with the voids at the fibre–matrix interfaces further affected the lowering of mechanical performances.

**Table 3.** Flexural and compressive strengths of the S, P and S/P mortars at 28, 60 and 90-days curing.

| Sample | Density (kg/m³) | $R_f$ (MPa) 28 Days | $R_f$ (MPa) 60 Days | $R_f$ (MPa) 90 Days | $R_C$ (MPa) 28 Days | $R_c$ (MPa) 60 Days | $R_c$ (MPa) 90 Days |
|---|---|---|---|---|---|---|---|
| S1 | 960 | 1.3 | 1.7 | 1.6 | 1.6 | 2.0 | 1.9 |
| S2 | 1100 | 1.7 | 2.0 | 2.1 | 2.4 | 2.7 | 2.8 |
| S3 | 1145 | 2.1 | 2.2 | 2.3 | 3.6 | 3.7 | 3.5 |
| S3A | 1220 | 2.5 | 2.6 | 2.4 | 4.1 | 4.3 | 4.3 |
| S3B | 990 | 1.9 | 2.0 | 2.3 | 3.1 | 3.4 | 3.3 |
| S3C | 900 | 1.7 | 2.0 | 2.0 | 2.4 | 2.5 | 2.7 |
| S4 | 1215 | 2.5 | 2.8 | 2.6 | 6.2 | 6.2 | 6.4 |
| P | 1250 | 3.5 | 4.3 | 4.5 | 18.8 | 19.3 | 19.7 |
| P/S1 | 1100 | 2.4 | 2.7 | 2.6 | 5.5 | 5.6 | 5.5 |
| P/S2 | 1180 | 2.6 | 2.9 | 3.0 | 9.8 | 10.1 | 10.4 |
| P/S3 | 1200 | 2.9 | 3.3 | 3.2 | 11.8 | 11.8 | 11.9 |
| P/S4 | 1240 | 3.2 | 3.6 | 3.8 | 15.1 | 15.3 | 15.2 |

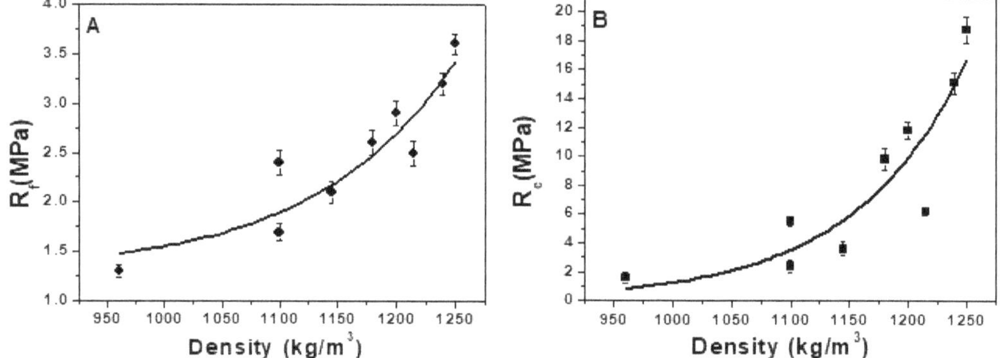


**Figure 8.** (**A**) Flexural and (**B**) compressive strengths of the S, P and S/P mortars with the same volume of aggregate (400 mL) as a function of the composite density (28 days ageing).

Figure 9A shows that these types of composites presented a good aggregate distribution and straw–matrix compatibility with the cement paste around and inside the fibres [21].

Moreover, it was observed that an increase in the resistances, with increase in straw length at the same volume of aggregate in the mixture, was also associated with increasing density and decreasing porosity of the conglomerates. The S1, S2, S3 and S4 specimens showed flexural strengths corresponding to 1.3 MPa, 1.7 MPa, 2.1 MPa and 2.5 MPa, respectively, while they showed compressive strengths corresponding to 1.6 MPa, 2.4 MPa, 3.6 MPa and 6.2 MPa, respectively.

The increase in straw content/volume in the composites decreased the flexural and compressive strengths, as shown in Table 3. Flexural strength at 28 days decreased from 2.5 MPa to 1.7 MPa, while compressive strength decreased from 6.2 MPa to 2.4 MPa for composites containing from 340 mL to 550 mL straw volume, respectively. This result was associated with decrease in encapsulation of the fibres in the cement matrix at increasing straw volume which increased the porosity of the samples. During the flexural load application, the interference of the nearby fibres induced a loss of the straw/matrix bonding.

Accordingly, the fibres were pulled out from the matrix and considerable energy was lost from the system in the form of frictional energy [22,23,66]. After the breakage, a separation of the two parts of the samples was not observed, with the two semi-prisms of the samples still connected by the fibres. This effect was associated with straw tensile strength [30,81] and was affected by the density decrease and by the structure of the fibres which showed a horizontal arrangement during the mortar preparation [30] that was confirmed after final separation of the two semi-prisms. In this respect, it was observed that the fibres were encapsulated in one of the two parts while pulling out from the other. Figure 9B,C show the holes derived from the pull-out of the cellulose fibres from the matrix, while Figure 10A,B show the horizontal arrangement of the aggregate after rupture [30]. Figure 10C shows that a real collapse of the specimen was not observed after breakage but only cracks ascribed to the plastic behaviour, mostly detected at high straw content.

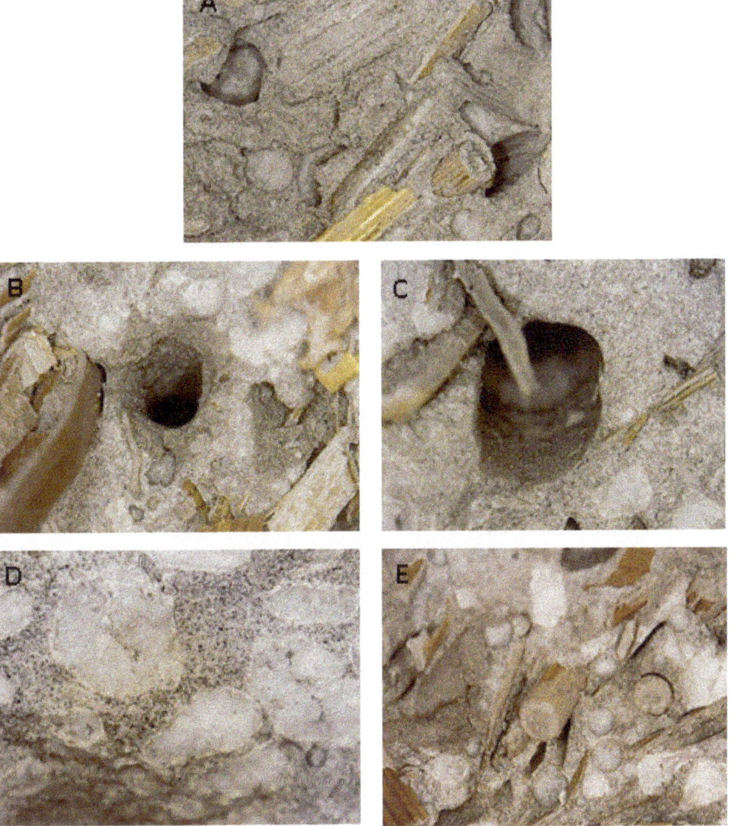


**Figure 9.** Failure plane of the (**A**) S1 sample, (**B**,**C**) holes caused by the pull-out of the straw, sections of the (**D**) P sample and of the (**E**) P/S1 sample.

The sample with bare perlite (P) showed the highest mechanical resistances ($R_f$ = 3.5 MPa and $R_c$ = 18.8 MPa) due to the higher stiffness of the silica aggregate with respect to straw and to the good adhesion to the ligand paste at the interface which increased the density of the mortar. Figure 9D shows an image of the surface of this composite after rupture and its good particle distribution. The straw/perlite samples (P/S1, P/S2, P/S3 and P/S4) showed intermediate values as a result of the intermediate values of density and porosity. Figure 9E shows that these composites, with a mixture of aggregates, also presented a good particle/fibre distribution and straw–matrix compatibility with the cement paste around and inside the fibres.

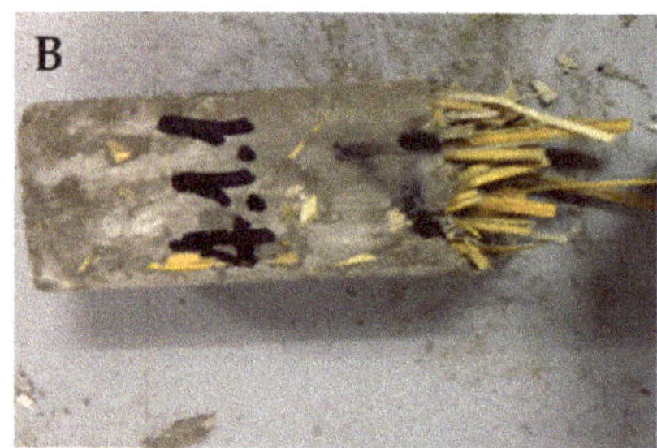


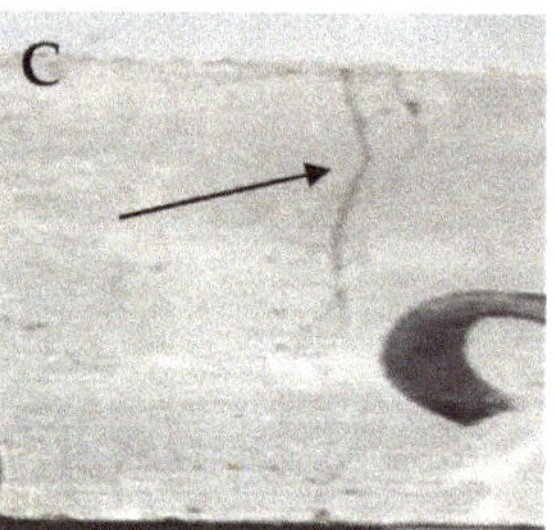


**Figure 10.** Fibres horizontal arrangement in (**A**) P/S3 and (**B**) S4 specimens. (**C**) Discrete cracks after rupture in the straw specimens (evidenced by the arrow), with the two parts of the sample still connected by the organic aggregate.

From a general point of view, the P/S3 and P/S4 samples can be considered interesting composites with good workability (similar to the normalised mortar), low thermal conductivities (~0.30 W/mK) and good mechanical properties ($R_f$ = 2.9 MPa and $R_c$ = 11.8 MPa, in the case of the P/S3 specimen, $R_f$ = 3.2 MPa and $R_c$ = 15.1 MPa, in the case of the P/S4 specimen).

The impact compression tests, obtained with the experimental apparatus of Figure 11A, showed that the straw samples (S1, S2, S3, S4, S3A, S3B and S3C) were characterised by high energy absorption capacity, with specific reference to the S1 (Figure 11B) and S3C (Figure 11C) samples. The toughness of these composites was improved as fibre length decreased and as fibre volume increased (increasingly low specific mass) and was characterised by a deep groove before complete failure (Figure 12A) [82,83]. The horizontal arrangement after pull-out was confirmed, as shown in Figure 12B relative to the S3 sample [30].

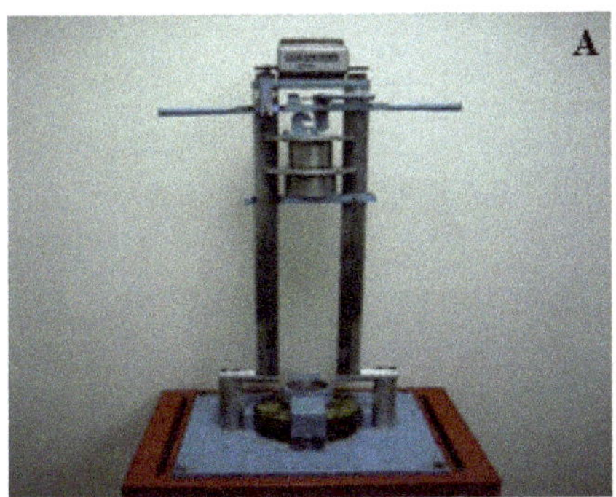


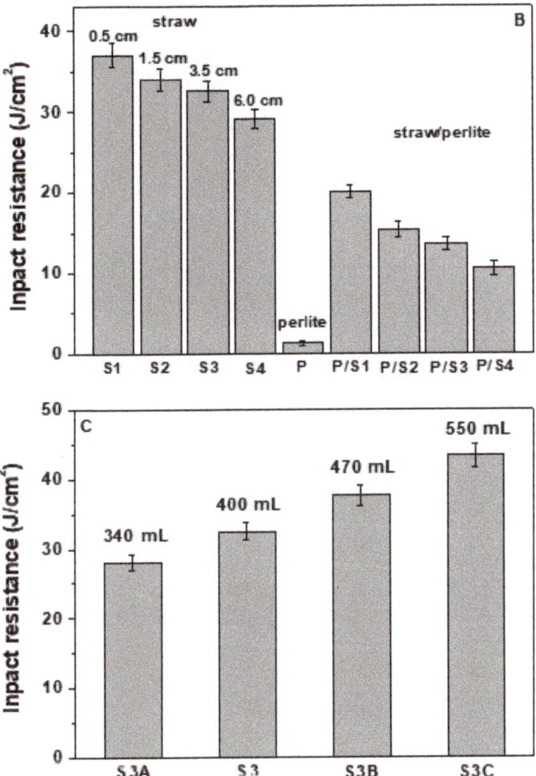


**Figure 11.** (**A**) Impact resistance apparatus. (**B**) Impact resistance of the specimens with the same volume of aggregates (400 mL). (**C**) Impact resistance of the S3 specimens with different volume of aggregates.

As observed in the flexural strength tests, a separation of the parts of the sample was not observed because the parts were still connected by the fibres (Figure 12A,B). The high straw tensile strength affected the formation of cracks instead of an evident breakage [81] and this effect was observed particularly in samples with high straw dosage, as in the case of the S3C specimen (Figure 12C).

The perlite sample (P) was fragile and breakage occurred after a few blows due to the presence of the brittle aggregate (Figure 12D,E), while the samples with 50% of straw and 50% of perlite represented a compromise between energy absorption capacity attributed to the natural fibres and mechanical resistance attributed to the inorganic aggregate (Figure 11B).

**Figure 12.** Samples after the impact test. (**A**,**B**) S3 sample, (**C**) S3C sample, (**D**,**E**) perlite sample.

### 3.4. *Stability of the Composites*

From visual inspections and optical microscopical observations, the straw surface structure, evidenced in Figures 9 and 10B, was similar to the pristine (inset Figure 1A) and also retained its original color, both indications of negligible degradation occurring after chemical interaction between the fibres and the ligand paste [30]. Specifically, Figure 9E is an optical microscope image of the P/S1 sample after more than one year from the breakage and after curing at room temperature and at 75–80% relative humidity.

It has been reported that the degradation of vegetable fibres is associated with the presence of calcium hydroxide in the matrix at the basic pH of the cement paste [42,45,46] due to the dissolution of water-soluble plant compounds. Accordingly, the concentration of soluble calcium compounds was much higher than the concentration of silicon compounds in the areas of the cement paste close to the straw fibres. As a result, drawbacks, such as shift of the setting time, delay in mechanical strength development, and stiffness increase were evidenced [30]. Figure 13B,C show EDX analyses of the cement paste close to the straw fibres and far from the fibres (Figure 13A) in order to have a chemical characterisation of two different zones, specifically the Ca/Si ratio [84,85]. The sample was withdrawn in a straw-based specimen cured for more than one year in a humid environment (75–80%). In the present case, the chemical compositions were similar to the Ca/Si ratios, which is a further indication of negligible degradation of these composites together with the optical observations and the stable values of mechanical resistance in the range of 28–90 days. The reported results can be considered preliminary investigations on the features of the composites during time and not durability studies, but they can be very useful for the preparation of sustainable indoor cement artifacts based on cellulose fibres and perlite.

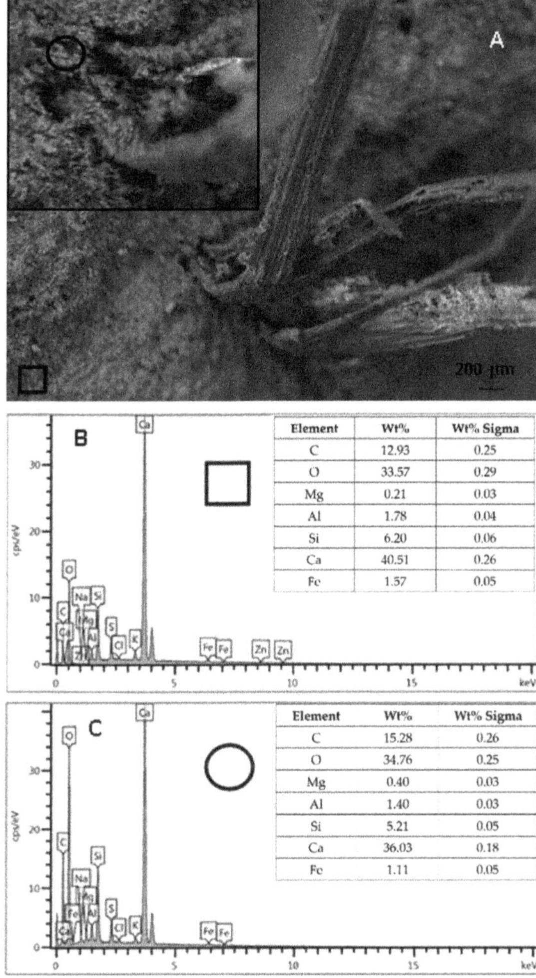


**Figure 13.** (**A**) SEM image of the cement/straw system with square spot showing the cement paste far from straw, in the inset: magnification of the cement paste close to straw and represented by circular spot. (**B**) EDX spectrum relative to the square spot (far from straw). (**C**) EDX spectrum relative to the circular spot (close to straw).

## 4. Conclusions

Eco-sustainable cement conglomerates containing untreated wheat straw and perlite beads as aggregates were prepared and characterised by rheological, thermal, acoustic, mechanical and microstructural measurements. A complete replacement of the conventional sand aggregate was carried out with straw cuttings of different length and dosage and with perlite beads. Composites with bare straw (S), perlite (P), and with a mixture of organic and inorganic aggregates (P/S), were characterised and compared with the properties of normalised sand mortar.

From the rheological characterisation, the fibres with the lowest length showed the highest specific surface which influenced water absorption with consequent reduction in workability (e.g., S1 and S2 samples, extremely dry). With an increase in straw length an increase in fluidity (S3 and S4) was observed because of the lower water absorption with respect to the former samples (i.e., plastic behaviour). Fresh composites also showed a sensible decrease in flow with increase in straw volume, with S3C extremely dry. The mixture with perlite (P) showed similar consistency to the control, while the straw/perlite mortars showed intermediate values between bare straw and bare perlite composites. All these samples had good workability useful for plastic castings. Specifically, the conglomerate with bare perlite and the P/S3 and P/S4 mixtures showed similar consistency to the control.

The thermal insulation of the straw mortars was extremely high compared to the sand reference. Specifically, the thermal conductivity quenching for the entire set of the straw-based composites was in the range of 86–91%. It was ascribed to the hollow lumen structure of the organic aggregate and to the poor adhesion of the straw fibres to the cement matrix which determined a reduction in the specific mass (39–52%) and an increase in porosity (45–54%) with respect to the reference. The results showed an enhancement of the thermal insulation with decrease in straw length and with increase in straw volume due to the increase in porosity of the composites. Lower thermal insulation was obtained with the mixtures of aggregates (P/S) because of the reduction in porosity associated with the presence of perlite which showed good adhesion to the cement paste.

The acoustic absorption of the straw mortars was extremely high compared to the sand reference, especially in the 500–1000 Hz range. These results were ascribed to the high porosity of these composites and showed an enhancement with decrease in straw length and with increase in straw volume.

A sensible decrease in mechanical performance with respect to the sand reference was obtained and the values increased with the length of the straw and decreased with the straw dosage. The addition of expanded perlite to the mixture allowed mortars to be obtained with an improvement in mechanical strength and negligible modification to thermal properties. Straw mortars showed discrete cracks after failure without separation of the two parts of the specimens due to the aggregate tensile strength.

The impact compression tests showed that the straw samples were characterised by high energy absorption capacity, with specific reference to the S1 and S3C samples. These parameters were improved as fibre length decreased and as fibre volume increased (increasingly low specific mass) and were characterised by a deep groove before complete failure.

Microscopical observations, after more than one year curing in 75–80% humidity, revealed negligible degradation of these composites, while mechanical tests showed stable values in the range of 28–90 days.

Based on the physical and mechanical results, non-structural indoor applications (e.g., panels, plasters) may be considered for these lightweight composites, with specific reference to conglomerates made of straw and perlite which can be considered a good compromise between thermo-acoustic and mechanical properties. It is important to underline the environmental advantages related to the recycling of agricultural waste adopting a safe and environmentally friendly process with respect to the circular economy.

**Author Contributions:** Conceptualization, M.N.; methodology, S.D.G.; validation, M.D.; formal analysis, A.P., M.E.D.C.; investigation, A.P., S.L., U.A.; resources, M.N.; data curation, F.T.; writing—original draft preparation, A.P., F.T.; writing—review and editing, A.P., M.N.; visualization, M.D.; supervision, M.N. All authors have read and agreed to the published version of the manuscript.

**Funding:** This research received no external funding.

**Institutional Review Board Statement:** Not applicable.

**Informed Consent Statement:** Not applicable.

**Data Availability Statement:** Data sharing not applicable.

**Acknowledgments:** Adriano Boghetich is acknowledged for SEM-EDX analysis and Regione Puglia (Micro X-ray Lab Project–Reti di Laboratori Pubblici di Ricerca, cod. n. 45 and 56). Acknowledgments to the DICATECh of the Polytechnic of Bari for SEM analyses. This article has been supported by the Polish National Agency for Academic Exchange under Grant No. PPI/APM/2019/1/00003.

**Conflicts of Interest:** The authors declare no conflict of interest.

## References

1. Hussain, Z.; Sajjad, W.; Khan, T.; Wahid, F. Production of bacterial cellulose from industrial wastes: A review. *Cellulose* **2019**, *26*, 2895–2911. [CrossRef]
2. Nayak, A.; Bhushan, B. An overview of the recent trends on the waste valorization techniques for food wastes. *J. Environ. Manag.* **2019**, *233*, 352–370. [CrossRef] [PubMed]
3. Kim, S.; Lee, Y.; Lin, K.Y.A.; Hong, E.; Kwon, E.E.; Lee, J. The valorization of food waste via pyrolysis: A review. *J. Clean. Prod.* **2020**, *259*, 120816. [CrossRef]
4. Sharma, P.; Gaur, V.K.; Kim, S.H.; Pandey, A. Microbial strategies for bio-transforming food waste into resources. *Bioresour. Technol.* **2020**, *299*, 122580. [CrossRef] [PubMed]
5. Antunes, A.; Faria, P.; Silva, V.; Brás, A. Rice husk-earth based composites: A novel bio-based panel for buildings refurbishment. *Constr. Build. Mater.* **2019**, *221*, 99–108. [CrossRef]
6. Spasiano, D.; Luongo, V.; Petrella, A.; Alfè, M.; Pirozzi, F.; Fratino, U.; Piccinni, A.F. Preliminary study on the adoption of dark fermentation as pretreatment for a sustainable hydrothermal denaturation of cement-asbestos composites. *J. Clean. Prod.* **2017**, *166*, 172–180. [CrossRef]
7. Petrella, A.; Petruzzelli, V.; Basile, T.; Petrella, M.; Boghetich, G.; Petruzzelli, D. Recycled porous glass from municipal/industrial solid wastes sorting operations as a lead ion sorbent from wastewaters. *React. Funct. Polym.* **2010**, *70*, 203–209. [CrossRef]
8. Todaro, F.; De Gisi, S.; Notarnicola, M. Contaminated marine sediment stabilization/solidification treatment with cement/lime: Leaching behaviour investigation. *Environ. Sci. Pollut. Res.* **2020**, *27*, 21407–21415. [CrossRef]
9. Gil, L.S.; Maupoey, P.F. An integrated approach for pineapple waste valorisation. Bioethanol production and bromelain extraction from pineapple residues. *J. Clean. Prod.* **2018**, *172*, 1224–1231.
10. Chintagunta, A.D.; Ray, S.; Banerjee, R. An integrated bioprocess for bioethanol and biomanure production from pineapple leaf waste. *J. Clean. Prod.* **2017**, *165*, 1508–1516. [CrossRef]
11. Guo, X.M.; Trably, E.; Latrille, E.; Carrere, H.; Steyer, J.P. Hydrogen production from agricultural waste by dark fermentation: A review. *Int. J. Hydrog. Energy* **2010**, *35*, 10660–10673. [CrossRef]
12. Tampio, E.; Marttinen, S.; Rintala, J. Liquid fertilizer products from anaerobic digestion of food waste: Mass, nutrient and energy balance of four digestate liquid treatment systems. *J. Clean. Prod.* **2016**, *125*, 22–32. [CrossRef]
13. Owamah, H.I.; Dahunsi, S.O.; Oranusi, U.S.; Alfa, M.I. Fertilizer and sanitary quality of digestate biofertilizer from the co-digestion of food waste and human excreta. *Waste Manag.* **2014**, *34*, 747–752. [CrossRef]
14. Kraiem, N.; Lajili, M.; Limousy, L.; Said, R.; Jeguirim, M. Energy recovery from Tunisian agri-food wastes: Evaluation of combustion performance and emissions characteristics of green pellets prepared from tomato residues and grape marc. *Energy* **2016**, *107*, 409–418. [CrossRef]
15. Pfaltzgraff, L.A.; Cooper, E.C.; Budarin, V.; Clark, J.H. Food waste biomass: A resource for high-value chemicals. *Green Chem.* **2013**, *15*, 307–314. [CrossRef]
16. Rizzi, V.; Gubitosa, J.; Fini, P.; Romita, R.; Nuzzo, S.; Cosma, P. Chitosan biopolymer from crab shell as recyclable film to remove/recover in batch ketoprofen from water: Understanding the factors affecting the adsorption process. *Materials* **2019**, *12*, 3810. [CrossRef]
17. Rizzi, V.; D'Agostino, F.; Gubitosa, J.; Fini, P.; Petrella, A.; Agostiano, A.; Semeraro, P.; Cosma, P. An alternative use of olive pomace as a wide-ranging bioremediation strategy to adsorb and recover disperse orange and disperse red industrial dyes from wastewater. *Separations* **2017**, *4*, 29. [CrossRef]
18. Ranieri, E.; Fratino, U.; Petrella, A.; Torretta, V.; Rada, E.C. Ailanthus Altissima and Phragmites Australis for chromium removal from a contaminated soil. *Environ. Sci. Pollut. Res.* **2016**, *23*, 15983–15989. [CrossRef] [PubMed]
19. Gorito, A.M.; Ribeiro, A.R.; Almeida, C.M.R.; Silva, A.M. A review on the application of constructed wetlands for the removal of priority substances and contaminants of emerging concern listed in recently launched EU legislation. *Environ. Pollut.* **2017**, *227*, 428–443. [CrossRef] [PubMed]
20. Stavrinou, A.; Aggelopoulos, C.A.; Tsakiroglou, C.D. A methodology to estimate the sorption parameters from batch and column tests: The case study of methylene blue sorption onto banana peels. *Processes* **2020**, *8*, 1467. [CrossRef]
21. Belhadj, B.; Bederina, M.; Makhloufi, Z.; Goullieux, A.; Quéneudec, M. Study of the thermal performances of an exterior wall of barley straw sand concrete in an arid environment. *Energy Build.* **2015**, *87*, 166–175. [CrossRef]

22. Ardanuy, M.; Claramunt, J.; Toledo Filho, R.D. Cellulosic fiber reinforced cement-based composites: A review of recent research. *Constr. Build. Mater.* **2015**, *79*, 115–128. [CrossRef]
23. Onuaguluchi, O.; Banthia, N. Plant-based natural fibre reinforced cement composites: A review. *Cem. Concr. Compos.* **2016**, *68*, 96–108. [CrossRef]
24. Yan, L.; Kasal, B.; Huang, L. A review of recent research on the use of cellulosic fibres, their fibre fabric reinforced cementitious, geo-polymer and polymer composites in civil engineering. *Compos. Part B Eng.* **2016**, *92*, 94–132. [CrossRef]
25. Mo, K.H.; Alengaram, U.J.; Jumaat, M.Z.; Yap, S.P.; Lee, S.C. Green concrete partially comprised of farming waste residues: A review. *J. Clean. Prod.* **2016**, *117*, 122–138. [CrossRef]
26. Aprianti, S.E. A huge number of artificial waste material can be supplementary cementitious material (SCM) for concrete production—A review part II. *J. Clean. Prod.* **2017**, *142*, 4178–4194. [CrossRef]
27. Paris, J.M.; Roessler, J.G.; Ferraro, C.C.; DeFord, H.D.; Townsend, T.G. A review of waste products utilized as supplements to Portland cement in concrete. *J. Clean. Prod.* **2016**, *121*, 1–18. [CrossRef]
28. Madurwar, M.V.; Ralegaonkar, R.V.; Mandavgane, S.A. Application of agro-waste for sustainable construction materials: A review. *Constr. Build. Mater.* **2013**, *38*, 872–878. [CrossRef]
29. Bederina, M.; Belhadj, B.; Ammari, M.S.; Gouilleux, A.; Makhloufi, Z.; Montrelay, N.; Quéneudéc, M. Improvement of the properties of a sand concrete containing barley straws–treatment of the barley straws. *Constr. Build. Mater.* **2016**, *115*, 464–477. [CrossRef]
30. Belhadj, B.; Bederina, M.; Makhloufi, Z.; Dheilly, R.M.; Montrelay, N.; Quéneudéc, M. Contribution to the development of a sand concrete lightened by the addition of barley straws. *Constr. Build. Mater.* **2016**, *113*, 513–522. [CrossRef]
31. Bentchikou, M.; Guidoum, A.; Scrivener, K.; Silhadi, K.; Hanini, S. Effect of recycled cellulose fibres on the properties of lightweight cement composite matrix. *Constr. Build. Mater.* **2012**, *34*, 451–456. [CrossRef]
32. Chabriac, P.A.; Gourdon, E.; Gle, P.; Fabbri, A.; Lenormand, H. Agricultural by-products for building insulation: Acoustical characterization and modeling to predict micro-structural parameters. *Constr. Build. Mater.* **2016**, *112*, 158–167. [CrossRef]
33. Mustapha, K.; Annan, E.; Azeko, S.T.; Kana, M.G.Z.; Soboyejo, W.O. Strength and fracture toughness of earth-based natural fiber-reinforced composites. *J. Compos. Mater.* **2016**, *50*, 1145–1160. [CrossRef]
34. Neithalath, N.; Weiss, J.; Olek, J. Acoustic performance and dumping behaviour of cellulose-cement composites. *Cem. Concr. Compos.* **2004**, *26*, 359–370. [CrossRef]
35. Roma, L.C., Jr.; Martello, L.S.; Savastano, H., Jr. Evaluation of mechanical, physical and thermal performance of cement-based tiles reinforced with vegetable fibers. *Constr. Build. Mater.* **2008**, *22*, 668–674. [CrossRef]
36. Toguyeni, D.Y.; Coulibaly, O.; Ouedraogo, A.; Koulidiati, J.; Dutil, Y.; Rousse, D. Study of the influence of roof insulation involving local materials on cooling loads of houses built of clay and straw. *Energy Build.* **2012**, *50*, 74–80. [CrossRef]
37. Xie, X.; Zhou, Z.; Jiang, M.; Xu, X.; Wang, Z.; Hui, D. Cellulosic fibers from rice straw and bamboo used as reinforcement of cement-based composites for remarkably improving mechanical properties. *Compos. B Eng.* **2015**, *78*, 153–161. [CrossRef]
38. Reddy, N.; Yang, Y. Biofibers from agricultural byproducts for industrial applications. *Trends Biotechnol.* **2005**, *23*, 22–27. [CrossRef] [PubMed]
39. Italian Organization for Standardization (UNI). Cement Composition, Specifications and Conformity Criteria for Common Cements. EN 197-1. Available online: http://store.uni.com/magento-1.4.0.1/index.php/en-197-1-2011.htmL (accessed on 2 December 2021).
40. Italian Organization for Standardization (UNI). Methods of Testing Cement-Part 1: Determination of Strength. EN 196-1. Available online: http://store.uni.com/magento-1.4.0.1/index.php/en-196-1-2016.htmL (accessed on 2 December 2021).
41. Petrella, A.; Spasiano, D.; Liuzzi, S.; Ayr, U.; Cosma, P.; Rizzi, V.; Petrella, M.; Di Mundo, R. Use of cellulose fibers from wheat straw for sustainable cement mortars. *J. Sustain. Cem. Based Mater.* **2018**, *8*, 161–179. [CrossRef]
42. Tonoli, G.H.D.; Santos, S.F.; Savastano, H., Jr.; Delvasto, S.; De Gutiérrez, R.M.; de Murphy, M.D.M.L. Effects of natural weathering on microstructure and mineral composition of cementitious roofing tiles reinforced with fique fibre. *Cem. Concr. Compos.* **2011**, *33*, 225–232. [CrossRef]
43. Ardanuy, M.; Claramunt, J.; García-Hortal, J.A.; Barra, M. Fiber-matrix interactions in cement mortar composites reinforced with cellulosic fibers. *Cellulose* **2011**, *18*, 281–289. [CrossRef]
44. Mohr, B.J.; Nanko, H.; Kurtis, K.E. Durability of kraft pulp fiber–cement composites to wet/dry cycling. *Cem. Concr. Compos.* **2005**, *27*, 435–448. [CrossRef]
45. Filho, R.D.T.; Scrivener, K.; England, G.L.; Ghavami, K. Durability of alkali-sensitive sisal and coconut fibres in cement mortar composites. *Cem. Concr. Compos.* **2000**, *22*, 127–143. [CrossRef]
46. Filho, R.D.T.; England, G.L. Development of vegetable fibre–mortar composites of improved durability. *Cem. Concr. Compos.* **2003**, *25*, 185–196. [CrossRef]
47. Mohr, B.J.; Biernacki, J.J.; Kurtis, K.E. Supplementary cementitious materials for mitigating degradation of kraft pulp fiber-cement composites. *Cem. Concr. Res.* **2007**, *37*, 1531–1543. [CrossRef]
48. Filho, R.D.T.; Silva, F.D.A.; Fairbairn, E.M.R.; Filho, J.D.A.M. Durability of compression molded sisal fiber reinforced mortar laminates. *Constr. Build. Mater.* **2009**, *23*, 2409–2420. [CrossRef]
49. Filho, J.D.A.M.; Silva, F.D.A.; Filho, R.D.T. Degradation kinetics and aging mechanisms on sisal fiber cement composite systems. *Cem. Concr. Compos.* **2013**, *40*, 30–39. [CrossRef]

50. Tonoli, G.H.D.; Santos, S.F.; Joaquim, A.P.; Savastano, H., Jr. Effect of accelerated carbonation on cementitious roofing tiles reinforced with lignocellulosic fibre. *Constr. Build. Mater.* **2010**, *24*, 193–201. [CrossRef]
51. Soroushian, P.; Won, J.P.; Hassan, M. Durability characteristics of $CO_2$-cured cellulose fiber reinforced cement composites. *Constr. Build. Mater.* **2012**, *34*, 44–53. [CrossRef]
52. Arsène, M.A.; Okwo, A.; Bilba, K.; Soboyejo, A.B.O.; Soboyejo, W.O. Chemically and thermally treated vegetable fibers for reinforcement of cement-based composites. *Mater. Manuf. Process.* **2007**, *22*, 214–227. [CrossRef]
53. Claramunt, J.; Ardanuy, M.; García-Hortal, J.A.; Tolêdo Filho, R.D. The hornification of vegetable fibers to improve the durability of cement mortar composites. *Cem. Concr. Compos.* **2011**, *33*, 586–595. [CrossRef]
54. Li, Z.; Wang, L.; Wang, X. Flexural characteristics of coir fiber reinforced cementitious composites. *Fibers Polym.* **2006**, *7*, 286–294. [CrossRef]
55. Sedan, D.; Pagnoux, C.; Smith, A.; Chotard, T. Mechanical properties of hemp fibre reinforced cement: Influence of the fibre/matrix interaction. *J. Eur. Ceram.* **2008**, *28*, 183–192. [CrossRef]
56. Bilba, K.; Savastano, H., Jr.; Ghavami, K. Treatments of non-wood plant fibres used as reinforcement in composite materials. *Mater. Res.* **2013**, *16*, 903–923.
57. Tonoli, G.H.D.; Belgacem, M.N.; Siqueira, G.; Bras, J.; Savastano, H., Jr.; Lahr, F.R. Processing and dimensional changes of cement based composites reinforced with surface-treated cellulose fibres. *Cem. Concr. Compos.* **2013**, *37*, 68–75. [CrossRef]
58. Blankenhorn, P.R.; Blankenhorn, B.D.; Silsbee, M.R.; DiCola, M. Effects of fiber surface treatments on mechanical properties of wood fiber–cement composites. *Cem. Concr. Res.* **2001**, *31*, 1049–1055. [CrossRef]
59. Juarez, C.; Duran, A.; Valdez, P.; Fajardo, G. Performance of "Agave Lecheguilla" natural fiber in Portland cement composites exposed to severe environment conditions. *Build. Environ.* **2007**, *42*, 1151–1157. [CrossRef]
60. Ferreira, S.R.; Silva, F.D.A.; Lima, P.R.L.; Filho, R.D.T. Effect of fiber treatments on the sisal fiber properties and fiber–matrix bond in cement based systems. *Constr. Build. Mater.* **2015**, *101*, 730–740. [CrossRef]
61. Ledhem, A.; Dheilly, R.M.; Queneudec, M. Reuse of waste oils in the treatment of wood aggregates. *Waste Manag.* **2000**, *20*, 321–326. [CrossRef]
62. Petrella, A.; Petrella, M.; Boghetich, G.; Petruzzelli, D.; Ayr, U.; Stefanizzi, P.; Calabrese, D.; Pace, L.; Guastamacchia, M. Thermo-acoustic properties of cement-waste-glass mortars. *Proc. Inst. Civ. Eng. Constr. Mater.* **2009**, *162*, 67–72. [CrossRef]
63. Petrella, A.; Petrella, M.; Boghetich, G.; Petruzzelli, D.; Calabrese, D.; Stefanizzi, P.; de Napoli, D.; Guastamacchia, M. Recycled waste glass as aggregate for lightweight concrete. *Proc. Inst. Civ. Eng. Constr. Mater.* **2007**, *160*, 165–170. [CrossRef]
64. Petrella, A.; di Mundo, R.; de Gisi, S.; Todaro, F.; Labianca, C.; Notarnicola, M. Environmentally sustainable cement composites based on end-of-life tyre rubber and recycled waste porous glass. *Materials* **2019**, *12*, 3289. [CrossRef] [PubMed]
65. Petrella, A.; di Mundo, R.; Notarnicola, M. Recycled expanded polystyrene as lightweight aggregate for environmentally sustainable cement conglomerates. *Materials* **2020**, *13*, 988. [CrossRef]
66. Savastano, H.; Warden, P.G.; Coutts, R.S.P. Brazilian waste fibres as reinforcement for cement-based composites. *Cem. Concr. Compos.* **2000**, *22*, 379–384. [CrossRef]
67. Petrella, A.; Notarnicola, M. Lightweight cement conglomerates based on end-of-life tire rubber: Effect of the grain size, dosage and addition of perlite on the physical and mechanical properties. *Materials* **2021**, *14*, 225. [CrossRef]
68. Petrella, A.; Petruzzelli, V.; Ranieri, E.; Catalucci, V.; Petruzzelli, D. Sorption of Pb(II), Cd(II) and Ni(II) from single- and multimetal solutions by recycled waste porous glass. *Chem. Eng. Commun.* **2016**, *203*, 940–947. [CrossRef]
69. Petrella, A.; Spasiano, D.; Rizzi, V.; Cosma, P.; Race, M.; De Vietro, N. Thermodynamic and kinetic investigation of heavy metals sorption in packed bed columns by recycled lignocellulosic materials from olive oil production. *Chem. Eng. Commun.* **2019**, *206*, 1715–1730. [CrossRef]
70. Petrella, A.; Spasiano, D.; Race, M.; Rizzi, V.; Cosma, P.; Liuzzi, S.; de Vietro, N. Porous waste glass for lead removal in packed bed columns and reuse in cement conglomerates. *Materials* **2019**, *12*, 94. [CrossRef]
71. Bageri, B.S.; Adebayo, A.R.; Al Jaberi, J.; Patil, S. Effect of perlite particles on the filtration properties of high-density barite weighted water-based drilling fluid. *Powder Technol.* **2020**, *360*, 1157–1166. [CrossRef]
72. Petrella, A.; Spasiano, D.; Rizzi, V.; Cosma, P.; Race, M.; de Vietro, N. Lead ion sorption by perlite and reuse of the exhausted material in the construction field. *Appl. Sci.* **2018**, *8*, 1882. [CrossRef]
73. International Organization for Standardization (ISO). Cement, Test Methods, Determination of Strength. ISO 679. Available online: http://store.uni.com/magento-1.4.0.1/index.php/iso-679-2009.htmL (accessed on 2 December 2021).
74. Italian Organization for Standardization (UNI). Determination of Consistency of Cement Mortars Using a Flow Table. 7044. Available online: http://store.uni.com/magento-1.4.0.1/index.php/uni-7044-1972 (accessed on 2 December 2021).
75. Gustafsson, S.E. Transient plane source techniques for thermal conductivity and thermal diffusivity measurements of solid materials. *Rev. Sci. Instrum.* **1991**, *62*, 797–804. [CrossRef]
76. Italian Organization for Standardization (UNI). Acoustics—Determination of Sound Absorption Coefficient and Impedance in Impedances Tubes—Method Using Standing Wave Ratio. EN ISO 10534-1. Available online: http://store.uni.com/magento-1.4.0.1/index.php/en-iso-10534-1-2001.htmL (accessed on 2 December 2021).
77. ACI Committee 544; ACI 544.2R-89. Measurement of properties of fibre reinforced concrete. In *ACI Manual of Concrete Practice, Part 5: Masonry, Precast Concrete and Special Processes*; American Concrete Institute: Farmington Hills, MI, USA, 1996.
78. Beranek, L.L. *Acoustic Measurements*; John Wiley & Sons: New York, NY, USA, 1949.

79. Tang, X.; Yan, X. Acoustic energy absorption properties of fibrous materials: A review. *Compos. Part A Appl. Sci. Manuf.* **2017**, *101*, 360–380. [CrossRef]
80. Chen, P.H.; Xu, C.; Chung, D.D.L. Sound absorption enhancement using solid-solid interfaces in a nonporous cement-based structural material. *Compos. Part B-Eng.* **2016**, *95*, 453–461. [CrossRef]
81. Merta, I.; Tschegg, E.K. Fracture energy of natural fibre reinforced concrete. *Constr. Build. Mater.* **2013**, *40*, 991–997. [CrossRef]
82. Khalil, E.; Abd-Elmohsen, M.; Anwar, A.M. Impact resistance of rubberized self-compacting concrete. *Water Sci.* **2015**, *29*, 45–53. [CrossRef]
83. Mastali, M.; Dalvand, A.; Sattarifard, A. The impact resistance and mechanical properties of the reinforced self-compacting concrete incorporating recycled CFRP fiber with different lengths and dosages. *Compos. Part B Eng.* **2017**, *112*, 74–92. [CrossRef]
84. Santos, S.F.; Schmidt, R.; Almeida, A.E.; Tonoli, G.H.; Savastano, H., Jr. Supercritical carbonation treatment on extruded fibre–cement reinforced with vegetable fibres. *Cem. Concr. Compos.* **2015**, *56*, 84–94. [CrossRef]
85. John, V.M.; Cincotto, M.A.; Sjöström, C.; Agopyan, V.; Oliveira, C.T.A. Durability of slag mortar reinforced with coconut fibre. *Cem. Concr. Compos.* **2005**, *27*, 565–574. [CrossRef]

*Article*

# Reuse of the Materials Recycled from Renewable Resources in the Civil Engineering: Status, Achievements and Government's Initiatives in Taiwan

**Chi-Hung Tsai [1], Yun-Hwei Shen [1,*] and Wen-Tien Tsai [2,*]**

1 Department of Resources Engineering, National Cheng Kung University, Tainan 701, Taiwan; ap29fp@gmail.com

2 Graduate Institute of Bioresources, National Pingtung University of Science and Technology, Pingtung 912, Taiwan

* Correspondence: yhshen@mail.ncku.edu.tw (Y.-H.S.); wttsai@mail.npust.edu.tw (W.-T.T.); Tel.: +886-6-2757575 (Y.-H.S.); +886-8-7703202 (W.-T.T.)

**Abstract:** Growing concerns about the circular economy and sustainable waste management for civil applications of non-hazardous mineral industrial waste have increased in recent years. Therefore, this study presents a trend analysis of industrial waste generation and treatment during the years of 2010–2020, and focused on promotion policies and regulatory measures for mandatory renewable resources from industrial sources in Taiwan, including reclaimed asphalt pavement (RAP) material, water-quenched blast furnace slag, and ilmenite chlorination furnace slag. According to the official database of the online reported statistics during the period of 2010–2020, approximately three million metric tons per year of renewable resources were totally reused in civil engineering or related cement products, reflecting a balanced supply chain in the domestic market. Among these, water-quenched blast furnace slag accounted for about 90% (about 2.7 million metric tons) in Taiwan. Currently, the legislative framework of sustainable waste management in Taiwan is based on the Waste Management Act and the Resource Recycling Act, but there are some problems with them. In order to effectively reduce environmental loadings and conserve natural resources to mitigate climate change, some recommendations are addressed from different points of view.

**Keywords:** industrial waste; renewable resource; reuse; civil engineering material; regulatory promotion

**Citation:** Tsai, C.-H.; Shen, Y.-H.; Tsai, W.-T. Reuse of the Materials Recycled from Renewable Resources in the Civil Engineering: Status, Achievements and Government's Initiatives in Taiwan. *Materials* **2021**, *14*, 3730. https://doi.org/10.3390/ma14133730

Academic Editors: Andrea Petrella and Michele Notarnicola

Received: 27 May 2021
Accepted: 30 June 2021
Published: 2 July 2021



## 1. Introduction

Situated to the east of mainland China, and as an export-dependent economy of 23.4 million people, Taiwan has shown rapid industrial development over the past few decades and now occupies a strategic position in global supply chains, such as those of microchips and information and communication technology (ICT) products. However, the enormous economic development in Taiwan has resulted in a greater complexity of industrial waste management during this period. Before the Taiwan government revised the Waste Management Act in 1999, businesses either had to dispose of their waste on their own or commission government departments. At that time, industrial waste was often abandoned illegally and/or exported to other Asian countries due to incompetent government administrations, insufficient capacity of waste treatment facilities, and the high cost of hazardous waste treatment [1–3]. In order to efficiently reduce waste generation and also follow international trends in a hierarchy of waste management [4,5], the Taiwan Environmental Protection Administration (EPA) revised the Act in 1999, and formed a legal frameworks for industrial waste management, including online reporting, strict penalties and treatment methods [6,7]. Since 2000, the EPA has formulated the "National Industrial Waste Management Program" and the "Industrial Waste Control Center", as well as a tracking system that controls the life cycle of waste from generation to disposal [8]. These historical developments of industrial waste management in Taiwan were similar to those of other Asian countries, such as Japan [9], Korea [10], China [11], and India [12].

In order to significantly reduce environmental loading and effectively reuse available materials from waste resources for the targets of zero waste [13–15] and industrial and urban symbiosis [16], the EPA has been actively undertaking the 4-in-1 Recycling Program under the authorization of the Waste Management Act since the early 2000s. This program integrates community residents, responsible and recycling enterprises, local governments, and a non-profit recycling fund for the recycling of regulated recyclable wastes among municipal solid wastes, especially in electrical and electronic waste equipment (WEEE) [17]. On the other hand, the EPA newly enacted the Resource Recycling Act in 2002 [18], which aims at conserving natural resources, promoting the recycling and reuse of materials, and mitigating environmental loading. The act focused on promotional policies and measures on the recycling and reuse of renewable resources from industrial waste sources, which were announced by a competent central industry authority. In this regard, reclaimed asphalt pavement (RAP) material has been listed as a renewable resource by the Ministry of Interior (MOI) because it has been reused as raw material for asphalt concrete since the 1990s [19–21]. In addition, furnace slags, including quenched blast furnace slag and ilmenite chlorination furnace slag, have been listed as renewable resources by the Ministry of Economic Affairs (MOEA) due to their successful application in cement products of Taiwan [22–24] and other regions [25].

As reviewed above, it was seldom found in combination with a description of the current status and regulatory measures of renewable resources from industrial waste in Taiwan's engineering applications. Therefore, this paper firstly analyzed the trends of industrial waste generation and treatment during the period of 2010–2020; furthermore, an updated status of the announced renewable resources that can be reused in civil engineering was also discussed in the study. Finally, regulatory promotion of renewable resource reused in civil engineering was addressed to echo a case study in the establishment of Environmental Science and Technology Parks (ESTP).

## 2. Data Mining

The main purposes of this study were to analyze the updated status of industrial waste and announced renewable resources (i.e., reclaimed asphalt pavement material, water-quenched blast furnace slag, and ilmenite chlorination furnace slag), and further address regulatory measures for promoting them in civil engineering applications. Therefore, a statistical database, recycling and reuse, and regulatory measures relevant to the announced renewable resources are briefly summarized below.

- Activity (statistics and status) of industrial waste generation and treatment:

The updated data on the statistics and status of industrial waste generation and treatment in Taiwan were obtained from the official yearbook [26] and website [27], which were compiled by the EPA.

- Activity (statistics and status) of renewable resources reused in civil engineering:

In order to highlight the recycling and reuse of renewable resource in Taiwan, the updated data on the statistics and status of the announced items (i.e., reclaimed asphalt pavement material, water-quenched blast furnace slag, and ilmenite chlorination furnace slag) reused in civil engineering were also accessed on the official website [27].

- Regulatory measures for the renewable resources and the establishment of ESTP:

Information about the regulatory measures for the announced renewable resources was accessed on the relevant website [28]. In addition, an official plan for the establishment of ESTP was addressed to echo the regulatory promotion for the recycling and reuse of renewable resources [29].

## 3. Results and Discussion

### *3.1. Trend Analysis of Industrial Waste Generation and Treatment*

#### 3.1.1. Industrial Waste Generation

Based on the Waste Management Act in Taiwan, there are two categories of industrial waste (i.e., general industrial waste and hazardous industrial waste), which refer to waste generated from industry activities, but excludes waste generated by the employees of those industries themselves. Herein, the industry activities include agricultural, industrial and mining plants and sites, construction enterprises, medical/hospital/clinic organizations, public and private waste management (clearance, treatment, and disposal) organizations, joint industrial waste treatment organizations, laboratories of schools or agency groups, and other enterprises designated by the central competent authority. General (non-hazardous) industrial waste is composed of movable solid, liquid substances, or objects other than hazardous industrial waste. In contrast, hazardous industrial waste is legally identified by its toxic characteristics and hazardous substances due to their negative impacts on human health and the environment when it is not managed properly. The central competent authority as referred to in the act means the EPA. It should be noted that radioactive waste management is in accordance with the relevant atomic energy regulations, such as the Ionization Radiation Protection Act.

In July 1999, the act was revised to promulgate a legal framework for industrial waste management, thus formulating the "National Industrial Waste Management Program", the "Industrial Waste Control Center", and the online reporting system that has tracked the complete life cycle of industrial wastes since 2001 from the generation source to the final disposal. On the other hand, the Resource Recycling Act was newly enacted to further promote recycling and reuse of so-called renewable resource, which are produced or derived from general industrial waste. Currently, there are three renewable resources announced by the central industry authorities, reclaimed asphalt pavement material, quenched blast furnace slag, and ilmenite chlorination furnace slag. Table 1 lists the reported amounts of industrial waste generation during the decade of 2010–2020 in Taiwan by accessing the official database [27]. Figure 1 shows the generation percentages of general industrial waste, hazardous industrial waste, and renewable resource, in 2019. Obviously, the reported amounts indicated a slight increase from about 18.1 million metric tons in 2010 to a peak record of approximately 22.3 million metric tons in 2018. Thereafter, the reported amounts showed a decreasing trend from 2019. The significant decrease during the years of 2018–2019 can be attributable to the exclusion of waste generated by the employees themselves (since 2019). As compared to the amounts of industrial waste generation in 2005 (i.e., about 14.6 million metric tons), it reflects the ongoing implementation of industrial waste minimization (including source reduction and waste recycling and reuse) in the industrial sector since the early 2000s [26,27], thus retarding the expected increase during the decade. For example, Taiwan's semiconductor industries contributed significantly to the supply chain globally, but also generated large amounts of industrial waste, such as calcium fluoride ($CaF_2$) sludge. Currently, this sludge is reused as raw cement material in Taiwan [28].

**Table 1.** Reported amounts of industrial waste generation during the decade of 2010–2020 in Taiwan [1].

| Year | General Industrial Waste | Hazardous Industrial Waste | Renewable Resource | Total |
|---|---|---|---|---|
| 2010 | $1.375 \times 10^7$ | $0.122 \times 10^7$ | $0.312 \times 10^7$ | $1.809 \times 10^7$ |
| 2011 | $1.412 \times 10^7$ | $0.120 \times 10^7$ | $0.341 \times 10^7$ | $1.873 \times 10^7$ |
| 2012 | $1.392 \times 10^7$ | $0.125 \times 10^7$ | $0.278 \times 10^7$ | $1.795 \times 10^7$ |
| 2013 | $1.448 \times 10^7$ | $0.145 \times 10^7$ | $0.275 \times 10^7$ | $1.867 \times 10^7$ |
| 2014 | $1.424 \times 10^7$ | $0.160 \times 10^7$ | $0.300 \times 10^7$ | $1.884 \times 10^7$ |
| 2015 | $1.449 \times 10^7$ | $0.137 \times 10^7$ | $0.330 \times 10^7$ | $1.916 \times 10^7$ |
| 2016 | $1.420 \times 10^7$ | $0.136 \times 10^7$ | $0.342 \times 10^7$ | $1.897 \times 10^7$ |
| 2017 | $1.485 \times 10^7$ | $0.144 \times 10^7$ | $0.307 \times 10^7$ | $1.937 \times 10^7$ |
| 2018 | $1.774 \times 10^7$ | $0.146 \times 10^7$ | $0.313 \times 10^7$ | $2.233 \times 10^7$ |
| 2019 | $1.506 \times 10^7$ | $0.139 \times 10^7$ | $0.339 \times 10^7$ | $1.984 \times 10^7$ |
| 2020 | $1.680 \times 10^7$ | | $0.294 \times 10^7$ | $1.975 \times 10^7$ |

[1] Sources [26,27]; unit: metric ton.

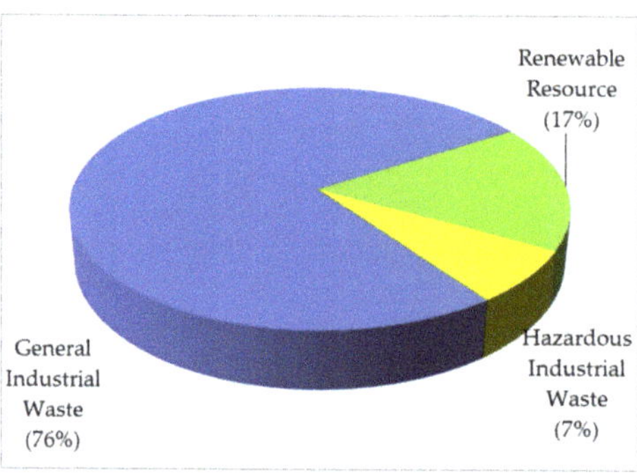


**Figure 1.** Pie chart of generation percentages of three categories of industrial waste in 2019.

3.1.2. Industrial Waste Treatment

As described above, industrial waste management in Taiwan started with the implementation of the Waste Management Act, which has been revised several times. According to the official definition of industrial waste treatment, the term "treatment" refers to three methods, immediate treatment (e.g., incineration, solidification), final disposal (e.g., sanitary landfill), and reuse. The term "reuse" is defined as "the reuse of industrial waste produced by an enterprise as raw material, materials, fuel, land reclamation fill, or other reuse methods recognized by the central industry competent authority via self-use, sale, transfer, or commissioning, and in compliance with the related regulations". Table 2 summarizes the reported amounts of industrial waste treatment during the period of 2010–2020 [26,27]. Figure 2 depicts the percentages of four methods of industrial waste treatment in 2020. Obviously, under the circular economy principle, focusing on waste recycling, the amounts of reused industrial waste indicated an increasing trend over the past decade, reaching an industrial waste reuse and recycling rate of over 80%. In fact, the EPA amended the Waste Management Act on 24 October 2001 to promote proper industrial waste management and reuse, leading to a higher willingness to reuse industrial waste and resource materials. Article 39 mandates that industrial waste reuse be implemented as required by the central industry competent authorities. Since then, there have been 10 agencies for formulating the corresponding regulations and management systems, which are relevant to industrial waste reuse under their jurisdictions.

**Table 2.** Reported amounts of industrial waste treatment during the decade of 2010–2020 in Taiwan [1].

| Year | Reuse | Self-Treatment | Commissioned or Joint Treatment | Exported Treatment | Total |
|---|---|---|---|---|---|
| 2010 | $1.458 \times 10^7$ | $4.943 \times 10^5$ | $2.625 \times 10^6$ | $3.329 \times 10^4$ | $1.773 \times 10^7$ |
| 2011 | $1.544 \times 10^7$ | $4.988 \times 10^5$ | $2.899 \times 10^6$ | $3.119 \times 10^4$ | $1.887 \times 10^7$ |
| 2012 | $1.451 \times 10^7$ | $5.007 \times 10^5$ | $2.880 \times 10^6$ | $3.160 \times 10^4$ | $1.792 \times 10^7$ |
| 2013 | $1.491 \times 10^7$ | $8.124 \times 10^5$ | $2.783 \times 10^6$ | $5.077 \times 10^4$ | $1.856 \times 10^7$ |
| 2014 | $1.521 \times 10^7$ | $8.658 \times 10^5$ | $2.753 \times 10^6$ | $4.950 \times 10^4$ | $1.888 \times 10^7$ |
| 2015 | $1.581 \times 10^7$ | $6.099 \times 10^5$ | $2.663 \times 10^6$ | $4.665 \times 10^4$ | $1.913 \times 10^7$ |
| 2016 | $1.469 \times 10^7$ | $6.358 \times 10^5$ | $2.587 \times 10^6$ | $1.582 \times 10^4$ | $1.793 \times 10^7$ |
| 2017 | $1.564 \times 10^7$ | $6.547 \times 10^5$ | $2.634 \times 10^6$ | $1.544 \times 10^4$ | $1.894 \times 10^7$ |
| 2018 | $1.680 \times 10^7$ | $6.889 \times 10^5$ | $2.615 \times 10^6$ | $0.771 \times 10^4$ | $2.011 \times 10^7$ |
| 2019 | $1.667 \times 10^7$ | $7.129 \times 10^5$ | $2.456 \times 10^6$ | $0.824 \times 10^4$ | $1.985 \times 10^7$ |
| 2020 | $1.676 \times 10^7$ | $7.442 \times 10^5$ | $2.343 \times 10^6$ | $0.882 \times 10^4$ | $1.985 \times 10^7$ |

[1] Sources [26,27]; unit: metric ton.

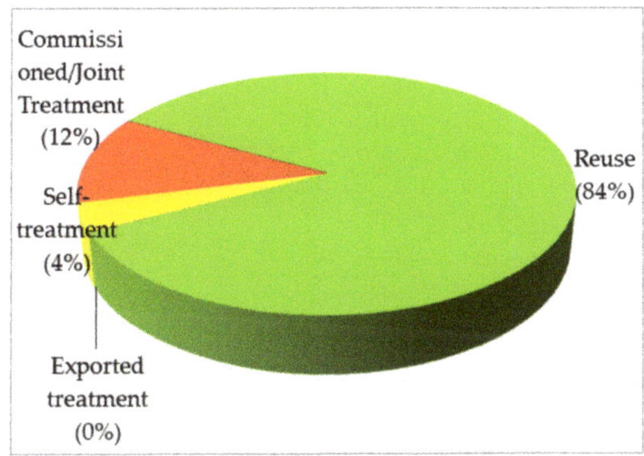


**Figure 2.** Pie chart of percentages of four methods of industrial waste treatment in 2020.

### 3.2. Status of Renewable Resource Reused in the Civil Engineering

Under the authorization of the Resource Recycling Act, a so-called "renewable resource" is defined as "the announced substances that have lost their original usefulness, are economically and technologically feasible to recycle, and may be recycled or reused" [28]. Herein, the term "reuse" means the practice of making direct and repeated use of renewable resources in their original form or using renewable resources after restoring some or all of their original functionalities. Another term, "recycling", means the practice of making renewable resources functional by altering the original form of substances, or combining them with other substances, so that they may serve as materials, fuel, fertilizers, animal feed, fillers, soil enhancers, or for other uses recognized by the central industry authorities. Regarding the differences between the terms reuse and recycling, they obviously have certain overlapping and ambiguous scopes, but reuse should have a preferred hierarchy. According to the "renewable resource" lists announced by the central competent authorities (i.e., MOEA, MOI, and EPA), they include water-quenched blast furnace slag, ilmenite chlorination furnace slag, cobalt-manganese compound precipitate, scrap masonry material, reclaimed asphalt pavement material, and electronic waste-derived materials (iron, copper, aluminum, glass, and plastic). Obviously, slag materials are typically referred to as industrial "by-products". These mining resources, once used to produce cement, are not renewable and circular since they are not used in the production of either fuel or cement again in the future. In this regard, the definition of "renewable resource" should be amended in the Resource Recycling Act. Table 3 summarizes the reported amounts of renewable resource generation during the period of 2010–2020 [26,27]. Figure 3 indicated the percentages of three categories of renewable resources in 2020, the majority being water-quenched blast furnace slag. As shown in Table 3, approximately 3 million metric tons of renewable resources were reused annually in civil engineering or in related cement products, reflecting a balanced supply chain in the domestic market. These announced renewable resources were completely reused or recycled in civil engineering or related products, such as cement materials, which will be further addressed in Section 3.3.

**Table 3.** Reported amounts of renewable resource generation during the decade of 2010–2020 in Taiwan [1].

| Year | Reclaimed Asphalt Pavement Material | Water-Quenched Blast Furnace Slag | Ilmenite Chlorination Furnace Slag | Total |
|---|---|---|---|---|
| 2010 | $0.203 \times 10^6$ | $2.739 \times 10^6$ | $0.178 \times 10^6$ | $3.120 \times 10^6$ |
| 2011 | $0.273 \times 10^6$ | $2.955 \times 10^6$ | $0.183 \times 10^6$ | $3.411 \times 10^6$ |
| 2012 | $0.146 \times 10^6$ | $2.590 \times 10^6$ | $0.041 \times 10^6$ | $2.777 \times 10^6$ |
| 2013 | $0.135 \times 10^6$ | $2.615 \times 10^6$ | - [2] | $2.750 \times 10^6$ |
| 2014 | $0.257 \times 10^6$ | $2.739 \times 10^6$ | - [2] | $2.996 \times 10^6$ |
| 2015 | $0.351 \times 10^6$ | $2.28 \times 10^6$ | $0.017 \times 10^6$ | $3.296 \times 10^6$ |
| 2016 | $0.364 \times 10^6$ | $3.034 \times 10^6$ | $0.022 \times 10^6$ | $3.420 \times 10^6$ |
| 2017 | $0.429 \times 10^6$ | $2.619 \times 10^6$ | $0.026 \times 10^6$ | $3.074 \times 10^6$ |
| 2018 | $0.399 \times 10^6$ | $2.702 \times 10^6$ | $0.027 \times 10^6$ | $3.128 \times 10^6$ |
| 2019 | $0.330 \times 10^6$ | $3.045 \times 10^6$ | $0.014 \times 10^6$ | $3.389 \times 10^6$ |
| 2020 | $0.239 \times 10^6$ | $2.685 \times 10^6$ | $0.019 \times 10^6$ | $2.943 \times 10^6$ |

[1] Sources [26,27]; unit: metric ton. [2] Not reported.

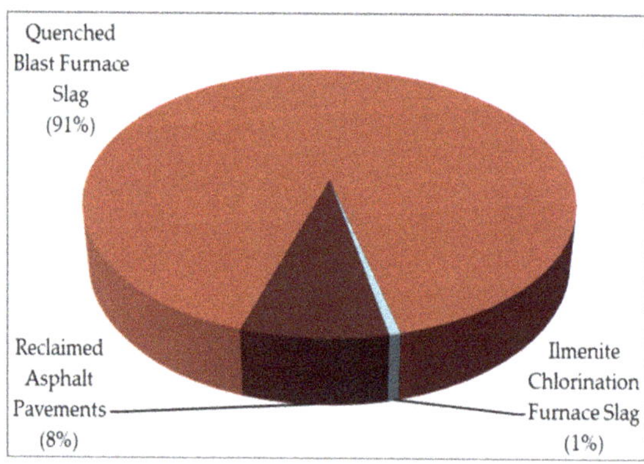


**Figure 3.** Pie chart of percentages of three categories of renewable resources in 2020.

In fact, there are many different slags from manufacturing industries in Taiwan, including electric arc furnace steelmaking, induction furnace steelmaking, cupola furnace steelmaking, water-quenched blast furnace steelmaking, basic oxygen furnace (BOF) steelmaking, ilmenite chlorination furnace titanium dioxide ($TiO_2$)-making, and rotary kiln in the thermal reuse treatment of steelmaking dust. Currently, slags from water-quenched blast furnace steelmaking and ilmenite chlorination furnace $TiO_2$-making have been listed as "renewable resources" by the central industry competent authority (i.e., MOEA). In Taiwan, water-quenched blast slag powder (after being ground) is blended with cement to replace general cement and is widely applied in various kinds of construction projects. On the other hand, RAP comes from oil-based tar (a fossil resource). RAP materials were derived from frequent milling and overlay activities for roads, and thus are produced in considerable amounts annually. In Taiwan, experience has indicated that the recycling of bitumen binder in RAP is a circular economy approach from technical, economical, and environmental viewpoints [19–21]. Therefore, RAP has been listed as one of the "renewable resources" by the central industry competent authority (i.e., MOI). In this regard, the Implementation Guidelines for the Outline Specifications for Public Construction—Chapter 02966 in Taiwan has allowed incorporating, at most, 40% of RAP into new hot-mix asphalt (HMA) since the late 1990s [21].

### 3.3. Regulatory Promotion for Renewable Resource Reused in the Civil Engineering

In Taiwan, the government established a renewable resource recycling system with the enactment of the Waste Management Act (also called the Waste Disposal Act) and the Resource Recycling Act [28]. The central industry authorities (i.e., MOEA and MOI) should be in consultation with the central competent authority (i.e., EPA) for announcing renewable resource items that must be recycled or reused according to the promulgated measures.

#### 3.3.1. Waste Management Act

In Taiwan, industrial waste management started with the promulgation of the Waste Management Act in 1974, which has been revised several times to comply with the domestic industrial structure and international trends. When established in 1987, the EPA began to put emphasis on industrial waste minimization (or cleaner production) and also set up a permit system concerning waste generation sources and waste treatment organizations. In recent years, the waste management concept has shifted to source reduction approaches, such as circular economy, sustainable materials management (SMM) and cradle-to-cradle (C2C). In order to promote industrial waste recycling and reuse, the regulatory measures and incentives have been incorporated into the act, which will be briefly described below.

The disposal of industrial waste, with the exception of that subject to reuse methods, shall be performed in accordance with the following methods:

1. Any product from industry activities shall be determined as waste when it is under any of the following circumstances:
    - The product is determined to be of no economic or market value by the EPA, and is intended to be disposed of illegally or harmful to the environment and human health.
    - The product is not lawfully stored or used, and is intended to be disposed of illegally or causing pollution.
    - The recycled/reused product is not used in accordance with this act, and is intended to be disposed of illegally or causing pollution.
2. Industrial waste management, with the exception of that subject to recycling/reuse methods, shall be performed in accordance with the methods of self-treatment, joint treatment and commissioned treatment.
3. Industrial waste reuse shall be processed in accordance with the regulations promulgated by the central industry competent authorities or central competent authority. As listed in Table 2, most industrial waste is currently reused or recycled as materials, fuel, land reclamation fill, and soil modifier according to the corresponding regulations by the ten authorities, especially in the Ministry of Economic Affairs (MOEA), the Council of Agriculture (COA), and the Ministry of Interior (MOI).
4. The expenses incurred by the enterprise to waste management should be partially exempted from tax. Furthermore, the enterprises that are in compliance with relevant waste management regulations and with excellent performance in the waste reduction, recycling, and reuse shall be rewarded by the EPA and the central industry competent authority.

3.3.2. Resource Recycling Act

In order to conserve natural resources, reduce waste generation, promote recycling and reuse of materials, and mitigate environmental loading, the Resource Recycling Act was passed on 21 January 2002. The act stipulates that businesses or enterprises must be in accordance with the management regulations when recycling or reusing the designated renewable resources designated by the central industry authority. The current promotion (assistance and incentive) measures for recycling or reusing renewable resources in industry were briefly addressed as follows:

1. To promote the recycling and reuse of renewable resources, the public organizations shall preferentially procure the government-certified environment-friendly products, domestic renewable resource, or recycled products in which contain a certain proportion of renewable resource. Therefore, the EPA requested relevant government agencies to consider the possibility of using products made from recyclable and reusable waste materials in their public construction projects. Through the efforts of these government agencies, the usage of such products in public construction has been increased gradually. As listed in Table 3, the three renewable resources were totally reused or recycled in the civil engineering or related products like cement and filling materials. It should be noted that the products recycled or recused from the renewable resources must comply with the operation management requirements like the national standards, international standards, engineering specifications, and other relevant regulations. For example, ilmenite chlorination furnace slag must comply with the Chapter 02726 of Taiwan's Construction Specifications (i.e., specific gravity $\geq 1.5$, water absorption $\leq 25\%$, and immersion swelling ratio $\leq 0.5\%$) when it is reused as bottom grade granular material in the pavement works.
2. The EPA shall regularly hold the awards for excellence in recycling/reusing technological developments and their actual achievements. The business or enterprise engaged in renewable resource recycling and reusing shall be granted tax incentives (tax deduction) for the cost of investment in research, facilities, tools, and equipment. In this regard, the "Award Ceremony for Excellent Performance of Industrial Waste and Resource Recycling and Reuse" has been held annually. In addition, the EPA held symposiums and exhibitions annually to further share the successful recycling experience with the general public and the industrial sector.
3. To promote the recycling and reuse of renewable resources, acquire advanced technologies and talents, and encourage innovative research and development (R&D) technologies by the domestic industry, the EPA in collaboration with the local governments established the environmental science and technology parks [29], which will be further addressed in the next section as a case study.

Table 4 lists the items, sources, and reuse types of renewable resource announced by the central industry competent authority [28]. In this regard, reclaimed asphalt pavement (RAP) material is widely used as a cheaper alternative to the conventional hot mix asphalt (HMA), but the mixing ratio shall not exceed 40% according to the commonly used practice [30]. As shown in Table 3, about 3 million metric tons of renewable resources were reused annually in the civil engineering or related cement products.

**Table 4.** Items and reuse types of renewable resource announced by the central industry competent authority.

| Central Industry Competent Authority | Item | Definition by Generation Source | Reuse Type |
|---|---|---|---|
| Environmental Protection Administration (EPA) [1] | Iron | Electronic waste (waste electrical and electronic equipment, WEEE) [4] | Raw material for steel making, ferric chloride, or reused to related chemical products |
| | Copper | WEEE | Raw material for copper/steel products, or reused to its chemical feedstock |
| | Aluminum | WEEE | Raw material for aluminum products, or reused to its chemical feedstock |
| | Glass | WEEE (without containing fluorescent powder or liquid crystal) | Raw material for glass/ceramic tile/cement products, glass, cement; additive for concrete/asphalt concrete; or reused to its chemical feedstock |
| | Plastic | Electronic waste (or waste electrical and electronic equipment) [4] | Raw material for plastic products and plastic pyrolysis; auxiliary fuel for cement/steel plants |
| Ministry of Interior (MOI) [2] | Reclaimed asphalt pavement material | By-product of construction project for asphalt concrete excavation | Raw material for asphalt concrete; or engineering filling material(note: when reusing as a hot-mix recycled asphalt concrete, the mixing ratio shall not exceed 40%) |
| Ministry of Economic Affairs (MOEA) [3] | Water-quenched blast furnace slag | By-product of steelmaking in wholly integrated steel mills where water-quenched blast furnace slag is formed by cooling slag in water | Raw material for quenched blast furnace slag powder, cement, cement products, ceramics, or fertilizer; concrete cementing material |
| | Ilmenite chlorination furnace slag | By-product of manufacturing titanium dioxide ($TiO_2$) in the ilmenite chlorination process | Raw material for recycled aggregate (for base or bottom grade granular material in the pavement works, controlled low-strength material, cement products, base filling material, embankment filling material only), or cement products |
| | Cobalt-manganese (Co/Mn) compound precipitate (content of Co $\geq$ 10 wt%) | By-product of manufacturing pure terephthalic acid (PTA) and isophthalic acid (ITA) | Raw material for cobalt-manganese catalyst |
| | Scrap masonry material | By-product of stone products manufacturing | Raw material for remade stone (board, brick or block), tile, ceramic clay powder, premixed concrete, cement, cement products, construction material, limestone (for marble trims only), recycled aggregate, fertilizer (for serpentine trims only) and craft; controlled low-strength material (CLSM); material for horticultural landscaping |

[1] Promulgated on 1 December 2006. [2] Promulgated on 23 April 2007. [3] Promulgated on 16 January 2004 (quenched blast furnace slag and ilmenite chlorination furnace slag) and 11 September 2020 (cobalt–manganese compound precipitate and scrap masonry material). [4] Registered WEEE according to the Article 18 of the Waste Management Act.

### 3.4. Case Study: Establishment of Environmental Science and Technology Parks

Since the 1992 United Nations Conference on Environment and Development (UNCED) held in Rio de Janeiro (Brazil), sustainable development and recycling-oriented societies have become global trends. Thereafter, one of the significant policies was to establish ecological industrial zones by the developed countries for incorporating industrial development into the natural ecological system. In 2001, the EPA started planning the establishment of Environmental Science and Technology Parks (ESTP) for jointly cooperating with local governments. The three main development axes of ESTP are "high-level resource recycling technology", "high-level environmental protection technology" and "ecological industries". The ESTP proposal was approved by the Executive Yuan (the Cabinet) on 9 September 2002, and was revised on 11 March 2004. The total budget for the ESTP planning and construction has reached 6.2 billion NT$ (about 0.22 billion US$). In order to successfully develop the ESTP, the EPA and local governments cooperated to construct so-called sustainable ecological communities. In this regard, the EPA was responsible for providing financial incentives to attract enterprises which needed to be reviewed by the examination mechanism and administrative measures. In addition, the EPA also provided tenants and research institutes with rewards for investing in technology development and industrial production of environment-friendly products. By contrast, the local governments were responsible for providing sufficient and suitable land, and administrative services for establishment, recruitment and operations.

By the end of 2011, the EPA had established four ESTP zones in Taiwan, as shown in Figure 4 [31]. They cover a total area of 123 acres; Benzhou ESTP (40 acres) in Kaohsiung City, Fenglin ESTP (22 acres) in Hualien County, Guanyin ESTP (31 acres) in Taoyuan City, and Liouying ESTP (30 acres) in Tainan City. In order to attractively promote recruitment, the government subsidized 50% of the land rent payment, and provided 25 million NT$ for total production and 50% of the investment in R&D. Currently, over 60 businesses have completed tenant procedures. It was estimated that 13.5 billion NT$ will be pooled from the private sector, and that these ESTP will create an annual revenue of about 3.0 billion NT$ (0.21 billion US$), in addition to over 2000 job opportunities and approximately 3 million metric tons of recyclable resources through mass production demonstration area in the ESTP. Furthermore, the recycling and reuse of renewable and recyclable resources from general waste (municipal solid waste) and industrial waste will form a complete green supply chain and a circular economy to assist businesses in lowering their operation costs and also achieve their corporate social responsibility (CSR) goals.

**Figure 4.** Locations of environmental science and technology parks in Taiwan [31].

## 4. Conclusions and Recommendations

With the implementation of renewable resources recycling and reuse in Taiwan under the authorization of the Resource Recycling Act in 2002, reclaimed asphalt pavement material, water-quenched blast furnace slag, and ilmenite chlorination furnace slag have been listed as "mandatory" recyclables by the EPA for the production of recycled aggregate materials in concrete and construction applications. During the period of 2010–2020, there were approximately three million metric tons of renewable resources reused annually in civil engineering or related cement products, reflecting a balanced supply chain in the domestic market. Currently, these announced renewable resources are completely reused in civil engineering or related products like cement material. This approach, not only reduces the amount of industrial waste disposal, but also recycles the valuable resources. More significantly, this sustainable waste management approach also raised green productivity and drove supply chain sustainability to develop a circular economy.

Currently, the legislative framework of sustainable waste management in Taiwan is based on the Waste Management Act and the Resource Recycling Act. However, there are some problems between them, including unclear definitions (e.g., reuse/recycling, waste/discard/resource, renewable resource/by-product), waste generation sources and central competent authorities. In the near future, major sources of industrial waste in Taiwan will be generated from the high-tech industries due to the global supply chains. To further enhance the recycling and reuse of WEEE and mineral industrial waste, the following measures were recommended:

- Combining the Waste Management Act and the Resource Recycling Act into a new act, which will incorporate the 5R (i.e., reduction, reuse, recycling, recovery, and reclamation) principles towards the ultimate goal of zero waste through total recycling.
- Promulgating the specific regulations for high-tech industries (e.g., semiconductor and opto-electronics manufacturing) to conduct an industrial symbiosis through industrial waste recycling and cleaner production.
- Adding several mineral waste sources (legally identified as non-hazardous industrial waste) to the lists of renewable resources like electric arc furnace slag, induced current furnace slag, coal ash, and scrap masonry material, which can be reused as available materials in civil engineering.
- Providing sufficient economic and financial (tax) incentives in the accounting/cost system of enterprises or businesses based on the performances of sustainable goals (SDGs) or corporate social responsibility (CSR).

**Author Contributions:** Conceptualization, C.-H.T. and W.-T.T.; data collection, C.-H.T.; data analysis, W.-T.T.; writing—original draft preparation, W.-T.T.; writing—review and editing, Y.-H.S.; supervision, Y.-H.S. and W.-T.T. All authors have read and agreed to the published version of the manuscript.

**Funding:** This research received no external funding.

**Institutional Review Board Statement:** Not applicable.

**Informed Consent Statement:** Not applicable.

**Data Availability Statement:** Not applicable.

**Conflicts of Interest:** The authors declare no conflict of interest.

## References

1. Houng, H.J. Industrial waste management in Taiwan. *Environ. Pract.* **1999**, *1*, 196–199. [CrossRef]
2. Wei, M.S.; Huang, K.H. Recycling and reuse of industrial wastes in Taiwan. *Waste Manag.* **2001**, *21*, 93–97. [CrossRef]
3. Hsing, H.J.; Wang, F.K.; Chiang, P.C.; Yang, W.F. Hazardous wastes transboundary movement management: A case study in Taiwan. Resour. *Conserv. Recycl.* **2004**, *40*, 329–342. [CrossRef]
4. Sakai, S.; Sawell, S.E.; Chandler, A.J.; Eighmy, T.T.; Kasson, D.S.; Vehlow, J.; van der Sloot, H.A.; Hartien, J.; Hjelmar, O. World trends in municipal solid waste management. *Waste Manag.* **1996**, *16*, 341–350. [CrossRef]
5. Wilson, D.C. Stick or carrot?: The use of policy measures to move waste management up the hierarchy. *Waste Manag. Res.* **1996**, *14*, 385–398. [CrossRef]
6. Tsai, W.T.; Chou, Y.H. A review of environmental and economic regulations for promoting industrial waste recycling in Taiwan. *Waste Manag.* **2004**, *24*, 1061–1069. [CrossRef] [PubMed]
7. Tsai, W.T.; Chou, Y.H. Government policies for encouraging industrial waste reuse and pollution prevention in Taiwan. *J. Clean. Prod.* **2004**, *12*, 725–734. [CrossRef]

8. Houng, H.; Cheng, Y.W. Electronic tracking and management of industrial waste in Taiwan. *J. Mater. Cycles Waste Manag.* **2013**, *15*, 146–153. [CrossRef]
9. Tanaka, M. Recent trends in recycling activities and waste management in Japan. *J. Mater. Cycles Waste Manag.* **1999**, *1*, 10–16.
10. Yang, W.S.; Park, J.K.; Park, S.W.; Seo, Y.C. Past, present and future of waste management in Korea. *J. Mater. Cycles Waste Manag.* **2015**, *17*, 207–217. [CrossRef]
11. Huang, Q.; Wang, Q.; Dong, L.; Xi, B.; Zhou, B. The current situation of solid waste management in China. *J. Mater. Cycles Waste Manag.* **2006**, *8*, 63–69. [CrossRef]
12. Narayana, T. Municipal solid waste management in India: From waste disposal to recovery of resources? *Waste Manag.* **2009**, *29*, 1163–1166. [CrossRef] [PubMed]
13. Shekdar, A.V. Sustainable solid waste management: An integrated approach for Asian countries. *Waste Manag.* **2009**, *29*, 1438–1448. [CrossRef] [PubMed]
14. Young, C.Y.; Ni, S.P.; Fan, K.S. Working towards a zero waste environment in Taiwan. *Waste Manag. Res.* **2010**, *28*, 236–244. [CrossRef] [PubMed]
15. Song, Q.; Li, J.; Zeng, X. Minimizing the increasing solid waste through zero waste strategy. *J. Clean. Prod.* **2015**, *104*, 199–210. [CrossRef]
16. Berkel, R.V.; Fujita, T.; Hashimoto, S.; Geng, Y. Industrial and urban symbiosis in Japan: Analysis of the Eco-Town program 1997–2006. *J. Environ. Manag.* **2009**, *90*, 1544–1556. [CrossRef]
17. Tsai, W.T. Recycling of waste electrical & electronic equipment (WEEE) and its toxics-containing management in Taiwan—A case study. *Toxics* **2020**, *8*, 48.
18. Lin, C.C.; Lin, C.H. What substances or objects should be recycled? The recycling legislative experience in Taiwan. *J. Mater. Cycles Waste Manag.* **2005**, *7*, 1–7. [CrossRef]
19. Chen, J.S.; Huang, C.C.; Chu, P.Y.; Lin, K.Y. Engineering characterization of recycled asphalt concrete and aged bitumen mixed recycling agent. *J. Mater. Sci.* **2007**, *42*, 9867–9876. [CrossRef]
20. Lin, P.S.; Wu, T.L.; Chang, C.W.; Chou, B.Y. Effects of recycling agents on aged asphalt binders and reclaimed asphalt concrete. *Mater. Struct.* **2011**, *44*, 911–921. [CrossRef]
21. Yang, S.H.; Lee, L.C. Characterizing the chemical and rheological properties of severely aged reclaimed asphalt pavement materials with high recycling rate. *Contr. Build. Struct.* **2016**, *111*, 139–146. [CrossRef]
22. Chen, H.J.; Huang, S.S.; Tang, C.W.; Malek, M.A.; Ean, L.W. Effect of curing environments on strength, porosity and chloride ingress resistance of blast furnace slag cement concretes: A construction site study. *Contr. Build. Struct.* **2012**, *35*, 1063–1070. [CrossRef]
23. Kuo, W.T.; Wang, H.Y.; Shu, C.Y. Engineering properties of cementless concrete produced from GGBFS and recycled desulfurization slag. *Contr. Build. Struct.* **2014**, *63*, 189–196. [CrossRef]
24. Ho, H.L.; Huang, R.; Hwang, L.C.; Lin, W.T.; Hsu, H.M. Waste-based pervious concrete for climate-resilient pavements. *Materials* **2018**, *11*, 900. [CrossRef] [PubMed]
25. Al-Hamrani, A.; Kucukvar, M.; Alnahhal, W.; Mahdi, E.; Onat, N.C. Green concrete for a circular economy: A review on sustainability, durability, and structural properties. *Materials* **2021**, *14*, 351. [CrossRef] [PubMed]
26. Environmental Protection Administration (EPA, Taiwan). *Yearbook of Environmental Protection Statistics 2019*; EPA: Taipei, Taiwan, 2020.
27. Industrial Waste Reporting and Management Information System (EPA, Taiwan). Available online: https://waste.epa.gov.tw/RWD/Statistics/?page=Month1 (accessed on 16 May 2021).
28. Laws and Regulation Retrieving System (Ministry of Justice, Taiwan). Available online: https://law.moj.gov.tw/Eng/index.aspx (accessed on 13 May 2021).
29. Establishment of Environmental Science and Technology Parks (ESTP) (EPA, Taiwan). Available online: https://www.epa.gov.tw/eng/EBE8CA1A17AF2F1D (accessed on 13 May 2021).
30. Rafiq, W.; Musarat, M.A.; Altaf, M.; Napiah, M.; Sutanto, M.H.; Alaloul, W.S.; Javed, M.F.; Mosavi, A. Life cycle cost analysis comparison of hot mix asphalt and reclaimed asphalt pavement: A case study. *Sustainability* **2021**, *13*, 4411. [CrossRef]
31. Environmental Science and Technology Parks (Ministry of Economic Affairs, Taiwan). Available online: https://investtaiwan.nat.gov.tw/showPagecht449?lang=eng&search=449 (accessed on 13 May 2021).

MDPI

*Article*

# Recycled Cellulose Fiber Reinforced Plaster

**Nadezda Stevulova [1,*], Vojtech Vaclavik [2], Viola Hospodarova [1] and Tomáš Dvorský [2]**

1 Faculty of Civil Engineering, Institute of Environmental Engineering, Technical University of Kosice, Vysokoskolska 4, 04200 Kosice, Slovakia; viola.hospodarova@gmail.com
2 Department of Environmental Engineering, Faculty of Mining and Geology, VSB-Technical University of Ostrava, 17. Listopadu 2172/15, 708 00 Ostrava-Poruba, Czech Republic; vojtech.vaclavik@vsb.cz (V.V.); tomas.dvorsky@vsb.cz (T.D.)
* Correspondence: nadezda.stevulova@tuke.sk; Tel.: +421-55-602-4126

**Abstract:** This paper aims to develop recycled fiber reinforced cement plaster mortar with a good workability of fresh mixture, and insulation, mechanical and adhesive properties of the final hardened product for indoor application. The effect of the incorporation of different portions of three types of cellulose fibers from waste paper recycling into cement mortar (cement/sand ratio of 1:3) on its properties of workability, as well as other physical and mechanical parameters, was studied. The waste paper fiber (WPF) samples were characterized by their different cellulose contents, degree of polymerization, and residues from paper-making. The cement to waste paper fiber mass ratios (C/WPF) ranged from 500:1 to 3:1, and significantly influenced the consistency, bulk density, thermal conductivity, water absorption behavior, and compressive and flexural strength of the fiber-cement mortars. The workability tests of the fiber-cement mortars containing less than 2% WPF achieved optimal properties corresponding to plastic mortars (140–200 mm). The development of dry bulk density and thermal conductivity values of 28-day hardened fiber-cement mortars was favorable with a declining C/WPF ratio, while increasing the fiber content in cement mortars led to a worsening of the water absorption behavior and a lower mechanical performance of the mortars. These key findings were related to a higher porosity and weaker adhesion of fibers and cement particles at the matrix-fiber interface. The adhesion ability of fiber-cement plastering mortar based on WPF samples with the highest cellulose content as a fine filler and two types of mixed hydraulic binder (cement with finely ground granulated blast furnace slag and natural limestone) on commonly used substrates, such as brick and aerated concrete blocks, was also investigated. The adhesive strength testing of these hardened fiber-cement plaster mortars on both substrates revealed lime-cement mortar to be more suitable for fine plaster. The different behavior of fiber-cement containing finely ground slag manifested in a greater depth of the plaster layer failure, crack formation, and in greater damage to the cohesion between the substrate and mortar for the observed time.

**Keywords:** waste paper fiber; fiber-cement plaster mortar; physical and mechanical performance; limestone; granulated blast furnace slag; adhesive strength

**Citation:** Stevulova, N.; Vaclavik, V.; Hospodarova, V.; Dvorský, T. Recycled Cellulose Fiber Reinforced Plaster. *Materials* **2021**, *14*, 2986. https://doi.org/10.3390/ma14112986

Academic Editors: Andrea Petrella and Michele Notarnicola

Received: 13 May 2021
Accepted: 28 May 2021
Published: 31 May 2021



## 1. Introduction

Buildings have huge untapped potential to become a key part of the solution to urgent sustainability challenges. One of the important ways to improve the sustainability of buildings is to produce environmentally friendly materials and to design green composites for eco-friendly building constructions meeting the requirements of low-energy, zero-carbon green buildings with their structural components being highly thermally insulating and breathable, ensuring effective climatic control. In this context, the construction industry needs to take a circular approach considering the entire life cycle of its production chain [1].

Recently, significant efforts to develop new sustainable composite materials have been mainly related to the replacement of traditional cement binders in cement mortars and concrete products through alternative cementitious materials [2] and the utilization of natural renewable fibrous materials [3]. Cementitious-based composites containing cellulose fibers meeting the sustainability requirements have become increasingly important for economic, environmental, and technical reasons [3–5]. A driving force for using natural vegetable fibers as reinforcing materials in structural applications is their ecological nature, biodegradability, good mechanical properties, and ability to exhibit good tensile strengths in the composite [6–10]. The incorporation of plant-based fibers into cementitious composites has been found to be an effective and low-cost solution for improving the brittle nature of cementitious composites [11]. According to [12], plant fiber-reinforced cementitious composites have a lower environmental impact than traditional materials, require less industrial and technology transformations, and improve the life cycle assessment of the final product [13] when compared with classic or enhanced concretes [14]. However, the adhesion between the fiber and matrix plays a significant role in the final mechanical properties of the composites. The stress transfer between the matrix and fibrous filler determines the reinforcement efficiency [15]. As demonstrated in [16], poor fiber-matrix bonding, aggregation of fibers, and increased porosity lead to an inferior strength and stiffness for plant fiber-reinforced cementitious composites. Another dominating factor affecting the mechanical properties is their reciprocal compatibility, leading to a homogeneous distribution of the reinforcing fibers in the matrix [17].

The main progress in the current building sector lies in the use of cellulosic nano-/micro-fibers/-fibrils extracted from diverse plant non-wood sources, including hemp, jute, flax, sisal, bamboo, bagasse, date-palm, etc., into composite materials of high-quality performance, service ability, durability, and reliability standards [18–22]. This strategy also plays a significant role in increasing the mechanical performance of mortars. Cement mortars reinforced by plant fibers are widely studied for various uses in the construction of new buildings and in the restoration of historical structures. Their use as plaster—one of the important parts of construction, covering the outdoor and indoor surfaces of buildings—is conditioned by the development of such compositions with improved functional properties. While external plaster provides protection from weather effects, it mainly offers a protective surface against the penetration of rain water and other atmospheric agents, improves the appearance of the structure, and gives decorative effects; internal plasters, in addition to their aesthetic function, mainly ensure the best possible diffusion properties, and thus they allow for improving indoor air quality, positively impacting the health of the occupants [23].

In recent years, a dramatic increase in the incorporation of cellulose fibers into cement-based, lime-based, or gypsum-based plasters/mortars has been reported, in line with the trend of sustainability in construction [24–28]. Research efforts in this area have confirmed that the application of cellulose fibers in quasi-brittle matrices of plaster mortars remarkably improves their mechanical, thermal, and acoustic properties [29–32]. The minimization or prevention of cracks propagation as a result of their crack-bridging ability has been proven [33–35]. The addition of natural fibers in plaster strengthens the plaster coating and thus enhances ductility. Over the last decades, lignocellulosic or cellulosic fibers derived/extracted from various plants (abaca, coir, cork, flax, hemp, rice straw, and sisal) have been used in plaster mortars/composites [6,8,36–43]. The use of a combination of two natural fibers (kenaf and sisal) for the reinforcement of lime plaster was studied in [26]. The developed multifunctional plaster contained $TiO_2$ nanoparticles (1 wt.%) and 0–4 wt.% cellulose fibers of two distinct sizes (1 mm < d < 2 mm; 2 mm < d < 4 mm), and showed superior thermal insulation properties, as well as the ability to control the indoor relative humidity and nitrogen oxide (NOx) photocatalytic degradation [44].

A key aspect of the application of lignocellulosic fibers in mortars is the nature of the inorganic matrix. As is known [45,46], the presence polymers like lignin and hemicelluloses in lignocellulosic fibers affected the setting process, and reduced the mechanical and durability properties of cement paste. Therefore, the pre-treatment of lignocellulosic fibers (hemp and flax) by chemical methods regarding the increase in the mechanical performance of lime and for mitigating the degradation of fibers in cement reinforced mortars was investigated [47–49]. As the quality of kraft fibers and pulp obtained through the chemical process of removing amorphous lignin and hemicelluloses from primary plant/wood sources is different from that of the cellulosic fibers obtained by recycling from secondary sources of waste paper, the integration of waste paper fibers (WPF) in the cement matrix requires more attention from researchers so that their usefulness can explicitly be examined. Waste paper containing a high amount of cellulose is an important source of cellulosic fibers, and because of the presence inorganic residues from the mineral fillers (calcite and kaolinite) used in paper making, is considered to be the most prominent recycled waste material, expanding the opportunities for its utilization in cement-based materials [50]; it supports the European Union circular economy goals [51,52].

As is shown in [24,53–55], the incorporation of waste paper fiber (WPF) has improved the properties of hardened cement-based composites. Thanks to the well-known homogeneity and composition of recycled fibers from waste paper, they are increasingly added in small amounts to cement mortars, providing a high-performance mortar/plaster with an insulating function and low environmental impact [56,57]. According to the results in the literature [58], the WPFs addition (0.2% by mass of cement) was found to show a significant self-shrinkage cracking control, while providing some internal curing. In [59], the effect of the addition of recycled cellulose fibers (up to 16 wt.% by mass of cement), obtained from waste packaging boxes and paper on the mechanical and thermal properties of lightweight cement paste was studied. The results revealed a reduction in the compressive strength of cement composite with the increasing fiber content; however, the thermal insulation properties were improved. Fibers derived from waste paper in various contents were applied in cement-based mortar mixed in a different way [25]. New ecological lightweight plastering mortars based on waste cellulosic fibers (newsprint or copy paper) using a technology with minimal embodied energy with very good thermal insulation characteristics, compressive strength assigned to class CS I or CS II, and water absorption by capillarity classified as W0 or W1 were developed [60]. Another effective approach to improving the properties of fiber-cement mortar is to partially replace the cement with supplementary cementitious materials (SCMs) [61,62]. SCMs positively influence the hydration kinetic of cement and lead to a considerable consumption of $Ca(OH)_2$.

Our research was devoted to the extensive investigation of the properties of multiple types of kraft wood pulp and recycled fibers from waste paper [63,64], as well as their implementation into cement mortars [55,65–70]. As the valorization of waste paper fibers in cement mortars has been relatively unexplored in current literature, a research emphasis was placed on the identification of a suitable type of cellulose fibers from the six samples of wood pulp and waste paper fibers for their use in cement fiber reinforced mortar. The originality of this work is connected with the development of recycled fiber reinforced plaster mortar with a good workability for the fresh mixture, and insulation, mechanical and adhesive properties for the final hardened product for indoor application. The research focus was on studying the influence of the mass ratio of cement and waste paper fibers (C/WPF) in fiber-cement mortars containing three types of WPF (different in cellulose content and degree of polymerization) on technically important properties of fresh mixture and compact test specimens after 28 days of hardening. Another investigation was aimed at testing the adhesion of fiber-cement mortar with a modified composition to two commonly used substrates (brick and aerated concrete block). The specific goal of this work is to provide a comprehensive overview of the key findings related to the use of recycled cellulosic fibers in cement-based mortars and plasters.

## 2. Materials and Methods

### 2.1. Materials

The basic components of the fiber-cement mortar used here were as follows: Ordinary Portland cement CEM I 42.5 N, standard natural silica sand, tap water, and waste paper cellulosic fibers (recycled). The addition of a super plasticizer and thickener to selected mixtures was applied for an improvement in the workability and water retention. Other mineral substances such as limestone and blast furnace granulated slag were also applied into a set of fiber-cement mortars for testing their adhesion to the substrate.

#### 2.1.1. Cement

Portland cement CEM I 42.5 N supplied by Cement Factory Ltd. (Povazska Cementaren Ladce, Ladce, Slovakia) in accordance with the European standard [71] and was used in all of the cement mortar samples. Its chemical composition is in Table 1.

**Table 1.** Chemical composition of Portland cement CEM I 42.5 N.

| Oxides/Element | Content (%) |
|---|---|
| MgO | 0.81 |
| $Al_2O_3$ | 2.13 |
| $SiO_2$ | 12.50 |
| $P_2O_5$ | 0.35 |
| $SO_3$ | 2.86 |
| Cl | 0.09 |
| $K_2O$ | 0.80 |
| CaO | 49.79 |
| $TiO_2$ | 0.17 |
| $V_2O_5$ | 0.02 |
| MnO | 0.03 |
| $Fe_2O_3$ | 2.29 |
| ZnO | 0.02 |
| SrO | 0.04 |

The mean particle size of the cement (calculated from the particle size analysis data) was 24.03 µm, and the specific surface area value measured using the B.E.T method was 0.968 $m^2/g$.

#### 2.1.2. Silica Sand

CEN standardized silica sand (fraction 0–2.0 mm) packed in bags (1350 g $\pm$ 5 g) supplied by Filtracni pisky (Chlum, Czech Republic) with a silica content of at least 98% as a filler in fiber-cement mixtures in accordance with the standard [71] was used.

#### 2.1.3. Water

Tap water was used for the preparation and treatment of fiber-cement mixtures in accordance with the standard [72].

#### 2.1.4. Waste Paper Cellulose Fiber (WPF)

Three types of WPF (A, B, and C), different in chemical composition, provided by Greencel, Ltd. (Hencovce, Slovakia), were used in the cement mortars (Table 2).

**Table 2.** Chemical composition of waste paper cellulose fiber (WPF) samples.

| Components of WPF (%) | Component Content (%) in WPF Samples | | |
|---|---|---|---|
| | A | B | C |
| Holocellulose | 71.03 | 71.98 | 81.30 |
| Cellulose | 47.40 | 46.95 | 56.97 |
| Hemicellulose | 23.63 | 25.03 | 24.33 |
| Lignin | 17.05 | 20.05 | 20.11 |
| Ash | 19.91 | 22.80 | 16.54 |
| Extractives | 1.39 | 1.69 | 1.72 |
| Moisture | 1.61 | 1.50 | 1.99 |

Their gray color corresponded to the remaining inks used in newspaper printing. The samples contained two inorganic substances, namely calcite and kaolinite, which are used as mineral fillers in the paper-making process [64]. The crystallinity index, one of the most significant crystalline structure parameters of the studied WPF samples, ranged 41–51%, and the lowest value was reached with sample C. The polymeric characteristics of the main component of cellulose, indicating the average polymer chain length (DP—degree of polymerization) and distribution of its molecular weights in the cellulose polymer (PDI—polydispersity index), are summarized in Table 3. Both parameters are numerical outputs of the molecular mass distribution of the cellulose obtained using the size exclusion chromatography method. PDI represents the ratio of the mass-average molecular mass (Mw)/number-average molecular mass (Mn). DP was calculated by dividing the molecular mass by the monomer equivalent mass of anhydroglucose [64]. DP is an important structural parameter that significantly influences the mechanical properties of cellulose.

**Table 3.** Polymeric characteristics of cellulose in the WPF samples. DP—degree of cellulose polymerization; PDI—polydispersity index.

| Cellulose Polymer Characteristic | WPF Samples | | |
|---|---|---|---|
| | A | B | C |
| DP | 1176 | 1190 | 1285 |
| PDI | 6.92 | 6.16 | 6.30 |

The characterization of the selected physical properties of the WPF samples (a more detailed description of the methods for determining these properties is in paper [64]) is given in Table 4. The morphology of the fibers is shown in Figure 1. Heterogeneity on the fiber dimensions and the roughness of the fiber surface due to the presence of fiber fragments and impurities is clearly observed in the WPF samples. The fragments are formed as a result of the mechanical damage to the fibers due to the recycling process of waste paper. The impurities originated from the paper-making processes.

**Table 4.** Physical properties of WPF samples.

| Characteristics of WPF Samples | WPF Sample | | |
|---|---|---|---|
| | A | B | C |
| Density ($kg/m^3$) | 1843 | 1966 | 1943 |
| Apparent density ($kg/m^3$) | 75 | 55 | 40 |
| Max. fiber length (μm) | 400 | 600 | 1200 |
| Dry matter (%) | 93 | 93 | 93 |
| pH | $7 \pm 1$ | $7 \pm 1$ | $7.5 \pm 1$ |
| Average fiber width D (μm) | 29.5 | 30.9 | 29.0 |
| Average fiber length L (μm) | 556 | 701 | 796 |
| Aspect ratio L/D (-) | 18.8 | 22.7 | 27.5 |
| BET specific surface area ($m^2/g$) | 9.48 | 8.63 | 7.60 |
| NLDFT cumulative pore volume ($10^3$ $cm^3/g$) | 20.64 | 12.98 | 22.06 |
| NLDFT pore radius (nm) | 2.71 | 1.66 | 2.08 |
| Thermal conductivity (W/m·K) | 0.0634 | 0.0599 | 0.0595 |
| Volume heat capacity $\times 10^{-6}$ ($J/m^3{\cdot}K$) | 0.2097 | 0.1785 | 0.1709 |
| Thermal diffusivity $\times 10^6$ ($m^2/s$) | 0.3024 | 0.3354 | 0.3478 |

2.1.5. Thickener

For the improvement and controlling of water retention, influencing the rheological properties, and adhesion of the fiber-cement mortar to the substrate, a thickener (mecellose) as a modified water-soluble cellulose ether derivative (hydroxypropylmethylcellulose (HPMC) manufactured by Lotte Fine Chemicals Ltd., Seoul, South Korea) in an amount of 1% by weight of the cement was used.

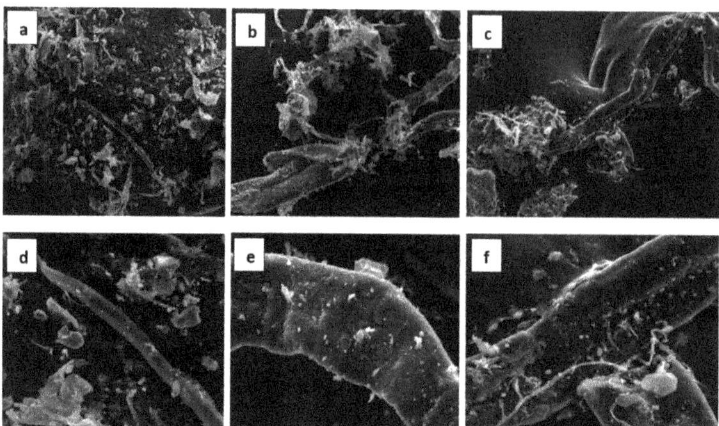

**Figure 1.** SEM micrographs of WPF samples (**a**,**d**) A, (**b**,**e**) B, and (**c**,**f**) C at magnifications of 1500× and 6000×.

2.1.6. Mixed Hydraulic Binders for Plastering Mortar

For experiments aimed at studying the adhesive ability of the fiber-cement mortar on the substrate, the binder component of the cement mortar was modified by using very fine ground products of natural limestone and metallurgical waste of granulated blast furnace slag (from KOTOUČ ŠTRAMBERK Ltd., Štramberk, Czech Republic). The fine products of the granulated slag as a powdered latent hydraulic binder met the requirements of the standard for [73], while ground limestone complied with the requirements of the standard for [74]. The chemical composition of both mortar components is in Table 5. The mean particle diameter of both mortar components was almost the same, 12.82 μm and 12.25 μm, respectively; however, the specific surface area of the ground slag was lower (0.96 $m^2/g$) than the value corresponding to the ground limestone (1.46 $m^2/g$).

**Table 5.** Chemical composition of the ground limestone and granulated slag.

| Oxides/Elements | Limestone | Slag |
|---|---|---|
| | Content (%) | |
| MgO | 0.59 | 0.29 |
| $Al_2O_3$ | 0.15 | 4.17 |
| $SiO_2$ | 0.91 | 25.00 |
| $P_2O_5$ | 0.18 | 0.52 |
| $SO_3$ | 0.07 | 1.04 |
| Cl | 0.01 | 0.02 |
| CaO | 61.26 | 35.70 |
| $K_2O_3$ | - | 0.22 |
| $TiO_2$ | 0.01 | 0.22 |
| $V_2O_5$ | 0.01 | 0.02 |
| MnO | - | 0.62 |
| $Cr_2O_3$ | 0.10 | - |
| $Fe_2O_3$ | 0.43 | 0.33 |
| SrO | 0.02 | 0.02 |

*2.2. Preparation of Fresh Fiber-Cement Mixtures*

Two sets of experimental cement mortar mixtures with different contents of all WPF samples, expressed as the cement to WPF material mass ratios (C/WPF), were prepared (I and II), as shown in Table 6. Based on the results of testing the fresh mixtures and hardened mortars of the sets, fiber sample C as a sand replacement and representative of the group of WPF with the highest content of cellulose and degree of polymerization was selected for the production of fiber-cement plaster mortars (set III) with a modified binder composition (see Section 2.1.6). The addition of 1 wt.% thickener (mecellose) was used in the mortars. The water to cement ratio (W/C) was varied because of the changing amount of fibers used in the mixture and taking into consideration the hydrophilic nature of WPF (Table 6).

**Table 6.** Formulations for fiber-cement mortars.

| Mortar Sample | C/WPF | W/C |
|---|---|---|
| | Set I | |
| R1 | - | 0.50 |
| R2 | - | 0.55 |
| A1; B1; C1 | 500:1 | 0.55 |
| A2; B2; C2 | 330:1 | 0.55 |
| A3; B3; C3 | 200:1 | 0.55 |
| | Set II | |
| R3 | - | 0.75 |
| A4; B4; C4 | 33:1 | 1.55 |
| A5; B5; C5 | 17:1 | 1.55 |
| A6; B6; C6 | 7:1 | 1.55 |
| C7 | 3:1 | 1.55 |
| | Set III | |
| C8 | 6:1 | 2.08 |
| | 6:1 | 2.17 |

The design of the composition of the sets of fiber-cement mortars was based on the standard for [71]; the mass ratio of cement/sand (C/S) was 1:3. Control (reference) cement mortars (R1, R2, and R3) without WPF addition were prepared. R1 was prepared according to the standard with W/C = 0.5. Reference samples R2 and R3 with varied W/C ratios corresponding to fiber-cement mortars were prepared so as to compare the properties of the fiber-cement mortars. The mass ratio of the components in the mortar mixture of set III (cement:limestone (slag):WPF:mecellose) was equal 1:2.04:0.17:0.01 (mortar based on limestone (C8) and slag (C9), respectively). The preparation of fiber-cement mixtures (I-II) consisted of two steps. The soaking of recycled fibers and manual mixing in approximately 50 wt.% of water was the first step in the preparation of the mixture. Next, the remaining water, the required amount of sand, and the cement were mixed by mechanical stirring in a mixer. Mixing took place automatically in a laboratory mixer BS MI-CM5AX (BETON SYSTEM, Brno, Czech Republic), according to the standard for [71] for 5.5 min. The procedure for preparing fiber-cement plasters (III) was as follows: the weighed individual dry materials (cement, limestone/slag, and WPF) were mixed, and water was gradually added during mechanical stirring. After sufficient mixing of the components, a thickener was added and incorporated into the mixture.

### 2.3. Production of Test Specimens

Standard steel molds with dimensions of 40 × 40 × 160 mm were used for the production of the mortar test specimens. The filling of the molds with fresh fiber-cement mixtures (I and II) took place in two layers; each layer being compacted by 60 strokes of the compaction table (according to the standard for [71]). The compaction of each layer of mixtures (set III) was performed by tapping (20 times) from a height of 3 cm on both sides of the mold. The excess layer of the mixture was wiped off with a saw motion and smoothed with a metal ruler. The filled molds were covered with PVC foil, which served to prevent water evaporation from the fresh mixture needed to hydrate the cement. Curing of these mix sets (I and II) took place for 2 days in molds at room temperature and humidity (+20 °C; 50% RH). After that time, the bodies were removed from the molds and were placed in a water bath (sets I and II), and the fiber-cement mixes (III) were held under PVC foil in a laboratory for 26 days.

In total, about 800 test specimens were prepared for testing the properties of the fresh fiber-cement mixtures and the technically significant parameters of the composites. Errors bars in the histograms represent the standard deviation of the three replicates.

### *2.4. Methods for Testing Properties of Fresh Mixtures and Compact Specimens*

#### 2.4.1. Consistency

The flow behavior of the fresh fiber-cement mixtures (indicating its workability) was detected through the determination of consistency using a flow table test in accordance with the standard of [75]. The mean diameter of a test sample placed on a flow table was measured before being impacted vertically after the release of a standard slump cone. The measurement was repeated twice with an accuracy of ±1 mm.

#### 2.4.2. Physical Properties of Compact Specimens

All of the physical properties of the compact mortar specimens were measured on prismatic bodies after 28 days of hardening.

##### Dry Bulk Density

Dry bulk densities of the 28-day hardened fiber-cement composite specimens were calculated according to the standard for [76].

##### Water Absorbency and Capillary Water Absorption

The water absorbency of mortars was determined according to [77]. The weighed specimens were immersed in a water bath at room temperature (20 °C) until their weight stabilized. The composites were then weighed and placed in an oven where they were dried at 80 °C until reaching a constant weight.

The standard for [78] was used for the determination of the capillary water absorption coefficient of the hardened fiber-cement mortars. A sealing compound (paraffin wax) was applied to the longitudinal sides of the test prisms, and they were broken in two halves. The prismatic bodies that were prepared were inserted with the fracture surface downwards into a container with water, the level of which was 5–10 mm (see Figure 2). Washers were inserted between the bottom of the vessel and the test specimen (its fracture surface) to ensure perfect contact with the water. The water level was kept constant throughout the test. The capillary water absorption coefficient (C) was determined from the difference of the two weight values measured after 10 (M1) and 90 min (M2), according to Equation (1)

$$C = 0.1 \times (M2 - M1) \tag{1}$$

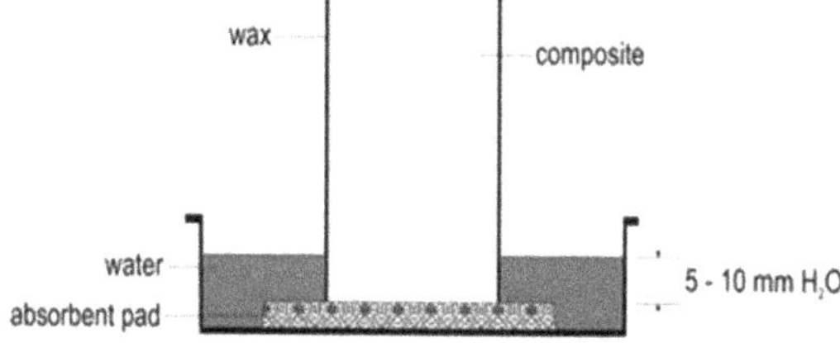


**Figure 2.** Capillary water absorption testing scheme.

The resulting average value of C is given with an accuracy of 0.05 $kg/m^2 \cdot min^{0.5}$.

Thermal Conductivity

The coefficient of the thermal conductivity of the hardened reference samples and fiber-cement mortar specimens was measured on their surface using the commercial device ISOMET 2114. The tests were performed using a measuring surface probe IPS 1100 placed on the surface of each wall of the prismatic specimen, and the mean value was reported.

### 2.4.3. Mechanical Properties of Compact Specimens

The compressive strength of the fiber-cement mortars (40 mm × 40 mm × 160 mm) after 28 days of hardening was determined using a compression test machine (FORM + TEST Seidner and Co. GmbH, Riedlingen, Germany) with a maximum load force of 300 kN and a loading rate of 2400 ± 200 N/s, in accordance with [79]. A three points bending test was used for the determination of the flexural strength on the aforementioned testing machine, with a maximum load force of 10 kN and a loading rate of 50 ± 10 N/s. The resulting parameter values were the average of six measurements.

### 2.4.4. Adhesive Strength of Plastering Mortars on the Substrates

The adhesive strength of the hardened plaster cement mortars based on the WPFs for two selected substrates (brick and aerated concrete blocks) was measured according to the standard for [80], which expresses the maximum tensile stress derived from a load perpendicular to the surface of the mortar applied to the substrate. The tensile load was determined using a Comtest OP3/4 (COMING Plus Ltd., Prague, Czech Republic) and a rigid tear-off target (diameter of 50 ± 0.1 mm; height at least 10 mm) made of corrosion-resistant steel, and it adhered to the tested surface of the mortar surface with Sikadur-31 CF Rapid adhesive (Sika Slovakia, Ltd., Bratislava, Slovakia), according to standard [80]. The mortar sample in a thickness of 30 ± 1 mm was applied to the substrate of ceramic fittings (375 mm × 80 mm × 238 mm) and aerated concrete blocks (50 mm × 249 mm × 599 mm). The substrate was kept vertical during application. The specimens were tested after 28 days of hardening. Five test circular surfaces with a diameter of 50 mm were drilled into the mortar so that the substrate was cut to a depth of at least 2 mm. Tear-offs were performed using glued test targets. The testing of the adhesion of the mortar to the substrate is demonstrated in Figure 3. The adhesion $R_{fu}$ (in MPa) is expressed by Equation (2) as the ratio of the derived load F (in N) and test area A (mm$^2$).

$$R_{fu} = F/A \tag{2}$$

(a)

(b)

**Figure 3.** Application of fiber-cement mortar on the substrate of an aerated concrete block (**a**); demonstration of testing the adhesion of fiber-cement mortar to the substrate (**b**).

## 3. Results and Discussion

### 3.1. Effect of C/WPF Ratio on Properties of Fresh Mixtures and Hardened Fiber-Cement Mortars

#### 3.1.1. Consistency of Fresh Fiber-Cement Mixtures

The consistency of fresh cement mixtures plays a vital role in the determination of the workability for mortars and for the compressive strength of hardened mortars. The water to cement ratio as a parameter determining the normal consistency of mortars and ensuring a complete hydration reaction was necessary in order to modify, in the case of fiber implementation into a cement mixture, in order to make a paste with an appropriate plasticity (Table 6). The effect of the WPF content in the cement mixes (expressed by C/WPF ratio) on the flow behavior of fresh fiber-cement mixtures was investigated. As shown in Figures 4–6, a decrease in the resulting mean values of the spill diameter of the fresh fiber-cement mixtures with an increasing amount of cellulose fibers in the mixture and decreasing C/WPF ratio can be observed. This phenomenon corresponds to properties of incorporated WPF as a material with a rougher surface, voids, high water absorption ability, and low apparent density, which significantly affect the mortar consistency. Fibers absorb larger amounts of mixing water and consequently significantly reduce the resulting consistency of the fresh mixture [48]. All of the experimental fiber-cement mortars reached mean spill diameter values in the range of 188–133 mm. The course of the mean spill diameter values with an increasing fiber content (0.2–10 wt.% from weight cement) for sample C (the highest cellulose content) clearly demonstrated their strong reduction in consistency (Figure 6). We used a simple moving average for the spill diameter as a tool of technical analysis to identify the trend. According to the measured values of consistency, fiber-cement mortars with a fiber content ≤2 wt.% can be classified as plastic mortars (140–200 mm). This value represents the limit for the fiber content in fiber-cement mortars in terms of their good workability. A 5–7% loss of workability from the lower limit (140 mm) was recorded for mortars with an increasing content of short fibers (C/WPF of 7:1 and 3:1). The presented results demonstrate the role of fibers in inducing the plasticity and reducing the flowability of fresh mortars. Similar results have been reported for paper [17,81].

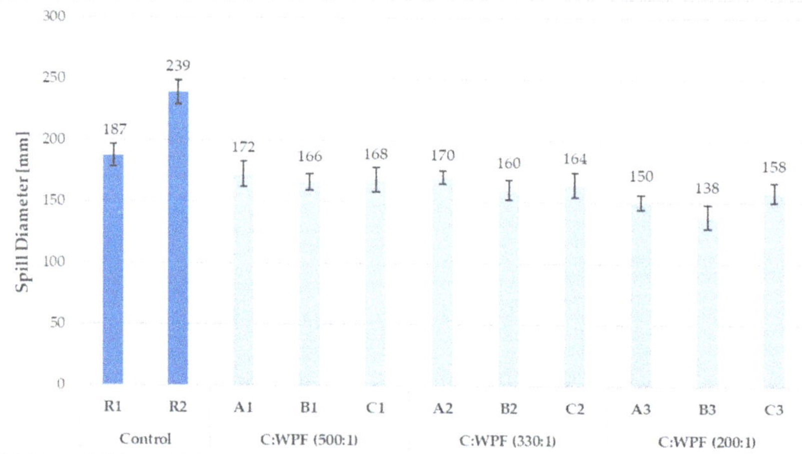


**Figure 4.** Mean spill diameter values for reference and fiber-cement fresh mixtures based on WPF samples (A, B, and C) with different C/WPF ratios (set I).

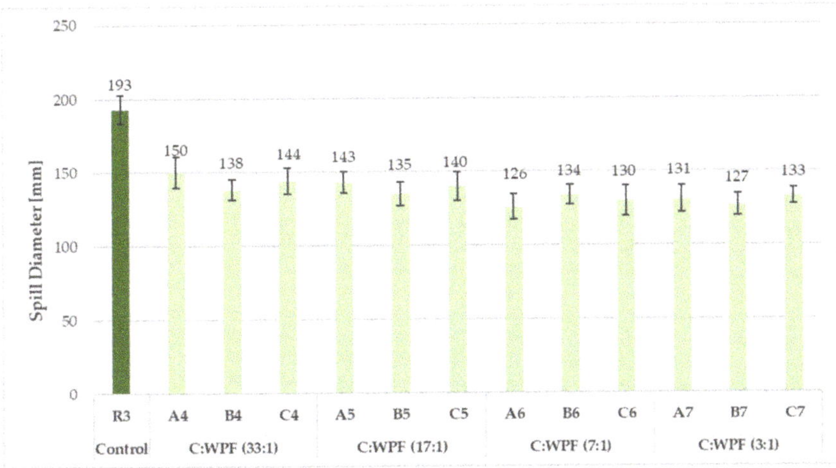


**Figure 5.** Mean spill diameter values for reference and fiber-cement fresh mixtures based on WPF samples (A, B, and C) with different C/WPF ratios (set II).

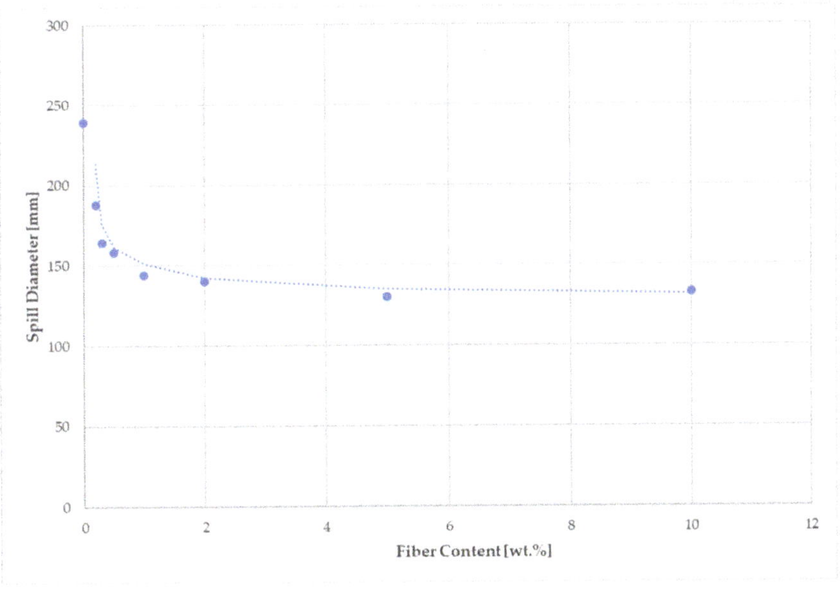


**Figure 6.** Effect of fiber content (sample C) on the consistency of the fiber-cement fresh mixture.

3.1.2. Bulk Density and Thermal Conductivity

From a comparison of the results of the bulk densities and thermal conductivity coefficients of fiber-cement mortars after 28 days of hardening composites (Figures 7–10), it is clear that as the content of cellulose fibers in the mixture increased (C/WPF ratio decreased), the bulk density and conductivity of the mortars had a decreasing trend. The values of the bulk density ranged 2160 to 1955 kg/m$^3$, and thermal conductivities were found in the range of 1.863–2.330 W/m·K. The values of both parameters found for references samples R1, R2, and R3 under the same laboratory conditions were likely mainly influenced by the ongoing phase transformation of the cement paste from a more liquid state to a solid state of mortar, and by the microstructure of hydrated phases formed under the conditions with different W/C ratios in the mixture during the hardening process. Another factor influencing the thermal conductivity can be related to the quality of the thermal contact of the measuring element and measured object.

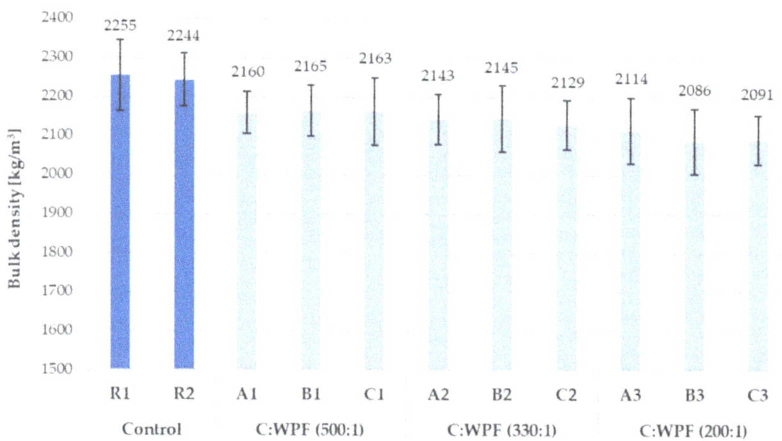


**Figure 7.** Development of bulk density with C/WPF ratios in fiber-cement mortars (set I) after 28 days of hardening.

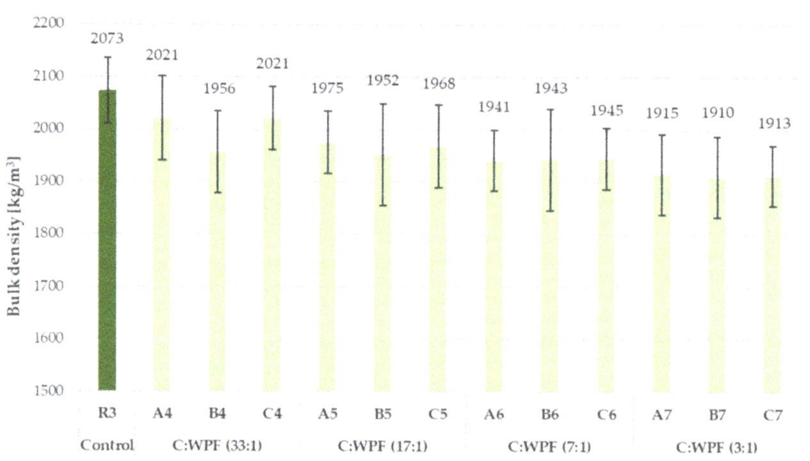


**Figure 8.** Development of bulk density with C/WPF ratios in fiber-cement mortars (set II) after 28 days of hardening.

A comparison of the thermal conductivity values of the fiber-containing mortars with those found for the reference mortars showed an almost equal reduction by 24–35% due to the low apparent density of the cellulose fibers and the increasing proportion of fibers in the mixture. Based on the values of the bulk densities of the mortar samples in the dried state, they belonged to the ordinary mortars [82] without thermal insulation properties. The measured thermal conductivities for 28-day hardened reference mortars R2 and R3 (2.519 and 2.699 W/m·K, respectively) were very close to the value (2.7 W/m·K) found for the cement mortar in [83]. Figure 11 illustrates the trend in the simple moving average of the thermal conductivity coefficient for hardened mortar samples with an increasing content of fiber sample C in the mixture. Using recycled cellulose fibers led to a better thermal insulation behavior for the cement matrix. The best thermal conductivity (1.783 W/m·K) among the tested mortars based on fiber sample C was reached with 10 wt.% (C7). The short cellulose fibers bring more pores and cavities into the cement matrix, which leads to a reduction in the thermal conductivity of the samples. A similar behavior has been observed on cement composites based on recycled waste paper and paper packaging [59], and waste fiber from young coconut and durian [84].

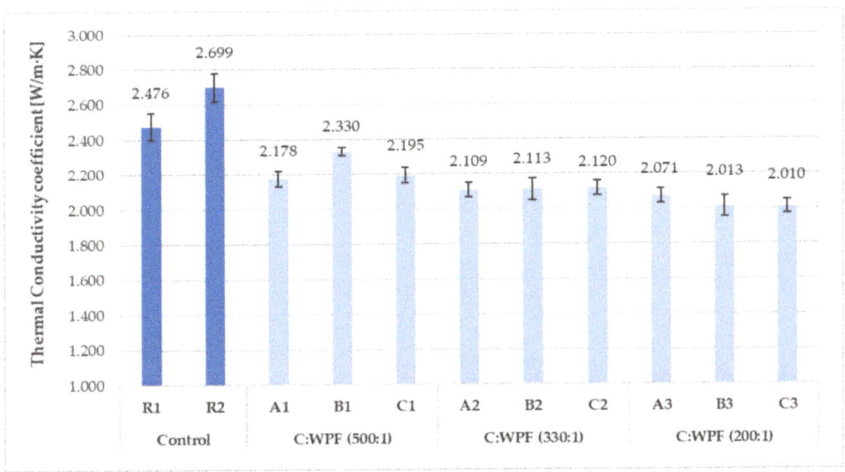


**Figure 9.** Thermal conductivity coefficients for 28-day hardened fiber-cement mortars (set I) with different C/WPF ratios.

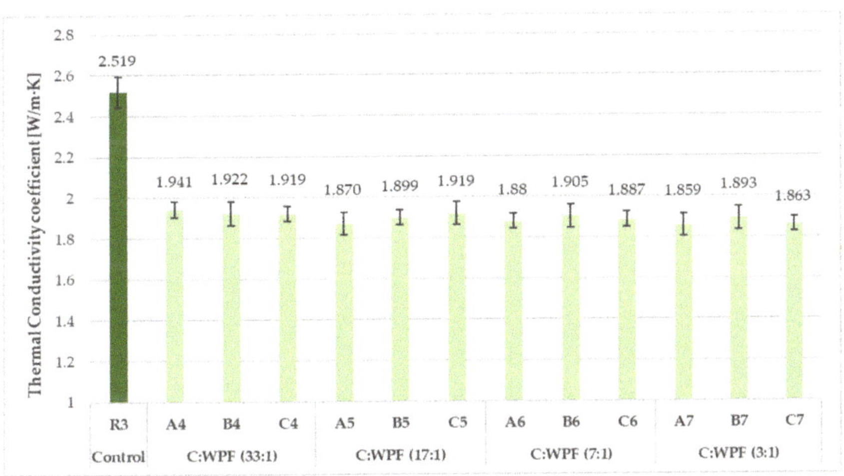


**Figure 10.** Thermal conductivity coefficients for 28-day hardened fiber-cement mortars (set II) with different C/WPF ratios.

As is known, the thermal conductivity of a material is determined not only by its chemical composition and structure, but also by the porosity, the nature of the pores, the moisture content of the material, and the temperature at which heat is transferred. The porosity of the fiber-cement material plays a crucial role, not only in thermal conductivity, but also in its water behavior.

The high value of the correlation coefficient (R = 0.9906) for the dependence shown in Figure 12 confirms the linear relationship between the thermal conductivity coefficient and the bulk density of the studied fiber-cement mortar of sample C with different C/WPF after 28 days of hardening.

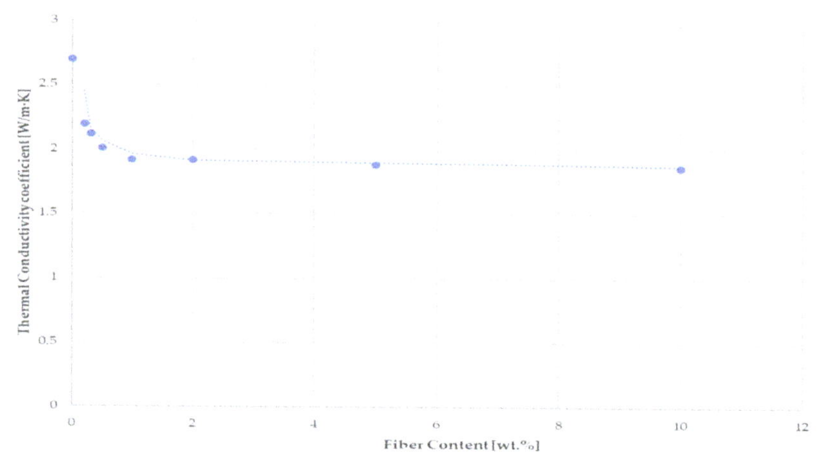


**Figure 11.** Dependence of the thermal conductivity coefficient of 28-day hardened fiber-cement mortars on the increased content of fiber sample C.

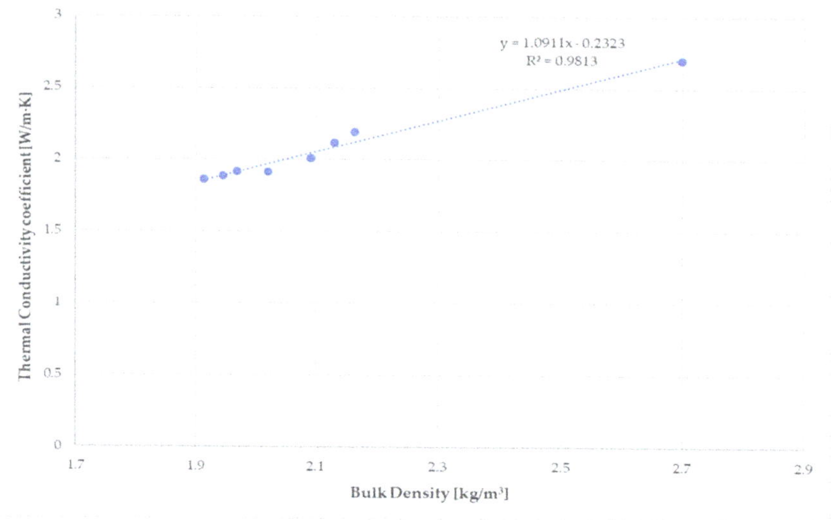


**Figure 12.** Relationship between the thermal conductivity coefficient and bulk density of 28-day fiber-cement mortar sample C with different C/WPF ratios.

3.1.3. Water Absorbency and Capillary Absorption

As shown in Table 7, an increase in the absorbency of 28-day hardened fiber-cement mortars by 6.3–21.1% and 17–32.4% dependent on the WPF content was observed in comparison with the absorbency value of the control samples R2 and R3. The values of the capillary water absorption coefficient of the cement mortars containing recycled fibers ranged from 0.21 to 0.39 $kg/m^2{\cdot}min^{0.5}$ dependent on the C/WPF ratio, and were also higher than the value of the control samples. The water behavior of the fiber-cement mortars is of a complex nature and is affected by a number of factors [85]. The main cause of such a water behavior for fiber-cement mortars is the implementation of cellulosic fibers to the matrix of cement mortars, with their typical absorbency ranging from 7% to 20% [86].

**Table 7.** Values of absorbency and capillary water absorption for 28-day hardened fiber-cement mortars.

| C/WPFs | Mortar Sample | Absorbency (%) | Capillary Water Absorption ($kg/m^2 \cdot min^{0.5}$) |
|---|---|---|---|
| | R2 | 8.66 | 0.23 |
| 500:1 | A1 | 9.21 | 0.21 |
| | B1 | 9.30 | 0.24 |
| | C1 | 9.35 | 0.22 |
| 330:1 | A2 | 9.61 | 0.25 |
| | B2 | 10.25 | 0.23 |
| | C2 | 10.35 | 0.24 |
| 200:1 | A3 | 10.38 | 0.25 |
| | B3 | 10.29 | 0.24 |
| | C3 | 10.49 | 0.26 |
| | R3 | 9.75 | 0.23 |
| 33:1 | A4 | 11.41 | 0.26 |
| | B4 | 12.24 | 0.36 |
| | C4 | 12.62 | 0.38 |
| 17:1 | A5 | 12.20 | 0.34 |
| | B5 | 12.65 | 0.37 |
| | C5 | 12.68 | 0.36 |
| 7:1 | A6 | 12.82 | 0.38 |
| | B6 | 12.91 | 0.39 |
| | C6 | 12.75 | 0.37 |

The fiber content in the matrix significantly affects the formation of the pore structure in these materials. Therefore, water absorption is considered a measure of the porosity of fiber-cement composites/mortars, and provides useful information on the permeable volume of pores inside the sample and the potential interconnection between these pores [33]. It is worth noting that the volume of voids and the number of open and closed pores and their size significantly affect the water absorption of fiber-cement materials.

Figure 13 shows the linear relationships between both parameters of water behavior for fiber-cement composites based on a particular type of cellulose fibers with changing the C/WPF ratio. All of the measured values of capillary water absorption correspond to class W1 ($\leq 0.40\ kg/m^2 \cdot min^{0.5}$), in accordance with standard for [82]. The lower fiber content in the cement mortar led to lower values of absorbency in the range of capillary absorption of 0.21–0.25 $kg/m^2 \cdot min^{0.5}$ because of the low porosity of the fiber-cement mortar, as well as the fibers themselves, while a higher fiber amount in cement paste (higher porosity) at almost equal capillary absorption values caused a higher absorbency. This relation between the measured values is most likely related to the number of open pores with different sizes, whereas the number of closed pores probably appears to be constant because they hardly exert capillary forces. Another possibility of this behavior could be related to the very small pore sizes in the mortar structure into which water molecules do not penetrate. As is evident from Figure 13, incorporating a higher proportion of WPF into mortars led to significantly higher values for both quantities, where the capillarity absorption values shifted to values higher than 0.26 up to 0.40 $kg/m^2 \cdot min^{0.5}$, and are related to the formation of a larger number of pores and to their size. Figure 14 documents the texture of the pores formed in fiber-cement mortar samples A3, B3, and C3 (C/WPF = 200:1). The pores are evenly distributed in the matrix and only occasionally are small agglomerates and clumps of fibers observed. The size of the pores, which are mostly spherical in shape, ranges from a few tenths of a millimeter to a millimeter. A small number of pores have an irregular shape, being slightly elongated. In some cases, the pores may be filled with fibers or clumps of fibers.

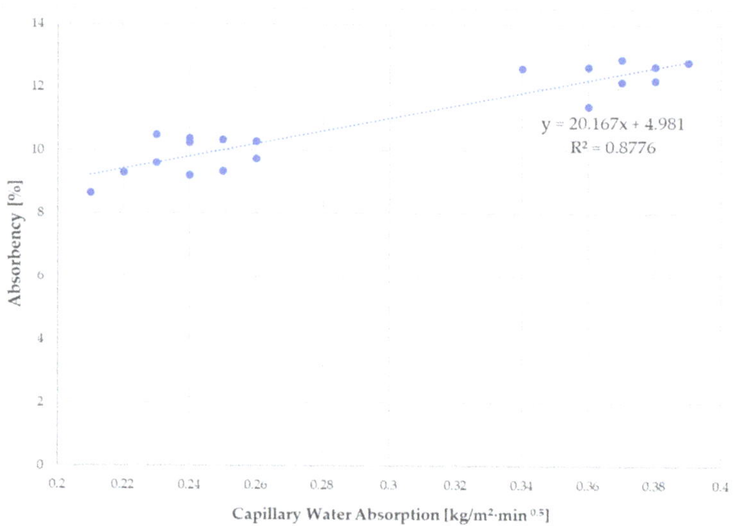


**Figure 13.** Relationship of the absorbency and capillary water absorption of fiber-cement mortars with increasing fiber content.

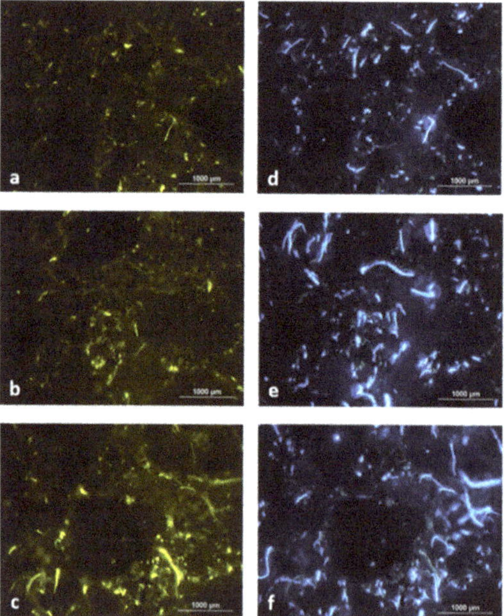


**Figure 14.** Micrographs of the texture of 28-day fiber-cement mortars samples A3 (**a**,**d**), B3 (**b**,**e**), and C3 (**c**,**f**) for C/WPF with a ratio of 200:1 under UV light using WB (left) and WU filter (right).

### 3.1.4. Compressive and Flexural Strength

The results of the compressive and flexural strength of the fiber-cement mortars with varied amounts of fiber content after 28 days of hardening are illustrated in Figures 15–18. The C/WPF ratio had a significant effect on both of the strength characteristics of the mortars.

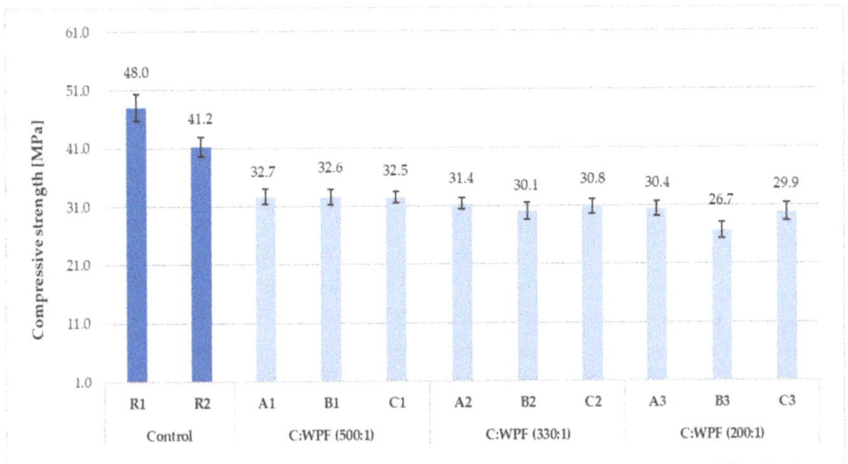


**Figure 15.** Compressive strength of fiber-cement mortars with different C/WPF ratios (set I).

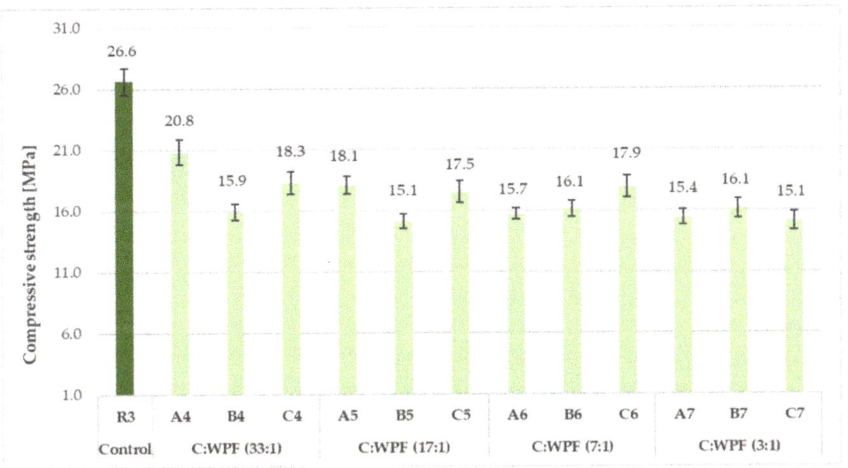


**Figure 16.** Compressive strength of fiber-cement mortars with different C/WPF ratios (set II).

As the C/WPF ratio decreased, the values of the compressive and flexural strength of the fiber-cement mortars decreased significantly. The fiber-cement mortars reached values of compressive and flexural strength in the range of 15.1–32.7 MPa and 3.01–5.92 MPa, respectively, depending on the WPF content, and they represented a lower level of the strength parameter found for reference samples R2 and R3. The higher the bulk density, the higher the value of the compressive strength of the fiber-cement mortar depending on the WPF content of sample C is observed (Figure 19; R = 0.8836).

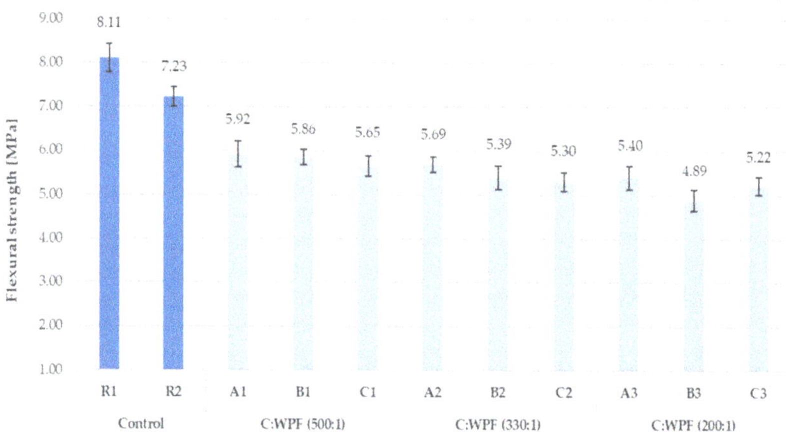


**Figure 17.** Flexural strength of fiber-cement mortars with different C/WPF ratios (set I).

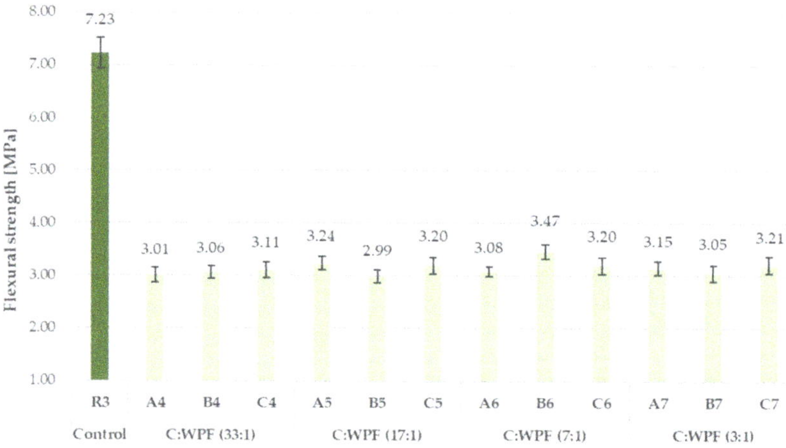


**Figure 18.** Flexural strength of fiber-cement mortars with different C/WPF ratios (set II).

A slightly higher correlation coefficient was calculated for the linear dependence between the flexural strength and bulk density (R = 0.8927). The presented decreasing trends of both strength parameters with an increasing fiber content for the mortar samples (C1–C7) were observed. A higher fiber content in the cement mortar induces the formation of larger interfacial pores/voids that make a more lightweight material, and it should be pointed out that a weaker adhesion at the matrix-fiber interface owing to the hydrophilic nature of WPFs leads to a decrease in the compressive and flexural strength. Similar conclusions were also found by Bentchikou et al. [59] upon investigation of the effect of recycled waste paper fiber contents on the mechanical properties of lightweight cement composites. As shown in Figure 20, there is a proportionality between the experimental determined levels of flexural and compressive strength for fiber-cement mortar samples (C1–C7). Two equations (linear and exponential) with almost equal values for the correlation coefficients (0.9973 and 0.9952) were found for the relationships of these strength parameters. This means that both models included factors affecting the strength of the fiber-cement mortars after 28 days in the range of the measured strength levels, and the flexural and compressive strengths had a very good correlation.

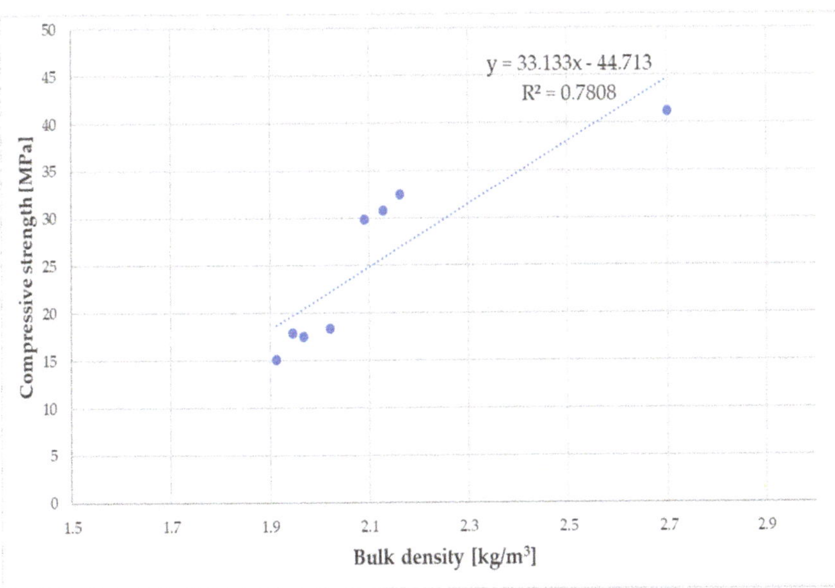


**Figure 19.** Linear dependence of compressive strength on bulk density of fiber-cement mortar samples (C1–C7).

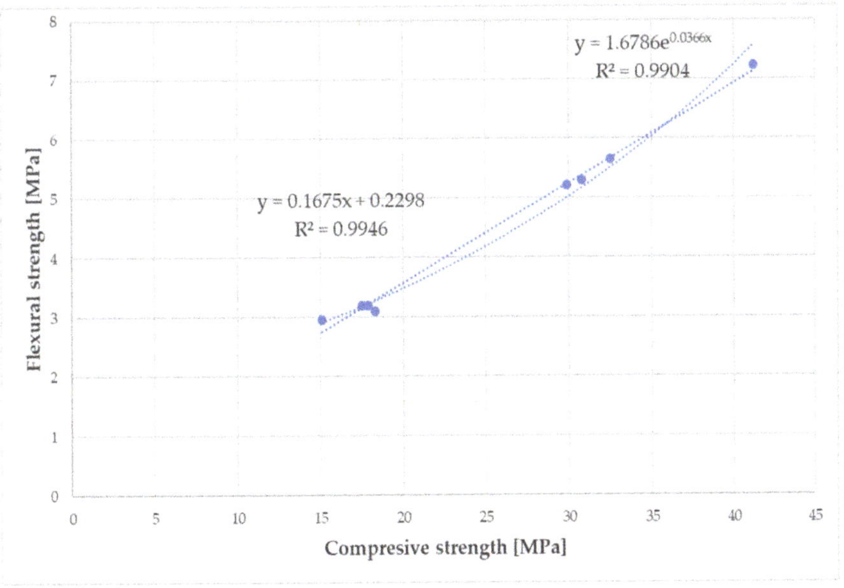


**Figure 20.** Relationship between flexural and compressive strength of fiber-cement mortars (C1–C7).

### 3.2. Modification of Fiber-Cement Plastering Mortar Composition and Its Influence on Technically Important Properties

In terms of determining the practical implementation of recycled cellulose fibers into cement plastering mortar for internal use and improving its plasticity behavior, two very finely ground materials, namely, limestone and granulated blast furnace slag, were selected. The effect of the modified composition of the fiber-cement plastering mortar on the important parameters of the fresh mixtures and hardened mortars was investigated.

#### 3.2.1. Physical Properties of Fresh Fiber-Cement Mixtures and Hardened Mortars

Table 8 summarizes the results obtained from the measurement of the consistency of fresh fiber-cement mixtures with the addition of very finely ground binder components of limestone (C8) and granulated slag (C9), and other properties of hardened mortars for application as interior plasters. As can be seen, the consistency values of the fiber-cement mixtures and the properties of the hardened specimens were influenced by the nature of the fine components in the cement mixture. The mean spill diameter of the fresh fiber-cement mixture with the addition of limestone (C8) reached a slightly higher consistency value than such a mixture with a slag addition (C9). This showed the same plasticity for both mortars.

**Table 8.** Properties of fresh fiber-cement mixtures, hardened mortars, and adhesive strength on two substrates of brick (B) and aerated concrete block (ACB).

| Set III | | |
|---|---|---|
| **Property** | **C8** | **C9** |
| Mean spill diameter (mm) | 157 | 150 |
| Bulk density (kg/m$^3$) | 1083 | 1112 |
| Thermal conductivity coefficient (W/kg·m) | 0.224 | 0.222 |
| Absorbency (wt.%) | 50.7 | 47.8 |
| Capillary water absorption coefficient (kg/m$^2$·min$^{0.5}$) | 2.71 | 1.59 |
| Compressive strength (MPa) | 4.86 | 10.07 |
| Flexural strength (MPa) | 1.70 | 3.32 |
| Adhesive strength on brick (MPa) | 0.215 | 0.347 |
| Adhesive strength on aerated concrete block (MPa) | 0.205 | 0.228 |

The bulk density value of the hardened mortar with finely ground limestone is evidently related to its lower density (2710 kg/m$^3$) compared with cement mortar containing granulated blast furnace slag (3100 kg/m$^3$). The thermal conductivity coefficient values of both mortar samples were almost the same (0.224 W/m·K and 0.222 W/m·K), and they were almost at the same level as the data (0.28 W/m·K) for fiber-cement mortar containing 16 wt.% fiber published in [59]. The measured conductivities were significantly lower than those declared for lime (0.9 W/m·K), lime-cement (1.0 W/m·K), and cement plaster (1.2 W/m·K).

Although the porosity of fiber-cement mortar plays a decisive role in this case, the different nature of the mineral components and the presence of waste paper fibers in the cement mixes affected the water absorption behavior of the mortar specimens (C8 and C9). The measured values of the absorbency and capillary water absorption were most likely related to the number of closed pores, which hardly exerted capillary forces; therefore, sample C9 reached lower values. As above mentioned, the water behavior of fiber-cement mortars is of a complex nature and is determined by a number of factors. Therefore, our assumption of a lower porosity for the C9 sample was not reflected in the thermal conductivity values.

#### 3.2.2. Mechanical Properties of Fiber-Cement Mortars

The results of the compressive and flexural strength of fiber-cement mortars containing two components after 28 days are presented in Table 8, and showed two times higher values

(10.07 MPa and 3.32 MPa, respectively) for both strength parameters for mortar C9 than for sample C8 (4.86 MPa; 1.70 MPa). The values of compressive strength found for the fiber-cement mortar samples of C8 and C9 corresponded to the minimum strength value required for the mortar of the CSIII (5 MPa) and CSII (2.5 MPa) classes [82]. The measured values for the flexural strength of the mortar samples were higher than the approximate flexural strength after 28 days of CS III (0.7 MPa). As the strength of such fiber-based materials depends on several factors, a higher value for the strength parameter (C9) could be related to a lower proportion of pores in the material. Another reason among the other factors influencing the strength of fiber-cement mortar might be attributed to better filling of the interparticle spaces in the mortar structure because of the presence fine particles of slag. Certainly, the chemical composition of this latent hydraulic material ($CaO/SiO_2$ = 1.43) significantly contributes to the strength, unlike fiber-cement mortar with fine limestone. An unreasonably high $CaO/SiO_2$ ratio in limestone (67.3), not supporting C-S-H phase formation in cement mortar, resulted in a decrease in strength properties.

### 3.2.3. Adhesive Strength of Fiber-Cement Plastering Mortar on Substrates

The adhesive strength of plaster mortar on a substrate (brick, concrete, and masonry) like the base layer is one of the important criteria for assessing the material quality, as well as its execution [87]. The plasticity of mortar directly depends on its ability to hold water against the suction of the surface to which it is applied.

Testing the adhesion of the fiber-cement plastering mortars with finely ground limestone (C8) and granulated blast furnace slag (C9) on two absorbent substrates of brick (B) and aerated concrete blocks (ACBs) after 28 days indicated higher values of this parameter for the fiber-cement formulation containing finely ground slag (C9) on both substrates (0.347 MPa and 0.228 MPa, respectively) compared with the cement mortar sample based on finely ground limestone (0.215 MPa and 0.205 MPa, respectively). Our results exceeded the minimum adhesive strength of the mortar to the substrate (0.18 MPa) declared by the manufacturer in the technical data sheets. According to the standard for [79], the failure of mortar adhesion on the substrates was also monitored. The failure occurred mostly in the surface of the plaster layer applied to the substrate. Different failure depths for the C8 and C9 plaster mortar samples on the two substrates after 28 days were found. While this depth was the same, about 0.5 mm, for the C8 mortar on both substrate surfaces (B and ACB), mortar sample C9 had a breakdown depth ranging 22–43 mm [69]. In addition, the cohesion between the substrate and the plaster mortar was broken, cracks formed, and the substrates were finally damaged, as illustrated in Figures 21 and 22. The damage of substrates (crack formation) was probably caused by the volume changes of the very finely ground granulated slag used in the proposed formulation (C9) because of its high tensile stress.

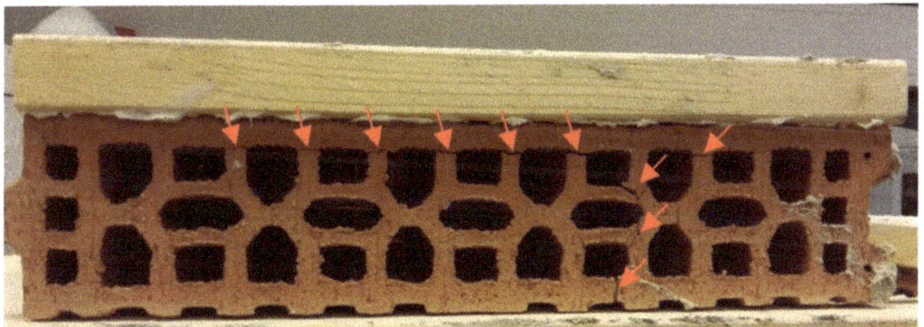

**Figure 21.** Damage of the brick substrate with C9 plaster mixture.

**Figure 22.** Damage of an aerated concrete block substrate with the C9 plaster mixture.

These deficiencies caused the formation of cracks in the plaster mortars and the subsequent damage to the brick and aerated concrete block substrates due to applied layer of plaster mortar formulations, which could be eliminated by modifying the recipes to better fix the amount of water needed for hydration, as well as for processing fresh mixes in the mix.

## 4. Conclusions

This work is focused on the valorization of waste paper fibers in cement mortars presenting an economic, technically feasible, and ecological approach to waste management and promoting a cleaner environment. The study presented the performance of sustainable cement-bonded mortars containing cellulose fibers from waste paper recycling. Based on the research results presented in two parts of this work, which were aimed at studying the effect of waste paper fiber content expressed by the cement to waste paper fiber (C/WPF) mass ratio on the important properties of fresh fiber-cement mixtures and 28-day hardened mortars, as well as a comparative investigation of the adhesive strength of fiber-cement plasters based on finely ground limestone and slag with two commonly used substrates, the following conclusions can be drawn:

1. The C/WPF ratio had a significant effect on all of the studied properties of fresh mixtures and hardened fiber-cement mortars. The consistency values of the fiber-cement mortars containing fibers with different cellulose contents in amounts less 2 wt.% were related to the plaster mortars. The optimum C/WPF ratio for achieving values of a mean spill diameter corresponding to the standard plastic mortars (140–200 mm) within the scope of the study is ranged up to 17:1, representing a fiber content $\leq$2 wt.% in cement mortar with fiber sample C. The higher fiber content in cement mortar and the higher porosity meant that the mortar became lighter, leading to lower values for the bulk density and thermal conductivity after 28 days of hardening. The study of the water absorption behavior of the fiber-cement mortars confirmed that incorporating a higher portion of WPF into mortars led to significantly higher values of absorbency and a capillarity absorption coefficient. This fact is related to the formation of a larger number of pores and their sizes because the low C/WPF ratio in cement mortars caused a decrease in the compressive and flexural strength. This is also a consequence of the weaker adhesion of the fibers and cement particles at the matrix–fiber interface because of the hydrophility of the cellulose fibers.
2. The influence of the fiber-cement mixture composition according to its modification of finely ground limestone and granulated slag in terms of achieving a better plasticity

and satisfactory adhesive strength for two absorbent substrates (brick and aerated concrete block) was investigated. Monitoring the consistency and the physical and mechanical properties of 28-day hardened fiber-cement plastering mortars, as well as the adhesion of these mortars based on the high content of fiber type C (C/WPF = 6:1) for both substrates showed their different behavior, which was influenced not only by the fiber amount, but also by the nature of the fine components in the cement mixture. Higher values for the adhesive strength of both substrates for brick (by 61%) and aerated concrete blocks (by 11%) were obtained for the plaster mortar with granulated slag (C9) compared with the fiber-cement mortar containing finely ground limestone (C8). It manifested in a greater depth of plaster layer failure, crack formation, and in greater cohesion damage between the substrate and mortar in the case of sample C9. The damage observed on both substrates was probably caused by the volume changes of the very finely ground granulated slag used in the proposed formulation (C9) because of its high tensile stress. The results of the comparative study of the physical and mechanical properties of the two cement plasters incorporated with recycled cellulosic fibers and a designed composition (a partial cement replacement by the ground materials—limestone and granulated blast furnace slag) provided complex information about their development. In addition, this study showed a future research direction towards optimizing the composition of the plaster mixture using waste material in order to better fix the plaster mortar to the substrate and to minimize its damage. A key aspect will also be to examine the long-term durability of the fiber-cement plaster mortar in terms of fiber degradation.

The results presented in this article demonstrate to researchers and designers the importance of fiber-cement plaster mortar composition and its influence on the properties of cement. It also highlights the benefits of these fiber-based plaster mortars for providing healthy living solutions, thanks to the fibers' ability to regulate humidity inside buildings by absorbing and/or releasing water molecules, depending on the air conditions.

**Author Contributions:** Conceptualization, N.S. and V.V.; methodology, N.S. and V.V.; validation, N.S., V.V., T.D. and V.H.; formal analysis, N.S. and V.H.; investigation, N.S., V.V., V.H. and T.D.; writing—original draft preparation, N.S., V.V., V.H. and T.D.; writing—review and editing, N.S., V.V., V.H. and T.D.; supervision, N.S. All authors have read and agreed to the published version of the manuscript.

**Funding:** This research was funded by the Scientific Agency of the Ministry of the Education, Science, Research, and Sport of the Slovak Republic (grant number VEGA 1/0222/19).

**Institutional Review Board Statement:** Not applicable.

**Informed Consent Statement:** Not applicable.

**Data Availability Statement:** The data presented in this study are available upon request from the corresponding author.

**Acknowledgments:** The authors thank the Greencell Ltd. in Hencovce for the provision of recycled cellulosic fibers used for experiments.

**Conflicts of Interest:** The authors declare no conflict of interest.

## References

1. Ellen MacArthur Foundation. Towards the Circular Economy. Economic and Business Rationale for an Accelerated Transition. 2013. Available online: https://www.ellenmacarthurfoundation.org/assets/downloads/publications/Ellen-MacArthur-Foundation-Towards-the-Circular-Economy-vol.1.pdf (accessed on 3 May 2021).
2. Shiny Brintha, G.; Sakthieswaran, N. Improving the performance of concrete by adding industrial by products as partial replacement for binder and fine aggregate. *Int. J. Adv. Eng.* **2016**, *7*, 932–936.
3. Gholampour, A.; Ozbakkaloglu, T. A review of natural fiber composites: Properties, modification and processing techniques, characterization, applications. *J. Mater. Sci.* **2020**, *55*, 829–892. [CrossRef]
4. Madurwar, M.V.; Ralegaonkar, R.V.; Mandavgane, S.A. Application of agro-waste for sustainable construction materials: A review. *Constr. Build. Mater.* **2013**, *38*, 872–878. [CrossRef]

5. Yan, L.; Kasal, B.; Huang, L. A review of recent research on the use of cellulosic fibres, their fibre fabric reinforced cementitious, geo-polymer and polymer composites in civil engineering. *Compos. B Eng.* **2016**, *92*, 94–132. [CrossRef]
6. Dalmay, P.; Smith, A.; Chotard, T.; Sahay-Turner, P.; Gloaguen, V.; Krausz, P. Properties of cellulosic fibre reinforced plaster: Influence of hemp or flax fibres on the properties of set gypsum. *J. Mater. Sci.* **2010**, *45*, 793–803. [CrossRef]
7. Onuaguluchi, O.; Banthia, N. Plant-based natural fibre reinforced cement composites: A review. *Cem. Concr. Res.* **2016**, *68*, 96–108. [CrossRef]
8. Xie, X.; Zhou, Z.; Jiang, M.; Xu, X.; Wang, Z.; Hui, D. Cellulosic fibers from rice straw and bamboo used as reinforcement of cement-based composites for remarkably improving mechanical properties. *Compos. B Eng.* **2015**, *78*, 153–161. [CrossRef]
9. Faruk, O.; Bledzki, A.K.; Fink, H.P.; Sain, M. Biocomposites reinforced with natural fibers: 2000–2010. *Prog. Polym. Sci.* **2012**, *37*, 1552–1596. [CrossRef]
10. Anandamurthy, A.; Guna, V.; Ilangovan, M.; Reddy, N. A review of fibrous reinforcements of concrete. *J. Reinf. Plast. Compos.* **2017**, *36*, 519–552. [CrossRef]
11. Fu, T.; Moon, R.J.; Zavattieri, P.; Youngblood, J.; Weiss, W.J. Cellulose Nanomaterials as Additives for Cementitious Materials. In *Cellulose-Reinforced Nanofiber Composites: Production, Properties and Applications*, 1st ed.; Jawaid, M., Boufi, S., Abdul Khalil, H.P.S., Eds.; Elsevier: Amsterdam, The Netherlands, 2017; pp. 455–482. [CrossRef]
12. Galan-Marin, C.; Rivera-Gomez, C.; Garcia-Martinez, A. Use of natural-fiber biocomposites in construction versus traditional solutions: Operational and embodied energy assessment. *Materials* **2016**, *9*, 465. [CrossRef]
13. Cabeza, L.F.; Rincón, L.; Vilariño, V.; Pérez, G.; Castell, A. Life cycle assessment (LCA) and life cycle energy analysis (LCEA) of buildings and the building sector: A review. *Renew. Sustain. Energy Rev.* **2014**, *29*, 394–416. [CrossRef]
14. D'Alessandro, A.; Pisello, A.L.; Fabiani, C.; Ubertini, F.; Cabeza, L.F.; Cotana, F. Multifunctional smart concretes with novel phase change materials: Mechanical and thermo-energy investigation. *Appl. Energy* **2018**, *212*, 1448–1461. [CrossRef]
15. Filho, A.S.; Parveen, S.; Rana, S.; Vanderlei, R.; Fangueiro, R. Micro-structure and mechanical properties of microcrystalline cellulose-sisal fiber reinforced cementitious composites developed using cetyltrimethylammonium bromide as the dispersing agent. *Cellulose* **2021**, *28*, 1663–1686. [CrossRef]
16. Fujiyama, R.; Darwish, F.; Pereira, M.V. Mechanical characterization of sisal reinforced cement mortar. *Theor. Appl. Mech. Lett.* **2014**, *4*, 061002. [CrossRef]
17. Chakraborty, S.; Kundu, S.P.; Roy, A.; Basak, R.K.; Adhikari, B.; Majumder, S.B. Improvement of the mechanical properties of jute fibre reinforced cement mortar: A statistical approach. *Constr. Build. Mater.* **2013**, *38*, 776–784. [CrossRef]
18. Fan, M.; Fu, F. A Perspective—Natural Fibre Composites in Construction. In *Advanced High Strength Natural Fibre Composites in Construction*, 1st ed.; Fan, M., Fu, F., Eds.; Elsevier: Amsterdam, The Netherlands, 2017; pp. 1–20. [CrossRef]
19. Mohammadkazemi, F.; Doosthoseini, K.; Ganjian, E.; Azin, M. Manufacturing of bacterial nano-cellulose reinforced fiber−cement composites. *Constr. Build. Mater.* **2015**, *101*, 958–964. [CrossRef]
20. Balea, A.; Fuente, E.; Blanco, A.; Negro, C. Nanocelluloses: Natural-based materials for fiber-reinforced cement composites. A critical review. *Polymers* **2019**, *11*, 518. [CrossRef]
21. De Pellegrin, M.Z.; Acordi, J.; Montedo, O.R.K. Influence of the lenght and the content of cellulose fibres obtained from sugarcane bagasse on the mechanical properties of fiber-reinforced mortar composites. *J. Nat. Fibers* **2021**, *18*, 111–121. [CrossRef]
22. Benaniba, S.; Driss, Z.; Djendel, M.; Raouache, E.; Boubaaya, R. Thermo-mechanical characterization of a bio-composite mortar reinforced with date palm fiber. *J. Eng. Fibers Fabr.* **2020**, *15*, 1–9. [CrossRef]
23. Dylewski, R.; Adamczyk, J. The comparison of thermal insulation types of plaster with cement plaster. *J. Clean. Prod.* **2014**, *83*, 256–262. [CrossRef]
24. Ashori, A.; Tabarsa, T.; Valizadeh, I. Fiber reinforced cement boards made from recycled newsprint paper. *Mater. Sci. Eng. A* **2011**, *528*, 7801–7804. [CrossRef]
25. Wang, Z.; Li, H.; Jiang, Z.; Chen, Q. Effect of waste paper fiber on properties of cement based mortar and relative mechanism. *J. Wuhan Univ. Technol. Mater. Sci. Ed.* **2018**, *33*, 419–426. [CrossRef]
26. Di Bella, G.; Fiore, V.; Galtieri, G.; Borsellino, C.; Valenza, A. Effects of natural fibres reinforcement in lime plasters (kenaf and sisal vs. polypropylene). *Constr. Build. Mater.* **2014**, *58*, 159–165. [CrossRef]
27. Sakthieswaran, N.; Sophia, M. Prosopis juliflora fibre reinforced green building plaster materials—An eco friendly weed control technique by effective utilization. *Environ. Technol. Innov.* **2020**, *20*, 101158. [CrossRef]
28. Jia, R.; Wang, Q.; Feng, P. A comprehensive overview of fibre-reinforced gypsum-based composites (FRGCs) in the construction field. *Compos. Part B Eng.* **2021**, *205*, 108540. [CrossRef]
29. D'Alessandro, F.; Asdrubali, F.; Mencarelli, N. Experimental evaluation and modelling of the sound absorption properties of plants for indoor acoustic applications. *Build. Environ.* **2015**, *94*, 913–923. [CrossRef]
30. Qamar, F.; Thomas, T.; Ali, M. Use of natural fibrous plaster for improving the out of plane lateral resistance of mortarless interlocked masonry walling. *Constr. Build. Mater.* **2018**, *174*, 320–329. [CrossRef]
31. Menna, C.; Asprone, D.; Durante, M.; Zinno, A.; Balsamo, A.; Prota, A. Structural behaviour of masonry panels strengthened with an innovative hemp fibre composite grid. *Constr. Build. Mater.* **2015**, *100*, 111–121. [CrossRef]
32. Sair, S.; Mandili, B.; Taqi, M.; El Bouari, A. Development of a new eco-friendly composite material based on gypsum reinforced with a mixture of cork fibre and cardboard waste for building thermal insulation. *Compos. Commun.* **2019**, *16*, 20–24. [CrossRef]

33. Gil, L.; Berant-Masó, E.; Cañavate, F.J. Changes in properties of cement and lime mortars when incorporating fibers from end-of-life tires. *Fibers* **2016**, *4*, 7. [CrossRef]
34. Yousefieh, N.; Joshaghani, A.; Hajibandeh, E.; Shekarchi, M. Influence of fibres on drying shrinkage in restrained concrete. *Constr. Build. Mater.* **2017**, *148*, 833–845. [CrossRef]
35. Fu, T.; Moon, R.J.; Zavattieri, P.; Youngblood, J.; Weiss, W.J. Cellulose Nanomaterials as Additives for Cementitious Materials. In *Composites Science and Engineering, Cellulose-Reinforced Nanofibre Composites*; Jawaid, M., Boufi, S., Abdul Khalil, H.P.S., Eds.; Woodhead Publishing Series; Woodhead Publishing: Sawston/Cambridge, UK, 2017; pp. 455–482. [CrossRef]
36. Iucolano, F.; Caputo, D.; Leboffe, F.; Liguori, B. Mechanical behavior of plaster reinforced with abaca fibers. *Constr. Build. Mater.* **2015**, *99*, 184–191. [CrossRef]
37. Andic-Cakir, O.; Sarikanat, M.; Tufekci, H.B.; Demirci, C.; Erdogan, U.H. Physical and mechanical properties of randomly oriented coir fiber-cementitious composites. *Compos. Part B Eng.* **2014**, *61*, 49–54. [CrossRef]
38. Ouedraogo, M.; Dao, K.; Millogo, Y.; Aubert, J.-E.; Messan, A.; Seynou, M.; Zerbo, L.; Gomina, M.M. Physical, thermal and mechanical properties of adobes stabilized with fonio (*Digitaria exilis*) straw. *J. Build. Eng.* **2019**, *23*, 250–258. [CrossRef]
39. Iucolano, F.; Boccarusso, L.; Langella, A. Hemp as eco-friendly substitute of glass fibres for gypsum reinforcement: Impact and flexural behaviour. *Compos. Part B Eng.* **2019**, *175*, 107073. [CrossRef]
40. Boccarusso, L.; Durante, M.; Iucolano, F.; Mocerino, D.; Langella, A. Production of hemp-gypsum composites with enhanced flexural and impact resistance. *Constr. Build. Mater.* **2020**, *260*, 120476. [CrossRef]
41. Gregoire, M.; de Luycker, E.; Bar, M.; Musio, S.; Amaducci, S.; Ouagne, P. Study of solutions to optimize the extraction of hemp fibers for composite materials. *SN Appl. Sci.* **2019**, *1*, 1293. [CrossRef]
42. Iucolano, F.; Liguori, B.; Aprea, P.; Caputo, D. Evaluation of bio-degummed hemp fibers as reinforcement in gypsum plaster. *Compos. Part B Eng.* **2018**, *138*, 149–156. [CrossRef]
43. Charai, M.; Sghiouri, H.; Mezrhab, A.; Karkri, M. Thermal insulation potential of non-industrial hemp (*Moroccan cannabis sativa* L.) fibers for green plaster-based building materials. *J. Clean. Prod.* **2021**, *292*, 126064. [CrossRef]
44. Senff, L.; Ascensão, G.; Ferreira, V.M.; Seabra, M.P.; Labrincha, J.A. Development of multifunctional plaster using nano-$TiO_2$ and distinct particle size cellulose fibers. *Energy Build.* **2018**, *158*, 721–735. [CrossRef]
45. Fan, M.; Ndikontar, M.K.; Zhou, X.; Ngamveng, J.N. Cement-bonded composites made from tropical woods: Compatibility of wood and cement. *Constr. Build. Mater.* **2012**, *36*, 135–140. [CrossRef]
46. Kochova, K.; Schollbach, K.; Gauvin, F.; Brouwers, H.J.H. Effect of saccharides on the hydration of ordinary Portland cement. *Constr. Build. Mater.* **2017**, *150*, 268–275. [CrossRef]
47. Le Troëdec, M.; Dalmay, P.; Patapy, C.; Peyratout, C.; Smith, A.; Cotard, T. Mechanical properties of hemp-lime reinforced mortars: Influence of the chemical treatment of fibers. *J. Compos. Mater.* **2011**, *45*, 2347–2357. [CrossRef]
48. Sawsen, C.; Fouzia, K.; Mohamed, B.; Moussa, G. Effect of flax fibers treatments on the rheological and the mechanical behavior of a cement composite. *Constr. Build. Mater.* **2015**, *79*, 229–235. [CrossRef]
49. Tonoli, G.H.D.; Belgacem, M.N.; Siqueira, G.; Bras, J.; Savastano, H.; Rocco Lahr, F.A. Processing and dimensional changes of cement based composites reinforced with surface-treated cellulose fibres. *Cem. Concr. Compos.* **2013**, *37*, 68–75. [CrossRef]
50. Aigbomian, E.P.; Fan, M. Development of wood-crete materials from sawdust and waste paper. *Constr. Build. Mater.* **2013**, *40*, 361–366. [CrossRef]
51. Merli, R.; Preziosi, M.; Acampora, A.; Lucchetti, M.C.; Petrucci, E. Recycled fibers in reinforced concrete: A systematic literature review. *J. Clean. Prod.* **2020**, *248*, 119207. [CrossRef]
52. Mikhailidi, A.; Kotelnikova, N. Chemical recycling of wastepaper to valuable products. *Bull. Polytech. Inst. Jassy* **2021**, *67*, 1–8.
53. Sangrutsamee, V.; Srichandr, P.; Poolthong, N. Re-pulped waste paper-based composite building materials with low thermal conductivity. *J. Asian Archit. Build. Eng.* **2012**, *11*, 147–151. [CrossRef]
54. Mármol, G.; Santos, S.F.; Savastano, H.; Borrachero, M.V.; Monzó, J.; Payá, J. Mechanical and physical performance of low alkalinity cementitious composites reinforced with recycled cellulosic fibres pulp from cement kraft bags. *Ind. Crops Prod.* **2013**, *49*, 422–427. [CrossRef]
55. Stevulova, N.; Hospodarova, V.; Junak, J. Potential utilization of recycled waste paper fibres in cement composites. *Chem. Technol.* **2016**, *67*, 30–34. [CrossRef]
56. Hospodarova, V.; Stevulova, N.; Vaclavik, V.; Dvorsky, T. Implementation of recycled cellulosic fibres into cement based composites and testing their influence on resulting properties. *IOP Conf. Ser. Earth Environ. Sci.* **2017**, *92*, 012019. [CrossRef]
57. Andrés, F.N.; Beltramini, L.B.; Guilarducci, A.G.; Romano, M.S.; Ulibarrie, N.O. Lightweight concrete: An alternative for recycling cellulose pulp. *Procedia Mater. Sci.* **2015**, *8*, 831–838. [CrossRef]
58. Jiang, Z.; Guo, X.; Li, W.; Chen, Q. Self-shrinkage behaviours of waste paper fiber reinforced cement paste considering its self-curing effect at early-ages. *Int. J. Polym. Sci.* **2016**, 8690967. [CrossRef]
59. Bentchikou, M.; Guidom, A.; Scrivener, K.; Silhadi, K.; Hanini, S. Effect of recycled cellulose fibres on the properties of lightweight cement composite matrix. *Constr. Build. Mater.* **2012**, *34*, 451–456. [CrossRef]
60. Aciu, C.; Ilu̦tiu-Varvara, D.A.; Cobirzan, N.; Balog, A. Recycling of paper waste in the composition of plastering mortars. *Procedia Technol.* **2014**, *12*, 295–300. [CrossRef]
61. Ardanuy, M.; Claramunt, J.; Toledo Filho, R.D. Cellulosic fiber reinforced cement-based composites: A review of recent research. *Constr. Build. Mater.* **2015**, *79*, 115–128. [CrossRef]

62. Mohr, B.J.; Biernacki, J.J.; Kurtis, K.E. Supplementary cementitious materials for mitigating degradation of kraft pulp fiber-cement composites. *Cem. Concr. Res.* **2007**, *37*, 1531–1543. [CrossRef]
63. Hospodarova, V.; Singovszka, E.; Stevulova, N. Characterization of cellulosic fibers by FTIR spectroscopy for their further implementation to building materials. *Am. J. Analyt. Chem.* **2018**, *9*, 303–310. [CrossRef]
64. Stevulova, N.; Hospodarova, V.; Estokova, A.; Singovszka, E.; Holub, M.; Demcak, S.; Briancin, J.; Geffert, A.; Kacik, F.; Vaclavik, V.; et al. Characterization of manmade and recycled cellulosic fibers for their application in building materials. *J. Renew. Mater.* **2019**, *7*, 1121–1145. [CrossRef]
65. Hospodarova, V.; Stevulova, N.; Vaclavik, V.; Dvorsky, T.; Briancin, J. Cellulose fibres as a reinforcing element in building materials. In Proceedings of the Environmental Engineering 10th International Conference, Vilnus, Lithuania, 27–28 April 2017; pp. 1–8.
66. Sicakova, A.; Hospodarova, V.; Stevulova, N.; Vaclavik, V.; Dvorsky, T. Effect of selected cellulosic fibers on the properties of cement based composites. *Adv. Mater. Lett.* **2018**, *9*, 606–609. [CrossRef]
67. Hospodarova, V.; Stevulova, N.; Briancin, J.; Kostelanska, K. Investigation of waste paper cellulosic fibers utilization into cement based building materials. *Buildings* **2018**, *8*, 43. [CrossRef]
68. Stevulova, N.; Hospodarova, V.; Vaclavik, V.; Dvorsky, T.; Danek, T. Characterization of cement composites based on recycled cellulosic waste paper fibres. *Open Eng.* **2018**, *8*, 363–367. [CrossRef]
69. Stevulova, N.; Hospodarova, V.; Vaclavik, V.; Dvorsky, T. Use of Cellulosic Fibers from Wood Pulp and Waste Paper for Sustainable Cement Based Mortars. In *Sustainable Industrial Processing Summit and Exhibition (SIPS 2019): New and Advanced Materials, Technologies, and Manufacturing*; Kongoli, F., Marquis, F., Chikhradze, N., Prikhna, T., Eds.; Flogen Stars Outreach: Quebec, QC, Canada, 2019; Volume 11, pp. 133–142.
70. Stevulova, N.; Hospodarova, V.; Vaclavik, V.; Dvorsky, T. Physico-mechanical properties of cellulose fiber-cement mortars. *Key Eng. Mater.* **2020**, *838*, 31–38. [CrossRef]
71. European Committee for Standardization. *Methods of Testing Cement, Part 1: Determination of Strength*; STN EN 196-1; Slovak Office of Standards, Metrology and Testing: Bratislava, Slovakia, 2016.
72. European Committee for Standardization. *Mixing Water Concrete: Specification for Sampling, Testing and Assessing the Suitability of Water, Including Water Recovered from Processes in the Concrete Industry, as Mixing Water for Concrete*; STN EN 1008; Slovak Office of Standards, Metrology and Testing: Bratislava, Slovakia, 2003.
73. European Committee for Standardization. *Cement, Part 1: Composition, Specifications and Conformity Criteria for Common Cements*; STN EN 197-1; Slovak Office of Standards, Metrology and Testing: Bratislava, Slovakia, 2012.
74. European Committee for Standardization. *Limestone, Dolomite: Quality*; STN 72 1217; Slovak Office of Standards, Metrology and Testing: Bratislava, Slovakia, 1992.
75. European Committee for Standardization. *Methods of Test for Mortar for Masonry, Part 3: Determination of Consistence of Fresh Mortar (by Flow Table)*; STN EN 1015-3; Slovak Office of Standards, Metrology and Testing: Bratislava, Slovakia, 2004.
76. European Committee for Standardization. *Methods of Test for Mortar for Masonry, Part 10: Determination of Dry bulk Density of Hardened Mortar*; STN EN 1015-10/A1; Slovak Office of Standards, Metrology and Testing: Bratislava, Slovakia, 2007.
77. European Committee for Standardization. *Determination of Moisture, Absorbency and Capillarity of Concrete*; STN 73 1316; Slovak Office of Standards, Metrology and Testing: Bratislava, Slovakia, 1989.
78. European Committee for Standardization. *Methods of Test for Mortar for Masonry, Part 18: Determination of Water Absorption Due to Capillary Action of Hardened Mortar*; STN EN 1015-18; Slovak Office of Standards, Metrology and Testing: Bratislava, Slovakia, 2003.
79. European Committee for Standardization. *Methods of Test for Mortar for Masonry, Part 11: Determination of Flexural and Compressive Strength of Hardened Mortar*; STN EN 1015-11; Slovak Office of Standards, Metrology and Testing: Bratislava, Slovakia, 2020.
80. European Committee for Standardization. *Methods of Test for Mortar for Masonry, Part 12: Determination of Adhesive Strength of Hardened Rendering and Plastering Mortars on Substrates*; STN EN 1015-12; Slovak Office of Standards, Metrology and Testing: Bratislava, Slovakia, 2016.
81. Pimentel, M.G.; das Chagas Borges, J.P.; de Souza Picanço, M.; Ghavami, K. Bending answer and toughness analysis of mortar reinforced with Curauá fibers. *Revista Matéria* **2016**, *21*, 18–26. [CrossRef]
82. European Committee for Standardization. *Specification for Mortar for Masonry, Part 1: Rendering and Plastering Mortar*; STN EN 998-1; Slovak Office of Standards, Metrology and Testing: Bratislava, Slovakia, 2019.
83. Shafigh, P.; Asadi, I.; Akhiani, A.R.; Mahyuddin, N.B.; Hashemi, M. Thermal properties of cement mortar with different mix proportions. *Materiales De Construcción* **2020**, *70*, e224. [CrossRef]
84. Mostefai, N.; Hamzaoui, R.; Guessasma, S.; Aw, A.; Nouri, H. Microstructure and mechanical performance of modified hemp fibre and shiv mortars: Discovering the optimal formulation. *Mater. Des.* **2015**, *84*, 359–371. [CrossRef]
85. Garbalińska, H.; Wygocka, A. Microstructure modification of cement mortars: Effect on capillarity and frost-resistance. *Constr. Build. Mater.* **2014**, *51*, 258–266. [CrossRef]
86. Stokke, D.D.; Wu, Q.; Han, G. *Introduction to Wood and Natural Fiber Composites*; John Wiley & Sons: West Sussex, UK, 2014.
87. Svoboda, L.; Bažantová, Z.; Myška, M.; Novák, J.; Tobolka, Z.; Vávra, R.; Vimmrová, A.; Výborný, J. *Building Materials*, 4th ed.; Czech Technical University in Prague: Prague, Czech Republic, 2018; Available online: http://people.fsv.cvut.cz/~{}svobodal/sh/SH4v1.pdf (accessed on 3 May 2021).

*Article*

# Mineral Addition and Mixing Methods Effect on Recycled Aggregate Concrete

**Hasan Dilbas [1,*] and Mehmet Şamil Güneş [2]**

1 Department of Civil Engineering, Yuzuncu Yil University, Van 65080, Turkey
2 Department of Statistics, Yildiz Technical University, Istanbul 34220, Turkey; msgunes@yildiz.edu.tr
* Correspondence: hasandilbas@yyu.edu.tr

**Abstract:** This paper presents influence of treatment and mixing methods on recycled aggregate concretes (RAC) designed regarding various techniques. Absolute Volume Method (AVM) according to TS 802, Equivalent Mortar Volume Method (EMV), silica fume (SF) as a mineral addition were considered in the design of concretes. In total, four groups of concretes were produced in the laboratory: (1) natural aggregate concrete (NAC) designed with AVM as control concrete, (2) RAC designed with AVM as control RAC, (3) RAC with SF as a mineral addition designed with AVM as treated RAC and (4) RAC designed with EMV as treated RAC. The tests were performed at 28th days and the statistical analysis were made on the test results. According to the results, EMV and SF increased the compressive strength of concretes and this resulted an increase in the strength class of concrete. A significant statistical difference between the concretes were determined. According to multiple comparison analysis, it was found that especially there was a significant relationship among NAC, RAC and RAC-EMV. In addition, it was recommended that EMV and AVM with 5% SF could be used in the design of RAC rather than AVM only to achieve the target strength class C30/37.

**Keywords:** recycled aggregate concrete; silica fume; mixing process; compressive strength

**Citation:** Dilbas, H.; Güneş, M.Ş. Mineral Addition and Mixing Methods Effect on Recycled Aggregate Concrete. *Materials* **2021**, *14*, 907. https://doi.org/10.3390/ma14040907

Academic Editor: Andrea Petrella

Received: 16 January 2021
Accepted: 10 February 2021
Published: 14 February 2021



## 1. Introduction

Demolition of concrete structures and waste concrete products have been mainly discussed in the countries. Authorities were worked on identifying a struggle process with the huge mass of waste concrete and offered environmental solutions for the use of waste concrete in newly manufactured concrete as recycled aggregate (RA) taking measures such as regulation and standards for RA [1]. Environmental approaches giving zero harm to the nature, recycling materials, and preserving natural resources so to the economic development have been defined as main aim by countries [2]. However, the authorities faced difficulties to act laws, regulations, and measures on the use of RA in concrete and it is emerged as a need to form a new code for recycled aggregate concrete (RAC) while present concrete codes have been developed for natural aggregate concrete (NAC). On the other hand, RA has attached old mortar (AOM) and natural aggregate (NA) phases and the ambiguity in the properties of RA due to its heterogeneity limits the use of RA in concrete. Hence, absence of reliability does not give hope to mix designer for structural concrete.

To solve the complexity of the use of RA in new concrete safely, some researchers proposed new mixture prescriptions including mineral additions (i.e., silica fume (SF) [3,4], ground granulated blast furnace slag (GGBS) [5,6], metakaolin [7], fly ash (FA) [8–11], fibers with/without mineral addition (i.e., polypropylene [9], polypropylene + SF [12], steel fiber + SF [4], basalt fiber [13,14], basalt fiber + nano-silica [15], waste-plastic strip + FA [16], steel fiber + FA [17], steel fiber + GGBS [18], glass fiber + FA [19] or redefining mixing method (i.e., Equivalent Mortar Volume Method (EMV) [20,21], two stage mixing method [22]. The suggestions presented satisfactory results and gave confidence to mix designers of RAC. However, despite the diversity of the suggestions, the most useful technique is not clear and comparison of them is required. Thus, the designer should compare the useful ones. In this perspective, Mineral Addition Treatment (MAT) and Equivalent Mortar Volume Methods (EMV) were the well-known and widely considered methods in the literature [3,20,23].

In this experimental research, the properties of concretes (NAC, RAC, RAC included SF as a mineral addition and RAC designed with EMV) were compared to observe the effect of treatment methods (Figure 1). Here, NAC, RAC and RAC with silica fume were designed with AVM. 120 concrete specimens for four group of concrete were produced in the laboratory. Then, 28th day compressive strength of concretes was determined, and the statistical analysis were conducted.

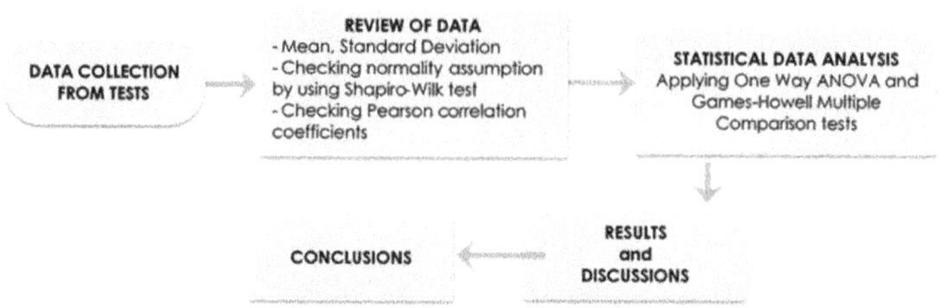


**Figure 1.** Methodology flow chart.

## 2. Materials and Methods

### 2.1. Materials

General purpose CEM I cement suitable with TS EN 197-1 [24] was used in the concrete mixes. The properties of cement and silica fume (SF) are given in Table 1.

**Table 1.** Properties of cement and silica fume.

| Contents | Cement | SF |
|---|---|---|
| $SiO_2$ (%) | 18.9 | 91.42 |
| CaO (%) | 64.7 | 0.52 |
| $SO_3$ (%) | 3.42 | 0.37 |
| $Al_2O_3$ (%) | 4.8 | 0.72 |
| $Fe_2O_3$ (%) | 3.4 | 1.66 |
| MgO (%) | 1.4 | 0.92 |
| $K_2O$ (%) | 0.4 | 1.21 |
| $Na_2O$ (%) | 0.7 | 0.38 |
| Density (g/cm$^3$) | 3.11 | 0.642 |
| Chlorine ratio (%) | 0.0241 | 0.04 |
| Specific surface area (m$^2$/kg) | 3840 | 21290 |
| Loss on ignition (%) | 1.82 | 1.72 |
| Activity index (%) | - | 118 |

In the concrete mixes, natural coarse aggregate and recycled coarse aggregate were used as the coarse ones and the granulometry of the mixes are the same. Natural gravel was crushed, calcareous aggregate and also sand was utilized as fine aggregate in the mixes (Table 2). Super plasticizer was used to enhance the low workability of fresh concretes (Table 3). The slump class is set to S2 for all mixes [25].

**Table 2.** The properties of natural aggregates.

| Notation | Density, g/cm$^3$ | Water Absorption, % | LA Abrasion value, % | Residual Content, % |
|---|---|---|---|---|
| Sand | 2.81 | 1.31 | - | - |
| NA (11.2–22.4 mm) | 2.70 | 0.75 | 24 | - |
| NA (4–11.2 mm) | 2.73 | 0.72 | - | - |
| RA (11.2–22.4 mm) | 2.00 | 8.95 | 55 | 52.5 |
| RA (4–11.2 mm) | 2.06 | 8.80 | - | 39.2 |

**Table 3.** The properties of super plasticizer.

| Content | Super Plasticizer |
|---|---|
| Structure of material | Polycarboxylic ether |
| Color | Amber |
| Density (kg/l) | 1.08–1.14 |
| Alkaline ratio (%) | <3 |
| Chlorine ratio (%) | <0.1 |

### 2.2. *Concrete Design Method and Data Evaluation Approachs*

Four concrete mixes were produced in the laboratory with the target strength class C30/37 (Table 4). Absolute Volume Method (AVM) and Equivalent Mortar Volume Method (EMV) were considered to design the mixes (Table 5). According to AVM a unit volume (it is generally 1 m$^3$) of concrete is filled with the components of concrete (Equation (1)) [23]:

$$V_{1m}^{3} = V_{agg} + V_{cem} + V_{w} + V_{ch} + V_{air} \tag{1}$$

**Table 4.** Ingredients of mixes.

| Components | NAC | RAC | RAC-SF | RAC-EMV |
|---|---|---|---|---|
| Cement, kg/m$^3$ | 340 | 340 | 323 | 255 |
| Silica fume, kg/m$^3$ | - | - | 17 | - |
| Water, kg/m$^3$ | 163 | 163 | 163 | 123 |
| Super plasticizer, % | 0.75 | 0.85 | 0.95 | 1.55 |
| Sand, kg/m$^3$ | 806 | 806 | 806 | 608 |
| Aggregate 4–11.2 mm, kg/m$^3$ | 392 | - | - | - |
| Recycled aggregate 4–11.2 mm, kg/m$^3$ | - | 296 | 296 | 368 |
| Aggregate 11–22.4 mm, kg/m$^3$ | 775 | - | - | - |
| Recycled aggregate 11–22.4 mm, kg/m$^3$ | - | 574 | 574 | 774 |

**Table 5.** Compression test results of specimens.

| Number of Specimen | NAC | RAC | RAC-SF | RAC-EMV |
|---|---|---|---|---|
| 1 | 36.73 | 31.05 | 34.20 | 48.00 |
| 2 | 35.98 | 34.87 | 35.78 | 42.89 |
| 3 | 35.70 | 34.59 | 32.53 | 40.94 |
| 4 | 35.16 | 35.29 | 37.01 | 38.37 |
| 5 | 35.56 | 31.33 | 34.42 | 38.26 |
| 6 | 38.07 | 23.83 | 36.32 | 42.73 |

**Table 5.** *Cont.*

| Number of Specimen | NAC | RAC | RAC-SF | RAC-EMV |
|---|---|---|---|---|
| 7 | 37.59 | 31.88 | 36.99 | 40.21 |
| 8 | 35.46 | 29.14 | 36.32 | 41.43 |
| 9 | 38.25 | 31.63 | 36.57 | 48.35 |
| 10 | 37.20 | 35.52 | 37.33 | 49.02 |
| 11 | 39.15 | 36.41 | 35.72 | 43.64 |
| 12 | 37.84 | 33.33 | 37.40 | 45.20 |
| 13 | 36.76 | 30.91 | 35.33 | 40.30 |
| 14 | 39.96 | 35.75 | 36.76 | 40.19 |
| 15 | 37.22 | 36.18 | 32.80 | 46.84 |
| 16 | 41.04 | 31.86 | 37.29 | 42.74 |
| 17 | 39.36 | 34.02 | 36.78 | 44.97 |
| 18 | 36.20 | 33.86 | 33.84 | 39.98 |
| 19 | 38.71 | 30.34 | 36.60 | 41.89 |
| 20 | 30.70 | 36.50 | 36.60 | 48.22 |
| 21 | 39.47 | 33.24 | 35.12 | 40.81 |
| 22 | 36.78 | 31.83 | 28.96 | 45.62 |
| 23 | 38.69 | 30.52 | 35.17 | 39.86 |
| 24 | 35.54 | 33.84 | 37.08 | 45.29 |
| 25 | 39.50 | 31.98 | 38.60 | 45.71 |
| 26 | 32.78 | 34.87 | 36.37 | 41.45 |
| 27 | 36.50 | 35.34 | 37.15 | 49.05 |
| 28 | 37.61 | 34.84 | 33.30 | 45.43 |
| 29 | 34.85 | 30.90 | 33.01 | 44.44 |
| 30 | 37.36 | 32.15 | 38.23 | 45.82 |

Here, $V_{agg}$ is volume of aggregate, $V_{cem}$ is volume of cement, $V_w$ is volume of water, $V_{ch}$ is volume of chemicals and $V_{air}$ is volume of air in concrete.

EMV requires that recycled aggregate concrete has same amount of total mortar volume with control concrete so to constant aggregate volume. Hence the residual content should be determined for RA [20]. HCl solution can be used to determine the amount of residual on RA [26] (Table 2). The remaining part over 4mm sieve was determined in the residual defining test after 0.1 M HCl solution attack to RA in a container. EMV requires the constant volume of aggregate as (Equation (3)) [21]:

EMV requires the constant volume of aggregate as (Equation (3)) [20]:

$$V_{RCA}^{RAC} = \frac{V_{NA}^{NAC} x(1-R)}{(1-RMC)x\frac{SG_b^{RCA}}{SG_b^{OVA}}} \quad (2)$$

Here, $V_{RCA}^{RAC}$ is the volume ratio of coarse in RAC, $V_{NA}^{NAC}$ is the volume ratio of fresh natural aggregate in control concrete, $SG_b^{RCA}$ and $SG_b^{OVA}$ are the bulk specific gravity of RA and original virgin aggregate, respectively, $RMC$ is the residual mortar content of RA and $R$ is the volume fraction of fresh natural aggregate content of RAC to fresh natural aggregate content of control mix.

The cement quantity and water-to-cement ratio was kept constant for all concrete mixes. Concrete was cast incompatible with ASTM C192/C192M–13a [27] and vibration was applied on the fresh concrete. For each concrete group, 30 cube specimens (15 × 15 × 15 cm) were produced and cured in lime saturated water for 28 days. At the end of the time (28th days), 120 concrete specimens were tested in 3000 kN compression machine in accordance with TS EN 12390-3 [28] and the results are given in Table 5.

#### 2.2.1. Strength Class Determination

95% confidence interval was considered, and strength class of concrete groups were determined using Equations (3) and (4) [25]:

$$f_{c,avg} \geq f_{ck} + 1.96\,\sigma \tag{3}$$

$$f_{c,min} \geq f_{ck} - 4.0 \tag{4}$$

Here, $f_{ck}$ characteristic compressive strength of group (MPa), $f_{c,\ avg}$ is the average compressive strength of group (MPa), $\sigma$ standard deviation, and $f_{c,\ min}$ is the minimum compressive strength of group (MPa).

#### 2.2.2. Statistical Analysis Method

In this study, Shapiro-Wilk normality test was performed on NAC, RAC, RAC-SF, RAC-EMV [29]. The Pearson correlation coefficient was used to test the degree of relationship between concrete types and to obtain information about the general structure of the results. Afterwards, the variance analysis (ANOVA) was used to measure whether the compressive strength values had a significant effect on the concrete types at 5% significance level. Also, the Games-Howell multiple comparison test was used to measure whether there was a significant difference between the concrete types of compressive strengths (where group variances were not equal). The analysis was performed using IBM SPSS 22 at 5% significance level.

## 3. Results

### *3.1. Strength Class of Concretes*

According to the results given in Table 6, the target strength class C30/37 was achieved for NAC and RAC-SF. The strength class of RAC-EMV was found as C35/45 and however, it was found as C25/30 for RAC. Poor properties of RA influenced the concrete properties and decreased the compressive strength of RAC [3,30–35]. Attached old mortar (AOM) content in RA had an important role on the decrease of compressive strength and AOM had porous structure with lower strength characteristics [36]. However, silica fume (SF) use in concrete mix increased the compressive strength and also the strength class of concrete giving satisfactory results. Here, SF showed two significant behaviors: 1) Causing extra C-S-H gels in matrix bounding free $Ca(OH)_2$ in the cement paste, 2) Filler effect (closing concrete pores) [37]. In addition, EMV, also, gave a comparable result to RAC-EMV and caused an increase in the compressive strength and the strength class [20]. This success was sourced by aggregate concentration consideration in the mix. Besides, the similar findings with the current literature are achieved observing the lower compressive strength and higher standard deviation values of compressive strength compared to control ones [3,38–43].

**Table 6.** Compressive strength and strength class of concretes.

| Parameters | NAC | RAC | RAC-SF | RAC-EMV |
|---|---|---|---|---|
| Average compressive strength, MPa | 44.12 | 39.20 | 42.44 | 51.89 |
| Minimum compressive strength, MPa | 36.55 | 28.37 | 34.48 | 45.55 |
| Std. deviation of compressive strength, MPa | 2.63 | 2.52 | 1.49 | 3.95 |
| Strength class of concrete | C30/37 | C25/30 | C30/37 | C35/45 |

### 3.2. Comparison of the Methods

As given in Table 6, the control concrete (NAC) that was designed with AVM had C30/37 strength class and the consideration of EMV as a mixing approach in the production of concrete ensured C35/45 but also C30/37 (the upper strength class covers and ensures the lower ones). Besides, SF treatment gave approximately close compressive strength and strength class with control concrete (NAC). However, increase in the strength class of RAC from C25/30, which is for RAC, to C35/45 which is for RAC-EMV, due to the consideration of EMV was not similar with the increase in the strength class of RAC from C25/30 which is for RAC to C30/37 which is for RAC-SF due to the consideration of SF. Although EMV seemed to be a potential to increase the strength class of RAC, EMV increased the standard deviation values of RAC-EMV. On the other hand, it was clear that SF decreased the standard deviation of the test results and the minimum standard deviation was calculated for RAC-SF and RA marginally changed the standard deviation values of RAC [3,33]. More tests should be conducted to observe the exact behavior of RAC.

### 3.3. Statistical Results

In the Table 7, the statistical values of concretes are given as standard deviation, mean, standard error and 95% confidence interval with histograms. According to Table 7, the lower bound of NAC crossed with upper bound of RAC-SF although the means of NAC and RAC-SF were different. Besides, the lower and upper bounds of RAC and RAC-EMV did not cross with the lower and upper bounds of NAC.

**Table 7.** Statistical parameters for concretes (with histograms).

| Concretes | Mean | Std. Deviation | Std. Error | 95% Confidence Interval | |
|---|---|---|---|---|---|
| | | | | Lower Bound | Upper Bound |
| NAC | 44.08 | 2.63 | 0.49 | 43.06 | 45.10 |
| RAC | 39.63 | 2.52 | 0.47 | 38.65 | 40.61 |
| RAC-SF | 42.70 | 1.97 | 0.37 | 41.93 | 43.46 |
| RAC-EMV | 51.84 | 3.95 | 0.74 | 50.30 | 53.37 |

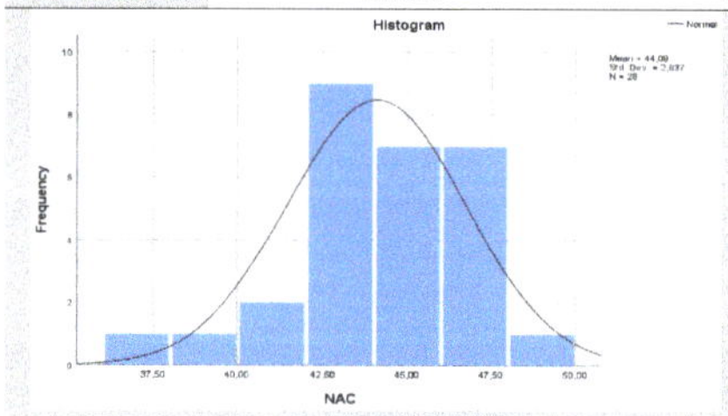

(NAC) 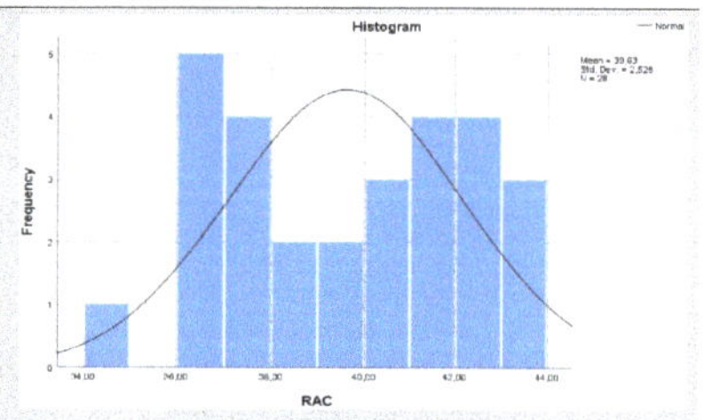 (RAC)

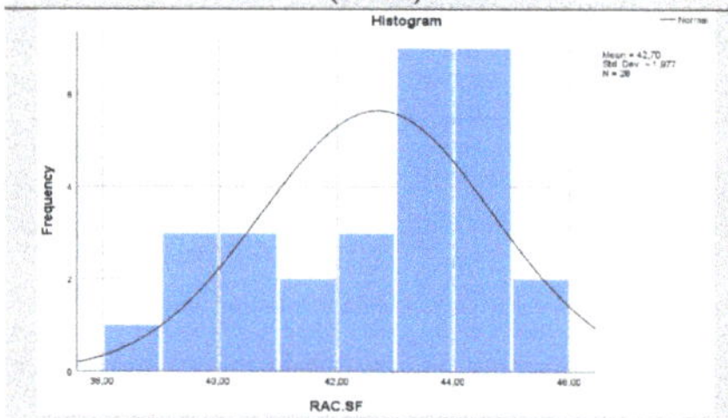

(RAC-SF) 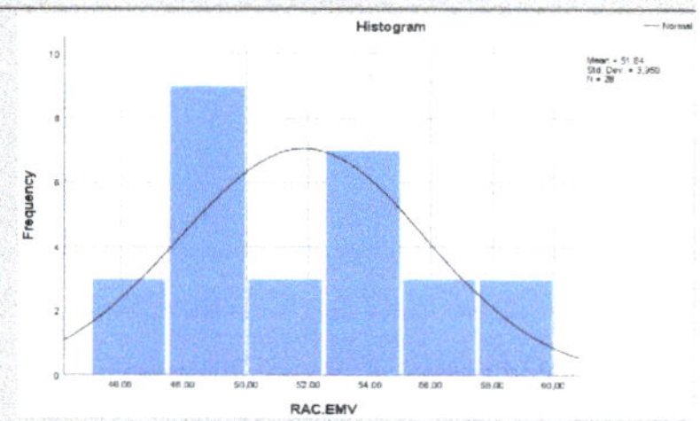 (RAC-EMV)

Shapiro-Wilk normality, skewness, kurtosis ratio to standard error results are given in Table 8. The values obtained from the standard error division of the observed kurtosis and skewness values for all variables varies between (−2, 2) indicating that the data was distributed normally. Furthermore, the Shapiro-Wilk normality statistics gave significant results for all qualifications at 5% significance level.

**Table 8.** Shapiro-Wilk normality test results.

| Evaluations | CONCRETES | | | |
|---|---|---|---|---|
| | NAC | RAC | RAC-SF | RAC-EMV |
| Skewness to std. Error ratio | −1.73 | −0.31 | −1.39 | 0.38 |
| Kurtosis to std. Error ratio | 1.59 | −1.41 | −0.64 | −1.38 |
| Shapiro-Wilk *p*-value | 0.302 | 0.154 | 0.058 | 0.119 |

The Pearson Correlation coefficient was a measure of the variation of two or more variables. The conducted correlation analysis showed how a change in interrelated variables affected the other variables and the relationship among them. For the correlation coefficient (it takes values between −1 and 1), "0" is the non-correlation, "−1" represents the perfect negative relationship and,"1" represents the perfect positive relationship. The interpretation of the correlation coefficient is given in the Figure 2. This figure showed samples of what vary correlations remind, in terms of the strength and direction of the relationship with histograms and fit lines. Pearson correlation coefficient between NAC and RAC was equal −0.20 which showed that there was a very weak relation between those variables. On the other hand, Pearson correlation coefficient between NAC and RAC-SF was equal 0.20 and that of NAC and RAC-EMV was equal −0.059.

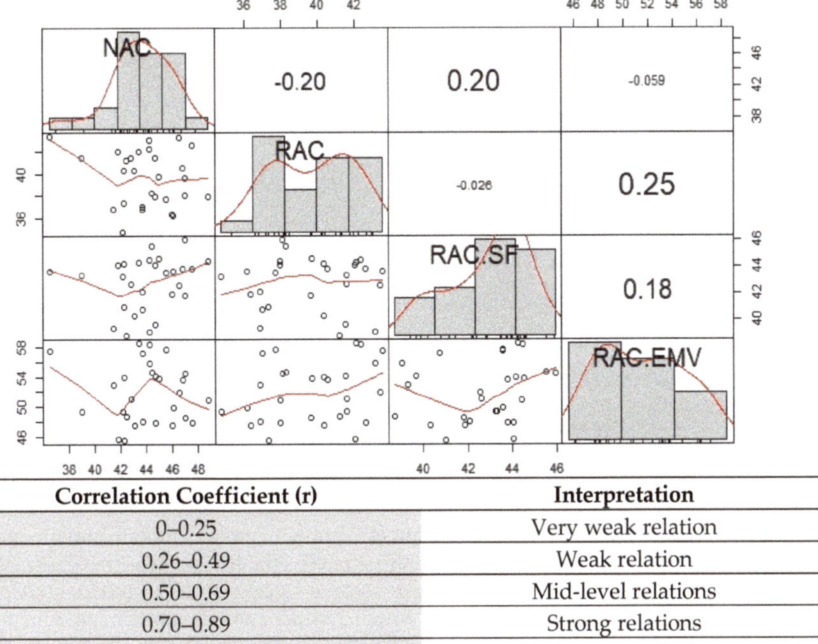


| Correlation Coefficient (r) | Interpretation |
|---|---|
| 0–0.25 | Very weak relation |
| 0.26–0.49 | Weak relation |
| 0.50–0.69 | Mid-level relations |
| 0.70–0.89 | Strong relations |
| 0.90–1 | Very strong relations |

**Figure 2.** The Pearson Correlation coefficient results.

A scatterplot gives the relations between two variables measured for the dataset. Each individual in the data appears as a point on the graph. It is convenient to use scatter plots with correlation test results. As shown in Figure 2, correlation coefficient between

NAC and RAC was equal −0.20. As it could be seen from the scatter plot in Figure 3, relationship between NAC and RAC had a negative line (downhill) with confidence interval (CI) which indicated the same interpretation with correlation coefficient.

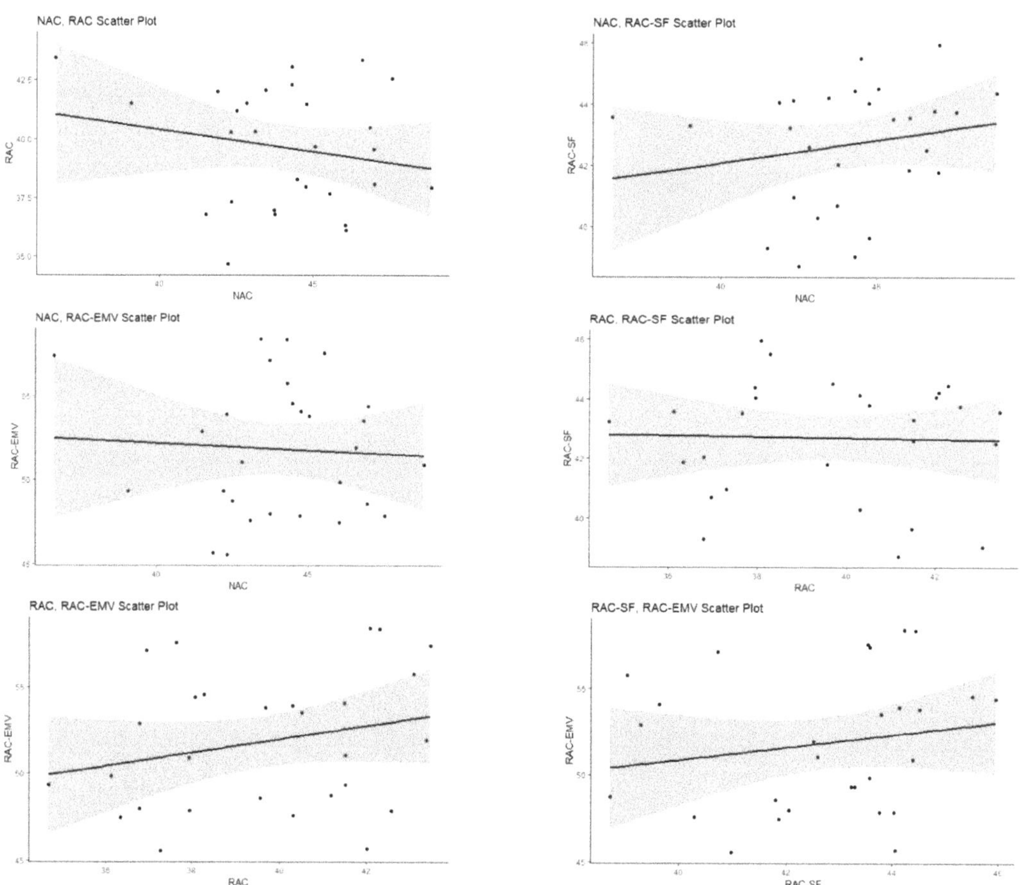


**Figure 3.** Scatter plots between concrete types with confidence intervals.

Variance analysis (ANOVA) measures the significance of compressive strengths on concrete types and the results of ANOVA is given in Table 9. Statistically, the relationship between the groups were significant at %5 significance level ($p$-value = 0.00 < 0.05). It is known that TS-500 [42] considered %10 significance level with higher tolerance compared to 5% significance level [43]. Also, a multiple comparison test was used to determine which concrete types were significant (Table 10) and here, if the confidence intervals for the multiple comparison test contained a value of "0", the bilateral relationship was not meaningful. Accordingly, when Table 10 was examined, it was found that the relationships of (NAC)-(RAC), (NAC)-(RAC-EMV), (RAC-SF)-(RAC), (RAC-EMV)-(RAC) and (RAC-EMV)-(RAC-SF) were significant at 5% significance level. However, the relationship between (NAC)-(RAC-SF) was not significant.

**Table 9.** ANOVA results with means and interval plots for concretes.

| Interactions | Sum of Squares | df | Mean Square | F | Sig. |
|---|---|---|---|---|---|
| Between Groups | 2267.729 | 3 | 755.910 | 91.849 | 0.000 |
| Within Groups | 888.833 | 108 | 8.230 | | |
| Total | 3156.561 | 111 | | | |

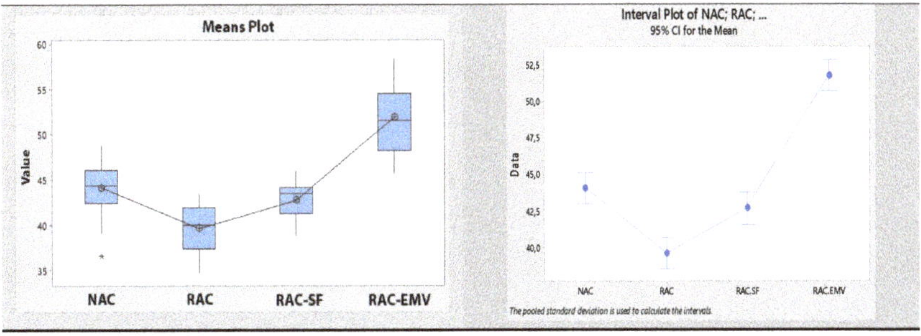


**Table 10.** Games-Howell Multiple Comparison test results and test graph.

| (I) kod | (J) kod | Mean Difference (I–J) | Std. Error | Sig. | 95% Confidence Interval | |
|---|---|---|---|---|---|---|
| | | | | | Lower Bound | Upper Bound |
| NAC | RAC | 40.45 * | 0.69 | **0.000** | 20.62 | 60.28 |
| | RAC-SF | 10.38 | 0.62 | 0.131 | −00.27 | 30.03 |
| | RAC-EMV | −70.75 * | 0.89 | **0.000** | −100.15 | −50.36 |
| RAC | NAC | −40.45 * | 0.69 | **0.000** | −60.28 | −20.62 |
| | RAC-SF | −30.06 * | 0.60 | **0.000** | −40.67 | −10.45 |
| | RAC-EMV | −120.20 * | 0.88 | **0.000** | −140.57 | −90.84 |
| RAC-SF | NAC | −10.38 | 0.62 | 0.131 | −30.03 | 00.27 |
| | RAC | 30.06 * | 0.60 | **0.000** | 10.45 | 40.67 |
| | RAC-EMV | −90.14 * | 0.83 | **0.000** | −110.38 | −60.89 |
| RAC-EMV | NAC | 70.75 * | 0.89 | **0.000** | 50.36 | 100.15 |
| | RAC | 120.20 * | 0.88 | **0.000** | 90.84 | 140.57 |
| | RAC-SF | 90.14 * | 0.83 | **0.000** | 60.89 | 110.38 |

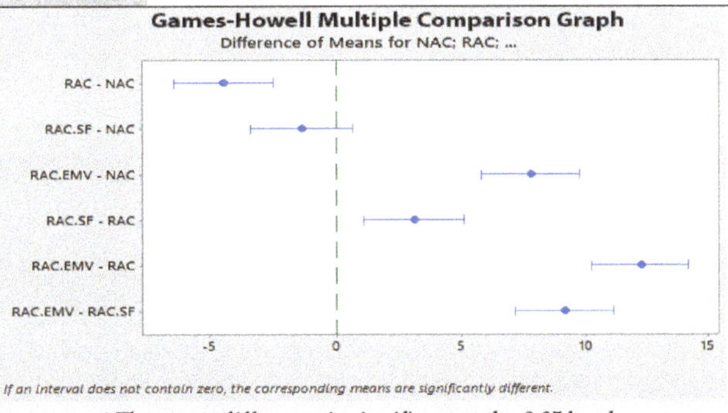


* The mean difference is significant at the 0.05 level.

## 4. Conclusions and Discussions

In this paper, a statistical study was conducted, and the compressive strength test results of concretes designed with Absolute Volume Method (AVM) and Equivalent Mortar Volume Method (EMV) and included natural aggregate (NA), recycled aggregate (RA) and silica fume (SF) were investigated. Based on the results, the following conclusions were made:

- Mineral Addition Treatment Method with SF and EMV gives convincing results eliminating the negative effect of attached old mortar (AOM) in RA.
- The target strength class C30/37 is achieved for NAC, RAC-SF and RAC-EMV, and however, RAC cannot achieve C30/37. SF addition and EMV facilitate to obtain the target strength class.
- The compressive strength of all concretes distributes normally according to the Shapiro-Wilk normality test and Skewness and Kurtosis to standard error ratios.
- It is found that the correlation analysis and the scatter plots give compatible results. The correlation analysis and the scatter plots indicate that the relation between concretes behavior pattern is observed at low level. Especially, for instance, the relation between behaviors of NAC and RAC is equal −0.20 and it means that there is very weak relation due to RA.
- According to the results of ANOVA, the relationship between the concretes are significant at 95% reliability level, although generally concrete standards consider 90% reliability level.
- Games-Howell Multiple Comparison Test demonstrates that the most significant bilateral relationships between (RAC-EMV)-(RAC) and (RAC)-(NAC) are found. (It is noted that the correlation test measures whether there is a relation between variables while ANOVA and comparison tests measure significance grades. These evaluation approaches should not be confused.)

In summary, according to the statistical evolutions, there is a major difference between the concretes and this phenomenon depends on the utilized components and considered mixing methods generally. However, Mineral Addition Treatment Method (here it is SF) and EMV are useful to improve the performance of RAC, and especially, EMV is strongly recommended for RAC mix design by the authors instead of AVM. If AVM is considered, SF addition use in mixes is recommended by the authors.

In addition, the following discussions were made after the evaluation of the results and conclusions:

- The data used in the statistical approaches were collected from experiments and at first strength class of concrete series were determined. In this point it was thought whether the concretes were in the required strength class (C30/37), and the results were checked in consideration of the helpful evaluation equations given in the related codes. However, it is well-known that the concrete, commonly used in the engineering area, includes the natural aggregate and, also the compressive strength results of natural aggregate concrete are distributed compatible with the normal distribution function. Here, it is expected that the concretes included recycled aggregate and designed with different mixing methods would show a similar behavior with natural aggregate concrete. However, the truth of recycled aggregate concrete was different. For instance, as a result of the normality test evaluations (skewness, kurtosis, etc.), when recycled aggregate was considered, the distribution of the test results of recycled aggregate concrete presented a non-similarity with natural ones although the use of silica fume changed a bit the behavior of recycled aggregate concrete from recycled aggregate concrete to natural aggregate concrete. According to this, it could be concluded that despite the consideration of silica fume in recycled aggregate concrete, natural aggregate concrete and recycled aggregate concrete had different characteristics and it was thought that the observed difference depended on the components such as recycled aggregate. To present the difference/similarity of concretes behavior, in addition, the comparison techniques were employed and hence the difference between the concrete types was obviously seen. The first comparison technique was made in consideration of Person Correlation Coefficient and the most suitable similarity between recycled aggregate concrete and natural aggregate concrete was found as 0.20 (the higher is good up to 1.0 and down to −1.0). The second comparison technique was made in consideration of ANOVA with Games-Howell Multiple Comparison Test and Games-Howell Comparison Test had an interrelation assessment approach.

As expected, the first and the second approaches demonstrated the similar results: There was a specific difference between the concretes included natural and recycled aggregates in dependent of mixing methods such as AVM and EMV and, also mineral addition such as silica fume.

- The critics and discussions on the results canalized the authors to think that the evaluation of test results of different types of concretes (i.e., natural aggregate concrete, heavy concrete, geopolymer concrete, recycled aggregate concrete) could mislead decision makers and the evaluation of test results of different concrete types may be separated in the standards and the evaluation equations for each concrete type could be proposed in the codes after several trial-and-error tests.
- Considering the various studies in the literature, Tukey's Test (it is a comparison test) was mostly used together with ANOVA analysis (i.e., Refs. [40,44]). In general, the crucial and inadequate points in the literature are the assumptions of Tukey's Test which was not properly considered for the situations examined in the study and, the lack of explanations of Multiple Comparison Tests (i.e., Refs. [45,46]). Therefore, the data were properly examined and discussed in detail in consideration of many statistical approaches in the current paper and one of them was Games-Howell Test. In the test, the assumption is made as there is no equal variance that is provided in the determination of the relationships between the groups. In addition, the application of the test on the recycled aggregate concretes designed with many methods was one of the novelty parts of the study and it clearly ensured the difference of the concretes revealing the characteristics of concretes.

## 5. Future Aspects

Also, there is a lack of knowledge in RAC application included many mineral additions (metakaolin, fly ash, granulated blast furnace slag, etc.) and fibers (basalt fiber, polypropylene fiber, steel fiber, etc.) and also those under the different curing conditions. In this paper only SF as a mineral addition, AVM and EMV as a design method is considered and compared evaluating statistical data obtained various stochastic approaches. In addition, it is though that aggregate types (light, normal and heavy ones), water-to-binder ratio, chemical admixtures, etc. are the other components effecting the statistical results and the approaches considered in this paper can be applied to those. Besides, in the future studies, clustering methods, the analysis of discriminant and regression and multivariate statistical methods etc. are able to be utilized to evaluate the effect of materials, methods etc. on concretes' relations and behaviors.

**Author Contributions:** Conceptualization, H.D. and M.Ş.G.; methodology, H.D.; software, M.Ş.G.; data curation, H.D.; writing—original draft preparation, H.D. and M.Ş.G.; writing—review & editing, H.D. and M.Ş.G.; visualization, M.Ş.G.; supervision, H.D. All authors have read and agreed to the published version of the manuscript.

**Funding:** This research received no external funding.

**Institutional Review Board Statement:** Not applicable.

**Informed Consent Statement:** Not applicable.

**Data Availability Statement:** The data and code related to the analysis presented in this paper will be made available upon request.

**Conflicts of Interest:** The authors declare no conflict of interest.

## References

1. Rao, M.C.; Bhattacharyya, S.K.; Barai, S.V. *Systematic Approach of Characterisation and Behaviour of Recycled Aggregate Concrete*, 1st ed.; Springer: Singapore, 2019. [CrossRef]
2. Bidabadi, M.S.; Akbari, M.; Panahi, O. Optimum mix design of recycled concrete based on the fresh and hardened properties of concrete. *J. Build. Eng.* **2020**, *32*, 101483. [CrossRef]

3. Dilbas, H.; Şimşek, M.; Çakır, Ö. An investigation on mechanical and physical properties of recycled aggregate concrete (RAC) with and without silica fume. *Constr. Build. Mater.* **2014**, *61*, 50–59. [CrossRef]
4. Xie, J.; Fang, C.; Lu, Z.; Li, Z.; Li, L. Effects of the addition of silica fume and rubber particles on the compressive behaviour of recycled aggregate concrete with steel fibres. *J. Cle. Prod.* **2018**, *197*, 656–667. [CrossRef]
5. Mansur, M.; Özgür, T.; Tüfekçi, M.; Çakır, Ö. An Investigation on Mechanical and Physical Properties of Recycled Coarse Aggregate (RCA) Concrete with GGBFS. *Int. J. Civil Eng.* **2017**, *15*, 549–563. [CrossRef]
6. Oner, A.; Akyuz, S. An experimental study on optimum usage of GGBS for the compressive strength of concrete. *Cem. Conc. Comp.* **2007**, *29*, 505–514. [CrossRef]
7. Ramyar, K.; Mardani-Aghabaglou, A.; Sezer, G.İ.; Ramyar, K. Comparison of fly ash, silica fume and metakaolin from mechanical properties and durability performance of mortar mixtures view point. *Constr. Build. Mater.* **2014**, *70*, 17–25. [CrossRef]
8. Kou, S.C.; Poon, C.S. Properties of self-compacting concrete prepared with coarse and fine recycled concrete aggregates. *Cem. Conc. Comp.* **2009**, *31*, 622–627. [CrossRef]
9. Xu, F.; Wang, S.; Li, T.; Liu, B.; Li, B.; Zhou, Y. Mechanical properties and pore structure of recycled aggregate concrete made with iron ore tailings and polypropylene fibers. *J. Build. Eng.* **2021**, *33*, 101572. [CrossRef]
10. Kurda, R.; de Brito, J.; Silvestre, J.D. A comparative study of the mechanical and life cycle assessment of high-content fly ash and recycled aggregates concrete. *J. Build. Eng.* **2020**, *29*, 101173. [CrossRef]
11. Zhou, Q.; Lu, C.; Wang, W.; Wei, S.; Lu, C.; Hao, M. Effect of fly ash and sustained uniaxial compressive loading on chloride diffusion in concrete. *J. Build. Eng.* **2020**, *31*, 101394. [CrossRef]
12. Jian-hua, W.; Shi-cheng, M.; Zhao-qing, T.; Xiao-quan, L.; Xiao-gen, L. Intensified test research on the influence of polypropylene fiber & silica fume on recycled concrete. *Concrete* **2006**, *11*, 36–38.
13. Dilbas, H.; Çakır, Ö. Influence of basalt fiber on physical and mechanical properties of treated recycled aggregate concrete. *Constr. Build. Mater.* **2020**, *254*, 119216. [CrossRef]
14. Katkhuda, H.; Shatarat, N. Improving the mechanical properties of recycled concrete aggregate using chopped basalt fibers and acid treatment. *Constr. Build. Mater.* **2017**, *140*, 328–335. [CrossRef]
15. Wang, Y.; Hughes, P.; Niu, H.; Fan, Y. A new method to improve the properties of recycled aggregate concrete: Composite addition of basalt fiber and nano-silica. *J. Clean. Prod.* **2019**, *236*, 117602. [CrossRef]
16. Sobhan, K.; Mashnad, M. *Fatigue Durability of Stabilized Recycled Aggregate Base Course Containing Fly Ash and Waste-Plastic Strip Reinforcement*; Final Report Submitted to the Recycled Materials Resource Centre; University of New Hampshire: Durham, NH, USA, 2000.
17. Sobhan, K.; Krizek, R.J. Fatigue Behavior of Fiber-Reinforced Recycled Aggregate Base Course. *J. Mat. Civil. Eng.* **1999**, *11*, 124–130. [CrossRef]
18. Kim, S.; Kim, Y.; Usman, M.; Park, C.; Hanif, A. Durability of slag waste incorporated steel fiber-reinforced concrete in marine environment. *J. Build. Eng.* **2021**, *33*, 101641. [CrossRef]
19. Mlv, P.; Pancharathi, R.K. Strength studies on glass fiber reinforced recycled aggregate concrete. *Asian J. Civil. Eng.* **2015**, *8*, 677–690.
20. Abbas, A.; Fathifazl, G.; Isgor, O.B.; Razaqpur, A.G.; Fournier, B.; Foo, S. Durability of recycled aggregate concrete designed with equivalent mortar volume method. *Cem. Conc. Comp.* **2009**, *31*, 555–563. [CrossRef]
21. Kim, J.; Sadowski, Ł. The equivalent mortar volume method in the manufacturing of recycled aggregate concrete. *Czas. Tech.* **2019**, *11*, 123–140. [CrossRef]
22. Abd Elhakam, A.; Awad, E. Influence of self-healing, mixing method and adding silica fume on mechanical properties of recycled aggregates concrete. *Constr. Build. Mater.* **2012**, *35*, 421–427. [CrossRef]
23. TSI, TS 802 (Turkish Standards Institution). *Design of Concrete Mixes*; TSI: Ankara, Turkey, 2016.
24. EN TS 197-1. *Cement—Part 1: Composition, Specification and Conformity Criteria for Common Cements*; Turkish Standards Institution: Ankara, Turkey, 2012.
25. EN TS 206-1. *Concrete—Part 1: Feature, Performance, Fabrication and Suitability*; Turkish Standards Institution: Ankara, Turkey, 2002.
26. Dilbas, H.; Çakır, Ö.; Atiş, C.D. Experimental investigation on properties of recycled aggregate concrete with optimized Ball Milling Method. *Constr. Build. Mater.* **2019**, *212*, 716–726. [CrossRef]
27. ASTM. *Standard Practice for Making and Curing Concrete Test Specimens in the Laboratory*; C192/C192M-13; ASTM International: West Conshohocken, PA, USA, 2013.
28. TS EN 12390-3. In *Testing Hardened Concrete—Part 3: Compressive Strength of Test Specimens*; Turkish Standards Institution: Ankara, Turkey, 2003.
29. Shapiro, S.S.; Wilk, M.B. An Analysis of Variance Test for Normality (Complete Samples). *Biometrika* **1965**, *52*, 591. [CrossRef]
30. Cantero, B.; Sáez del Bosque, I.F.; Matías, A.; Medina, C. Statistically significant effects of mixed recycled aggregate on the physical-mechanical properties of structural concretes. *Constr. Build. Mater.* **2018**, *185*, 93–101. [CrossRef]
31. Dilbas, H.; Çakır, Ö. Fracture and Failure of Recycled Aggregate Concrete (RAC)–A Review. *Int. J. Conc. Tech.* **2016**, *1*, 31–48.
32. Dilbas, H.; Çakır, Ö.; Şimşek, M. Recycled Aggregate Concretes (RACs) for Structural Use: An Evaluation on Elasticity Modulus and Energy Capacities. *Int. J. Civil Eng.* **2017**, *15*, 247–261. [CrossRef]
33. Dilbas, H. An Examination on Mechanical Behaviour of a Cantilever Beam Produced with Recycled Aggregate Concrete. Master's Theiss, Graduate School of Natural and Applied Science, Yildiz Technical University, Istanbul, Turkey, 2014.

34. Duan, Z.H.; Poon, C.S. Properties of recycled aggregate concrete made with recycled aggregates with different amounts of old adhered mortars. *Mater. Des.* **2014**, *58*, 19–29. [CrossRef]
35. Matias, D.; de Brito, J.; Rosa, A.; Pedro, D. Mechanical properties of concrete produced with recycled coarse aggregates—Influence of the use of superplasticizers. *Constr. Build. Mater.* **2013**, *44*, 101–109. [CrossRef]
36. Wardeh, G.; Ghorbel, E. Mechanical Properties of Recycled Concrete: An Analytical Study. In Proceedings of the International Symposium on Eco-Crete, Reykjavik, Iceland, 13–15 August 2019; pp. 1–8.
37. de Brito, J.; Saikia, N. *Recycled Aggregate in Concrete Use of Industrial, Construction and Demolition Waste*; Springer: Berlin, Germany, 2013. [CrossRef]
38. Silva, R.V.; De Brito, J.; Dhir, R.K. Properties and composition of recycled aggregates from construction and demolition waste suitable for concrete production. *Constr. Build. Mater.* **2014**, *65*, 201–217. [CrossRef]
39. Akça, K.; Çakır, Ö.; İpek, M.; Ipek, M. Properties of polypropylene fiber reinforced concrete using recycled aggregates. *Constr. Build. Mater.* **2015**, *98*, 620–630. [CrossRef]
40. Xiao, J.; Li, W.; Poon, C. Recent studies on mechanical properties of recycled aggregate concrete in China—A review. *Sci. China Technol. Sci.* **2012**, *55*, 1463–1480. [CrossRef]
41. Xiao, J.; Li, W.; Fan, Y.; Huang, X. An overview of study on recycled aggregate concrete in China (1996–2011). *Constr. Build. Mater.* **2012**, *31*, 364–383. [CrossRef]
42. TS 500 (Turkish Standards Institution). *Design and Construction Rules of Reinforced Concrete Structures*; Turkish Standards Institution: Ankara, Turkey, 2000.
43. Dilbas, H.; Erkılıç, E.; Saraylı, S.; Haktanır, T. Beton Mukavemet Deneyleri Değerlendirmesi için Bir İstatistiksel Çalışma (A Statistical Study for the Evaluation of Concrete Strength Tests). *Yapı Dünyası Dergisi* **2011**, *178*, 25–31. (In Turkish)
44. Brand, A.S.; Roesler, J.R.; Salas, A. Initial moisture and mixing effects on higher quality recycled coarse aggregate concrete. *Constr. Build. Mater.* **2015**, *79*, 83–89. [CrossRef]
45. Güneyisi, E.; Gesoğlu, M.; Kareem, Q.; İpek, S. Effect of different substitution of natural aggregate by recycled aggregate on performance characteristics of pervious concrete. *Mater. Struc.* **2016**, *49*, 521–536. [CrossRef]
46. Carneiro, J.A.; Lima, P.R.L.; Leite, M.B.; Toledo Filho, R.D. Compressive stress–strain behavior of steel fiber reinforced-recycled aggregate concrete. *Cem. Concr. Comp.* **2014**, *46*, 65–72. [CrossRef]

Article

# Experimental Investigation on the Shear Behaviour of Stud-Bolt Connectors of Steel-Concrete-Steel Fibre-Reinforced Recycled Aggregates Sandwich Panels

Arash Karimipour [1], Mansour Ghalehnovi [2,*], Mohammad Golmohammadi [3] and Jorge de Brito [4,*]

1 Department of Civil Engineering, University of Texas at El Paso (UTEP) and the Member of Center for Transportation Infrastructure Systems (CTIS), El Paso, TX 79968, USA; akarimipour@miners.utep.edu
2 Department of Civil Engineering, Ferdowsi University of Mashhad, Mashhad 9177948944, Iran
3 Civil Engineering Department, Faculty of Engineering, University of Torbat Heydarieh, Torbat Heydarieh 9516168595, Iran; m.golmohammadi@torbath.ac.ir
4 Department of Civil Engineering, Architecture and Georresources, Instituto Superior Técnico, Universidade de Lisboa, 1649-004 Lisbon, Portugal
* Correspondence: Ghalehnovi@um.ac.ir (M.G.); jb@civil.ist.utl.pt (J.d.B.)

**Citation:** Karimipour, A.; Ghalehnovi, M.; Golmohammadi, M.; de Brito, J. Experimental Investigation on the Shear Behaviour of Stud-Bolt Connectors of Steel-Concrete-Steel Fibre-Reinforced Recycled Aggregates Sandwich Panels. *Materials* **2021**, *14*, 5185. https://doi.org/10.3390/ma14185185

Academic Editors: Andrea Petrella and Michele Notarnicola

Received: 4 August 2021
Accepted: 6 September 2021
Published: 9 September 2021



**Abstract:** Steel-concrete-steel (SCS) sandwich panels are manufactured with two thin high-strength steel plates and a moderately low-density and low-strength thick concrete core. In this study, 24 specimens were produced and tested. In these specimens, a new stud-bolt connector was used to regulate its shear behaviour in sandwich panels. The bolts' diameter, concrete core's thickness and bolts' spacing were the parameters under analysis. Furthermore, the concrete core was manufactured with normal-strength concrete and steel fibres concrete (SFC). Steel fibres were added at 1% by volume. In addition, the recycled coarse aggregate was used at 100% in terms of mass instead of natural coarse aggregate. Therefore, the ultimate bearing capability and slip of the sandwich panels were recorded, and the failure mode and ductility index of the specimens were evaluated. A new formula was also established to determine the shear strength of SCS panels with this kind of connectors. According to this study, increasing the diameter of the stud-bolts or using SFC in sandwich panels improve their shear strength and ductility ratio.

**Keywords:** bolt connectors; fibre-reinforced concrete; steel-concrete-steel sandwich; shear behaviour; ultimate load

## 1. Introduction

Nowadays, waste construction materials have become a major challenge in the field of environmental pollution. As per prior reports, between 2012 and 2014, nearly 370 million tonnes of waste construction materials were dumped in nature, which accounts for approximately 70% of the total waste construction materials generated in each year [1]. Therefore, if the production of construction waste maintains this tendency, 430 million tonnes of waste will be dumped annually. In addition, some natural phenomena such as tornadoes, earthquakes, and floods play an effective role in increasing construction wastes [2–5]. While the recycling of old building rubble has become very popular, much of it is still dumped in nature. Velay-Lizancos et al. [6] declared that reusing waste materials to manufacture new concrete can play a crucial role in keeping environment safe and clean. Furthermore, the use of coarse recycled aggregates (CRA) as a substitute of coarse natural aggregates (CNA) substantially decreases $CO_2$ production by about 15–20% [7]. As a result, the importance of using recycled materials and the increasing use of CRA concrete is clear, and understandable. Many investigations have been performed on using of CRA to produce conventional recycled aggregate concrete. Thus, so far, a large number of disadvantages of CRA use on the general behaviour of a concrete mixes have been detected, largely because old mortar is adhered to the CRA surfaces [8–10]. Hansen and Narud [11] showed that, based on the aggregates size, 25% to 60% of the mortar can attach to the aggregates surface. Additionally, with an increase in the RCA content, the mechanical properties of concrete decreased [12–14]. Amer et al. [15] found the same observations and declared that raising RCA substitution content led to reducing the compressive strength of concrete. On the other hand, some research indicated an enhancement effect of CRA on the structural behaviour of concrete beams due to their larger broken surface, relative to that of CNA. Therefore, the behaviour of concrete members still needs evaluation.

On the other hand, prior investigations indicate that using fibres in concrete mixes substantially enhances the mechanical and rheological characteristics of concrete. Fibres decrease cracks width as well as their expansion [16,17]. In addition, the bridging role of fibres interacts with the paste at the level of micro-cracks and postpones cracks propagation as well as improving the resistance of concrete when enough fibres content is used [18]. Therefore, when tensile strength of concrete increases, micro-cracks accumulate and convert into large cracks, and fibres prevent them from opening and propagating by efficiently bridging them. This post-peak macro-crack bridging is the main feature improving the properties of fibre-reinforced concrete (FRC). Using a low to intermediate fibres content does not improve the tensile and flexural strengths of concrete and only slightly enhances the energy absorption and durability in the post-cracking state. Conversely, the addition of high fibres fractions improves the tensile strength, strain-hardening performance before localization and toughness beyond crack localization [19–23]. Therefore, once FRC is conducted under a flexural load, fibres enhance the bending behaviour. Additionally, the use of fibres at the cracking location leads to delaying in crack propagation, reducing the decrease caused by high CRA substitution of CNA on the structural behaviour of concrete [24,25]. In another research, Niu et al. [26] evaluated the structural behaviour of FRC in water and salt freezing conditions. Experiments showed that the use of SF can substantially enhance the micro structure and tensile strength of samples. Olivito and Zuccarello [27] investigated the influence of SF on the mechanical characteristics of concrete and its classification considering SF fraction and mix-design variations. They found that the SF contents and their geometric characteristics are vital features and crack width and tensile strength of concrete went up by raising the length of SF. Köksal et al. [28] assessed the effect of water/cement ratio, tensile strength and volume of SF on the properties of concrete and provided solutions for FSC design based on the maximum fracture energy. They also found that the tensile strength of SF and the water/cement ratio directly affect the structural behaviour of FRC.

On the other hand, steel-concrete-steel (SCS) panels comprise two outer steel plates and a concrete core. An interrelated material or connectors are generally used to guarantee an adequate connection between the concrete core and steel plates. Besides, using connectors improves the ductility and the biaxial performance of SCS members [29,30]. In recent years, SCS sandwich plates have been widely used to manufacture oil and gas silos in the Arctic region and shear walls for offshore structures. This structure can withstand the high pressures of ice up to 45 MPa [29]. The different kinds of connectors that have been proposed for SCS sandwich panels in the past decades [31–36]. To manufacture slim lightweight concrete (LWC) SCS panels resistant to impact and blast loads, connectors were proposed in previous studies. SCS panels with connectors showed promising behaviour when subjected to static and impact loads [37]. Connectors act in couples and interlock the concrete core in SCS panels. By this interlocking mechanism, the up-lift as well as local buckling of the steel faceplates are prevented. In SCS panel members with steel connectors, the transverse shear force is withstood by concrete core and connectors. Connectors act as shear links to avoid inclined shear cracks in the concrete core. Moreover, the tensile strength of the connectors contributes to the lateral shear resistance of the SCS panels. Another function of the connector is that their tensile strength can be used to reduce local buckling of the steel faceplates and avoid the rupture of steel plates from the concrete core. Local buckling of the steel plates can happen when the SCS panels are tested under compression, or the compression region of the steel plate is subjected to a bending load.

A significant separation between the steel plate and concrete happens when a SCS panel is tested under a lateral impact or blast loading [37]. Therefore, connectors can sufficiently connect concrete core and steel plate, avoiding tensile separation and keeping the overall structural performance. As a result, the tensile resistance of connectors is a vital feature that will influence the structural behaviour of the SCS panels. This offers the motivation for the current investigation to assess the tensile strength of this novel type of connectors.

Punching shear rupture can happen in the steel plates of SCS members under different loads, and so many experimental studies have been done by Huang et al. [38] and Sohel and Liew [39] on this mechanism. However, these researchers only reported limited evidence on the SCS sandwich panels. In these panels, after flexural yielding, a membrane action is developed in the slab due to the effectiveness of connectors in maintaining composite action that further increases its load-carrying capacity after flexural yielding. The main benefit of the connectors is providing the shear capacity to restrict the shear cracks resulting from the shear force between the concrete core and the steel plates under an external lateral load. Lately, diverse sorts of connectors have been invented to enhance the shear performance of SCS members [40,41]. Furthermore, different types of the concrete core, such as LWC and high-strength concrete, were assessed in previous studies [42]. In the initial investigations, normal weight concrete (NWC) was used to produce the concrete core [43]. After that, lightweight concrete (LWC) was used to decrease the weight of SCS members [44]. Sohel et al. [45] tested eight SCS slabs with J-shaped connectors and LWC core. Furthermore, the membrane effect after yielding was evaluated. The results demonstrated that the failure modes and cracks propagation in SCS slabs with J-hook connectors are very comparable to those of reinforced concrete (RC) slabs. In another research, Yan et al. [46] investigated the punching shear strength of LWC slabs and panels. They established a new model to estimate the shear capability of RC shells, but not valid for SCS shell members. Subsequently, a new formula was established to anticipate the bending and shear capability of SCS shells [47,48].

Nevertheless, there are inadequate design procedures to estimate the punching capacity of the SCS members and some standards such as Eurocode [49] and ACI 318 [50] have established general schemes for RC slabs. To enhance the behaviour of SCS panels in offshore structures, J-hook connectors were tested by Liew et al. [51]. According to this study, SCS sandwich panels with J-hook connectors exhibited outstanding performance under the impulse and long-term loads. In 2009, Liew and Sohel [52] proposed a new technique to design SCS sandwich panels with LWC core. In their study, J-hook connectors were utilized to raise the shear capability. The obtained experimental outcomes showed that the J-hook connector is an efficient element to withstand the shear stress between steel plates and concrete core. In 2015, Yan et al. [53] measured the bending performance of SCS LWC beams experimentally and numerically. The results of the failure mode and shear strength of structures show the influence of the thickness of the steel skin shell, curvature, spacing of the connectors, depth of the cross-section, the strength of the concrete core, and boundary conditions on the ultimate strength behaviour of the curved SCS sandwich beam. In 2011, Leekitwattana [54] employed corrugated-strip connectors (CSC) to produce SCS panels. These connectors are placed normally to the inclined crack of concrete. Consequently, a limitation should be defined for the concrete core's thickness. The stud-bolt connector is one of the cheapest and simplest types of connectors. In 2017, Yousefi and Ghalehnovi [55] studied the impact of one-end welded corrugated-strip connectors on the shear behaviour of SCS panels. The specimens were produced with different connectors' angles and tested under the push-out test. The results indicated that, by increasing the connectors' strength, the shear strength considerably increased. In another study, Yousefi and Ghalehnovi [56] proposed a FEM scheme to foresee the shear capability of SCS panels with one-end welded corrugated-strip connectors. In that study, a formula was proposed to control the interlayer performance of a double-skin SCS structure.

## 2. Research Significance

Based on the results of the previous studies, using diverse connectors enhances the shear capability of SCS panels, but in this study, a new type of stud-bolt connector was applied, which is cheap and easy to use. To manufacture the previous connectors, some equipment such as a welding motor and electrode is needed to weld the connectors to the steel faceplates. Additionally, the heat produced to weld is going to deform the molecular structure of the steel plates and results in a reduction of the strength of plates and an increase of stress concentration in the welding points. In addition, to present a new low-cost connector, CRA were sorted from a demolished old building and used as a substitute of CNA at 100% in terms of weight to consider solving environmental problems. According to previous investigations, the use of CRA decreased the load-bearing capacity of concrete members. Therefore, to mitigate the negative effect of CRA on the load-carrying of SCS panels, SF were added to specimens. Furthermore, a new model is presented to estimate the shear capability of SCS panels with stud-bolts connectors. Figure 1 shows the experimental program and novelty carried out in this study.

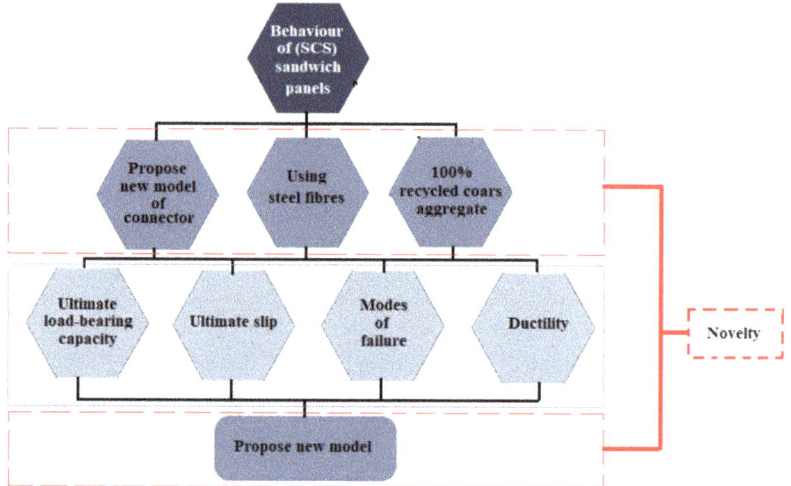


**Figure 1.** General flowchart of experimental program and novelty of this study.

## 3. Materials and Specimens' Specifications

### 3.1. Steel Plates

Square steel faceplates with 300 mm sides and 6 mm thickness were cut at the factory and holes were created using computer numerical control (CNC), as seen in Figure 2. In addition, in order to determine the characteristics of the steel faceplates, three specimens were tested under the directional tensile test according to ASTM A517 [57], and the average properties were considered and those of the steel plates and presented in Table 1.

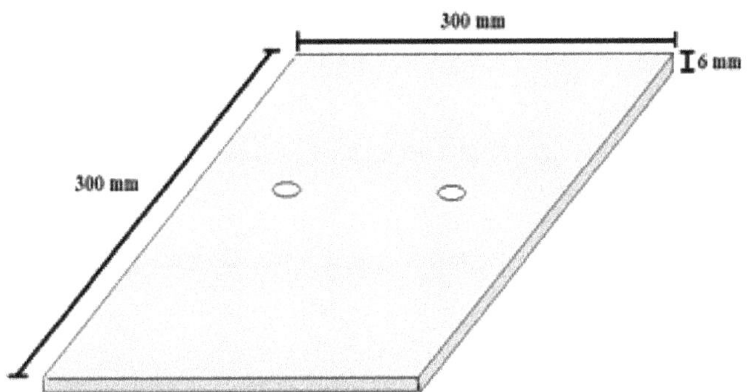


**Figure 2.** Geometry of the steel plates.

**Table 1.** Properties of steel plates with 6 mm thickness.

| Materials | Yield Stress (MPa) | Ultimate Stress (MPa) | Modulus of Elasticity (GPa) | Ultimate Strain |
|---|---|---|---|---|
| Steel plate | 283 | 491 | 201 | 0.0024 |
| Coefficient of variation (%) | 0.7 | 0.8 | 0.4 | 4.0 |

*3.2. Bolts and Nuts*

In this study, bolts of three diameters (8 mm, 10 mm and 12 mm) were used. To manufacture, the SCS sandwich panels, two bolts and four nuts were used. The mechanical characteristics of the bolts are presented in Table 2. The shear bearing capacity of the bolts with diameters of 8 mm, 10 mm and 12 mm was obtained (30,000 N, 55,000 N and 58,000 N, respectively), according to the properties presented in Table 2. There were two bolts' spacings: 100 mm and 150 mm. Figure 3 presents the reinforced plates with stud-bolt connectors. The bolts were selected according to ASTM C293 [58–60].

**Table 2.** Mechanical characteristics of the bolts.

| | Service Classes | | | | | | | | |
|---|---|---|---|---|---|---|---|---|---|
| **ASTM C293 [58]** | **B5** | **B6** | **B7** | **B8** | **B8C** | **B8M** | **B8T** | **B16** | **B7M** |
| | Chemical Analysis | | | | | | | | |
| Carbon | > 0.15 | < 0.20 | 0.4–0.15 | > 0.10 | > 0.10 | < 0.10 | < 0.10 | 0.35–0.45 | 0.40–0.50 |
| Manganese | < 15 | < 2 | 0.70–1 | < 3 | < 3 | < 3 | < 3 | 0.50–0.75 | 0.70–1 |
| Phosphorous mud | 0.05 | 0.05 | 0.05 | 0.05 | 0.05 | 0.05 | 0.05 | 0.05 | 0.05 |
| Sulphur mud | 0.04 | 0.09 | 0.05 | 0.09 | 0.09 | 0.09 | 0.09 | 0.05 | 0.05 |
| Silicon | < 1 | < 1 | 0.20–0.35 | < 1 | < 1 | < 1 | < 1 | 0.20–0.35 | 0.25–0.35 |
| Nickle | – | – | – | 8–20 | 9–13 | 10–14 | 9–22 | – | – |
| Chromium | 4.5–6.5 | 12–14 | 0.85–1.2 | 5–20 | 15–20 | 15–20 | 15–20 | 0.8–1.2 | 0.85–1.15 |
| Molybdenum | 0.05–0.7 | - | 0.2–0.3 | – | – | 2.5–3.5 | – | 0.55–0.70 | 0.2–0.3 |
| Vanadium | – | – | – | – | – | – | – | 0.3–0.4 | – |
| Tanium mini | – | – | – | – | – | – | 6 | – | – |
| Columbium + Titanium | – | – | – | – | – | – | – | – | – |

**Table 2.** Cont.

| Tensile Requirements | | | | | | | | | | |
|---|---|---|---|---|---|---|---|---|---|---|
| Minimum tensile strength | Lbs/psi | 100.0 | 110.0 | 125.0 | 75.0 | 75.0 | 75.0 | 75.0 | 125.0 | 100.0 |
| | kg/mm$^2$ | 70.5 | 77.5 | 88.0 | 53.0 | 53.0 | 53.0 | 53.0 | 88.0 | 70.5 |
| Minimum yield strength | Lbs/psi | 80.0 | 85.0 | 105.0 | 50.0 | 30.0 | 30.0 | 30.0 | 105.0 | 80.0 |
| | kg/mm$^2$ | 56.5 | 60.0 | 74.0 | 60.0 | 21.0 | 21.0 | 21.0 | 74.0 | 50.5 |
| Elongation in 2 inches (%) | | 16.0 | 15.0 | 16.0 | 55.0 | 30.0 | 30.0 | 30.0 | 18.0 | 18.0 |
| Reduction of area (%) | | 50.0 | 50.0 | 50.0 | 50.0 | 50.0 | 50.0 | 50.0 | 50.0 | 50.0 |
| Internal Molecules Equilibrium | | | | | | | | | | |
| AlSi | | 501 | 410 | 4140–4142 | 304 | 347 | 321 | 316 | – | 4142–4145 |
| AFNOR | | Z12CO5 | Zr12cr13 | 4co4 | Z6CN | Z6CNN | – | Z6CND | 40CDV 4.06 | 42CD4 |
| WERISTOFF | | 12 crMo 19.05 | X10cr13 | 4crMo4 | XSCNi 18.09 | X10CNiNb 18.90 | – | 25cr NiMO 18.10 | 40crMoN 48 | 42CrMo4 |
| B. S | | 15.06–625 | 15.06–713 | 15.06–624Gr.A | 15.06–801Gr.A | – | – | 150.6–845 | 150.6–661 | 150.6–62 GrA |
| **Recommended Temperature Range (°C)** | | | | | | | | | | |
| Minimum | | – | – | −45 | −198 | −198 | −198 | −198 | −129 | – |
| Maximum | | – | – | −48.2 | 67 | 675 | 675 | 67 | 575 | – |

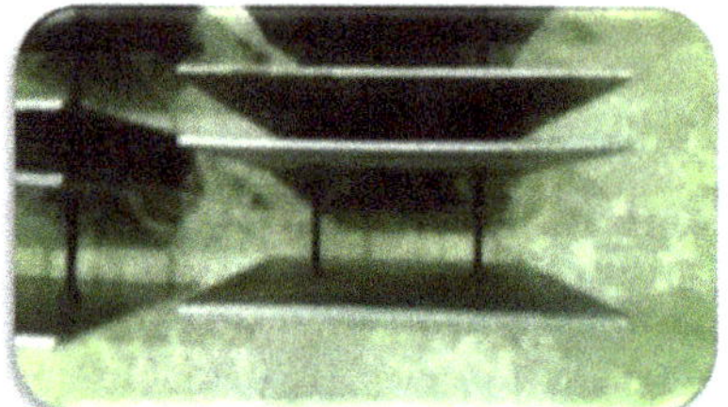

**Figure 3.** Reinforced plates with stud-bolt connectors.

### 3.3. Concrete Core

The spacing between plates is filled with concrete to make the concrete core, where both NWC and steel fibres concrete (SFC) were used. Furthermore, two thicknesses of the concrete core were considered: 80 mm and 100 mm. The width and length of the specimens were 300 mm and 250 mm, respectively, according to the standards [58–74]. The concrete mixes' characteristics are represented in Table 3. In addition, SF are used at 1% (by volume). In addition, CRA were used as a substitute of CNA at 100% in terms of weight. To evaluate the concrete's compressive strength, three cylindrical specimens with 150 mm diameter and 300 mm height were produced and tested under a hydraulic jack. For each mix, the average compressive strength of three samples was considered [62–66]. The results are explained in Table 4. The specimens were tested after 28 days. Figure 4 presents a filled specimen.

**Table 3.** Concrete mix composition (kg/m$^3$).

| Materials | Water | Artificial Sand | Natural Sand | Recycled Coarse Gregates | Cement | Weight |
|---|---|---|---|---|---|---|
| Concrete | 200 | 450 | 630 | 720 | 400 | 2400 |

**Table 4.** Compressive strength of the various mixes (MPa).

| NWC | | SFC | |
|---|---|---|---|
| Compressive strength | Average strength | Compressive strength | Average strength |
| 29.6<br>31.4<br>30.5 | 30.6 | 34.9<br>34.3<br>32.6 | 33.9 |

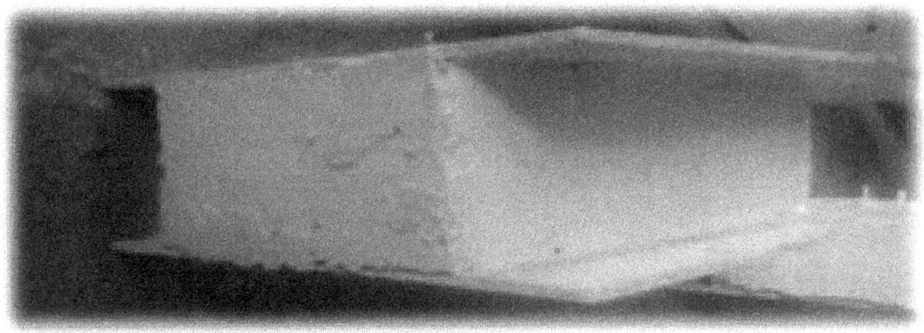

**Figure 4.** SCS sandwich panel.

### 3.4. Steel Fibres

In this study, two bent ends are used as illustrated in Figure 5. The tensile strength, elasticity modulus and failure strain of the fibres are 200 GPa, 2 GPa and 3%, correspondingly. Additionally, the length and equivalent diameter of SF are 60 mm and 0.9 ± 0.03, respectively. SF are employed in order to produce concrete at 1% volumetric content.

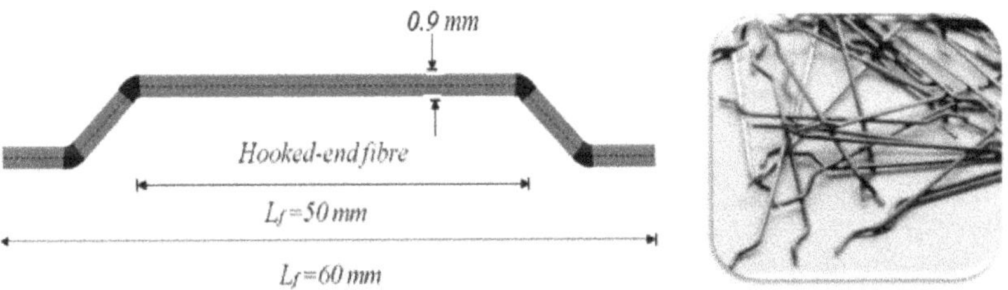


**Figure 5.** Steel fibres.

### 3.5. Coarse Recycled Aggregate

In this study, CRA were sourced from building demolition and replaced CNA at 100% weight ratio. The physical and chemical characteristics of CRA are provided in Tables 5 and 6, respectively. In addition, in Figure 6, the chemical composition of CRA achieved by XRD is illustrated. There, each component is drawn with a colour spectrum.

**Table 5.** Physical properties of CRA.

| Apparent Density (g/cm$^3$) | Bulk Density (g/cm$^3$) | Water Absorption (wt%) | Crushing Index (%) | Porosity (%) |
|---|---|---|---|---|
| 2.66 | 2.56 | 1.519 | 46.1 | 3.76 |

**Table 6.** Chemical properties of CRA.

| Chemical Composition | CRA |
|---|---|
| Ca Mg($CO_3$) (%) | 100 |
| Overall diffraction profile (%) | 100 |
| Background radiation (%) | 25.12 |
| Diffraction peaks (%) | 74.88 |
| Peak area belonging to selected phases (%) | 47.15 |
| Peak area of Phase A (calcium magnesium carbonate) | 47.15 |

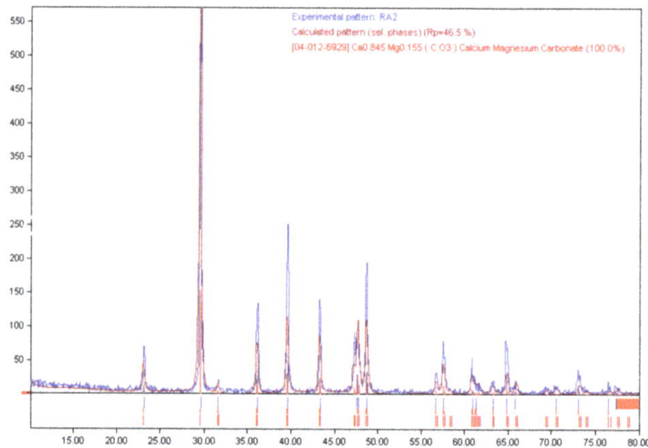

**Figure 6.** XRD patterns of CRA.

## 3.6. Specimens' Properties

In this study, 24 SCS sandwich panels were manufactured and tested. In these specimens, the diameter of bolts, bolts' spacing, thickness and type of concrete core were variable. The geometric characteristics of the specimens are illustrated in Table 7.

**Table 7.** Compressive strength of the various mixes (MPa).

| Specimen | Bolts' Diameter (mm) | Thickness of the Concrete Core (mm) | Bolts' Spacing (mm) | Specimen | Bolts' Diameter (mm) | Thickness of the Concrete Core (mm) | Bolts' Spacing (mm) |
|---|---|---|---|---|---|---|---|
| ND8T80S100 | 8 | 80 | 100 | FD8T80S100 | 8 | 80 | 100 |
| ND8T100S100 | 8 | 100 | 100 | FD8T100S100 | 8 | 100 | 100 |
| ND8T80S150 | 8 | 80 | 150 | FD8T80S150 | 8 | 80 | 150 |
| ND8T100S150 | 8 | 100 | 150 | FD8T100S150 | 8 | 100 | 150 |
| ND10T80S100 | 10 | 80 | 100 | FD10T80S100 | 10 | 80 | 100 |
| ND10T100S100 | 10 | 100 | 100 | FD10T100S100 | 10 | 100 | 100 |
| ND10T80S150 | 10 | 80 | 150 | FD10T80S150 | 10 | 80 | 150 |
| ND10T100S150 | 10 | 100 | 150 | FD10T100S150 | 10 | 100 | 150 |
| ND12T80S100 | 12 | 80 | 100 | FD12T80S100 | 12 | 80 | 100 |
| ND12T100S100 | 12 | 100 | 100 | FD12T100S100 | 12 | 100 | 100 |
| ND12T80S150 | 12 | 80 | 150 | FD12T80S150 | 12 | 80 | 150 |
| ND12T100S150 | 12 | 100 | 150 | FD12T100S150 | 12 | 100 | 150 |

In the designation of the specimens in Table 7, N, F, D, T and S specify normal concrete, SFC, the bolts' diameter, the thickness of the concrete core and spacing of the bolts.

## 4. Test Setup

Specimens were produced and measured under a hydraulic jack for external pressure loading, as represented in Figure 7a. To evaluate deformations, two LVDTs were set up at the top and bottom of the specimens. The test was done under displacement control circumstances at a 0.5 mm rate, and the stoppage condition was set to be a failure. The purpose of this test was to determine the ultimate load and slip of the sandwich panels. Additionally, Figure 7b shows the force distribution in the specimens under lateral load.

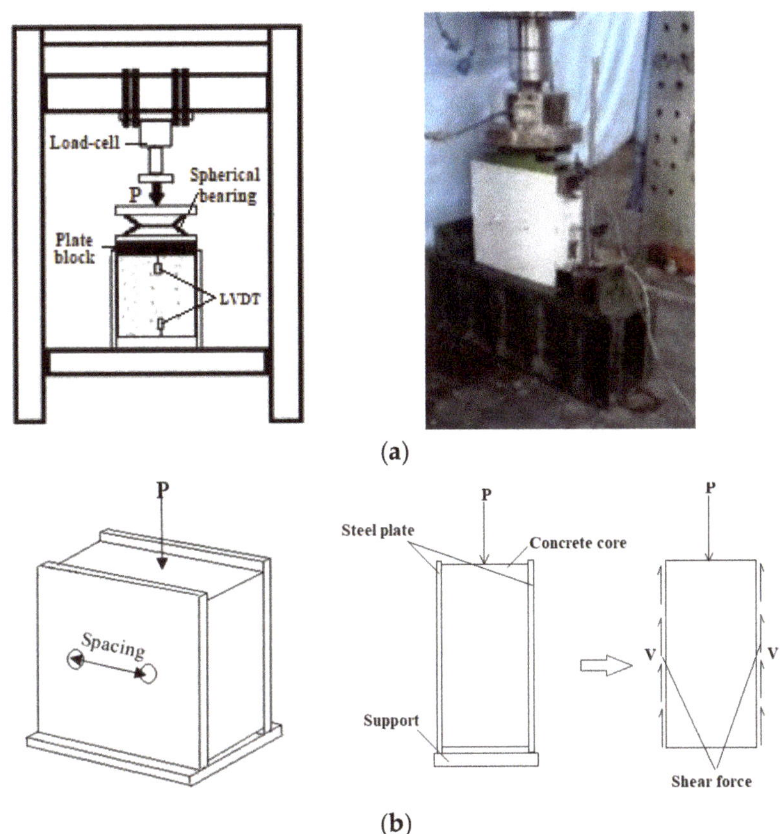


**Figure 7.** (**a**) Test setup and loading condition; (**b**) shear force distribution.

## 5. Results and Discussion

To consider the impact of the stud-bolt connectors on the shear behaviour of SCS, with two sandwich panels, different outcomes were obtained and discussed in the following sections.

### *5.1. Shear Behaviour and Failure Modes*

The failure mode is an important factor to analyse the performance of SCS panels under different loads. In Table 8, the failure mode of the specimens is identified. Furthermore, in Figure 8, the ultimate loading capability of specimens is demonstrated. It should be stated that the loading was performed over the thickness of specimens in the lateral and shear direction, as illustrated in Figure 7b. As seen in Table 8, SF significantly increased the shear strength of the specimens, especially when 12 mm diameter bolts were used. The main reason for improving the shear strength of SCS sandwich panels due to using SF could be attributed to the bridging role of fibres. Adding SF keeps particles together and increases the stiffness of concrete paste. In addition, under the shear stress, stress is transferred through cracks using SF, which leads to increasing the concrete core's resistance exposed to a shear load. Additionally, the bridging role of SF led to a significant reduction in cracks width, which is one of the main reasons for bolt failure and results in reaching the highest load-bearing capacity of SCS sandwich panels.

**Table 8.** Ultimate bearing capacity and failure mode of specimens.

| Specimens | Ultimate Loading Capability (kN) | Failure Mode |
|---|---|---|
| ND8T80S100 | 52 | Bottom of bolts fracture |
| ND8T100S100 | 62 | Bottom of bolts fracture |
| ND8T80S150 | 70 | Bolts fail in the concrete core and bolts detachment from the bottom plate and the concrete core fractured |
| ND8T100S150 | 77 | Bottom of bolts fracture |
| ND10T80S100 | 85 | Concrete core fracture and bolts failure |
| ND10T100S100 | 94 | Concrete core fracture and bolts failure |
| ND10T80S150 | 79 | Concrete core fracture and bolts failure |
| ND10T100S150 | 85 | Concrete core fracture and bolts failure |
| ND12T80S100 | 115 | Concrete core fracture |
| ND12T100S100 | 156 | Concrete core fracture |
| ND12T80S150 | 151 | Concrete core fracture and bolts failure |
| ND12T100S150 | 158 | Concrete core fracture and bolts failure |
| FD8T80S100 | 68 | Bolt failed and bottom of bolts crushing |
| FD8T100S100 | 81 | Bolt failed and bottom of bolts crushing |
| FD8T80S150 | 82 | Bolt failed and bottom of bolts crushing |
| FD8T100S150 | 94 | Bolt failed and bottom of bolts crushing |
| FD10T80S100 | 70 | Bottom of bolts fracture |
| FD10T100S100 | 99 | Concrete core crushing and bolts failure |
| FD10T80S150 | 92 | Concrete core crushing and bolts failure |
| FD10T100S150 | 99 | Concrete core crushing and bolts failure |
| FD12T80S100 | 168 | Bolt failed and the concrete core crushed |
| FD12T100S100 | 204 | Bolt failed and the concrete core crushed |
| FD12T80S150 | 175 | Bolt failed and the concrete core crushed |
| FD12T100S150 | 180 | Bolt failed and the concrete core crushed |

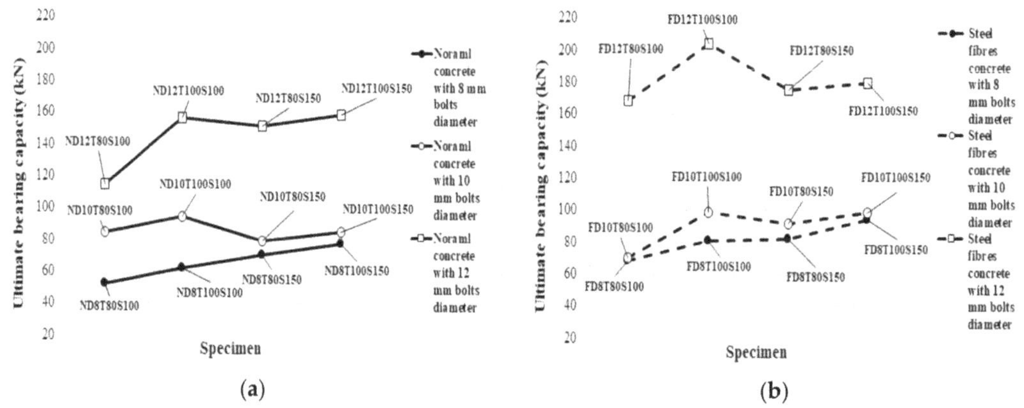


**Figure 8.** Impact of diverse factors on the critical bearing capacity: (**a**) NWC; (**b**) SFC.

Moreover, the bearing capacity increased by increasing the concrete core's thickness. Furthermore, simultaneously increasing both the bolts' spacing and core thickness had a negative effect on the maximum strength of SF reinforced concrete specimens when bolts with a higher diameter (>8 mm) are used, while, in specimens with 8 mm bolts' diameter, increasing both the bolts' spacing and core thickness causes the ultimate shear strength to rise considerably. Furthermore, in specimens with no SF, the failure modes changed from localized failure at the bottom end of the bolts to concrete core fracture by increasing the diameter of the bolt above 8 mm. The same results were reported by previous studies for other types of connectors which confirms the results presented in this study [68–70]. In contrast, the mode of failure was crushing of the concrete core as a result of bolts failure when SF was used because adding SF increases the compressive, tensile and shear strength of concrete and concentrates more stress on the bolts. These changes are visible in Figure 8. Increasing the bolts' diameter resulted in concrete core crushing and fracture mode in the specimens. Conversely, the concrete strength increased when SF was used. Therefore, in specimens with lower bolt diameter, the failure mode occurred by bolts failure; however, failure of the concrete core occurred by increasing the bolts' diameter. According to Figure 8, in specimens with 8 mm bolts' diameter, increasing both the core thickness and bolts' spacing increased the shear strength of the panels, while in specimens with 10 mm and 12 mm bolts' diameter, the maximum shear strength was achieved when the core thickness and bolts' spacing were 100 mm and 100 mm, respectively. The reason for this phenomenon is less contribution of the bolts with a diameter of 8 mm to shear strength and a significant contribution to the shear force of the concrete's strength; however, by increasing the bolts' diameter (10 mm and 12 mm), the role of bolts in providing shear strength increases, which resulted in better shear performance for specimens for lower bolts' spacing. Additionally, the improvement of the maximum shear strength of specimens with 100 mm bolts' spacing from raising the core thickness is more significant in those with 150 bolts' spacing.

Figure 9 illustrates the mode of failure samples. In NWC with 8 mm bolts' diameter, the cohesion between concrete core and faceplates dropped and the plates detached from the concrete core by increasing the load. Besides, the concrete core failed when the bolts' spacing increased because the confinement of concrete between bolts drops as a result of increasing the bolts' spacing. The concrete core was completely fractured by raising the bolts' diameter. In addition, the crack width increased when the bolts' diameter went up. Furthermore, cracks width decreased by using SF due to the bridging role of fibres that kept particles together in the concrete paste and increased the strength of the concrete matrix, and the ultimate bearing capacity significantly improved. Therefore, detachment between the concrete core and plates occurred by increasing the load. In addition, the crack width decreased, and cracks propagated more.

**ND8T80S100**

Bottom of bolts fracture

**ND8T100S100**

Bottom of bolts fracture

**ND8T80S150**

Bolts fail in the concrete core and bolts detachment from the bottom plate and concrete core fractured

**ND8T100S150**

Bottom of bolts fracture

**ND10T80S100**

Concrete core fracture and bolts failure

**ND10T100S100**

Concrete core fracture and bolts failure

**ND10T80S150**

Concrete core fracture and bolts failure

**ND10T100S150**

Concrete core fracture and bolts failure

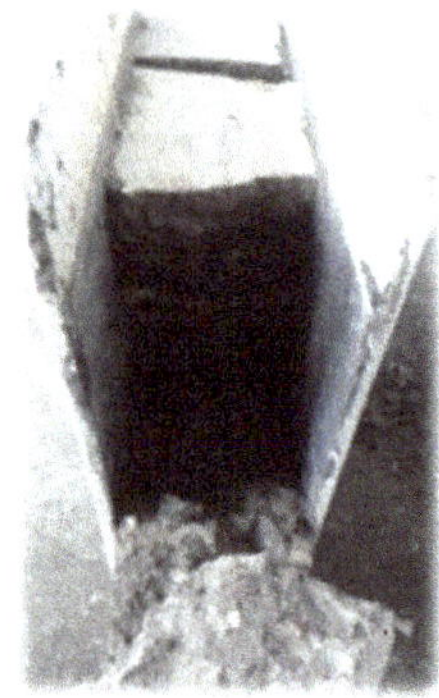

**ND12T80S100**

Concrete core fracture

**ND12T100S100**

Concrete core fracture

**ND12T80S150**

Concrete core fracture and bolts failure

**ND12T100S150**

Concrete core fracture and bolts failure

**Figure 9.** *Cont.*

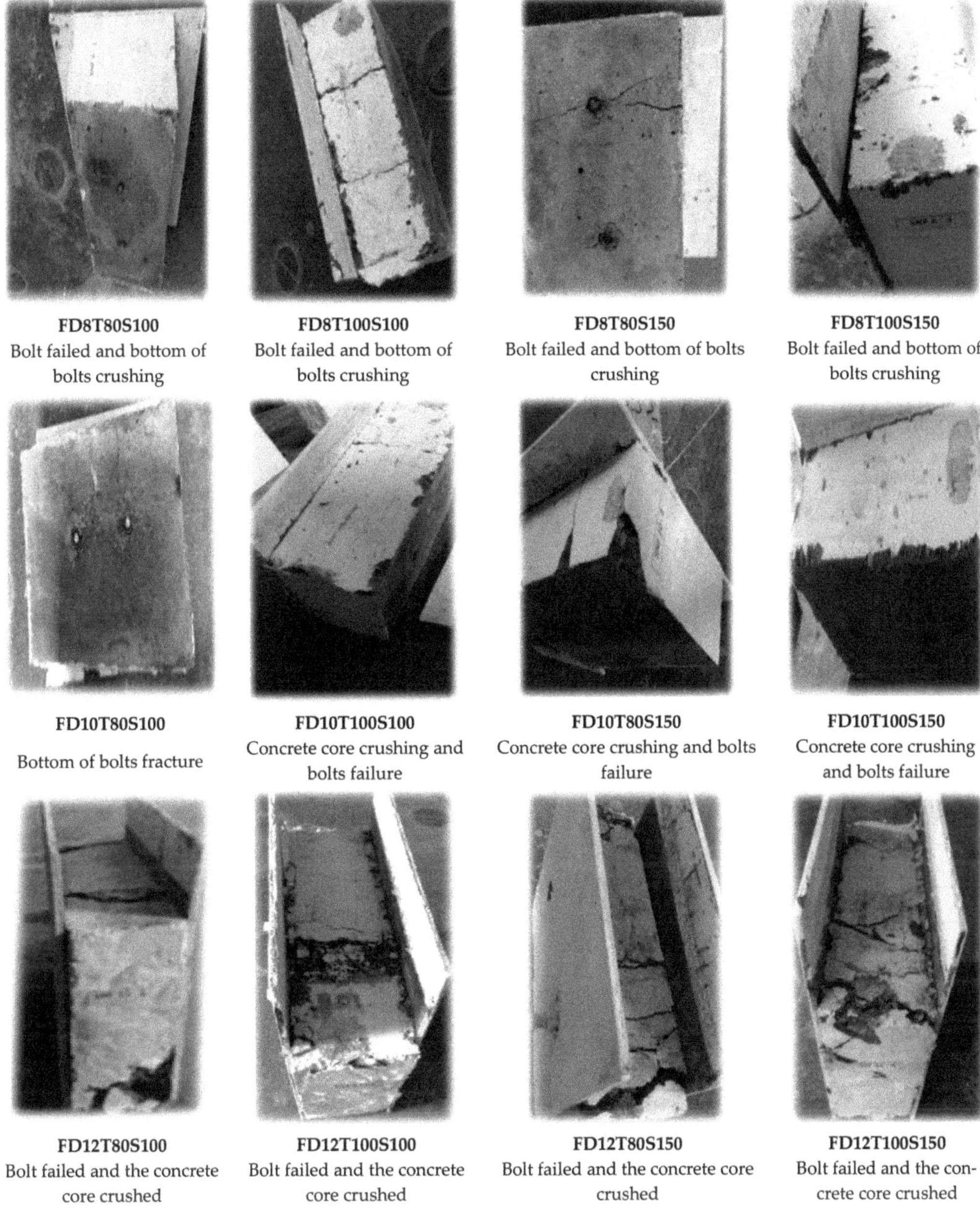


**Figure 9.** Typical static failure modes of specimens with and without SF.

In order to consider the impact of bolts' diameter, the load-slip relationship of specimens with the same concrete core's thickness and bolts' spacing was evaluated, as illustrated in Figures 10 and 11 for NWC and SFC, correspondingly. The ultimate shear strength substantially increased when bolts of 12 mm diameter were used, but there is no significant variance between the ultimate shear strength of specimens made with bolts of 8 mm and

10 mm diameter; however, the shear capacity of specimens with 10 mm bolts' diameter is slightly higher. This could be attributed to increasing the area of bolts, which leads to reducing the stress value over bolts' diameter and is the main reason for improving the shear strength of the specimens. There are two peaks in the load-slip behaviour of some specimens (Figures 10a and 11a). According to these figures, the shear strength of the specimens increased and then slightly dropped as a result of the concrete core's fracture and then went up and declined again after the second peak resulting from bolts' failure and detachment from the steel plate face. Additionally, there is no relevant difference in slip at ultimate strength by increasing the concrete core's thickness.

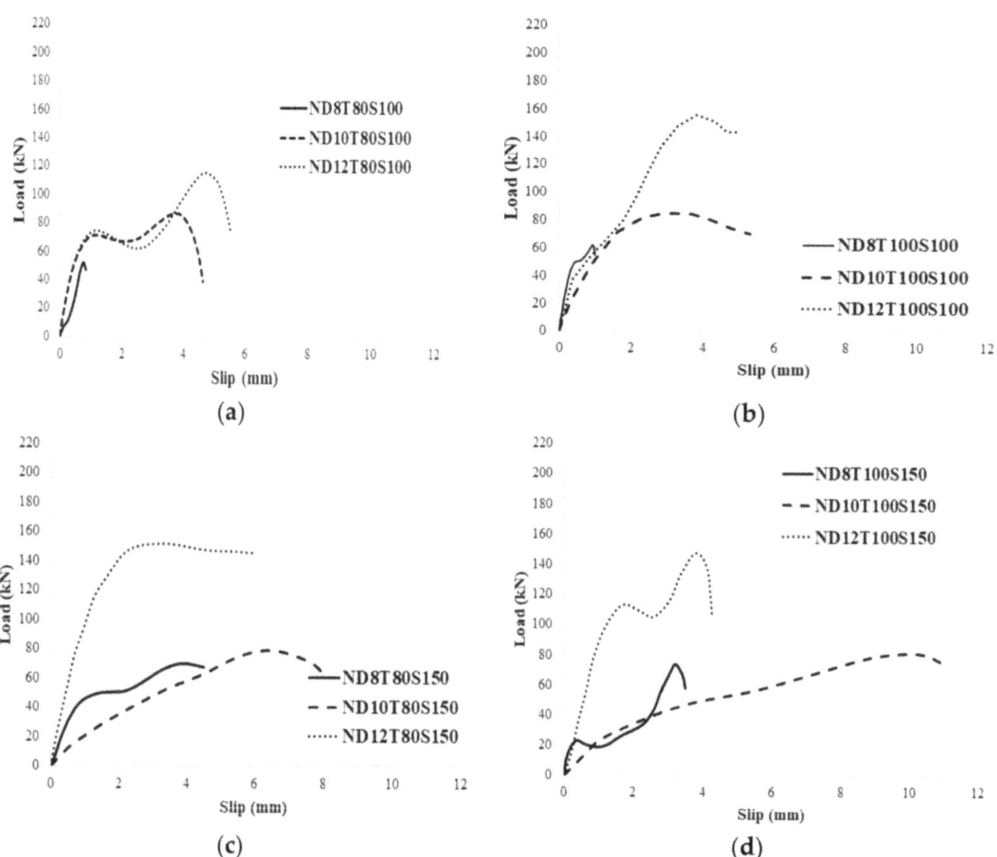


**Figure 10.** Impact of bolts' diameter on the load-slip behaviour of specimens with NWC. (**a**) 8 mm bolt's diameter with 100 mm spacing and 80 mm core thickness, (**b**) 8 mm bolt's diameter with 100 mm spacing and 100 mm core thickness, (**c**) 8 mm bolt's diameter with 150 mm spacing and 80 mm core thickness and (**d**) 8 mm bolt's diameter with 150 mm spacing and 100 mm core thickness.

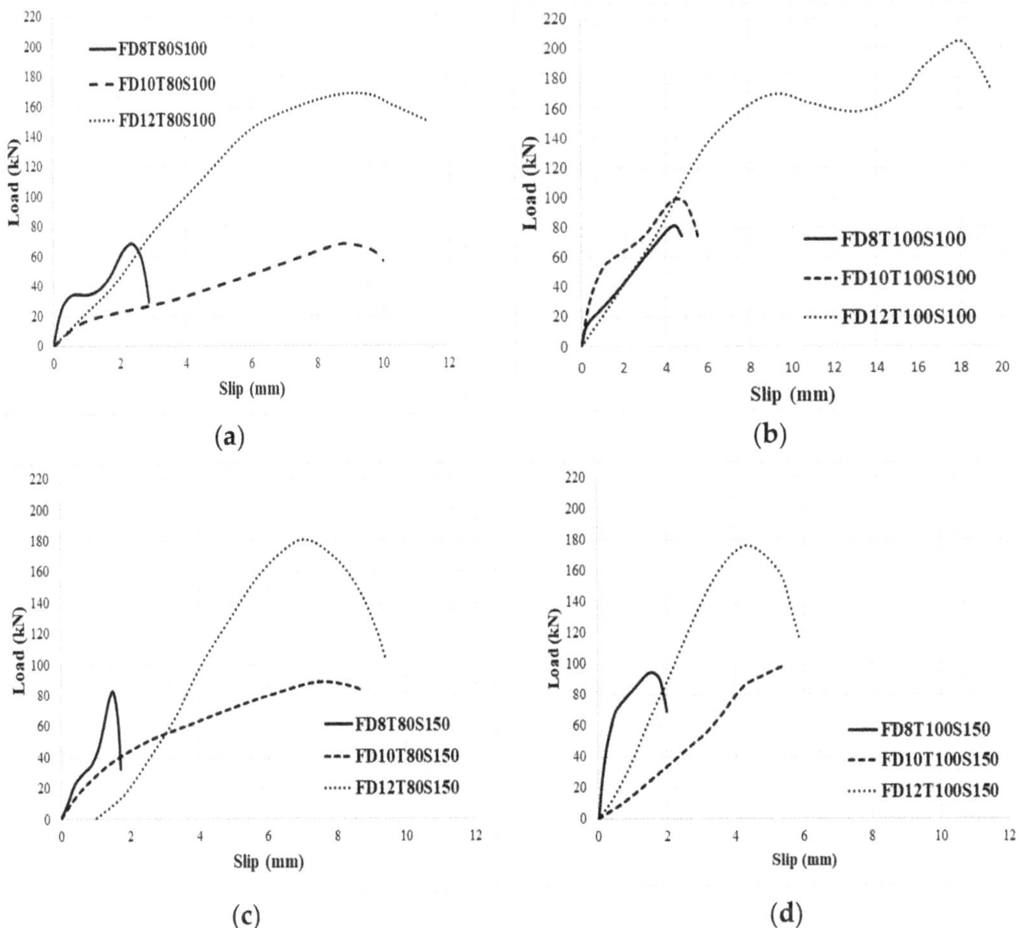


**Figure 11.** Impact of bolts' diameter on the load-slip behaviour of specimens with SFC. (**a**) 100 mm spacing and 80 mm core thickness, (**b**) 100 mm spacing and 100 mm core thickness, (**c**) 150 mm spacing and 80 mm core thickness and (**d**) 150 mm spacing and 100 mm core thickness.

As shown in Figure 11, in specimens with 80 mm core thickness and 100 bolts' spacing, increasing the bolts' diameter to 10 mm and 12 mm substantially increased the maximum slip by about 475% and 625% for NWC, and 233% and 283% for SFC, respectively. This could be attributed to increasing the deformation of the bolts with an increase of the bolts' diameter. According to Figures 10 and 11, the initial stage of the loading curves was linear-elastic followed by a sudden drop of the force due to tension cracking in normal concrete core specimens. In contrast to SFRC cored specimens, only a single crack was observed in the mid-section of the specimen indicating that the sandwich plate failed by core tension, and possibly core/face debonding in certain areas, which led to a sudden drop of the applied force after the maximum load-bearing capacity. Very similar results were reported by previous investigations for CNA concrete specimens [68]. On the other hand, in specimens with 150 mm bolts' spacing, increasing the bolts' diameter considerably increased the slip at ultimate strength; however, in NWC, the slip at ultimate strength dropped by increasing the bolts' diameter to more than 10 mm while the slip at ultimate strength of SFC specimens improved by increasing the bolts' diameter to more than 10 mm.

Moreover, specimens with 100 mm bolts' spacing and 8 mm, 10 mm and 12 mm bolts' diameter, the maximum shear strength is considerably increased by adding SF by approximately 40%, 11% and 44%, respectively, in specimens with 80 mm core thickness while in specimens with 100 mm core thickness, the improvement was about 33%, 25% and 30%, respectively. Furthermore, in specimens with 150 mm bolts' spacing and 8 mm, 10 mm and 12 mm bolts' diameter, the maximum shear capability improved by adding SF by about 17%, 13% and 20%, respectively, in specimens with 80 mm core thickness, while in specimens with 100 mm core thickness this improvement was nearby 25%, 26% and 20%, respectively. Therefore, the impact of SF was boosted for higher diameter of the bolts. Consequently, using SF considerably improves the ultimate shear strength.

In order to investigate the impact of concrete core's thickness on the performance of SCS sandwich panels, the influence of the type of concrete core (NWC and SFC) on the load-slip behaviour of specimens was assessed, as seen in Figures 12 and 13, respectively.

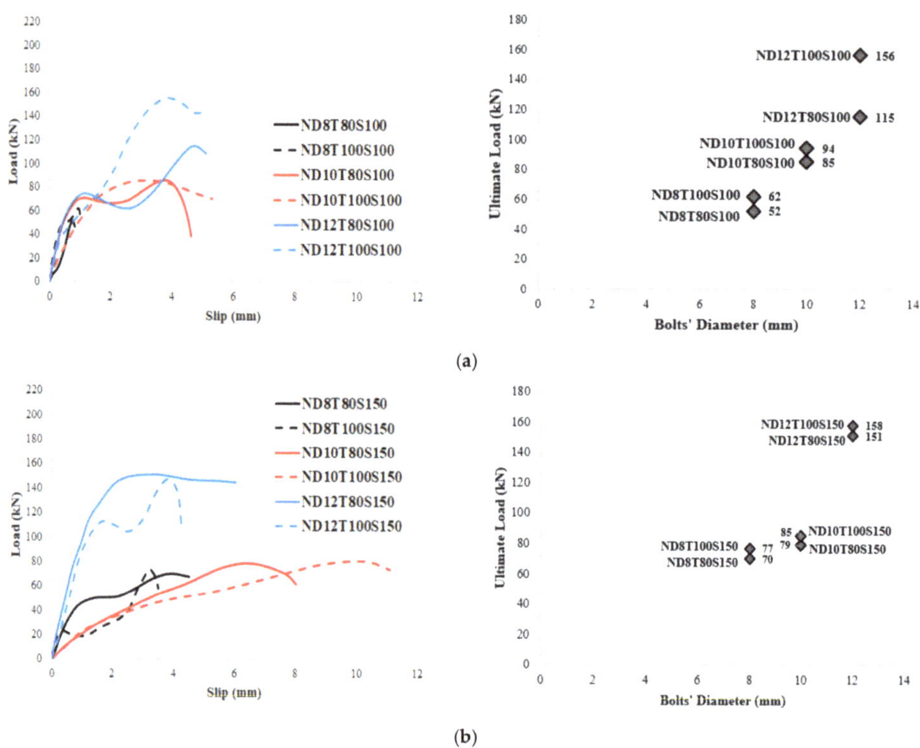


**Figure 12.** Impact of concrete core's thickness on the performance of SCS panels with no SF. (**a**) 100 mm bolts spacing and (**b**) 150 mm bolts spacing.

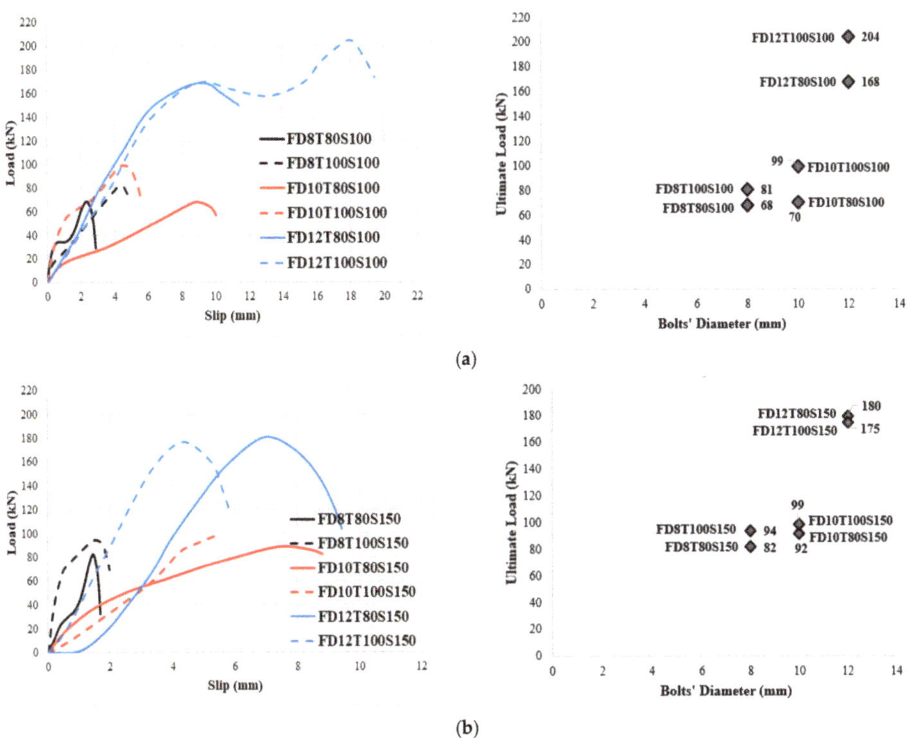


**Figure 13.** Impact of concrete core's thickness on the performance of SCS panels with SF. (**a**) 100 mm bolts spacing and (**b**) 150 mm bolts spacing.

In NWC, there is a significant difference between the slip values at ultimate shear strength. Additionally, increasing the concrete core's thickness did not improve the maximum shear strength of the specimens when the bolts' spacing is 100 mm and 8 mm and 10 mm bolts' diameter are used; however, it improves the maximum strength of 12 mm bolts' diameter reinforced SCS by about 36% (Figure 12a). Thus, increasing the concrete core's thickness is not a good way to increase the shear strength of the SCS when bolts with a smaller diameter (≤10 mm) are used. The same trend can be observed when the bolts' spacing is 150 mm (Figure 10b). Generally, raising the core thickness is not an effective way to improve the ultimate shear strength of the SCS panels; however, raising the bolts' diameter could be a useful manner to improve the shear performance of composite panels. Oppositely, in SFC samples, the maximum shear strength went up with the increase of concrete core's thickness when the bolts' spacing was 100 mm, while by increasing the bolts' spacing to 150 mm the ultimate shear strength is not enhanced by increasing the core thickness and the slip at ultimate strength declined as well (Figure 13a,b). As a result, it is recommended that, in order to improve the shear capacity of SCS sandwich panels, increasing the bolts' diameter and using SF play an important role; however, raising the bolts' spacing and the concrete core's thickness did not have a considerable influence on the maximum shear strength. In Figures 14 and 15, the impact of SF on both the ultimate strength and slip at ultimate strength was studied. As illustrated, SF significantly improve the performance of SCS sandwich panels.

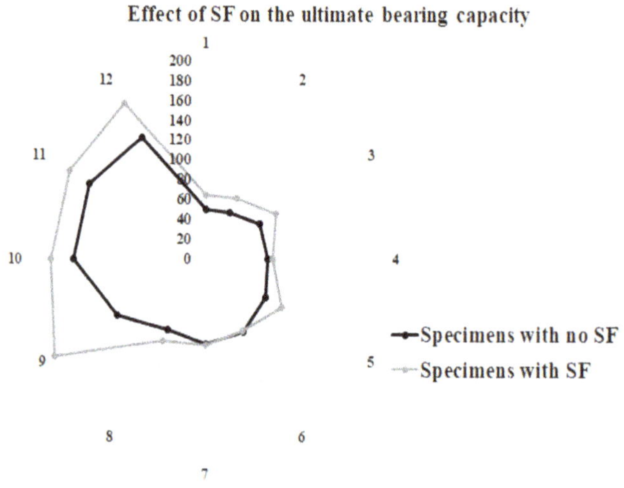

Figure 14. Impact of SF on the ultimate bearing capacity of specimens.

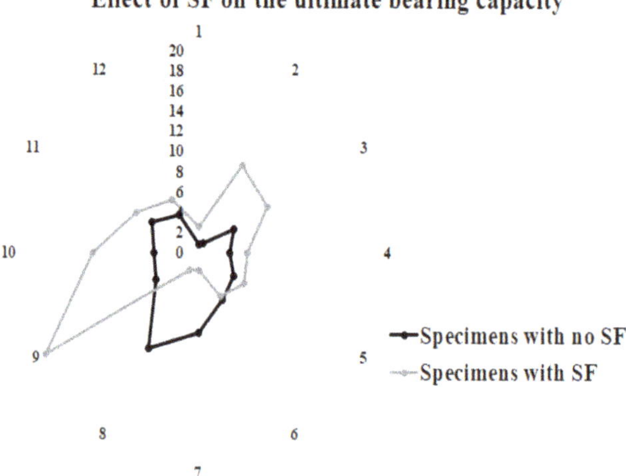


**Figure 15.** Impact of SF on the maximum slip of specimens.

### *5.2. Ductility*

In this section, the ductility index of specimens was evaluated. This index shows the deformation capability of specimens and is defined as the ratio between ultimate deformation and the corresponding displacement at yielding. Therefore, the value of ductility could be calculated using the following formula, as illustrated in Figure 16. The achieved experimental consequences of ductility are presented in Table 9 and Figure 17. As shown in Figure 17, raising the diameter of bolts and using SF resulted in considerable improvement of the value of ductility and deformation of specimens. The main reason for improving the ductility of SCS sandwich panels with SF could be attributed to increasing the strength of CRA concrete matrix obtained from the transferred stress through cracks using SF.

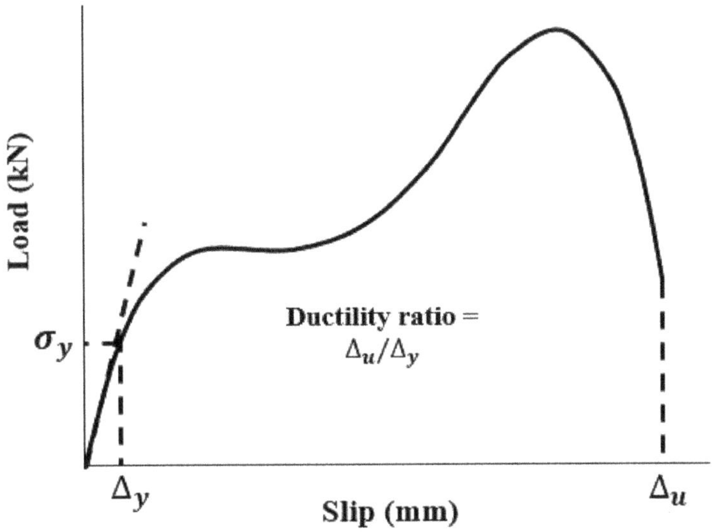


**Figure 16.** Ductility ratio.

**Table 9.** Ductility values of the laboratory samples.

| Specimens | $\Delta_u$ (mm) | $\Delta_y$ (mm) | Ductility (i) | Specimens | $\Delta_u$ (mm) | $\Delta_y$ (mm) | Ductility (i) |
|---|---|---|---|---|---|---|---|
| ND8T80S100 | 0.84 | 0.60 | 1.40 | FD8T80S100 | 2.89 | 0.32 | 9.03 |
| ND8T100S100 | 0.85 | 0.34 | 2.50 | FD8T100S100 | 4.76 | 0.95 | 5.00 |
| ND8T80S150 | 1.70 | 0.32 | 5.29 | FD8T80S150 | 4.51 | 0.60 | 7.51 |
| ND8T100S150 | 1.99 | 0.36 | 5.38 | FD8T100S150 | 3.51 | 0.33 | 10.51 |
| ND10T80S100 | 4.64 | 0.65 | 7.13 | FD10T80S100 | 10.00 | 0.94 | 10.63 |
| ND10T100S100 | 5.32 | 1.57 | 3.37 | FD10T100S100 | 5.54 | 0.86 | 6.44 |
| ND10T80S150 | 8.78 | 1.99 | 4.44 | FD10T80S150 | 8.04 | 1.64 | 4.89 |
| ND10T100S150 | 5.35 | 1.34 | 3.93 | FD10T100S150 | 11.10 | 1.15 | 9.65 |
| ND12T80S100 | 5.52 | 0.70 | 9.88 | FD12T80S100 | 11.40 | 0.91 | 12.52 |
| ND12T100S100 | 5.01 | 1.12 | 10.00 | FD12T100S100 | 19.40 | 1.96 | 9.89 |
| ND12T80S150 | 9.24 | 1.81 | 5.71 | FD12T80S150 | 6.04 | 1.03 | 5.86 |
| ND12T100S150 | 5.83 | 1.75 | 3.27 | FD12T100S150 | 4.28 | 0.82 | 5.22 |

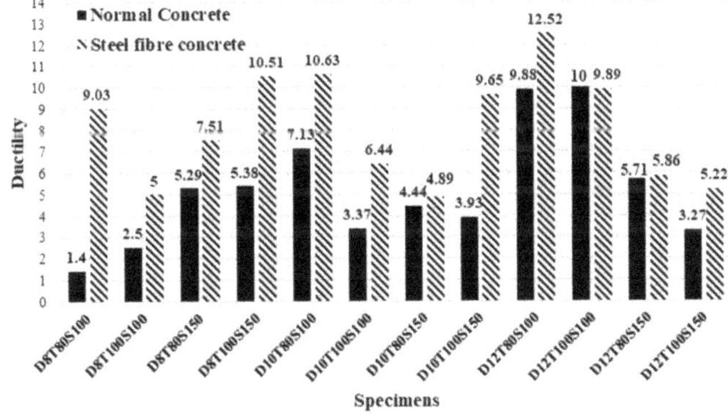


**Figure 17.** Ductility index of SCS sandwich panels.

The yield point is defined as the point where the behaviour of specimen starts to be nonlinear. The ductility index is a qualitative parameter that depends on the deformation of specimens and brittle or ductile failure mode of the concrete members. Therefore, the exact value of Δy is not of major importance. As seen in Table 9 and Figure 17, in addition to SF, increasing bolts' diameter plays an appropriate role to improve the ductility and deformation of specimens and prevents brittle shear failure of specimens.

$$i = \frac{\Delta_u}{\Delta_y} \quad (1)$$

*5.3. Compression with Other Types of Connectors*

In order to show the performance of the proposed connectors in this study, the results were compared with those presented in previous studies. Liew et al. [67] proposed a J-hook connector to raise the shear strength of SCS panels. In another research, Yousefi and Ghalehnovi [55] proposed one-end welded corrugated-strip connectors in SCS members. In Figure 18, the structure of J-hook and welded corrugated-strip connectors are illustrated. Moreover, the characteristics of the specimens are represented in Table 10. Therefore, the compression results are illustrated in Figures 19 and 20, and listed in Table 11. Even though not all the characteristics of the specimens from previous researches are the same as those from this study, they are similar, which allows an acceptable comparison. Stud-bolts connectors have been used to improve the shear of composite beams and shear walls, but there is no study about the shear behaviour of SCS sandwich panels with stud-bolts, which is the novelty of this study.

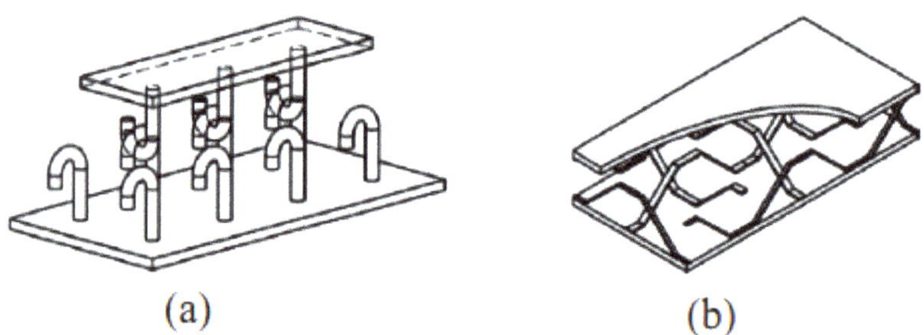


**Figure 18.** SCS sandwich constructions based on shear connector shape. (**a**) J-hook connectors and (**b**) welded end connectors.

**Table 10.** Characteristics of specimens proposed by Liew et al. [67] and Yousefi and Ghalehnovi [55].

| | Specimens | Failure Modes of These Samples | $f_c$ (MPa) | Thickness (mm) | Plate Thickness (mm) | Yield Strength of Plates (MPa) | Yield Strength of the Connectors (MPa) | Diameter of the Connectors (mm) |
|---|---|---|---|---|---|---|---|---|
| J-hook connectors [67] | N5 | Concrete embedment failure | 47.7 | 60 | 6 | 310 | 310 | 11.7 |
| | N7 | Concrete embedment failure | 47.7 | 75 | 6 | 310 | 310 | 11.7 |
| | N9 | Concrete herringbone shear crack | 47.7 | 100 | 6 | 310 | 310 | 11.7 |
| | HN1 | Concrete wedge splitting | 43.5 | 75 | 6 | 310 | 315 | 15.6 |
| | HN2 | Left-strip shear fracture and concrete herringbone shear crack | 43.5 | 75 | 6 | 310 | 315 | 15.6 |

**Table 10.** *Cont.*

| | Specimens | Failure Modes of These Samples | $f_c$ (MPa) | Thickness (mm) | Plate Thickness (mm) | Yield Strength of Plates (MPa) | Yield Strength of the Connectors (MPa) | Diameter of the Connectors (mm) |
|---|---|---|---|---|---|---|---|---|
| | HN3 | Concrete embedment failure | 43.5 | 40 | 6 | 310 | 315 | 15.6 |
| | HN4 | Left-strip shear fracture and concrete herringbone shear crack | 43.5 | 75 | 6 | 310 | 340 | 19.5 |
| | HN9 | Concrete wedge splitting | 43.5 | 75 | 6 | 310 | 340 | 19.5 |
| | HN11 | Concrete wedge splitting | 43.5 | 100 | 6 | 310 | 340 | 19.5 |
| Welded end connectors [56] | 6D-1 | Connectors shear fracture | 27.9 | 100 | 6 | 315 | 380 | 20 |
| | 8D-2 | Left-strip shear fracture and concrete herringbone shear crack | 27.9 | 100 | 8 | 315 | 380 | 20 |
| | 10D-3 | Top branch bent down and bottom branch straighten of right-strip and concrete wedge shear | 27.9 | 100 | 10 | 315 | 495 | 20 |
| | 12D-4 | Left-strip shear fracture and concrete wedge shear | 27.9 | 100 | 12 | 315 | 516 | 20 |
| | 6Db10-5 | Left-strip shear fracture | 27.9 | 100 | 6 | 315 | 380 | 10 |
| | 6Db70-6 | Concrete crushing and plate buckling | 27.9 | 100 | 6 | 315 | 495 | 70 |
| | 6Db140-7 | Concrete wedge splitting | 27.9 | 127 | 6 | 315 | 516 | 140 |
| | 6Db200-8 | Concrete wedge splitting | 27.9 | 186 | 6 | 315 | 615 | 200 |

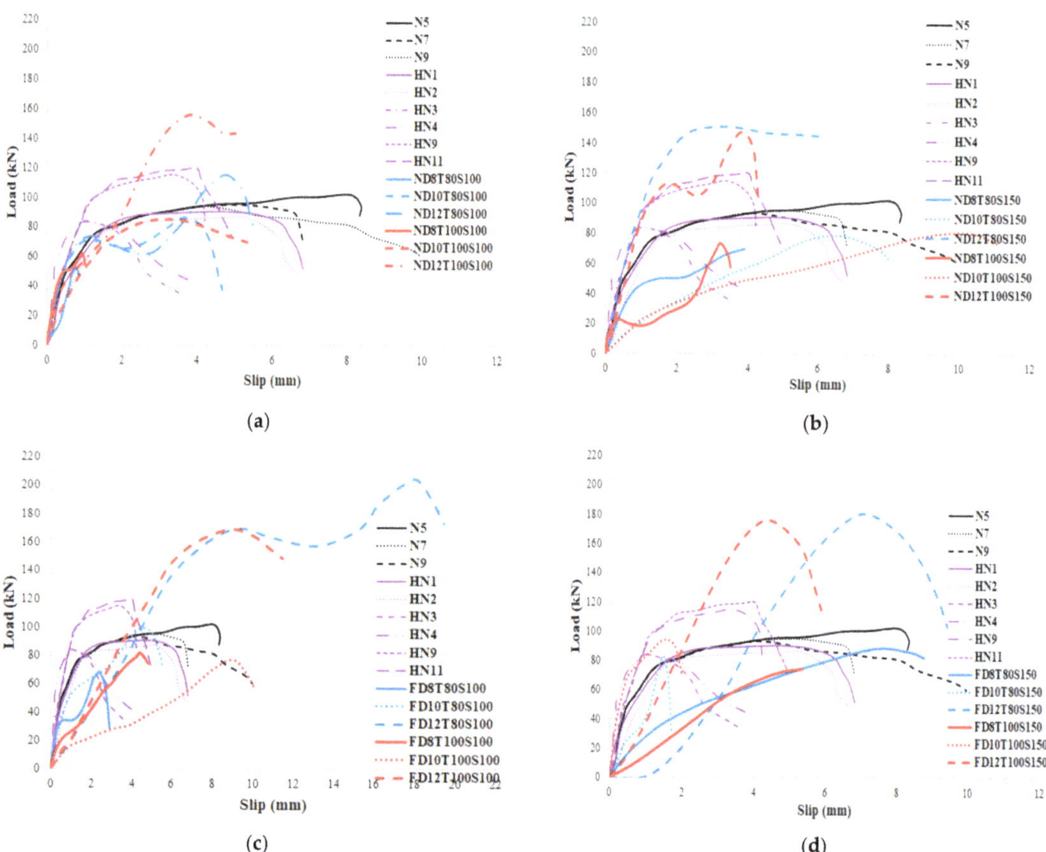

**Figure 19.** Compression between the proposed connector in this study and J-hook connectors. (**a**) normal concrete with 100 mm bolts' spacing, (**b**) normal concrete with 150 mm bolts' spacing, (**c**) fibre-reinforced concrete with 100 mm bolts' spacing and (**d**) fibre-reinforced concrete with 150 mm bolts' spacing.

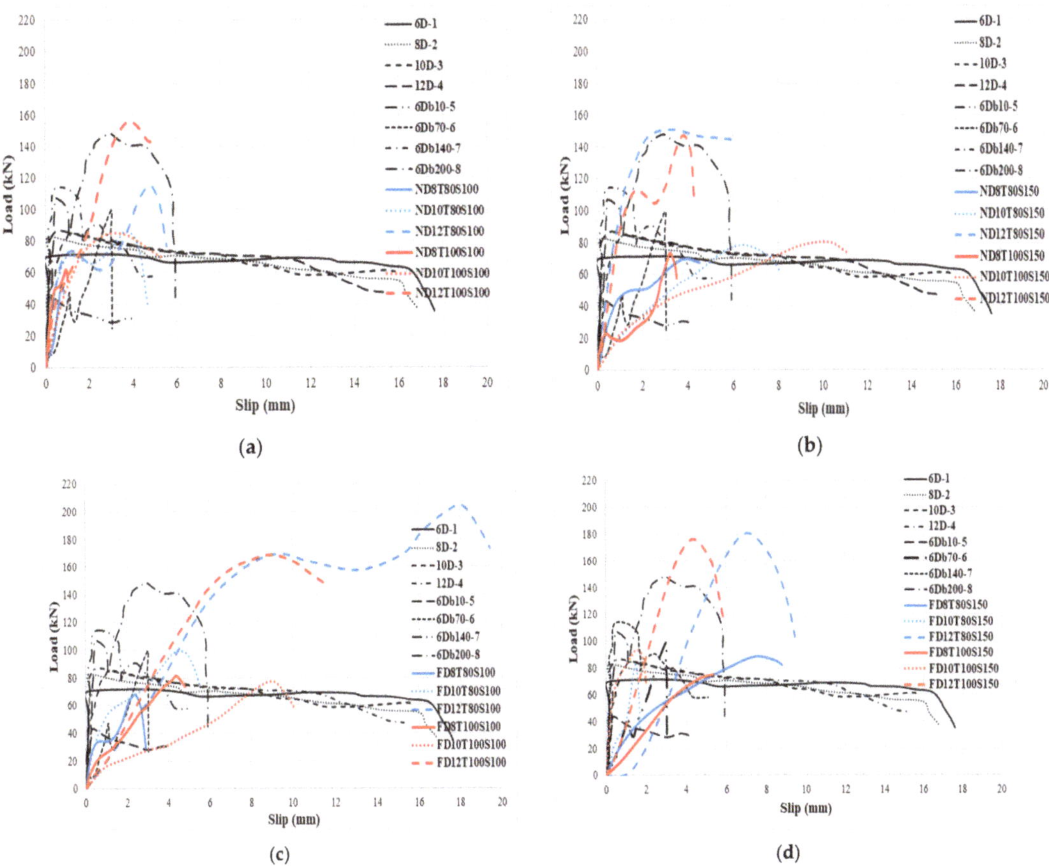


**Figure 20.** Compression between the proposed connector in this study and welded end connectors. (**a**) normal concrete with 100 mm bolts' spacing, (**b**) normal concrete with 150 mm bolts' spacing, (**c**) fibre-reinforced concrete with 100 mm bolts' spacing and (**d**) fibre-reinforced concrete with 150 mm bolts' spacing.

According to Figure 19, the bearing capacity of the specimens significantly increased when stud-bolt connectors were used relative to those produced with J-hook connectors. This could be attributed to connecting two ends of the stud-bolt while in the J-hook connector only one end was welded to the plate. The concrete compressive strength of specimens produced using J-hook connectors and tested by Liew et al. [67] was almost twice of the one of those manufactured with stud-bolts connectors and tested in this study, but other parameters of specimens are similar.

Consequently, the compressive strength of the concrete core noticeably dropped when stud-bolt connectors were employed, which is the main advantage of stud-bolts connectors compared with the J-hook type. Besides, using SF enhanced the shear behaviour of the SCS members due to the bridging role of SF, which improved the shear resistance of the concrete matrix. The maximum shear capability of the specimens was also substantially improved when stud-bolts connectors were used compared to the specimens with welded end connectors (Figure 20). It should be stated that the yield strength of steel faceplates used by Yousefi and Ghalehnovi [55] to manufacture welded end connectors SCS sandwich structures was greater than that of the plates employed in this study, which is the foremost benefit of stud-bolts connectors. Therefore, the steel plated with lower yield strength could be used when stud-bolts connectors are used.

**Table 11.** Comparison of the influence of stud-bolts connectors with J-hook and welded end connectors on the maximum shear strength.

| Connector | Specimens | Ultimate Strength (kN) | Connector | Specimens | Ultimate Strength (kN) |
|---|---|---|---|---|---|
| J-hook connectors [67] | N5 | 102.0 | Stud-bolt | ND10T80S100 | 78 |
| | N7 | 94.8 | | ND10T100S100 | 94 |
| | N9 | 93.8 | | ND10T80S150 | 79 |
| | HN1 | 90.5 | | ND10T100S150 | 85 |
| | HN2 | 86.8 | | ND12T80S100 | 115 |
| | HN3 | 84.1 | | ND12T100S100 | 156 |
| | HN4 | 83.7 | | ND12T80S150 | 151 |
| | HN9 | 115.0 | | ND12T100S150 | 158 |
| | HN11 | 120.0 | | FD8T80S100 | 68 |
| Welded end connectors [55] | 6D-1 | 71.6 | | FD8T100S100 | 81 |
| | 8D-2 | 82.6 | | FD8T80S150 | 82 |
| | 10D-3 | 85.2 | | FD8T100S150 | 94 |
| | 12D-4 | 86.9 | | FD10T80S100 | 70 |
| | 6Db10-5 | 43.6 | | FD10T100S100 | 99 |
| | 6Db70-6 | 99.4 | | FD10T80S150 | 92 |
| | 6Db140-7 | 114.0 | | FD10T100S150 | 99 |
| | 6Db200-8 | 148.0 | | FD12T80S100 | 168 |
| Stud-bolt | ND8T80S100 | 52 | | FD12T100S100 | 204 |
| | ND8T100S100 | 62 | | FD12T80S150 | 175 |
| | ND8T80S150 | 70 | | FD12T100S150 | 180 |
| | ND8T100S150 | 77 | | | |

### *5.4. Suggested Load-Slip Behaviour Model for SCS Sandwich Panels with Stud-Bolt Connectors*

In this subdivision, a new technical scheme is presented by using a regression technique using the experimental results. This method is mandatory to assess the load-slip relationship and estimating the maximum capability of SCS members with stud-bolt connectors. Previously, many formulas were proposed by Ollgaard et al. [31], Cederwall [68], Lorence and Kuica [69], Gattesco and Giuriani [32] and Xue et al. [70] to estimate the shear load-slip of SCS panels based on the experimental results as briefly presented in Table 12. The previously proposed models are appropriate tools to predict the shear behaviour of SCS sandwich panels for other types of connectors. Therefore, it is necessary to develop a model in order to estimate the shear strength of the SCS sandwich panels when stud-bolts are used.

**Table 12.** Previous models to predict the shear strength of SCS sandwich panels with different types of connectors.

| Study | Model |
|---|---|
| Ollgaard et al. [31] | $\frac{P}{P_u} = \left(1 - e^{-18\delta}\right)^{0.4}$ |
| Cederwall [68] | $\frac{P}{P_u} = \frac{2.24(\delta-0.058)}{1+1.98(\delta-0.058)}$ |
| Lorence and Kuica [69] | $\frac{P}{P_u} = \left(1 - e^{0.55\delta}\right)^{0.3}$ |
| Gattesco and Giuriani [32] | $\frac{P}{P_u} = \alpha\sqrt{1 - e^{-\beta\delta/\alpha}} + \gamma\delta$ |
| Xue et al. [70] | $\frac{P}{P_u} = \frac{\delta}{0.5+0.97\delta}$ |

In order to present a new technique to anticipate the shear behaviour of SCS panels, the bearing capability of specimens was normalized relative to $P/Pu$ is the ultimate bearing capacity and can be determined based on the load-slip relationship. The result is presented in Figure 21. The normalized bearing capability ($P/Pu$) and slip ($\delta$) are alienated into four categories (a, b, c and d) based on the type of concrete core and the diameter of bolts (Figure 21). According to Figure 22, the following equations estimate the shear performance of NWC SCS panels. Equations (2) and (3) differ in terms of the bolts' diameter. The $R^2$ values of Equations (2) and (3), 0.98 and 0.97 respectively, show a good fit with the experimental results.

$$\frac{P}{P_u} = 0.24ln(\delta) + 0.64 \quad D \leq 8\,mm \quad R2 = 0.98 \tag{2}$$

$$\frac{P}{P_u} = 0.26ln(\delta) + 0.42 \; D \geq 10\,mm \quad R2 = 0.97 \tag{3}$$

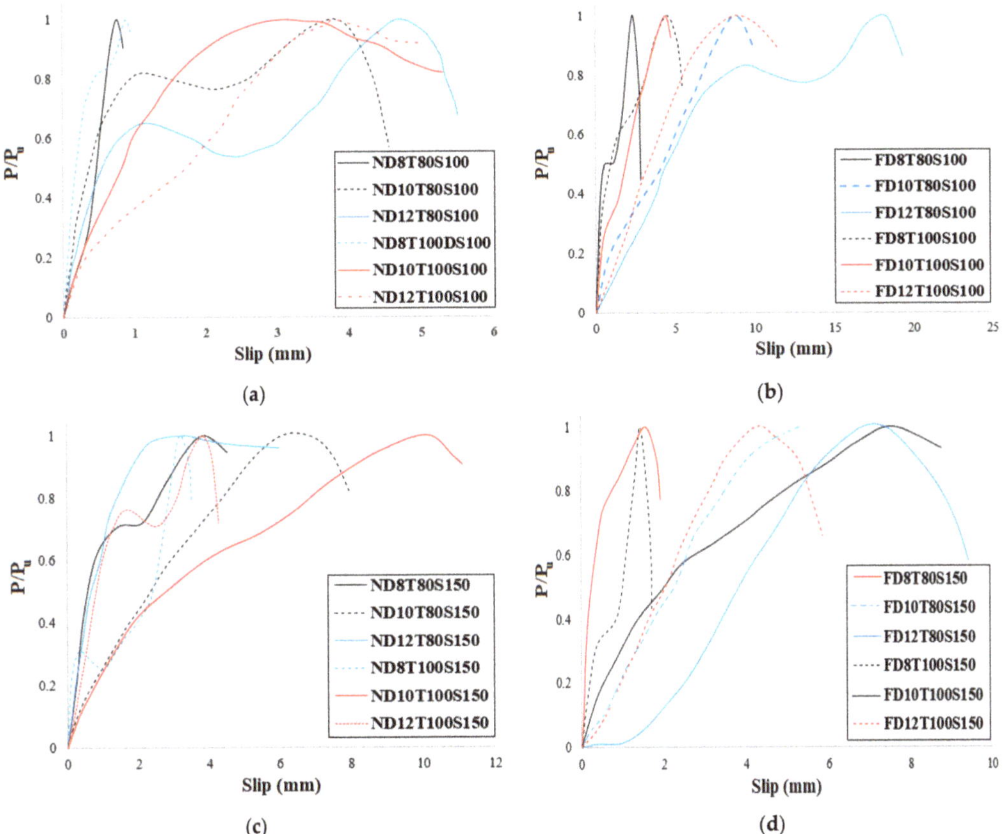


**Figure 21.** Normalized load-slip relationship of SCS specimens. (**a**) normal concrete with 100 mm bolts' spacing, (**b**) fibre-reinforced concrete with 100 mm bolts' spacing, (**c**) normal concrete with 150 mm bolts' spacing and (**d**) fibre-reinforced concrete with 150 mm bolts' spacing.

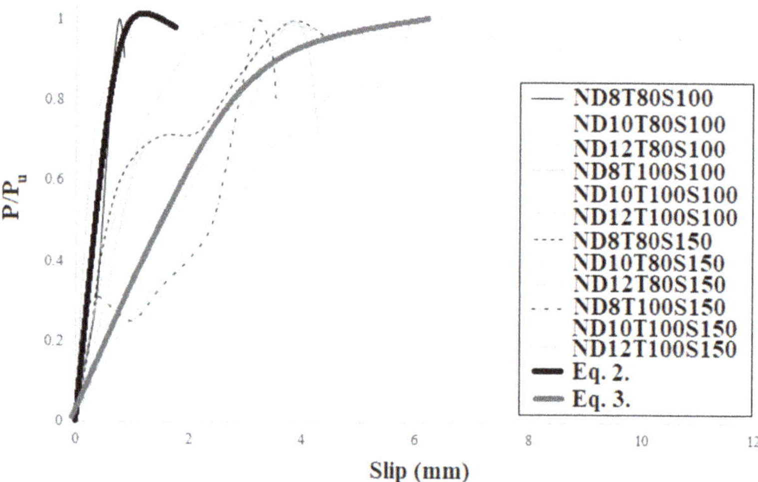


**Figure 22.** Comparison between the normalized load-slip relationship of the test results and the proposed model for NWC SCS panels with stud-bolt connectors.

For SFC specimens, a diverse scheme was proposed, as represented below. These formulas, with $R^2$ values of 0.95 and 0.91, respectively, also show a good fit with the experimental consequences. They also differ in terms of the bolts' diameter for SFC SCS panels. The comparison of the proposed formulas with the experimental results is illustrated in Figure 23.

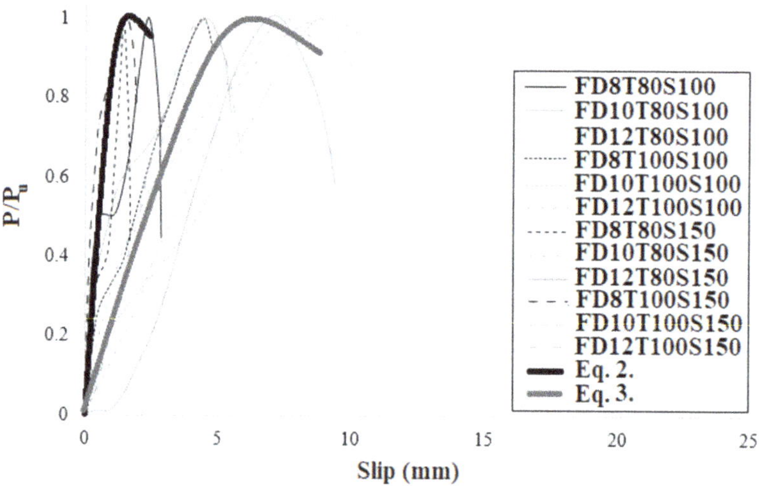


**Figure 23.** Comparison between the normalized load-slip relationship of the test consequences and the proposed technique for SFC SCS panels with stud-bolt connectors.

When the diameter of bolts is up to 8 mm, Equations (2)–(5) can be used to forecast the shear capability of NWC and SFC SCS panels, correspondingly. Alternatively, Equations (4) and (5) can be employed when the diameter of bolts is equal or greater than 10 mm. Besides, in Figure 24, the presented formulas were compared with the models proposed for previous techniques. This figure demonstrates the higher accuracy of the presented model in this study relative to the previous ones. Previous models were based on one equation that is not useful for the wide-range of SCS panels. In this research, a two-dimensional equation is proposed that covers a wide-range of specimens based on the bolts' diameter, which shows the high accuracy of the model proposed in this study.

$$\frac{P}{P_u} = 0.5\delta^{0.53} \quad D \leq 8\,mm \quad R2 = 0.88 \tag{4}$$

$$\frac{P}{P_u} = 0.19\delta^{0.8} \quad D \geq 10\,mm \quad R2 = 0.91 \tag{5}$$

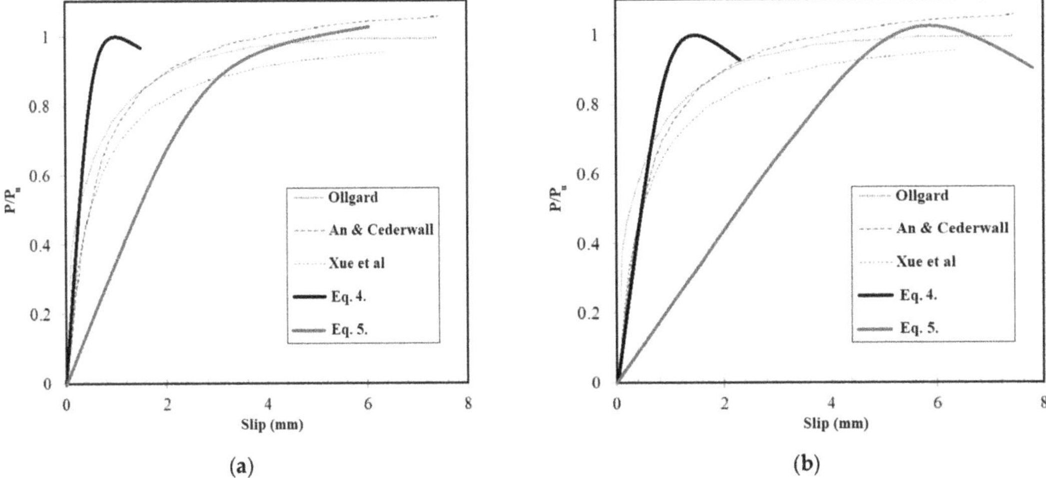


**Figure 24.** Comparison between the presented scheme and the proposed previous equations: (**a**) NWC (**b**) SFC.

## 6. Conclusions

In this study, a new model of stud-bolt connector was presented. The effect of this connector on the shear capability of SCS panels was the object of this study. Additionally, the bolts' diameter, concrete core's thickness, bolts' spacing and concrete type were the governing parameters. In total, 24 laboratory samples were produced and examined, and the shear performance of SCS panels was assessed. According to the obtained outcomes, the following conclusions could be discussed:

1. CRA could be used as an effective way to produce concrete members in order to decrease waste materials resources worldwide. In addition, the use of SF plays an effective role to mitigate the negative influence of high CRA incorporation on the structural performance of SCS sandwich panels;
2. The new proposed stud-bolt connector is efficient to improve the shear strength of concrete sandwich panels, and this type of connectors are low-cost and easy to design compared with others;
3. The failure mode was changed when the diameter of bolts is kept constant and SF were used. On the contrary, increasing the bolts' spacing increases the deformation of the SCS panels when bolts with a higher diameter (>8 mm) are used while, in specimens with 8 mm bolts' diameter, increasing both the bolts' spacing and the core thickness causes the ultimate shear strength to rise considerably. Moreover, in specimens with 8 mm bolts' diameter, increasing both the core thickness and the bolts' spacing increases the shear strength of the panels while, in specimens with 10 mm and 12 mm bolts' diameter, the maximum shear strength was achieved when the core thickness and bolts' spacing are 100 mm and 10 mm, respectively. The reason

for these phenomena is the low contribution of the bolts of 8 mm diameter to shear strength and significant increase of shear force with concrete strength; however, by increasing the bolts' diameter (10 mm and 12 mm), the contribution of the bolts to shear strength increases, which resulted in better shear performance for specimens with less spacing;

4. In specimens with no SF, the failure modes change from localized failure at the bottom end of the bolts to concrete core fracture by increasing the bolts' diameter above 8 mm. Furthermore, the concrete core fractured when the bolts' spacing increased because the confinement of concrete between bolts drops as a result of increasing bolts' spacing. In contrast, the failure mode is crushing of the concrete core when SF are used;
5. Increasing the bolts' spacing causes failure of the concrete core. In addition, the concrete core completely fractures, and the crack width increases by increasing the diameter of the bolts. Alternatively, the crack width dropped and the maximum bearing capability increased by using SF because fibres play a bridge role to keep particles close to each other;
6. The shear strength of the specimens increased and then slightly dropped as a result of the concrete core fracture and then raised and declined again after the second peak resulting from bolts' failure and detachment from the steel plate face. Additionally, there is no relevant difference in slip at ultimate strength by increasing the concrete core's thickness. Therefore, in specimens with 100 mm bolts' spacing and 8 mm, 10 mm and 12 mm bolts' diameter, the maximum shear strength considerably increased by adding SF by approximately 40%, 11% and 44% respectively in specimens with 80 mm core thickness, while in specimens with 100 mm core thickness the improvement was about by 33%, 25% and 30%, respectively. Furthermore, in specimens with 150 mm bolts' spacing and 8 mm, 10 mm and 12 mm bolts' diameter, the maximum shear capacity improved by adding SF by about 17%, 13% and 20% in specimens with 80 mm core thickness, respectively, while in specimens with 100 mm core thickness this improvement was nearby 25%, 26% and 20%, respectively;
7. Increasing the concrete core's thickness did not improve the maximum shear strength of specimen when the bolts' spacing was 100 mm and 8 mm and 10 mm bolts' diameter was used; however, it improved the maximum strength of 12 mm bolts' diameter reinforced SCS by about 36%. Therefore, increasing the concrete core's thickness is not a good way to increase the shear strength of SCS when bolts with a smaller diameter ($\leq$10 mm) are used;
8. In SFC samples, the maximum shear strength went up with the increase of concrete core's thickness when the bolts' spacing is 100 mm while, by increasing the bolts' spacing to 150 mm, the ultimate shear strength was not enhanced by increasing the core thickness and the slip at ultimate strength declined. As a result, it is recommended, in order to improve the shear capacity of SCS sandwich panels, to increase the bolts' diameter and use SF; however, increasing the bolts' spacing and concrete core's thickness did not have a considerable influence on the maximum shear strength;
9. Increasing the bolts' diameter plays an appropriate role to improve the ductility and deformation of specimens and prevents brittle shear failure of the specimens;
10. The proposed model is highly accurate in the estimation of the shear behaviour of SCS panels with stud-bolt connectors and can be used for both NWC and SFC.

It is stressed that this research investigated the effect of bolts' diameter, concrete core's thickness, bolts' spacing, RCA and SF on the performance of SCS sandwich panels using a novel proposed connector shape. In the experiments, SF and RCA contents were kept constant in all specimens. Therefore, it is recommended to measure the effect of various concrete, fibres and aggregate types as well as both coarse and fine aggregates on the performance of SCS sandwich panels with different kinds of connectors in future studies.

**Author Contributions:** Conceptualization, M.G. (Mansour Ghalehnovi) and J.d.B.; formal analysis, M.G. (Mohammad Golmohammadi); investigation, A.K. and M.G. (Mohammad Golmohammadi); methodology, A.K. and M.G. (Mohammad Golmohammadi); software, M.G. (Mansour Ghalehnovi); supervision, M.G. (Mansour Ghalehnovi) and J.d.B.; writing—original draft, A.K. and M.G. (Mohammad Golmohammadi); writing—review & editing, J.d.B. All authors have read and agreed to the published version of the manuscript.

**Funding:** This research received no external funding.

**Data Availability Statement:** The raw/processed data required to reproduce these findings cannot be shared at this time as the data also form part of an ongoing study.

**Acknowledgments:** The authors wish to thank the CERIS (Civil Engineering Research and Innovation for Sustainability) research centre and the FCT (Foundation for Science and Technology).

**Conflicts of Interest:** The authors declare that they have no conflicts of interest.

## Nomenclature

| | |
|---|---|
| CNA | coarse natural aggregates |
| CNC | computer numerical control |
| CRA | coarse recycled aggregates |
| CSC | corrugated-strip connectors |
| D | bolts' diameter |
| DSC | double skin composite |
| LWC | lightweight concrete |
| N | normal strength concrete |
| NWC | normal weight concrete |
| P | external shear force |
| $P_u$ | ultimate shear capacity of the connector |
| RC | reinforced concrete |
| S | bolts' spacing |
| SF | steel fibres |
| SCS | steel-concrete-steel |
| SFC | steel fibres concrete |
| T | concrete core's thickness |
| $\delta$ | slip induced by the applied load |

## References

1. USEPA. *Advancing Sustainable Materials Management: 2017 Fact Sheet*; Office of Land and Emergency Management: Washington, DC, USA, 2019.
2. Ahmari, S.; Ren, X.; Toufigh, V.; Zhang, L. Production of geopolymeric binder from blended waste concrete powder and fly ash. *Constr. Build. Mater.* **2012**, *35*, 718–729. [CrossRef]
3. Beckmann, B.; Schicktanz, K.; Reischl, D.; Curbach, M. DEM simulation of concrete fracture and crack evolution. *Struct. Concr.* **2012**, *13*, 213–220. [CrossRef]
4. Rezaiee-Pajand, M.; Rezaiee-Pajand, A.; Karimipour, A.; Abbadi, J.A.N.M. Particle swarm optimization algorithm to suggest formulas for the behaviour of the recycled materials reinforced concrete beams. *Int. J. Optim. Civ. Eng.* **2020**, *10*, 451–479.
5. Farhan, N.A.; Sheikh, M.N.; Hadi, M.N.S. Investigation of engineering properties of normal and high strength fly ash based geopolymer and alkali-activated slag concrete compared to ordinary Portland cement concrete. *Constr. Build. Mater.* **2019**, *196*, 26–42. [CrossRef]
6. Velay-Lizancos, M.; Martinex-Lage, M.; Azenha, J.; Granja, A.; Vazquez-Burgo, P. Concrete with fine and coarse recycled aggregates: E-modulus evolution, compressive strength and non-destructive testing at early ages. *Constr. Build. Mater.* **2018**, *193*, 323–331. [CrossRef]
7. Katz, A. Properties of concrete made with recycled aggregate from partially hydrated old concrete. *Cem. Concr. Res.* **2003**, *33*, 703–711. [CrossRef]
8. Karimipour, A.; Ghalehnovi, M. Comparison of the effect of the steel and polypropylene fibres on the flexural behaviour of recycled aggregate concrete beams. *Structures* **2021**, *29*, 129–146. [CrossRef]
9. Laskar, S.M.; Talukdar, S. Development of ultrafine slag-based geopolymer mortar for use as repairing mortar. *J. Mater. Civ. Eng.* **2017**, *29*, 04016292. [CrossRef]

10. Rezaiee-Pajand, M.; Abad, J.M.N.; Karimipour, A.; Rezaiee-Pajand, A. Propose new implement models to determine the compressive, tensile and flexural strengths of recycled coarse aggregate concrete via Imperialist Competitive Algorithm. *J. Build. Eng.* **2021**, *40*, 102337. [CrossRef]
11. Hansen, T.C.; Narud, H. Strength of recycled concrete made from crushed concrete coarse aggregate. *Concr. Int.* **1983**, *5*, 79–83.
12. Ghalehnovi, M.; Karimipour, A.; Anvari, A.; de Brito, J. Flexural strength enhancement of recycled aggregate concrete beams with steel fibre-reinforced concrete jacket. *Eng. Struct.* **2021**, *240*, 112325. [CrossRef]
13. Karimipour, A.; Edalati, M.; de Brito, J. Influence of magnetized water and water/cement ratio on the properties of untreated coal fine aggregates concrete. *Cem. Concr. Compos.* **2021**, *122*, 104121. [CrossRef]
14. Karimipour, A.; Rakhshanimerhr, M.; Ghalehnovi, M.; de Brito, J. Effect of different fibre types on the structural performance of recycled aggregate concrete beams with spliced bars. *J. Build. Eng.* **2021**, *37*, 102090. [CrossRef]
15. Amer, A.A.M.; Ezziane, K.; Bougara, A.; Adjoudj, M.H. Rheological and mechanical behavior of concrete made with pre-saturated and dried recycled concrete aggregates. *Constr. Build. Mater.* **2016**, *123*, 300–308. [CrossRef]
16. Borujerdi, A.S.; Mostofinejad, D.; Hwang, H.J. Cyclic loading test for shear-deficient reinforced concrete exterior beam-column joints with high-strength bars. *Eng. Struct.* **2021**, *237*, 112140. [CrossRef]
17. Karimipour, A.; Ghalehnovi, M. Influence of steel fibres on the mechanical and physical performance of self-compacting concrete manufactured with waste materials and fillers. *Constr. Build. Mater.* **2021**, *267*, 121806. [CrossRef]
18. Bernardo, L.F.A.; Filho, B.M.V.C.; Horowitz, B. Efficient softened truss model for prestressed steel fiber concrete membrane elements. *J. Build. Eng.* **2021**, *40*, 102363. [CrossRef]
19. Karimipour, A.; de Brito, J.; Edalati, M. Influence of polypropylene fibres on the thermal and acoustic behaviour of untreated coal coarse aggregates concrete. *J. Build. Eng.* **2021**, *37*, 102125. [CrossRef]
20. Tretyakov, A.; Tkalenko, I.; Wald, F. Fire response model of the steel fibre reinforced concrete filled tubular column. *J. Constr. Steel Res.* **2021**, *186*, 106884. [CrossRef]
21. Karimipour, A.; Edalati, M.; de Brito, J. Biaxial mechanical behaviour of polypropylene fibres reinforced self-compacting concrete. *Constr. Build. Mater.* **2021**, *278*, 122416. [CrossRef]
22. Yue, J.G.; Wang, Y.N.; Beskos, D.E. Uniaxial tension damage mechanics of steel fiber reinforced concrete using acoustic emission and machine learning crack mode classification. *Cem. Concr. Compos.* **2021**, *123*, 104205. [CrossRef]
23. Karimipour, A.; de Brito, J. Influence of polypropylene fibres and silica fume on the mechanical and fracture properties of ultra-high-performance geopolymer concrete. *Constr. Build. Mater.* **2021**, *283*, 122753. [CrossRef]
24. Gao, D.; Huang, Y.; Yuan, J.; Gu, Z. Probability distribution of bond efficiency of steel fiber in tensile zone of reinforced concrete beams. *J. Build. Eng.* **2021**, *43*, 102550. [CrossRef]
25. Karimipour, A.; Ghalehnovi, M.; de Brito, J. Effect of micro polypropylene fibres and nano TiO2 on the fresh- and hardened-state properties of geopolymer concrete. *Constr. Build. Mater.* **2021**, *300*, 124239. [CrossRef]
26. Niu, D.; Jiang, L.; Bai, M.; Miao, Y. Study of the performance of steel fibre reinforced concrete to water and salt freezing condition. *Mater. Des.* **2013**, *44*, 267–273. [CrossRef]
27. Olivito, R.S.; Zuccarello, F.A. An experimental study on the tensile strength of steel fibre reinforced concrete. *Compos. Part B Eng.* **2010**, *41*, 246–255. [CrossRef]
28. Köksal, F.; Şahin, Y.; Gencel, O.; Yiğit, İ. Fracture energy-based optimization of steel fibre reinforced concretes. *Eng. Fract. Mech.* **2013**, *107*, 29–37. [CrossRef]
29. Ameryan, A.; Ghalehnovi, M.; Rashki, M. Investigation of shear strength correlations and reliability assessments of sandwich structures by kriging method. *Compos. Struct.* **2020**, *253*, 112782. [CrossRef]
30. Remennikov, A.M.; Kong, S.Y. Numerical simulation and validation of impact response of axially-restrained steel–concrete–steel sandwich panels. *Compos. Struct.* **2012**, *94*, 3546–3555. [CrossRef]
31. Sohel, K.M.A.; Liew, J.Y.R.; Yan, J.B.; Zhang, M.H.; Chia, K.S. Behavior of steel-concrete-steel sandwich structures with lightweight cement composite and novel shear connectors. *Compos. Struct.* **2012**, *94*, 3500–3509. [CrossRef]
32. Hopkins, P.M.; Norris, T.; Chen, A. Creep behavior of insulated concrete sandwich panels with fiber-reinforced polymer shear connectors. *Compos. Struct.* **2017**, *172*, 137–146. [CrossRef]
33. Oduyemi, T.O.S.; Wright, H.D. An experimental investigation into the behaviour of double skin sandwich beams. *J. Constr. Steel Res.* **1989**, *14*, 197–220. [CrossRef]
34. Bergan, P.G.; Bakken, K. Sandwich design: A solution for marine structures. In Proceedings of the International Conference on Computational Methods in Marine Engineering—Eccomas Marine, Oslo, Norway, 27–29 June 2005.
35. Zuk, W. *Prefabricated Sandwich Panels for Bridge Decks*; Special Report, no. 148; Transportation Research Board: Washington, DC, USA, 1974; pp. 115–121.
36. Bowerman, H.; Gough, M.; King, C. *Bi-Steel Design & Construction Guide*; British Steel Ltd.: Scunthorpe, UK, 1999.
37. Fam, A.; Sharaf, T. Flexural performance of sandwich panels comprising polyurethane core and GFRP skins and ribs of various configurations. *Compos. Struct.* **2010**, *92*, 2927–2935. [CrossRef]
38. Huang, C.J.; Chen, S.J.; Leng, Y.B.; Qian, C.H.; Song, X.B. Experimental research on steel-concrete-steel sandwich panels subjected to biaxial tension-compression. *J. Constr. Steel Res.* **2019**, *162*, 105725. [CrossRef]
39. Sohel, K.; Liew, J.R. Steel–concrete–steel sandwich slabs with lightweight core—Static performance. *Eng. Struct.* **2011**, *33*, 981–992. [CrossRef]

40. Yan, J.B.; Zhang, W.; Liew, J.Y.R.; Li, Z.X. Numerical studies on the shear resistance of headed stud connectors in different concrete under Arctic low temperature. *Mater. Des.* **2016**, *112*, 184–196. [CrossRef]
41. Xie, M.; Chapman, J.C. Static and fatigue tensile strength of friction-welded bar–plate connections embedded in concrete. *J. Constr. Steel Res.* **2005**, *61*, 651–673. [CrossRef]
42. Zhang, P.; Cheng, Y.; Liu, J.; Wang, C.; Hou, H.; Li, Y. Experimental and numerical investigations on laser-welded corrugated-core sandwich panels subjected to air blast loading. *Mar. Struct.* **2015**, *40*, 225–246. [CrossRef]
43. Huang, Z.; Liew, J.Y.R. Compressive resistance of steel-concrete-steel sandwich composite walls with J-hook connectors. *J. Constr. Steel Res.* **2016**, *124*, 142–162. [CrossRef]
44. Joseph, J.D.R.; Prabakar, J.; Alagusundaramoorthy, P. Experimental studies on through-thickness shear behaviour of EPS based precast concrete sandwich panels with truss shear connectors. *Compos. Part B Eng.* **2019**, *166*, 446–456. [CrossRef]
45. Sohel, K.; Liew, R.; Alwis, W.; Paramasivam, P. Experimental investigation of low-velocity impact characteristics of steel-concrete-steel sandwich beams. *Steel Compos. Struct.* **2003**, *3*, 289–306. [CrossRef]
46. Yan, J.B.; Liew, J.Y.R.; Zhang, M.H.; Li, Z.X. Punching shear resistance of steel–concrete–steel sandwich composite shell structure. *Eng. Struct.* **2016**, *117*, 470–485. [CrossRef]
47. Anandavalli, N.; Lakshmanan, N.; Rajasankar, J.; Parkash, A. Numerical studies on blast loaded steel-concrete composite panels. *J. Community Engagem. Scholarsh.* **2012**, *1*, 102–108.
48. Eom, T.S.; Park, H.G.; Lee, C.H.; Kim, J.H.; Chang, I.H. Behavior of double skin composite wall subjected to in-plane cyclic loading. *J. Struct. Eng.* **2009**, *135*, 1239–1249. [CrossRef]
49. The European Union. *Eurocode 4: Design of Composite Steel and Concrete Structures Part 1.1: General Rules and Rules for Buildings*; BS EN 2004-1-1; The European Union: Maastricht, The Netherlands, 2004.
50. ACI Committee 318. *Building Code Requirement for Structural Concrete and Commentary*; ACI 318-08; American Concrete Institute: Farmington Hills, MI, USA, 2008.
51. Yan, J.B.; Qian, X.; Liew, J.Y.R.; Zong, L. Damage plasticity-based numerical analysis on steel-concrete-steel sandwich shells used in the Arctic offshore structure. *Eng. Struct.* **2016**, *117*, 542–559. [CrossRef]
52. Liew, J.R.; Sohel, K. Lightweight steel-concrete-steel sandwich system with J-hook connectors. *Eng. Struct.* **2009**, *31*, 1166–1178. [CrossRef]
53. Yan, J.B.; Liew, J.Y.R.; Qjan, X.; Wang, J.Y. Ultimate strength behaviour of curved steel-concrete-steel sandwich composite beams. *J. Constr. Steel Res.* **2015**, *115*, 316–328. [CrossRef]
54. Leekitwattana, M. Analysis of an Alternative Topology for Steel-Concrete-Steel Sandwich Beams Incorporating Inclined Shear Connectors. Ph.D. Thesis, University of Southampton, Southampton, UK, 2011.
55. Yousefi, M.; Ghalehnovi, M. Push-out test on the one end welded corrugated-strip connectors in steel-concrete-steel sandwich structure. *Steel Compos. Struct.* **2017**, *24*, 23–35. [CrossRef]
56. Yousefi, M.; Ghalehnovi, M. Finite element model for interlayer behavior of double skin steel-concrete-steel sandwich structure with corrugated-strip shear connectors. *Steel Compos. Struct.* **2018**, *27*, 123–133. [CrossRef]
57. ASTM International. *Standard Specification for Pressure Vessel Plates, Alloy Steel, High-Strength, Quenched and Tempered*; A517/A5M; ASTM International: West Conshohocken, PA, USA, 2017.
58. ASTM International. *Standard Test Method for Flexural Strength of Concrete (Using Simple Beam with Centre-Point Loading)*; ASTM C293-08; ASTM International: West Conshohocken, PA, USA, 2008.
59. Meghdadaian, M.; Ghalehnovi, M. Improving seismic performance of composite steel plate shear walls containing openings. *J. Build. Eng.* **2018**, *21*, 336–342. [CrossRef]
60. Meghdadian, M.; Gharaei-Moghaddam, N.; Arabshahi, A.; Mahdavi, N.; Ghalehnovi, M. Proposition of an equivalent reduced thickness for composite steel plate shear walls containing an opening. *J. Constr. Steel Res.* **2020**, *168*, 105985. [CrossRef]
61. Asadi Shamsabadi, E.; Ghalehnovi, M.; De Brito, J.; Khodabakhshian, A. Performance of concrete with waste granite powder: The effect of superplasticizers. *Appl. Sci.* **2018**, *8*, 1808. [CrossRef]
62. British Standards Institution. *Testing Hardened Concrete: Shape, Dimensions and Other Requirements for Specimens and Moulds*; BS EN 12390 1; British Standards Institution: London, UK, 2000.
63. British Standards Institution. *Testing Hardened Concrete: Compressive Strength of Test Specimens*; BS EN 12390-3; British Standards Institution: London, UK, 2009.
64. British Standards Institution. *Testing Hardened Concrete: Making and Curing Specimens for Strength Tests*; BS EN 12390-2; British Standards Institution: London, UK, 2000.
65. Chaboki, H.R.; Ghalehnovi, M.; Karimipour, A.; de Brito, J. Experimental study on the flexural behaviour and ductility ratio of steel fibres coarse recycled aggregate concrete beams. *Constr. Build. Mater.* **2018**, *186*, 400–422. [CrossRef]
66. Chaboki, H.R.; Ghalehnovi, M.; Karimipour, A.; de Brito, J.; Khatibinia, M. Shear behaviour of concrete beams with recycled aggregate and steel fibres. *Constr. Build. Mater.* **2019**, *204*, 809–827. [CrossRef]
67. Liew, J.Y.R.; Yan, J.B.; Sohel, K.M.A.; Zhang, M.H. Push-out tests on J-hook connectors in steel–concrete–steel sandwich structure. *Mater. Struct.* **2013**, *47*, 1693–1714. [CrossRef]
68. An, L.; Cederwall, K. Push-out tests on studs in high strength and normal strength concrete. *J. Constr. Steel Res.* **1996**, *36*, 15–29. [CrossRef]

69. Lorenc, W.; Kubica, E. Behaviour of composite beams prestressed with external tendons: Experimental study. *J. Constr. Steel Res.* **2006**, *62*, 1353–1366. [CrossRef]
70. Xu, W.; Ding, M.; Wang, H.; Luo, Z. Static behaviour and theoretical model of stud shear connectors. *J. Bridge. Eng.* **2008**, *13*, 623–634. [CrossRef]
71. Jahangir, H.; Khatibinia, M.; Kavousi, M. Application of Contourlet transform in damage localization and severity assessment of prestressed concrete slabs. *J. Soft Comput. Civ. Eng.* **2021**, *5*, 39–67. [CrossRef]
72. Jahangir, H.; Eidgahee, D.R. A new and robust hybrid artificial bee colony algorithm—ANN model for FRP-concrete bond strength evaluation. *Compos. Struct.* **2020**, *10*, 87–93. [CrossRef]
73. Jahangir, H.; Bagheri, M. Evaluation of seismic response of concrete structures reinforced by shape memory alloys. *Int. J. Eng.* **2020**, *33*, 410–418. [CrossRef]
74. Bagheri, M.; Chahkandi, A.; Jahangir, H. Seismic reliability analysis of RC frames rehabilitated by glass fiber-reinforced polymers. *Int. J. Civ. Eng.* **2019**, *17*, 1785–1797. [CrossRef]

 materials

*Article*

# Physical and Mechanical Performance of Coir Fiber-Reinforced Rendering Mortars

**Cinthia Maia Pederneiras [1,2], Rosário Veiga [2] and Jorge de Brito [1,*]**

[1] CERIS, Instituto Superior Técnico, University of Lisbon, 1049-001 Lisbon, Portugal; cinthiamaia@tecnico.ulisboa.pt
[2] National Laboratory for Civil Engineering, 1700-066 Lisbon, Portugal; rveiga@lnec.pt
* Correspondence: jb@civil.ist.utl.pt

**Abstract:** Coir fiber is a by-product waste generated in large scale. Considering that most of these wastes do not have a proper disposal, several applications to coir fibers in engineering have been investigated in order to provide a suitable use, since coir fibers have interesting properties, namely high tensile strength, high elongation at break, low modulus of elasticity, and high abrasion resistance. Currently, coir fiber is widely used in concrete, roofing, boards and panels. Nonetheless, only a few studies are focused on the incorporation of coir fibers in rendering mortars. This work investigates the feasibility to incorporate coir fibers in rendering mortars with two different binders. A cement CEM II/B-L 32.5 N was used at 1:4 volumetric cement to aggregate ratio. Cement and air-lime CL80-S were used at a volumetric ratio of 1:1:6, with coir fibers were produced with 1.5 cm and 3.0 cm long fibers and added at 10% and 20% by total mortar volume. Physical and mechanical properties of the coir fiber-reinforced mortars were discussed. The addition of coir fibers reduced the workability of the mortars, requiring more water that affected the hardened properties of the mortars. The modulus of elasticity and the compressive strength of the mortars with coir fibers decreased with increase in fiber volume fraction and length. Coir fiber's incorporation improved the flexural strength and the fracture toughness of the mortars. The results emphasize that the cement-air-lime based mortars presented a better post-peak behavior than that of the cementitious mortars. These results indicate that the use of coir fibers in rendering mortars presents a potential technical and sustainable feasibility for reinforcement of cement and cement-air-lime mortars.

**Keywords:** vegetable fiber; fiber-reinforced mortar; cement and cement-lime mortars; sustainability; render

**Citation:** Maia Pederneiras, C.; Veiga, R.; de Brito, J. Physical and Mechanical Performance of Coir Fiber-Reinforced Rendering Mortars. *Materials* **2021**, *14*, 823. https://doi.org/10.3390/ma14040823

Academic Editor: Lizhi Sun
Received: 15 January 2021
Accepted: 5 February 2021
Published: 9 February 2021



## 1. Introduction

Agricultural waste has been considered an environmental issue. Coir fiber is a by-product waste of the production of other coconut products [1,2], and the world production is approximately 250,000 tonnes a year [3]. In order to provide a proper disposal, many researchers seek different approaches to use the coir waste fiber. Concerning engineering applications, coir has been incorporated in concrete, roofing, boards, panels, and others building materials [4–10]. Coir fibers are extracted from between the outer husk of coconut and the internal shell [11], and their physical and mechanical properties are seen as great potential to improve the ductility, flexural toughness, and energy absorption capacity of the composites. The high toughness and flexibility of these fibers offer a better post-cracking behavior of the reinforced composites.

Regarding the incorporation of coir fibers in mortars, a few studies were carried out with the purpose of enhance their cracking performance [1,12–17]. In previous studies, the use of coir fiber was investigated in mortars with cement as the only binder (further referred as cement mortars) and in mortars using more than one binder. In general, the authors have reported that the addition of coir fibers decreased the workability of the mortars; thus, it was necessary to add more mixing water when compared to that of the mortar without fibers. This effect is attributed to the high water absorption and the retentive nature of the coir fibers [6]. A reduction in the mortar's density was found by increasing the coir fibers content and water/binder ratio.

Concerning the compressive strengths, there are contradictory outcomes presented in previous works. For cement mortars, Hwang et al. [1] found that the incorporation of coir fibers decreased their compressive strength when increasing the fiber content. The authors attributed this reduction to the fibers clustering inside the matrix; an increase of the volume of voids inside the composite was observed as a high volume fraction of coir fibers was added, which indicates a more porous structure. On the other hand, other researchers found an increase in compressive strength with the incorporation of coir fibers in cement mortars [12,15]. Al-Zubaidi [15] attributed this effect to the distribution of stresses by the fibers. In what concerns the addition of coir fiber in cement-lime mortars, Sathiparan et al. [14] found that the compressive strength of the mortars increased with the volume fraction of coir fibers up to 0.5%, whereas a higher content of coir fiber decreased the compressive strength of the mortar when compared to the reference mortar.

With regard to flexural properties, it is well known that the addition of fibers increases the flexural strength, fracture toughness, and ductility of the mortars. Hwang et al. [1] reported a significant improvement in flexural behavior of the coir fiber-reinforced cement-based mortars. The flexural strength increased by increasing the fiber content. The explanation given by the authors was that the fibers distribute the stresses before rupture. Additionally, the coir fibers surface seems rough, which provides a better interfacial adhesion between the fiber and the cementitious matrix. Andiç-Çakir et al. [12] also found an increase in flexural strength with increasing of the coir fiber amount in cement mortars.

For cement-lime mortars, the incorporation of coir fibers in the studies of Sathiparan et al. [14] presented improvements in flexural strength up to 0.5% of addition. The authors reported that the incorporation of 0.75% coir fiber decreased by 16.5% the flexural strength when compared to the reference mortar.

Previous studies have found an enhancement in mortar's toughness as increasing the addition of coir fibers [1,12,14]. Sathiparan et al. [14] reported that the flexural toughness indices of the cement-lime mortars were significantly higher than that of the control mortar, which reveals a higher energy absorption during post-peak. The authors calculated the toughness indices based on the total area under the load-deflection curves from the flexural strength test. Therefore, the authors verified that the flexural ductility of the mortars with coir fibers have increased by increasing the coir fiber content, i.e., the fiber-reinforced mortars showed a more ductile failure mode when compared to that of the mortars without fibers. Regarding the incorporation of coir fibers in cement-based mortars, they also presented improvements in terms of residual strength, ductility and toughness. Hwang et al. [1] found that the increase of coir fibers content presented a significant increase in mortars' toughness. This improvement was associated to the bridging mechanism of the fibers, which transfers the stress in the matrix across the opening cracks and withstands a residual load after achieving the maximum load. The results found by Andiç-Çakir et al. [12] also presented a remarkable increase in toughness values, which can be attributed to the bridge phenomena and interface between fibers and the matrix.

Cracking behavior of the mortars with vegetable fibers was also investigated in previous works [1,12,14,16]. It was clear that the effectiveness of the fibers in led to reducing the mortars' shrinkage. Toledo Filho et al. [16] evaluated the free, restrained, and drying shrinkage of coir fiber-reinforced cement mortars. The authors reported that the addition of vegetable fibers delayed the first crack opening and crack propagation in the mortars. This effect is mainly due to the bridging mechanism of the fibers across the cracks. Hwang et al. [1] and Sathiparan et al. [14] also found that the incorporation of coir fibers contributes to control the cracking opening and its propagation, due to the stresses distribution by the fibers. As a result, conversely to the mortars without fibers that present a brittle failure, the modified mortars present a ductile behavior and a gradual failure.

Notwithstanding these previous research studies, the study of coir fibers in rendering mortars with two different binders has not been found in the literature. Therefore, the aim of this work was to investigate the feasibility of the renders with the addition of coir fibers, in order to minimize cracking by improving the mortars' ductility. Cement and cement-lime mortars with compositions adequate for use as renders were produced and modified by adding 10% and 20% of coir fibers by the total mortar volume. Two common volumetric ratios were used to produce the mortars for render's application following the European Standard for specifications for rendering mortars (EN 998-1 [18]). Cement to aggregate ratio at 1:4 and cement: air-lime: sand at 1:1:6 volumetric ratio. Two fibers' length were chosen based on previous research studies that used coconut fiber as reinforcement in mortars. The coir fiber-reinforced mortars' properties were investigated at fresh and hardened state, and physical and mechanical behavior was evaluated through several tests.

## 2. Experimental Program

### *2.1. Materials*

The objective of the experimental program was to evaluate the physical and mechanical properties of cement and cement-air-lime-based mortars with coir fibers for non-structural uses, namely renders. The materials used in this study are the following:

- Cement (Secil, Portugal): CEM II/B-L 32.5 N, according to EN 197-1 [19];
- Calcium hydrated lime powder-air lime (Calcidrata S.A., Portugal): Class CL80-S, according to EN 459-1 [20];
- Sand (Areipor—Areias Portuguesas S.A., Portugal): Sieved river sand to obtain the size range previously defined;
- Coir fibers (waste from an insulation company—Amorim Cork Insulation, Portugal): With lengths of 1.5 cm and 3.0 cm.

The results of the tensile properties of the coir fibers used in this current work are: tensile strength of 237.26 ± 79.55 MPa, modulus of elasticity of 2.25 ± 1.75 GPa and elongation at break of 11.25%. The coir fiber's water absorption is 115%. Coir fibers consist of cellulose as crystalline microfibrils held together by amorphous lignin and hemicellulose fibrils [21]. In general, plant-based fibers present a similar morphology. The cellulose fibrils packed inside these bundles bonded with lignin forming an unidirectional filament [22].

The grading curve of the sand used in this work is presented in Figure 1. The sand was previously washed and calibrated by the producer. From the technical sheet of the producer, the sand is mainly composed by quartz (>98% silica). The sand was sieved to achieve the previously size distribution chosen. The opening of the sieves were 0.063, 0.15, 0.25, 0.50, 1.00, 1.70, and 2.00 mm.

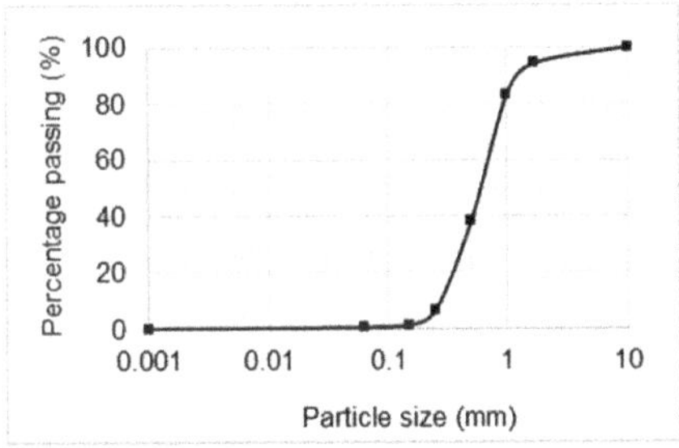


**Figure 1.** Size distribution of the sand used.

The fibers' length was obtained by manually cutting of the waste, which is presented in Figure 2. These fibers were previously washed with neutral detergent, in order to remove any impurities. Before the fibers incorporation in the mix, they were distributed inside a properly closed receptacle by blowing compressed air in order to achieve an adequate dispersion and disentangle the fibers.

**Figure 2.** Coir fibers used in this work.

A microscopic observation was performed using an Olympus SZH-10 optical microscope (Tokyo, Japan) in order to evaluate the fibers' surface and estimate their diameter. Figure 3 presents a micrograph of coir fiber, and the average of coir fibers diameter is 179 ± 3 μm. Therefore, the aspect ratios (length/diameter) were 83.80 and 167.60 for 1.5 and 3.0 cm, respectively.

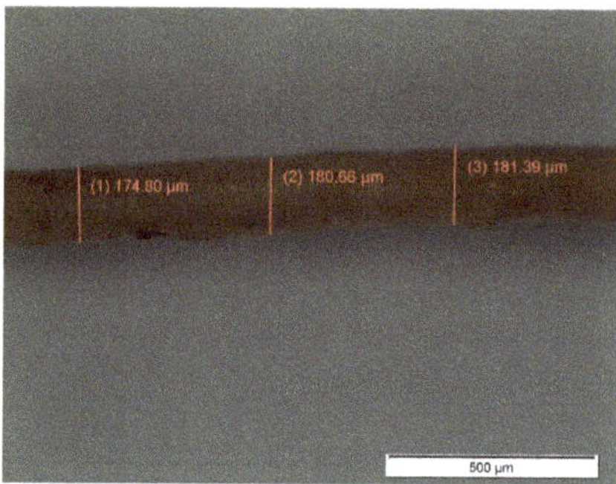


**Figure 3.** Optical microscope image of a coir fiber.

The bulk density of the constituents of the mortars produced is presented in Table 1.

**Table 1.** Bulk density of the constituents.

| Component | Apparent Bulk Density (kg/m$^3$) |
|---|---|
| Cement | 975.5 |
| Air-lime | 565.7 |
| Sand | 1230.8 |
| Coir 1.5 cm | 5.4 |
| Coir 3.0 cm | 2.6 |

### 2.2. Mix Design

The mortars were produced at two volumetric ratios: 1:4 (cement: aggregates) and 1:1:6 (cement: air-lime: aggregates). The water to binder ratio varied according to the amount of mixing water required in each mortar, since the consistency by flow table value was fixed at 140 ± 2 mm, which provides adequate workability for renderings. The composition of the mortars produced in this work is presented in Table 2. The incorporation of the fibers waste was analyzed at two ratios: 10% and 20% by total volume of the mortar;

since the coir fiber bulk density is low, the contents expressed in volume represent a considerable weight of incorporation, as seen in Table 2. Two different lengths were used: 1.5 cm and 3.0 cm.

**Table 2.** Composition of the mortars mixes by mass.

| Mortar | Water (mL) | Cement (g) | Air-Lime (g) | Sand (g) | Coir Fiber (g) | Incorporation |
|---|---|---|---|---|---|---|
| REF 1:4 | 445 | 487.8 | - | 2461.6 | 0 | 0% |
| C 1.5-10c | 415 | 439.1 | - | 2215.4 | 1.4 | 10% of 1.5 cm |
| C 3.0-10c | 430 | 439.1 | - | 2215.4 | 0.7 | 10% of 3.0 cm |
| C 1.5-20c | 370 | 390.2 | - | 1969.3 | 2.7 | 20% of 1.5 cm |
| C 3.0-20c | 400 | 390.2 | - | 1969.3 | 1.3 | 20% of 3.0 cm |
| REF 1:1:6 | 465 | 304.8 | 176.8 | 2307.8 | 0 | 0% |
| C 1.5-10cl | 425 | 274.4 | 159.1 | 2077.0 | 1.4 | 10% of 1.5 cm |
| C 3.0-10cl | 420 | 274.4 | 159.1 | 2077.0 | 0.7 | 10% of 3.0 cm |
| C 1.5-20cl | 396 | 243.9 | 141.4 | 1846.2 | 2.7 | 20% of 1.5 cm |
| C 3.0-20cl | 396 | 243.9 | 141.4 | 1846.2 | 1.3 | 20% of 3.0 cm |

### 2.3. Methods

The standards and number of specimens used for each test performed are listed below. The properties determined in the fresh and hardened mortars tests were:

- Consistency of fresh mortar (by flow table)—EN 1015-3 [23]. Three samples per mortar.
- Bulk density of fresh mortar—EN 1015-6 [24]. Three samples per mortar.
- Dry bulk density of hardened mortar—EN 1015-10 [25], at 28, 90, 180, and 365 days. Three prisms per mortar.
- Flexural strength of hardened mortar—EN 1015-11 [26], at 28, 90, 180, and 365 days. Three prisms per mortar.
- Compressive strength of hardened mortar—EN 1015-11 [26], at 28, 90, 180, and 365 days. Six prisms per mortar.
- Dynamic modulus of elasticity by resonance frequency of hardened mortar—EN 14146 [27], at 28, 90, 180, and 365 days. Three prisms per mortar.
- Ultrasound pulse velocity of hardened mortar—EN 12504-4 [28]. To measure this property, two methods were applied: direct and indirect. In the direct method, the electrodes are on opposite sides of the prisms and, in the indirect method, the electrodes are on the same surface of the prisms. The direct method measures the wave's propagation time between extremities and the indirect method makes the measurements in small increasing distances on the same surface. This test evaluates the mortar's compactness; a lower wave propagation velocity indicates a less compact material, since it means a greater volume of intercepted voids. Three prisms per mortar at 28 days.
- Open porosity—EN 1936 [29]. Three samples per mortar, resulting from the compressive strength test at 28 and 365 days.

Prismatic samples with dimensions of $160 \times 40 \times 40$ mm$^3$ were used for the hardened mortars tests. The mortars were cured as specified by EN 1015-11 [26], which establishes that the specimens should be kept inside the molds for two days at a temperature of $20 \pm 2$ °C and a relative humidity of $95 \pm 5\%$. Then, the prisms are demolded and the specimens kept in the same conditions for 5 days inside of plastic bags. After seven days of the mortars production, the specimens were kept in a room with temperature of $20 \pm 2$ °C and $65 \pm 5\%$ of relative humidity, until the day of the test.

## 3. Results and Discussion

### 3.1. Workability

The fresh behavior of the mortars was evaluated through the consistency by flow table test and bulk density. The workability was previously chosen by a proper consistency of the mortar to be applied on vertical surfaces. In order to ensure an adequate workability, an application on a brick was carried out. The consistency of the mortars was previously fixed at 140 ± 5 mm. Therefore, the amount of mixing water needed by each mortar was different. The water to binder ratios are presented in Table 3. It was noticed that the incorporation of coir fibers increased the water content required in order to achieve the intended workability. Nonetheless, it can be seen in Figure 3 that the fibers maintain the mortar agglutinated. The water/binder ratio increased by increasing the fibers length and volume fraction. C 3.0-20c presented the highest increase of about 12% compared to that of the REF 1:4. It is also stressed that longer fibers presented worse workability than the shorter ones. The finding of coir fibers addition in cement-based mortars reducing the workability was also presented by Hwang et al. [1] and Andiç-Çakir et al. [12]. The authors reported that as increasing the fibers volume fraction, the mortars workability decreases due to the fibers clustering.

**Table 3.** Fresh mortars properties.

| Mortar | Water/Binder Ratio | Bulk Density (kg/m$^3$) |
|---|---|---|
| REF 1:4 | 0.91 | 2005 ± 4 |
| C 1.5-10c | 0.94 | 1959 ± 16 |
| C 3.0-10c | 0.97 | 1971 ± 5 |
| C 1.5-20c | 0.94 | 1940 ± 30 |
| C 3.0-20c | 1.02 | 1989 ± 15 |
| REF 1:1:6 | 0.98 | 1999 ± 8 |
| C 1.5-10cl | 0.99 | 1989 ± 18 |
| C 3.0-10cl | 0.98 | 2000 ± 7 |
| C 1.5-20cl | 1.03 | 1986 ± 8 |
| C 3.0-20cl | 1.03 | 1993 ± 8 |

Mortars with 10% of coir fibers using a mix of cement and air-lime as binders presented similar water/binder ratio to the mortar without fibers, i.e., the addition of fibers did not affect the mortars workability. On the other hand, the addition of 20% of coir fibers showed an increase of the water required when compared to that of the REF 1:1:6. The results found in this work are in agreement with the one found in the technical literature. Sathiparan et al. [14] noticed that, as increasing the coir fibers content in cement-lime mortars, the amount of water needed has increased.

The bulk density of the mortars reduced with the addition of coir fibers, regardless of the type of binder used. However, this reduction is more significant for the modified cement-based mortars. This could be due to the lower bulk density of the fibers. Figure 4 shows the bulk density test's preparation.

**Figure 4.** Flow table and bulk density test for coir fiber-reinforced cement mortar.

In short, the incorporation of fibers reduced the mortars' workability. Therefore, a higher mixing water content was used in order to reach an intended flow table value. The mortars with fibers showed an increase in water to binder ratio when compared to the reference mortar, which can affect the hardened properties of the mortars. It was noticed that the workability decreased as the fibers' length and volume fraction increased.

### 3.2. Dry Bulk Density

Figure 5 presents the dry bulk density results of the mortars at 28, 90, 180, and 365 days. A similar trend of the fresh bulk density was observed. The addition of coir fibers reduced the bulk density of the mortars. It was found for C 3.0-10c a decrease of 4.5% in bulk density when compared to the reference mortar at 365 days. Regarding the cement-lime mortars, this reduction was not significant. Hwang et al. [1] also found a reduction in bulk density of the modified mortars, this effect was attributed to the low density of the natural fiber and the higher porosity of the modified mortars.

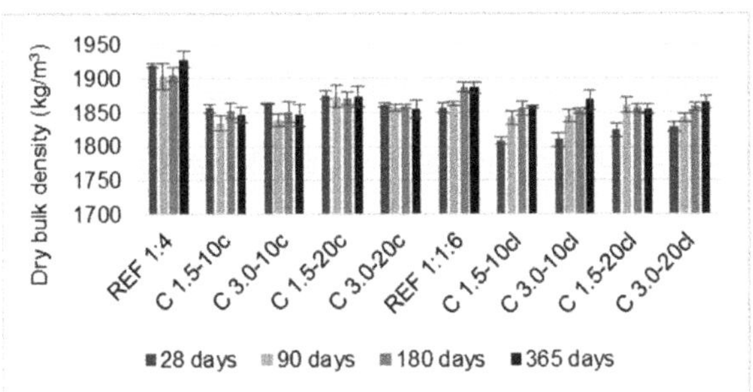


**Figure 5.** Dry bulk density of the hardened mortars.

### 3.3. Dynamic Modulus of Elasticity

The dynamic modulus of elasticity was determined by resonance frequency over time, from 28 days until 365 days, and the results are presented Figure 6. Rendering mortars should be able to deform and accommodate the stresses without cracking. The modulus of elasticity measures the mortar's ability to deform under stress. A low modulus of elasticity indicates that the mortars present a certain deformability, which may prevent cracking. Regardless of the binder used, it was noticed that the incorporation of fibers reduced the modulus of elasticity, which means that the modified mortars may behave in a more deformable way than the control mortar.

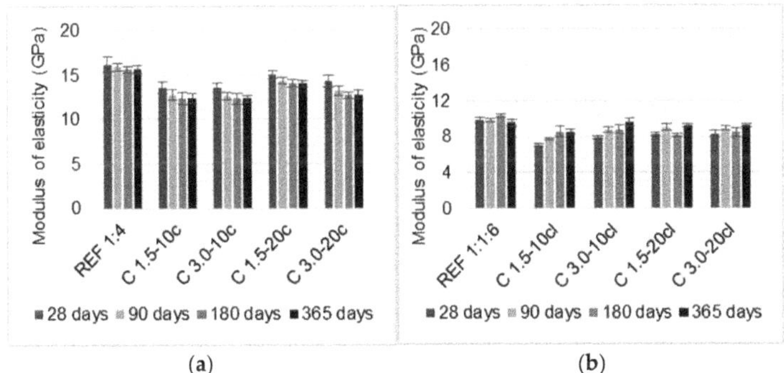


**Figure 6.** Dynamic modulus of elasticity of the: (**a**) cement mortars and (**b**) cement-lime mortars.

At 365 days, cement-based mortar with 10% of 3.0 cm coir fibers presented the highest decrease of approximately 21% when compared to the reference mortar. In what concerns the cement-lime mortars, the fibers' length and volume fraction seemed not to affect this reduction, since all the modified mortars presented similar values among them. The lowest modulus of elasticity was attributed to the C 1.5-10cl sample, which reduced approximately 19.5% compared to that of the REF 1:1:6, at 365 days.

A reduction in the modulus of elasticity with the addition of natural fibers in mortars was also found in the literature. For cement-based mortars, Maia et al. [30] reported a decrease in modulus of elasticity when natural sheep's wool fibers were added. Sathiparan et al. [14] also noticed a reduction in modulus of elasticity when coir fibers were incorporated in cement-lime mortars. Therefore, the authors stated that the coir fiber-reinforced mortars presented a more ductile behavior than that of the control mortar.

### 3.4. Ultra-Sound Pulse Velocity

The ultra-sound pulse velocity of the mortars is presented in Figure 7, at 28 days. Two methods were used: direct and indirect. The ultra-sound pulse velocity results, in both methods, showed that the incorporation of fibers reduced the pulse velocity through the mortar. This result indicates that the modified mortars present a higher volume of pores and suggests a decrease of the mortars modulus of elasticity with the addition of the coir fibers, which is consistent with the results of the resonance frequency test.

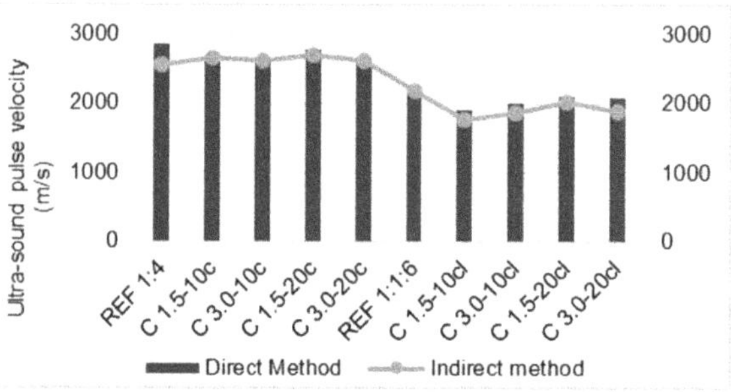


**Figure 7.** Ultra-sound pulse velocity of the mortars.

A reduction in ultra-sound pulse velocity was also observed by Maia et al. [30]. The authors incorporated sheep's wool fibers in cement and cement-lime mortars, and, regardless of the binder used, the natural fiber-reinforced mortars presented less compactness when compared to the reference mortar. These results corroborate the reduction in modulus of elasticity, since the modified mortars may deform more than the mortars without fibers.

### 3.5. Compressive and Flexural Strengths

Compressive and flexural strengths tests were performed at 28, 90, 180, and 365 days, and the results are presented in Figures 8 and 9.

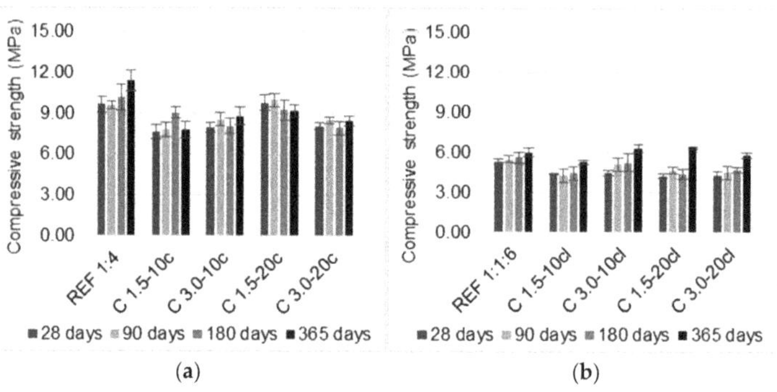


**Figure 8.** Compressive strength of the mortars: (**a**) cement mortars and (**b**) cement-lime mortars.

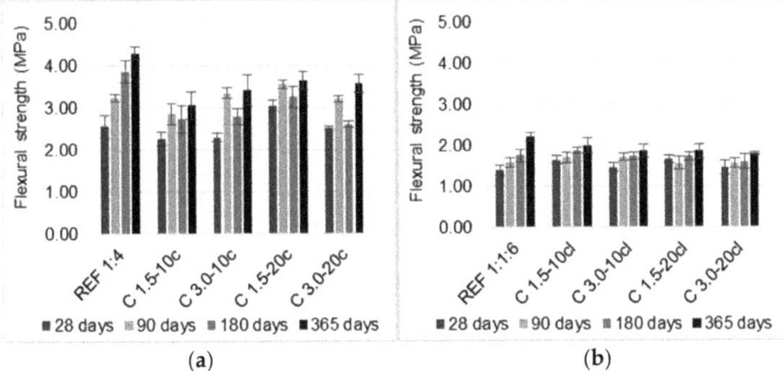


**Figure 9.** Flexural strength of the mortars: (**a**) cement mortars and (**b**) cement-lime mortars.

Rendering mortars should not exhibit a high compressive strength, which indicates high rigidity, since a brittle behavior may be more susceptible to cracking. Therefore, these results obtained in this work may be a positive factor for a render. On the other hand, the flexural strength is most requested and should withstand the building movements and thermal variations stresses without cracking. Flexural strength is strongly correlated to other characteristics, such as susceptibility to cracking and adhesive strength of rendering mortars.

For cement-based mortars, the incorporation of coir fibers slightly decreased the compressive strength, in general, with the exception of C 1.5-20c, which presented an increase in compressive strength when compared to the reference mortar until up to 180 days. C 1.5-10c presented the lowest compressive strength, which was a reduction of about 32% when compared to the reference mortar at 365 days. In terms of flexural strength, it can be noticed an increase in flexural strength with the coir addition until 90 days. After that, a reduction is evidenced for all the samples. C 1.5-10c had the major reduction of approximately 30% when compared to the REF 1:4, at 365 days. The volume fraction had more influence in the mechanical strengths than the length of the fibers, since C 1.5-20c and C 3.0-20c presented similar values at 28 and 365 days.

The mortars with cement and air-lime as a binary binder followed the same trend: the addition of coir fibers reduced the mortars' compressive strength in the first ages. However, at 365 days, the C 3.0-10cl and C 1.5-20cl mortars both obtained an increase of 6% compared to the reference mortar. This effect could be due to the improvement in interfacial transition zone between the matrix and the fibers over time, which provides a better distribution of the stresses when submitted to loading.

The major reduction was presented by C 1.5-10cl, which showed a decrease of 12% in relation with the REF 1:1:6, at 365 days. Concerning the flexural strength, the incorporation of coir fibers increased when compared with a reference mortar, at 28 days. This improvement lasted up to 180 days. Modified mortars presented a reduction in flexural strength at 365 days, as increasing the fiber content and fiber length. C 3.0-20cl presented major reduction of 19% compared to the REF 1:1:6. It can be seen that the reduction over time is more relevant for cement-based mortars. This could be due to a higher water to binder ratio in cement-mortars with coir fibers and the fiber's degradation over time. The use of air-lime in the binder improved the fibers performance in terms of flexural strength, since the cement-lime mortar with coir fibers required less water to achieve the intended workability than the reference mortar; thus, the cement-lime mortars presented higher improvements in mechanical strengths than the mortar without fibers.

Figure 10 presents the coir fiber-reinforced cement-based mortars specimens after compressive strength and flexural strength tests.

**Figure 10.** Coir fiber-reinforced cement-based mortars specimens after (**a**) compressive strength test and (**b**) flexural strength test.

In short, the compressive strength of the coir fiber-reinforced mortars showed a decrease. C 1.5-10c and C 1.5-10cl had the lowest compressive strength, a reduction of 32% and 12% compared to the respective reference mortars. Concerning flexural strength, the use of coir fibers enhanced the flexural strength up to 90 days of the cement-based mortars. The flexural strength of the cement-lime mortars showed an improvement until 180 days. It was noticed that, over time, the flexural strength of the modified mortars presented similar values to those of the mortars without fibers. It is stressed that the coir fiber may degrade inside the matrices of cement or cement-lime mortar due to their composition. Therefore, a reduction in the improvements provided by the fibers over time was noticed. A treatment for coir fibers could enhance their performance.

The findings of previous studies present different results based on the fibers lengths and volume fraction. The results presented in the literature referred to 28 days tests. Hwang et al. [1] followed the trend found in the results obtained in this paper, i.e., the authors reported a reduction in compressive strength of cement-based mortars as the coir fibers content increased. Hwang et al. [1] and Andiç-Çakir et al. [12] found, similarly to this research, that the coir fibers incorporation increased the flexural strength of cement-based mortars.

Regarding the cement-lime mortars, Rupasinghe et al. [31] and Sathiparan et al. [14] found that the incorporation of coir fibers increased the compressive strength up to 0.5% of incorporation. Mortars with higher volume fraction reduced the compressive strength. The flexural strength of mortars with 0.5% of coir fraction was 6% higher than the control mortar, whilst the addition of 0.75% reduced the flexural strength around 16.5%. The authors attributed this reduction to the fiber's clustering.

### 3.6. Cracking Behaviour

The cracking behavior of the mortars was evaluated through some parameters proposed by the Center Scientifique et Technique du Bâtiment (CSTB) [32] and by their fracture toughness, which are presented in Table 4. Rendering mortars should dissipate the tensile stresses without cracking, i.e., the mortar's capacity to absorb and accommodate the tensions correlates to its cracking resistance. In order to analyze the cracking susceptibility of the mortars, these parameters may indicate the mortars' ability to resist cracking. The former criterion is based on the ratio between the modulus of elasticity and flexural strength ($E/\sigma_f$). It is well known that a low modulus of elasticity contributes to a better deformation capacity, and a high flexural strength indicates a better mechanical resistance to support the load applied. Consequently, when the $E/\sigma_f$ ratio is high, the tendency of the mortar to crack is greater. The other factor is related to another ratio ($\sigma_f/\sigma_c$), which is between the flexural strength and compressive strength, which suggests the ductility of the material, i.e., when the $\sigma_f/\sigma_c$ ratio is closer to one, the mortar tends to be more ductile. The deformability of the mortar before failure is measured through the ductility of the material. The fracture toughness indicates the mortars ability to absorb energy during failure, and it can be measured through the area under the load-deflection curves from the results of flexural strength at 28 days [33]. These load-deflection curves are presented in Figure 11.

**Table 4.** Parameters related to cracking susceptibility of the mortars tested.

| Mortar | Dynamic Modulus of Elasticity (MPa) | Flexural Strength (MPa) | Compressive Strength (MPa) | $E/\sigma_f$ | $\sigma_f/\sigma_c$ | Fracture Toughness (N·mm) |
|---|---|---|---|---|---|---|
| REF 1:4 | 16,210 | 2.56 | 9.66 | 6332 | 0.27 | 195 |
| C 1.5-10c | 13,560 | 2.27 | 7.64 | 5974 | 0.30 | 225 |
| C 3.0-10c | 13,740 | 2.31 | 7.22 | 5948 | 0.32 | 192 |
| C 1.5-20c | 15,010 | 3.06 | 9.78 | 4905 | 0.31 | 393 |
| C 3.0-20c | 14,360 | 2.52 | 8.09 | 5698 | 0.31 | 217 |
| REF 1:1:6 | 9820 | 1.40 | 5.27 | 7014 | 0.27 | 135 |
| C 1.5-10cl | 7020 | 1.65 | 4.35 | 4255 | 0.38 | 155 |
| C 3.0-10cl | 7860 | 1.47 | 4.44 | 5347 | 0.33 | 149 |
| C 1.5-20cl | 8240 | 1.67 | 4.24 | 4934 | 0.39 | 212 |
| C 3.0-20cl | 8290 | 1.47 | 4.30 | 5639 | 0.34 | 223 |

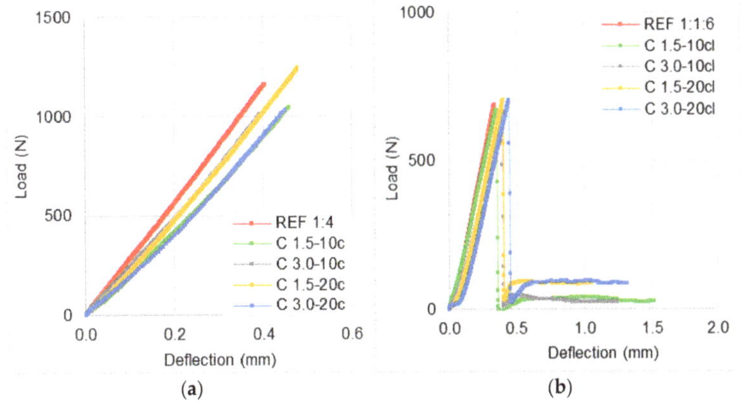


**Figure 11.** Load/deflection curves obtained from flexural strength of the mortars: (**a**) cement mortars and (**b**) cement-lime mortars.

From the results, it was noticed that the addition of coir fibers in mortars, considering these two ratios parameters, has shown a more ductile behavior and less susceptibility to cracking, regardless of the type of binder used. Moreover, it is clear, as expected, that the mortars with cement and air-lime exhibited higher ductility when compared to the cement-based mortars.

Taking into account the fracture toughness values, it can be seen that the modified mortars presented an increase in relation to the REF's. For cement-based mortars, shorter fibers seemed to be more effective in improving the fracture toughness of the mortars. C 1.5-20c attained up to 100% higher toughness values when compared to the reference mortar. On the other hand, for mortars with cement and air-lime this increment was not so high, since the highest value was of C 3.0-20cl, which was 65% higher than that of REF 1:1:6. It was observed that the fracture toughness has increased as increasing the fiber content and fiber length.

To conclude, mortars' toughness exhibited a remarkable enhancement when coir fibers were added. Moreover, according to the parameters analyzed, the coir fiber-reinforced mortars showed a more ductile behavior. Therefore, the mortars with incorporation of coir fibers revealed to be less susceptible to cracking. Furthermore, for cement-lime mortars, a higher volume fraction of fibers resulted in better load-carrying capacity after achieving the maximum peak load.

Other authors also reported this improvement in fracture toughness by adding vegetable fibers in mortars [14,31,34,35]. Xie et al. [36] found that the incorporation of 16 wt.% of rice and bamboo cellulosic fibers increased the fracture toughness in 37 and 45 times, respectively, compared to the mortar without fibers. Rupasinghe et al. [31] and Sathiparan et al. [14] found that the incorporation of coir fibres in cement-lime mortars presented a better performance in terms of fracture toughness, residual strengths, and ductility. The authors reported that the fracture toughness increased with an increase in the volume fraction of coir fibers. The sample with a higher fibers content has increased about 10 times over the control mortar [31].

From the load-deflection curves of the specimens, it can be seen the load-carrying capacity after the maximum peak load achieved. It was observed that the incorporation of coir fibers enhanced the post-peak behavior of the mortars. The cement-based mortars with coir fibers showed a lower decay of load after the mortar achieved its maximum force, since the reference mortars exhibited a brittle failure mode. It is evidenced that the coir fiber-reinforced mortars with cement and air-lime withstand a residual load during failure and sustain greater deformations. It is stressed that a higher volume fraction of fibers presented a better load-carrying capacity.

Other authors also analyzed the load-deflection curves of vegetable fiber-reinforced mortars and verified a similar trend obtained in this work [1,14,35,37]. Hwang et al. [1] reported that the area under the load-deflection curves of the coir fiber-reinforced cement mortars was greater than that of the mortar without fibers, which indicates an increase in toughness indices of the modified mortars. Xie et al. [35] and Benaimeche et al. [37] also followed a similar trend, showing that the addition of vegetable fibers in cement mortars, namely rice, bamboo, and date-mesh palm fibers improved the post-peak behavior by providing a not abrupt failure. This effect was attributed to the bridging mechanism of the fibers across the cracks, which enable the mortars to distribute the stresses in non-brittleness mode [14]. In the study of Pereira et al. [34], it was found that longer sisal fibers exhibited higher fracture toughness than the shorter ones.

It can be concluded that the binders used to produce rendering mortars with coir fibers addition affects their properties. In order to the fibers to be effectively requested to improve the mortars behavior in terms of failure, the modulus of elasticity of the fibers must be compatible with the modulus of elasticity of the mortar. Therefore, it is important to note that the modulus of elasticity of the coir fibers is more similar to the modulus of elasticity obtained in cement and air-lime mortars. For this reason, it is clear that the cement-lime mortars presented a better cracking behavior with coir fibers addition. It is also stressed that the volume fraction of the fibers has more influence than the fibers' length in improving specific properties, namely cracking susceptibility. In order to evaluate the durability of the coir fibers inside the matrix, the tests were performed over time. From the results, it was clear that the improvement of the fiber's addition in mortar's mechanical properties slightly decayed, over time. However, in general, the fibers improved the failure mode of the mortar, increasing their ductility that remains over time.

### 3.7. Open Porosity

The results of the open porosity test are presented in Figure 12. This test measures the volume of interconnected pores inside the mortars, and it is strongly correlated to the modulus of elasticity, ultra-sound pulse velocity, and mechanical strengths. From the results, it can be noticed that the incorporation of coir fibers increased the porosity of the mortars, regardless of the type of the binder used. At 28 days, the C 1.5-20cl and C 3.0-20cl both presented a reduction of the modulus of elasticity of about 16%, and an increase of the open porosity of 25% and 27%, respectively, relative to the reference mortar.

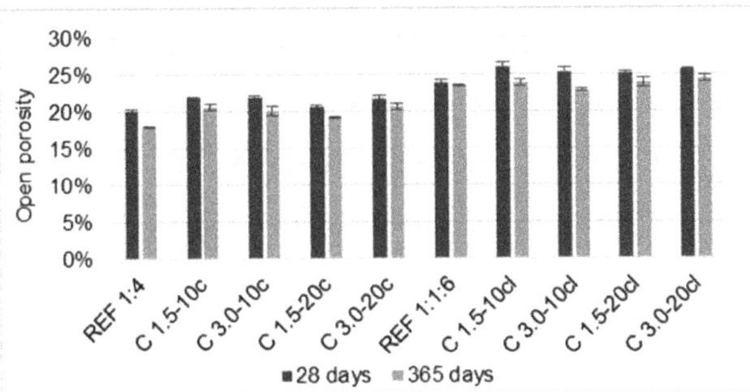


**Figure 12.** Open porosity test results of the mortars tested.

Mortars with longer and higher volume fraction of fibers exhibited higher porosity. At 365 days, C 3.0-20c and C 3.0-20cl increased 15% and 4% over their respective reference mortar. This increase in open porosity of the mortars with fibers is attributable to the fiber-matrix interfacial bond, which may generate some voids inside the matrix due to the fibers clustering. Previous studies are in agreement with those results found in this work [1,14,34]. The incorporation of coir fibers increased the porosity of the mortars. It can be seen that the mortars with higher porosity obtained the lower modulus of elasticity.

## 4. Conclusions

From the results obtained in this study, the following conclusions can be drawn:

- Coir fiber addition reduces the mortars' workability, regardless of the type of binder used. As increasing the fiber length and volume fraction, a higher mixing water content is required to achieve the intended consistency when compared to the reference mortars.
- The mortars with coir fibers presented a more ductile behavior and less susceptibility to cracking than that of the control mortars, since they presented lower modulus of elasticity and higher fracture toughness. The addition of coir fibers also increased the porosity of the mortars due to the fibers' clustering inside the matrix.
- Concerning the mechanical behavior of the mortars, the coir fiber addition improved in the first ages of the mortars. Over time, the coir fibers did not significantly affect their compressive and flexural strengths.

The findings of this current work show that the addition of coir fibers in rendering mortars led to improvements mainly in terms of cracking behavior. Furthermore, it provided an environmentally-friendly and low-cost product. The characterization of the coir fiber-reinforced mortars highlighted that the volume fraction of the fibers and the binder used are the most influencing factors to improve the brittleness of the mortar. A higher fiber content and cement-lime as a binary binder obtained the highest fracture toughness, according to the load-deflection curves.

**Author Contributions:** C.M.P. performed all the tests and established the correlation between all tests results. C.M.P. wrote the manuscript. J.d.B. and R.V. supervised the research and revised the manuscript. All authors have read and agreed to the published version of the manuscript.

**Funding:** This research was funded by Portuguese Foundation for Science and Technology (FCT), grant number PD/BD/135193/2017.

**Data Availability Statement:** The data presented in this study are available on request from the corresponding author.

**Acknowledgments:** The authors acknowledge Instituto Superior Técnico, CERIS Research Centre, the National Laboratory for Civil Engineering and the Portuguese Foundation for Science and Technology for the support given to this research.

**Conflicts of Interest:** The authors declare no conflict of interest.

## References

1. Hwang, C.L.; Tran, V.A.; Hong, J.W.; Hsieh, Y.C. Effects of short coconut fiber on the mechanical properties, plastic cracking behavior, and impact resistance of cementitious composites. *Constr. Build. Mater.* **2016**, *127*, 984–992. [CrossRef]
2. Ghavami, K.; Toledo Filho, R.D.; Barbosa, N.P. Behaviour of composite soil reinforced with natural fibres. *Cem. Concr. Compos.* **1999**, *21*, 39–48. [CrossRef]
3. Kicińska-Jakubowska, A.; Bogacz, E.; Zimniewska, M. Review of Natural Fibers. Part I-Vegetable Fibers. *J. Nat. Fibers* **2012**, *9*, 150–167. [CrossRef]
4. Ali, M.; Liu, A.; Sou, H.; Chouw, N. Mechanical and dynamic properties of coconut fibre reinforced concrete. *Constr. Build. Mater.* **2012**, *30*, 814–825. [CrossRef]
5. Holbery, J.; Houston, D. Natural-fiber-reinforced polymer composites in automotive applications. *JOM* **2006**, *58*, 80–86. [CrossRef]
6. Savastano, H.; Agopyan, V.; Nolasco, A.M.; Pimentel, L. Plant fibre reinforced cement components for roofing. *Constr. Build. Mater.* **1999**, *13*, 433–438. [CrossRef]
7. Syed, H.; Nerella, R.; Madduru, S.R.C. Role of coconut coir fiber in concrete. *Mater. Today Proc.* **2019**, *27*, 1104–1110. [CrossRef]
8. Ahmad, W.; Farooq, S.H.; Usman, M.; Khan, M.; Ahmad, A.; Aslam, F.; Yousef, R.A.; Abduljabbar, H.A.; Sufian, M. Effect of coconut fiber length and content on properties of high strength concrete. *Materials* **2020**, *13*, 1075. [CrossRef]
9. Agopyan, V.; Savastano, H.; John, V.M.; Cincotto, M.A. Developments on vegetable fibre-cement based materials in São Paulo, Brazil: An overview. *Cem. Concr. Compos.* **2005**, *27*, 527–536. [CrossRef]
10. Tolêdo Filho, R.D.; Scrivener, K.; England, G.L.; Ghavami, K. Durability of alkali-sensitive sisal and coconut fibres in cement mortar composites. *Cem. Concr. Compos.* **2000**, *22*, 127–143. [CrossRef]
11. Faruk, O.; Bledzki, A.K.; Fink, H.P.; Sain, M. Biocomposites reinforced with natural fibers: 2000–2010. *Prog. Polym. Sci.* **2012**, *37*, 1552–1596. [CrossRef]
12. Andiç-Çakir, Ö.; Sarikanat, M.; Tüfekçi, H.B.; Demirci, C.; Erdoğan, Ü.H. Physical and mechanical properties of randomly oriented coir fiber-cementitious composites. *Compos. Part B Eng.* **2014**, *61*, 49–54. [CrossRef]
13. Ramakrishna, G.; Sundararajan, T. Impact strength of a few natural fibre reinforced cement mortar slabs: A comparative study. *Cem. Concr. Compos.* **2005**, *27*, 547–553. [CrossRef]
14. Sathiparan, N.; Rupasinghe, M.N.; Pavithra, B.H. Performance of coconut coir reinforced hydraulic cement mortar for surface plastering application. *Constr. Build. Mater.* **2017**, *142*, 23–30. [CrossRef]
15. Al-Zubaidi, A.B. Effect of natural fibers on mechanical properties of green cement mortar. *AIP Conf. Proc.* **2018**, *1968*, 020003. [CrossRef]
16. Toledo Filho, R.D.; Ghavami, K.; Sanjuán, M.A.; England, G.L. Free, restrained and drying shrinkage of cement mortar composites reinforced with vegetable fibres. *Cem. Concr. Compos.* **2005**, *27*, 537–546. [CrossRef]
17. Pederneiras, C.M.; Veiga, R.; Brito, J. De Effects of the Incorporation of Waste Fibres on the Cracking Resistance of Mortars: A Review. *Int. J. Green Technol.* **2018**, *4*, 38–46.
18. European Committee for Standardization. *Specification for Mortar for Masonry—Part 1: Rendering and Plastering Mortar*; EN 998-1; European Committee for Standardization (CEN): Brussels, Belgium, 2010.
19. European Committee for Standardization. *Cement Part 1: Composition, Specifications and Conformity Criteria for Common Cements*; EN 197-1; European Committee for Standardization (CEN): Brussels, Belgium, 2011.
20. European Committee for Standardization. *Building Lime; Part 1: Definitions, Specifications and Conformity Criteria*; EN 459-1; European Committee for Standardization (CEN): Brussels, Belgium, 2015.
21. Bledzki, A.K.; Gassan, J. Composites reinforced with cellulose based fibres. *Prog. Polym. Sci.* **1999**, *24*, 221–274. [CrossRef]
22. Hamzaoui, R.; Guessasma, S.; Mecheri, B.; Eshtiaghi, A.M.; Bennabi, A. Microstructure and mechanical performance of modified mortar using hemp fibres and carbon nanotubes. *Mater. Des.* **2014**, *56*, 60–68. [CrossRef]
23. European Committee for Standardization. *Methods of Test for Mortar for Masonry—PART 3: Determination of Consistence of Fresh Mortar (by Flow Table)*; EN 1015-3; European Committee for Standardization (CEN): Brussels, Belgium, 1999.
24. European Committee for Standardization. *Methods of Test for Mortar for Masonry—Part 6: Determination of Bulk Density of Fresh Mortar*; EN 1015-6; European Committee for Standardization (CEN): Brussels, Belgium, 1998.
25. European Committee for Standardization. *Methods of Test for Mortar for Masonry—Part 10: Determination of Dry Bulk Density of Hardened Mortar*; EN 1015-10; European Committee for Standardization (CEN): Brussels, Belgium, 1999.
26. European Committee for Standardization. *Methods of Test for Mortar for Masonry—Part 11: Determination of Flexural and Compressive Strength of Hardened Mortar*; EN 1015-11; European Committee for Standardization (CEN): Brussels, Belgium, 1999.
27. European Committee for Standardization. *Natural Stone Test Methods. Determination of the Dynamic Elastic Modulus of Elasticity (by Measuring the Fundamental Resonance Frequency)*; EN 14146; European Committee for Standardization (CEN): Brussels, Belgium, 2004.

28. European Committee for Standardization. *Testing Concrete in Structures. Part 4: Determination of Ultrasonic Pulse Velocity*; EN 12504-4; European Committee for Standardization (CEN): Brussels, Belgium, 2007.
29. European Committee for Standardization. *Natural Stone Test Methods. Determination of Real Density and Apparent Density and Total and Partial Open Porosity*; EN 1936; European Committee for Standardization (CEN): Brussels, Belgium, 2007.
30. Pederneiras, C.M.; Veiga, R.; de Brito, J. Rendering mortars reinforced with natural sheep's wool fibers. *Materials* **2019**, *12*, 3648. [CrossRef]
31. Rupasinghe, M.N.; Sathiparan, N. Mechanical behavior of masonry strengthened with coir fiber reinforced hydraulic cement mortar as surface plaster. *J. Struct. Eng. Appl. Mech.* **2019**, *2*, 12–24. [CrossRef]
32. Centre scientifique et technique du bâtiment. *Certification CSTB des Enduits Monocouches D'imperméabilisation, Modalités D'essais*; Cahier 2669-4; Centre scientifique et technique du bâtiment: Marne-la-Vallée, France, 1993.
33. Coutts, R.S.P. Flax fibres as a reinforcement in cement mortars. *Int. J. Cem. Compos. Light. Concr.* **1983**, *5*, 257–262. [CrossRef]
34. Pereira, M.V.; Fujiyama, R.; Darwish, F.; Alves, G.T. On the Strengthening of Cement Mortar by Natural Fibers. *Mater. Res.* **2015**, *18*, 177–183. [CrossRef]
35. Xie, X.; Zhou, Z.; Jiang, M.; Xu, X.; Wang, Z.; Hui, D. Cellulosic fibers from rice straw and bamboo used as reinforcement of cement-based composites for remarkably improving mechanical properties. *Compos. Part B Eng.* **2015**, *78*, 153–161. [CrossRef]
36. Zhang, T.; Dieckmann, E.; Song, S.; Xie, J.; Yu, Z.; Cheeseman, C. Properties of magnesium silicate hydrate (M-S-H) cement mortars containing chicken feather fibres. *Constr. Build. Mater.* **2018**, *180*, 692–697. [CrossRef]
37. Benaimeche, O.; Carpinteri, A.; Mellas, M.; Ronchei, C.; Scorza, D.; Vantadori, S. The influence of date palm mesh fibre reinforcement on flexural and fracture behaviour of a cement-based mortar. *Compos. Part B Eng.* **2018**, *152*, 292–299. [CrossRef]

 materials

MDPI

*Article*

# Long Persistent Luminescent HDPE Composites with Strontium Aluminate and Their Phosphorescence, Thermal, Mechanical, and Rheological Characteristics

**Anesh Manjaly Poulose [1,*], Hamid Shaikh [1], Arfat Anis [1], Abdullah Alhamidi [1], Nadavala Siva Kumar [2], Ahmed Yagoub Elnour [2] and Saeed M. Al-Zahrani [1]**

1 SABIC Polymer Research Center, Department of Chemical Engineering, King Saud University, Riyadh 11421, Saudi Arabia; hamshaikh@ksu.edu.sa (H.S.); aarfat@ksu.edu.sa (A.A.); AKFHK90@hotmail.com (A.A.); szahrani@ksu.edu.sa (S.M.A.-Z.)
2 Department of Chemical Engineering, King Saud University, Riyadh 11421, Saudi Arabia; snadavala@ksu.edu.sa (N.S.K.); aelnour@ksu.edu.sa (A.Y.E.)
* Correspondence: apoulose@ksu.edu.sa

**Abstract:** In this work, HDPE/strontium aluminate-based auto glowing composites ($SrAl_2O_4$: Eu, Dy ($AG_1$) and $Sr_4Al_{14}O_{25}$: Eu, Dy ($AG_2$)) were prepared, and their phosphorescence studies were conducted. In HDPE/$AG_1$ composites, the green emission was observed at ~500 nm after the UV excitation at 320 nm. The HDPE/$AG_2$ has a blue emission at ~490 nm and, in both cases, the intensity of emission is proportional to the $AG_1$ and $AG_2$ content. The DSC data show that the total crystallinity of both the composites was decreased but with a more decreasing effect with the bulky $AG_2$ filler. The melting and crystallization temperatures were intact, which shows the absence of any chemical modification during high shear and temperature processing. This observation is further supported by the ATR-FTIR studies where no new peaks appeared or disappeared from the HDPE peaks. The tensile strength and modulus of HDPE, HDPE/$AG_1$, and HDPE/$AG_2$ composites were improved with the $AG_1$ and $AG_2$ fillers. The rheological studies show the improvement in the complex viscosity and accordingly the storage modulus of the studied phosphorescent HDPE composites. The SEM images indicate better filler dispersion and filler–matrix adhesion, which improves the mechanical characteristics of the studied HDPE composites. The ageing studies in the glowing composites show that there is a decrease in the intensity of phosphorescence emission on exposure to drastic atmospheric conditions for a longer period and the composites become more brittle.

**Keywords:** phosphorescent composites; thermal; mechanical; rheology

**Citation:** Poulose, A.M.; Shaikh, H.; Anis, A.; Alhamidi, A.; Kumar, N.S.; Elnour, A.Y.; Al-Zahrani, S.M. Long Persistent Luminescent HDPE Composites with Strontium Aluminate and Their Phosphorescence, Thermal, Mechanical, and Rheological Characteristics. *Materials* **2022**, *15*, 1142. https://doi.org/10.3390/ma15031142

Academic Editors: Andrea Petrella and Michele Notarnicola

Received: 16 November 2021
Accepted: 28 January 2022
Published: 1 February 2022

**Publisher's Note:** MDPI stays neutral with regard to jurisdictional claims in published maps and institutional affiliations.



## 1. Introduction

The known history of persistent luminescence started at the beginning of the 17th century. In 1602, an Italian shoemaker, V. Casciarolo, observed a solid luminescence from the barite ($BaSO_4$), a mineral known as the Bologna stone. The cause for the luminescent emission was unclear, but it was employed for many applications [1]. In the 20th century, the luminous paints were based on the luminescent emission from Cu- or Mn-doped zinc sulfide (green emission). Their application was later restrained by their shorter afterglow time (~30 min.) and lower brightness. These sulfide phosphors have an affinity towards moisture and $CO_2$ and are chemically unstable [2]. The researchers were in search of phosphorescent material that releases visible light for a longer period even after the excitation source (UV, X-ray, etc.) has been stopped. The detection of strontium aluminate-based rare-earth-doped phosphors initiated the modern luminescent materials era. When compared with sulfide phosphors, these phosphors (e.g., SrAlxOy: Eu, Dy) exhibited better afterglow, brightness, chemical stability, environmental safety, and photo-resistance [3]. These materials gain solar light energy, remain photo-luminescent for a longer time (~16 h), and find applications in roadway displays at night-time, glowing paints, fluorescent lamps, pavements, etc. [4–10]. The application side of rare earth metals doped alkaline-earth metals is rapidly expanding due to the faster growth in the nanotechnology field. Due to its long persistent nature and stability, it has also been utilized in areas such as medical imaging and fluorescent probes [11,12].

After the invention of long-lasting $SrAl_2O_4$:$Eu^{2+}$, $Dy^{3+}$ phosphor in 1996, only a limited number of phosphors were developed to date, which is suitable for practical applications. In recent years, different phases of rare-earth-doped ($Pr^{3+}$, $Ce^{3+}$, $Sm^{2+}$, $Nd^{3+}$, etc.) aluminates were developed, and their emission wavelength depends upon the crystalline structure of the alkaline earth metal aluminate phase [13–18]. Among these phosphors, $SrAl_2O_4$: Eu, Dy and $Sr_4Al_{14}O_{25}$: Eu, Dy have displayed a robust potential for phosphorescence applications and are available commercially [19,20]. These materials undergo hydrolysis in the presence of atmospheric moisture, and the encapsulation process is important to extend the glowing property and is reported in the literature [21–25]. Out of these methods, polymeric encapsulation attracts more due to its low cost, and the process enables the composites to be shaped easily and imparts better physical properties [26–29]. In recent years, research on polymer–phosphor material, composites have gained much attention due to their wide range of potential applications such as safety indication, emergency lighting, road signs, interior decorations, photovoltaics, and optoelectronics [30–32]. In one study, a facile way to synthesize a luminescent polymer nanocomposite of PMMA-($Sr_3B_2O_6$:$Dy^3$) is presented, and detailed structural, optical, and photoluminescence properties were investigated. The nanocomposites films consisting of ($Sr_3B_2O_6$:$Dy^3$) dispersed in polymethyl methacrylate (PMMA) matrix were prepared via solution casting method [31]. In another study, the effect of incorporating different volume loadings (ratios ranging from 0.05% to 5%) of the green-emitting $SrAl_2O_4$ phosphor into both the low-density polyethylene (LDPE) matrix and the (PMMA) matrix was investigated. The composites were produced by the melt-mixing process, and the results showed that LDPE can be used to build a suitable three-dimensional phosphor network for luminescence applications [33]. In recent studies, the incorporation of $SrAl_2O_4$:$Eu^{2+}$, $Dy^{3+}$ powder into polylactic acid (PLA) matrix, as an example of biodegradable polymers with good thermo-plasticity and machinability, was investigated. The results showed that the concentration of $SrAl_2O_4$:$Eu^{2+}$, $Dy^{3+}$ significantly impacts the fluorescence and mechanical properties of resulting composites and that the composites with 15 wt.% have the best fluorescence properties [34]. Furthermore, the modification of the $SrAl_2O_4$:$Eu^{2+}$, $Dy^{3+}$ powder with $SiO_2$ has brought about an improved filler dispersion and compatibility with the (PLA) matrix for 3D printing technologies [26].

Due to its excellent mechanical properties and ease of manufacturing, high-density polyethylene (HDPE) is recognized as one of the most versatile commodity thermoplastics, and it finds a wide acceptance in many industrial applications. The unique properties of HDPE combined with the attractive luminescent features will widen the window of its application in several other interesting areas such as nighttime display boards, sensing, etc. Furthermore, in these reported works the phosphorescence ageing studies exposing the luminescence film to drastic outside temperature conditions are missing. Therefore, in this work, two different strontium aluminate-based phosphorescent materials were melt-mixed with HDPE matrix, and adding to their luminescence, thermal, mechanical, rheological, and morphological studies, the phosphorescence ageing studies were also conducted. The solution casting method makes use of strong solvents, which will inversely affect the mechanical properties of the resultant composites and hence the melt-mixing process is preferred in the present study.

## 2. Materials and Methods

### *2.1. Materials*

High-density polyethylene (TASNEE HD F0455) was supplied by TASNEE, Jubail Industrial City, Saudi Arabia, having a density of 0.957 g/$cm^3$ (ISO 1183). It has a melt flow index (190 °C/5 kg) of 0.40 g/10 min. (ISO 1133). Two strontium aluminate-based phosphors

($SrAl_2O_4$: Eu, Dy (Mw = 209.11 g/mol) ($AG_1$) and $Sr_4Al_{14}O_{25}$: Eu, Dy (1139.55 g/mol) ($AG_2$), were bought from Sigma Aldrich Company, St. Louis, MO, USA.

### 2.2. *Methods*

#### 2.2.1. Composites Preparation

The strontium aluminate phosphors $AG_1$ and $AG_2$ (1, 3, 5, and 10 wt.%) were melt-blended with the HDPE matrix in DSM Xplore micro-compounder, Sittard, Netherlands, at a temperature of 200 °C at 100 rpm for a mixing time of 3 min. Thin films with an average thickness of 0.5 mm were prepared with the help of COLLIN Press, Maitenbeth, Germany (100 bar pressure, 200 °C), for the phosphorescence measurements. Dumb-bell-shaped tensile testing specimens ASTM, Type1, were prepared with a microinjection molding machine, DSM Xplore, Sittard, Netherlands (12 $cm^3$). The dumb-bell specimens were shaped in a mold maintained at 6 bar pressure and at room temperature.

#### 2.2.2. Composites Characterization

The thermal analysis of the samples (DSC) was carried out in Shimadzu DSC-60A, Kyoto, Japan, taking 5–10 mg material in an aluminum pan, and the temperature program was from 30 to 220 °C at 10 °C/min heating and cooling rate with 4 min. holding time in the melt.

The crystallinity percent was obtained as:

$$X_c\% = \frac{\Delta H_m}{(1-\Phi)\Delta H_m^{\circ}}$$

where ($\Phi$) is the weight fraction of filler in the composites, ($\Delta H_m$) is the enthalpy of melting obtained from the DSC melting peak, and ($\Delta H_m^{\circ}$) is the theoretical value of 100% crystalline HDPE, which is 293 J/g [35].

The chemical composition of prepared composites was investigated in an Attenuated Total Reflection-Fourier Transform Infrared Spectroscopy (ATR-FTIR), from Thermo-Scientific, Waltham, MA USA, Nicolet iN10 model with a germanium micro-tip. The FTIR scanning range was between 400–4000 $cm^{-1}$.

The morphology and dispersion of the filler particles were studied using SEM. A scanning Electron Microscope (SEM), JEOL JSM-6360A, (Tokyo, Japan), at an accelerating voltage of 20 kV, was used for these analyses. The samples for SEM examination were cryo-fractured after exposure to liquid nitrogen. The elemental detection was carried out with the help of the attached energy-dispersive X-ray spectroscopy (EDS) facility.

Tensile testing was carried out in a Hounsfield H100 KS (Salfords, UK) universal testing machine at a crosshead speed of 10 mm/min according to the ASTM D638 standard testing method, and an average of five tests is reported.

The rheological properties of HDPE/$AG_1$ and HDPE/$AG_2$ samples were carried out in a TA instruments, New Castle, DE, USA, ARG2 model, with parallel-plate geometry. The experimental temperature was set at 190 °C so that the composites were in a melt state, and the gap between the parallel plate was kept at 1000 μm for the analysis. The angular frequency sweep tests were performed from 0.01 to 628.3 rad/s under an oscillation of 3.259 Pa.

Phosphorescence measurements were performed on an Agilent Technologies Fluorescence Spectrophotometer, Santa Clara, CA, USA equipped with a Xe lamp as a source of UV. The wavelength of excitation used was 320 nm, and the emission spectra from these composites were monitored in the visible range.

#### 2.2.3. Phosphorescence Intensity vs. Time in HDPE/$AG_1$ and HDPE/$AG_2$ Composites

The phosphorescence decay in these composites was carried out in an Agilent Technologies Fluorescence Spectrophotometer with the delay time and gate time as 0.1 and 10,000 ms. The excitation and emission wavelength is chosen were 320 and 490 nm, respectively, and the decay in the intensity of emission was monitored for a period of 1400s.

2.2.4. Effect of Ageing on Phosphorescence and Mechanical Properties

HDPE/3AG$_1$ composites films and standard dumb-bell specimens were chosen for the ageing studies (tensile and phosphorescence) and were exposed to outside climatic conditions for 40 days (August-September; Riyadh, Saudi Arabia). The measurements of phosphorescence emission intensity and mechanical properties were monitored in a time interval of 10 days. The ageing phenomenon is mainly influenced by the exposed temperature and humidity. The average day and nighttime temperatures for every 10 days were 42.9, 43.1, 41.6, and 38.4 °C and 28.3, 27.6, 27.6, and 24.5 °C; respectively. The average humidity value was around 10% during that period.

## 3. Results and Discussion

### 3.1. Particles and Composites SEM Analysis

The SEM analysis of AG$_1$ and AG$_2$ powders are shown in Figure 1A,B, respectively. While that of HDPE/10AG$_1$ and HDPE/10AG$_2$ composites along with the EDS elemental analysis are shown in Figure 1C–E, respectively. From Figure 1A,B, the SEM images of powder samples, it can be seen that both samples have irregular (random) particle shapes with particle sizes ranging between 5 to 150 μm approximately.

Though agglomeration is visible on the SEM images in the highest filler loading composites (Figure 1C,D), the dispersion of the filler and the adhesion among fillers and HDPE is good enough to improve the tensile and storage modulus of the studied composites. The EDS analysis is carried out on the particle visible on the SEM images, and the analysis shows the presence of elements such as Sr, Al, and $O_2$, confirming the composition of the AG$_1$ fillers, as in Figure 1E.

### 3.2. DSC and FTIR Data for HDPE/AG$_1$ and HDPE/AG$_2$ Composites

The DSC results for the HDPE matrix and the composites are shown in Tables 1 and 2. The AG$_1$ and AG$_2$ filler do not affect the melting temperature of HDPE, showing that the mixing process is purely physical, and this observation is also supported by the ATR-FTIR data. As shown in Figure 2A,B, the HDPE and the composites with AG$_1$ and AG$_2$ have similar FTIR peaks, confirming the absence of any chemical modification of HDPE during the high-temperature mixing process.

**Table 1.** DSC results on HDPE and HDPE/AG$_1$ composites.

| Material | $T_c$ (°C) | $T_m$ (°C) | $\Delta H_m$ (J/g) | $X_c$ (%) |
|---|---|---|---|---|
| HDPE | 111.1 | 134.4 | 190.2 | 64.9 |
| HDPE/1AG$_1$ | 110.4 | 135.0 | 181.4 | 61.9 |
| HDPE/3AG$_1$ | 111.4 | 134.1 | 173.6 | 59.2 |
| HDPE/5AG$_1$ | 110.2 | 134.6 | 172.1 | 58.7 |
| HDPE/10AG$_1$ | 110.0 | 135.0 | 163.1 | 55.7 |

**Table 2.** DSC results on HDPE and HDPE/AG$_2$ composites.

| Material | $T_c$ (°C) | $T_m$ (°C) | $\Delta H_m$ (J/g) | $X_c$ (%) |
|---|---|---|---|---|
| HDPE | 111.1 | 134.4 | 190.2 | 64.9 |
| HDPE/1AG$_2$ | 111.5 | 134.3 | 174.1 | 59.4 |
| HDPE/3AG$_2$ | 110.1 | 135.4 | 160.4 | 54.7 |
| HDPE/5AG$_2$ | 110.5 | 134.5 | 160.0 | 54.6 |
| HDPE/10AG$_2$ | 110.6 | 134.3 | 154.1 | 52.6 |

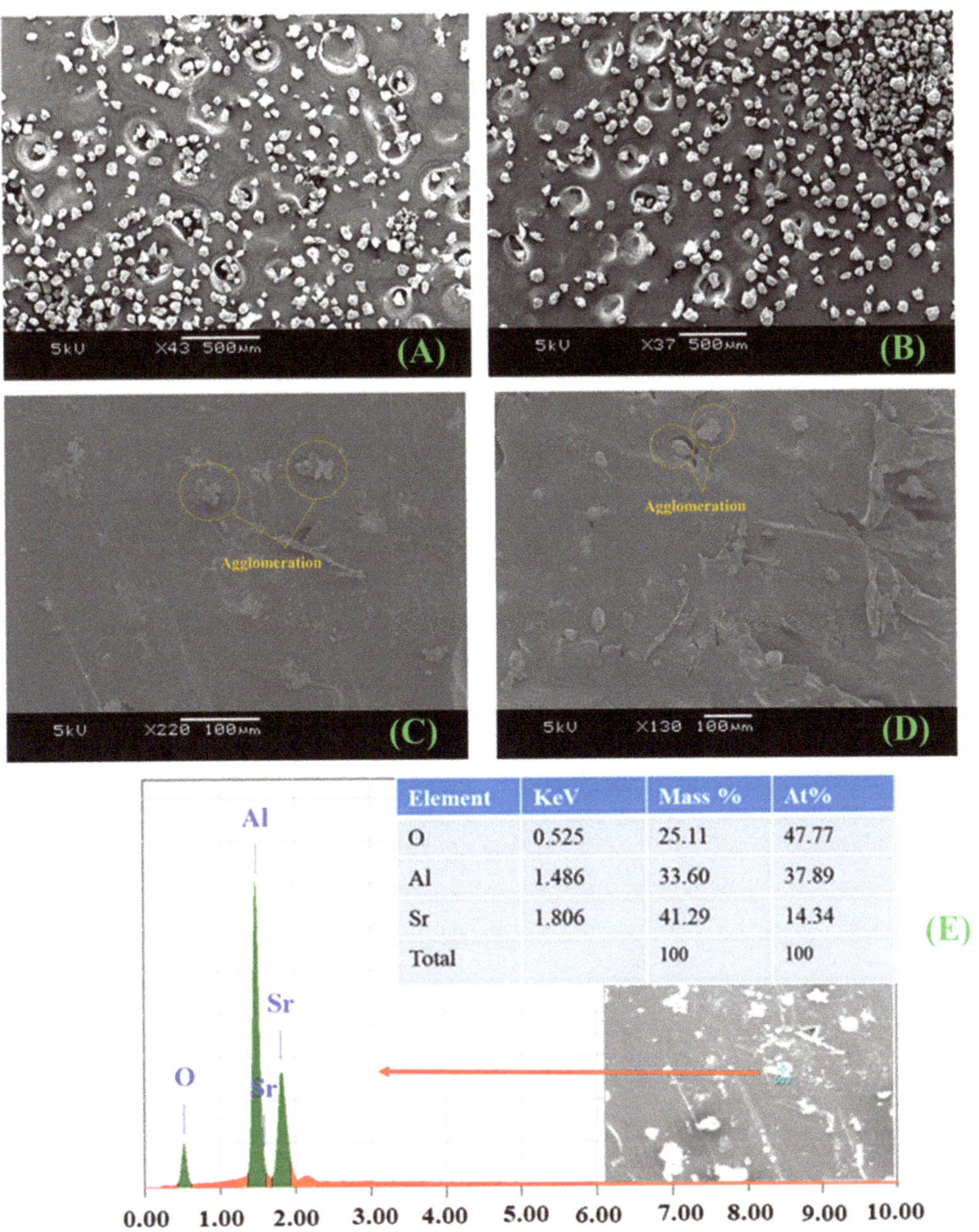


| Element | KeV | Mass % | At% |
|---|---|---|---|
| O | 0.525 | 25.11 | 47.77 |
| Al | 1.486 | 33.60 | 37.89 |
| Sr | 1.806 | 41.29 | 14.34 |
| Total | | 100 | 100 |

**Figure 1.** SEM images of $AG_1$ (**A**) and $AG_2$ (**B**) powders; and HDPE/10$AG_1$ (**C**), HDPE/10$AG_2$ (**D**), and EDS of HDPE/10$AG_1$ (**E**) composites.

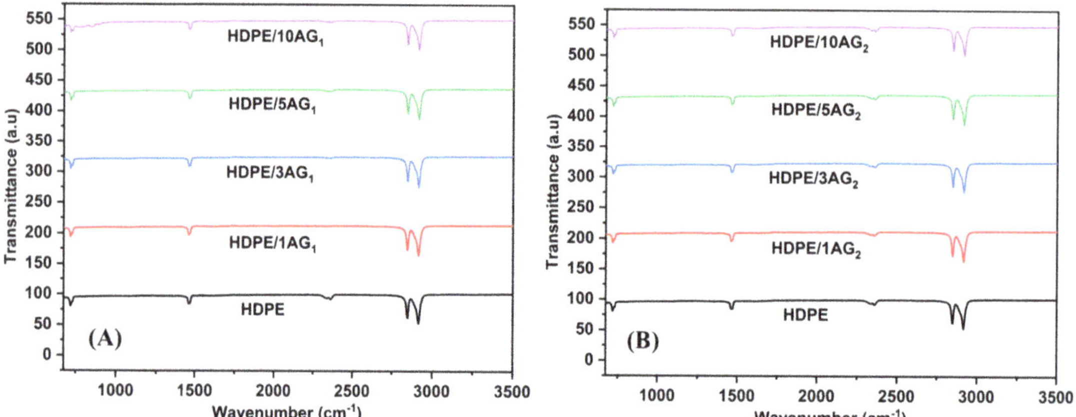


**Figure 2.** ATR-FTIR data of HDPE and HDPE/$AG_1$ (**A**) and HDPE/$AG_2$ (**B**) composites.

On the other hand, the $AG_1$ and $AG_2$ fillers have a significant effect on the crystallinity of HDPE, as shown in Tables 1 and 2, with a more pronounced effect on composites with $AG_2$ filler due to the bulky chemical structure of $AG_2$ than $AG_1$ [36].

### 3.3. Mechanical Characterization of HDPE/$AG_1$ and HDPE/$AG_2$

The mechanical properties of the composites are important for the application side of the auto-glowing composites. The tensile strength and modulus of HDPE, HDPE/$AG_1$, and HDPE/$AG_2$ composites are shown in Figure 3A,B, respectively. It is found that at lower filler loading (1–3 wt.%), the tensile strength was found to be improved and then decreases on further filler loading due to the agglomeration of the fillers, especially at higher filler loading concentration. The tensile modulus increases from 0.84 to 0.873 GPa, increasing the filler content from 0 to 10 wt.%.

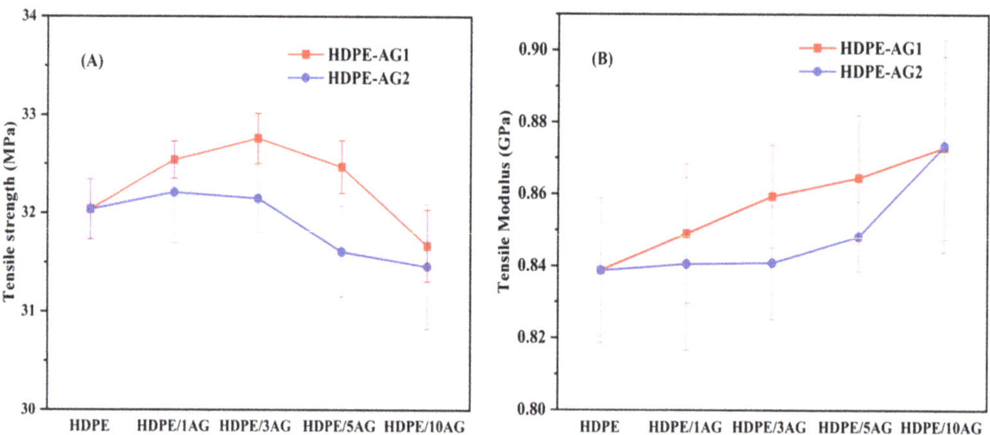


**Figure 3.** Tensile strength (**A**) and tensile modulus (**B**) in HDPE, HDPE/$AG_1$, and HDPE/$AG_2$ composites.

### 3.4. Rheological Studies in HDPE/$AG_1$ and HDPE/$AG_2$ Composites

The storage modulus and complex viscosity of HDPE/$AG_1$ and HDPE/$AG_2$ composites are shown in Figures 4 and 5, respectively. For both composites, the $AG_1$ and $AG_2$ fillers have a positive effect on both the complex viscosity and storage moduli of the composites, especially at a lower frequency range. The complex viscosity increment shows the good filler–HDPE interaction and improves the storage moduli of the studied composites. As shown in Figure 4, there is a gradual increase in the moduli value of the composites with the $AG_1$ and $AG_2$ content.

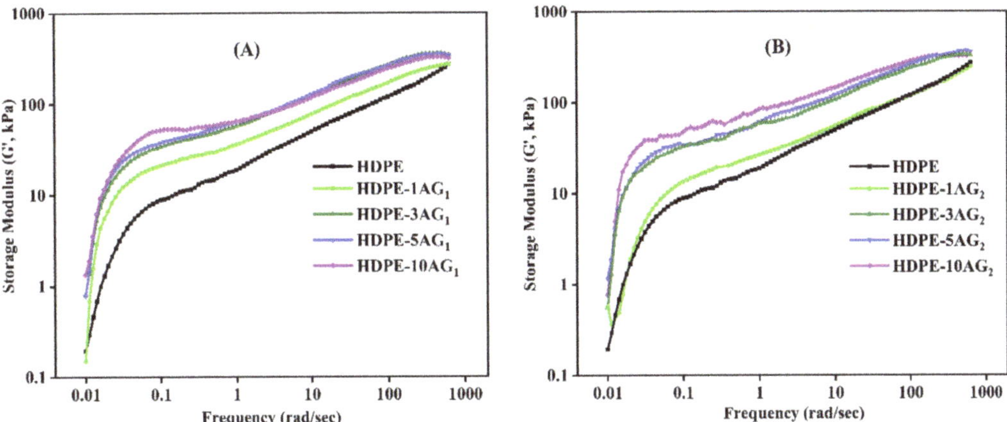


**Figure 4.** Storage modulus in HDPE/$AG_1$ (**A**) and HDPE/$AG_2$ (**B**) composites.

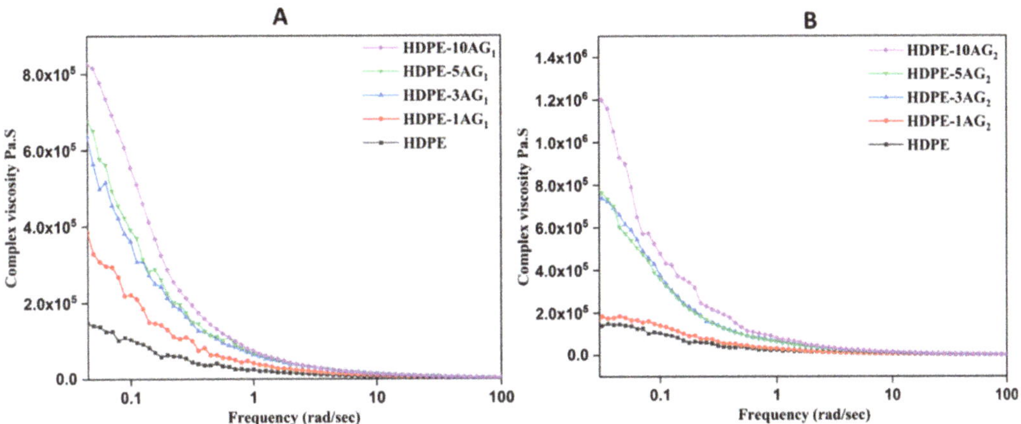


**Figure 5.** Complex viscosity data in HDPE/$AG_1$ (**A**) and HDPE/$AG_2$ (**B**) composites.

### *3.5. Phosphorescence Studies in HDPE/$AG_1$ and HDPE/$AG_2$ Composites*

The physical mechanism behind the phosphorescence phenomena is briefly divided into four processes: (i) when the phosphor is excited by external excitation illumination at specific wavelengths takes place upon the liberation of charge carriers (electrons and/or holes); (ii) the excited electrons or generated holes can be non-radiatively acquired by the electron or hole traps through a conduction band (CB) or valence band (VB), respectively, or by the quantum tunneling process through the forbidden band, known as the trapping process. The traps do not emit electromagnetic radiation but they store the excitation energy for a long time (optical battery). (iii) After stopping the excitation, the captured charge carriers can be released mainly by thermal stimulation energy (other stimulations like optical or mechano-ones are out of the scope in this paper), termed a de-trapping process. (iv) Finally, the released charge carriers move back to the emission center, yielding the delayed luminescence due to the electron–hole recombination [37–39].

The phosphorescence measurements for HDPE/$AG_1$ and HDPE/$AG_2$ are shown in Figure 6A,B, respectively. The UV excitation was carried out at 320 nm, and the intensity of emission was monitored in HDPE composites with $AG_1$ and $AG_2$ fillers. In HDPE/$AG_1$ composites, green emission was observed at ~500 nm, due to the $4f^65d^1$ to $4f^7$ transition of $Eu^{+2}$ ion of the $AG_1$ phosphor sample, and, as expected, the intensity of emission was found to increase with the increased loading percentage of the $AG_1$ phosphor, as shown in Figure 6A. For the HDPE/$AG_2$, a blue emission at ~490 nm was observed, and similar to the $AG_1$ sample, the emission intensity is proportional to the $AG_2$ content (Figure 6B).

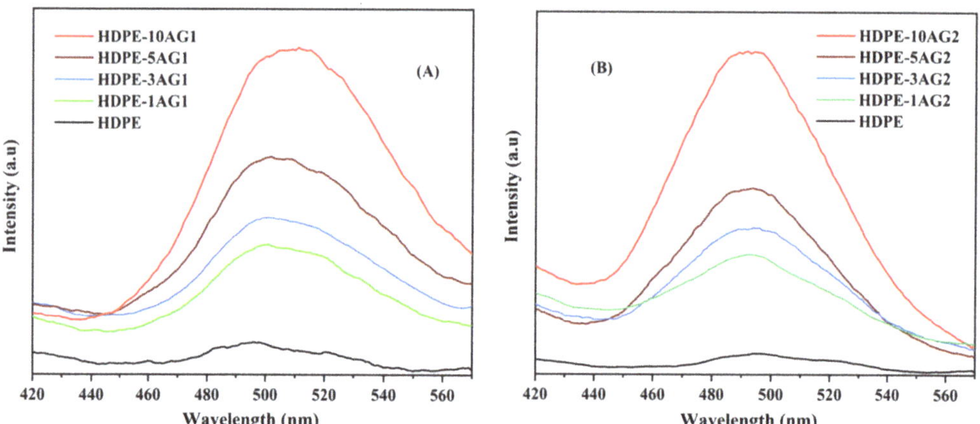


**Figure 6.** Phosphorescence emission in HDPE, HDPE/$AG_1$ (**A**), and HDPE/$AG_2$ (**B**) composites.

The HDPE/$AG_1$ (green) and HDPE/$AG_2$ (blue) emission under darkness can be visible from Figure 7, and the intensity of emission is proportional to the filler content ($AG_1$ and $AG_2$: 1 to 10 wt.%).

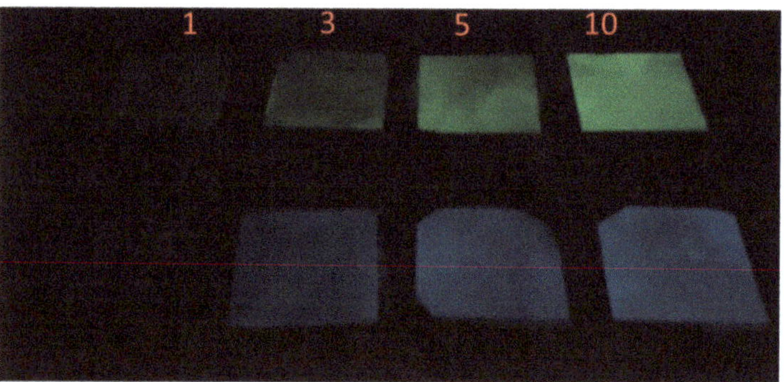


**Figure 7.** Phosphorescence emission under darkness; green for $AG_1$ composites and blue for $AG_2$ composites (1–10 wt.%).

*3.6. Phosphorescence Intensity Decay Studies in HDPE/$AG_1$ and HDPE/$AG_2$ Composites*

The phosphorescence decay for the studied composites is shown in Figure 8. In both composites, the phosphorescence intensity decay rate is inversely proportional to the $AG_1$ and $AG_2$ filler content, i.e., the composites with higher filler content decay slowly. One can see that, with the 10 and 5 wt.% $AG_1$ and $AG_2$ composites, the decay rate is very slow and the final intensity is not reaching a zero value as previously reported for similar materials [34,40]. The afterglow for these materials is considerably higher; therefore, no such behavior was observed within the studied timespan.

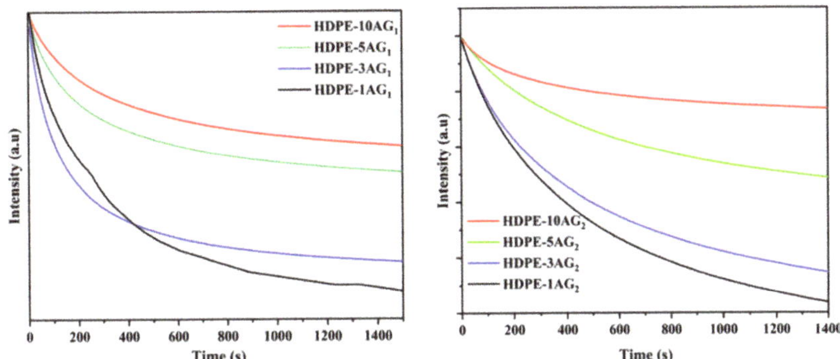


**Figure 8.** Phosphorescence intensity vs. time in HDPE/AG$_1$ and HDPE/AG$_2$ composites.

### *3.7. Effect of Ageing on Phosphorescence and Mechanical Properties*

The effect of aging on the phosphorescence and mechanical properties of the composite samples were conducted by the composite films to outside atmospheric conditions, for different exposure times (up to 40 days), and the phosphorescent and mechanical properties measurements were conducted in 10 days. For the sake of brevity, only the HDPE/3AG$_1$ composite sample was selected. Considering the phosphorescence characteristics of composites, it was found that there is a decrease in the emission intensity due to the ageing phenomena and that this decrease in intensity is directly proportional to the number of exposure days (Figure 9). On the other hand, the effect of aging on tensile strength was found to be negligible, and the tensile modulus values decrease on ageing. It was also observed that both tested samples, neat HDPE and HDPE/3AG$_1$ composite, became more brittle, as evidenced by the reduction in elongation percentages with an increased number of exposure days. The elongation at break for neat HDPE was found to be decreased from 58.5 to 18.82, while that of HDPE/3AG$_1$ composite was from 48 to 11 (Table 3). This latter observation is due to the physical aging phenomena that takes place for the sample at the outside atmospheric conditions [41].

**Table 3.** Tensile strength, tensile modulus, and elongation data on HDPE and HDPE/3AG$_1$ composites on ageing.

| Material | Tensile Strength MPa | SD | Tensile Modulus GPa | Elongation % | SD |
|---|---|---|---|---|---|
| HDPE | 32.04 | 0.31 | 0.84 | 58.5 | 4.58 |
| HDPE-10days | 33.63 | 0.49 | 0.70 | 51.22 | 5.17 |
| HDPE/20days | 34.12 | 1.48 | 0.67 | 39.51 | 6.31 |
| HDPE/30days | 33.51 | 0.20 | 0.66 | 20.28 | 7.5 |
| HDPE/40 days | 32.1 | 0.63 | 0.64 | 18.82 | 3.1 |
| HDPE/3AG$_1$ | 31.94 | 0.34 | 0.84 | 48.14 | 4.11 |
| HDPE/3AG$_1$/10days | 34.41 | 0.32 | 0.82 | 40.12 | 3.12 |
| HDPE/3AG$_1$/20days | 34.12 | 0.51 | 0.79 | 36.32 | 4.2 |
| HDPE/3AG$_1$/30days | 33.51 | 0.27 | 0.77 | 15.66 | 3.02 |
| HDPE/3AG$_1$/40days | 33.95 | 0.59 | 0.72 | 11.21 | 2.51 |

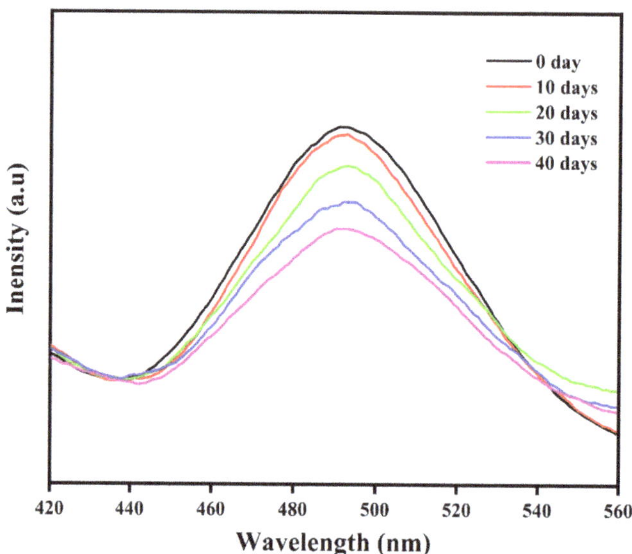


**Figure 9.** Phosphorescence decay for HDPE/3$AG_1$ composites every 10 days.

## 4. Conclusions

Long afterglow HDPE-based strontium aluminate composites were prepared and characterized for their application side. The HDPE-encapsulated $AG_1$ and $AG_2$ composites resulted in long afterglow composites, which lasted for many hours. The physical mixing was confirmed from both the DSC and ATR-FTIR studies. The green (~500 nm) and blue emission (~490 nm) were observed for the $AG_1$ and $AG_2$ composites, respectively, and the intensity of emission improves with the amount of $AG_1$ and $AG_2$ fillers. The DSC data show that the total crystallinity of both the composites was decreased but with a more decreasing effect with the bulky $AG_2$ filler without affecting the melting and crystallization temperature. The rheological results show the increase in complex viscosity and thereby the storage modulus values in the resulted composites. SEM pictures reveal good dispersion of the fillers in the HDPE matrix, and the tensile strength and modulus were found to be increased with the proportion of the fillers. Due to the better mechanical characteristics and long afterglow time, these composites can be found in applications in roadway nighttime displays, fluorescent lamps, etc. The ageing studies in the glowing composites show that there is a decrease in the intensity of phosphorescence emission on exposure to outside temperature for a longer period, and the composites become more and more brittle on ageing, .

**Author Contributions:** Conceptualization, A.M.P.; investigation, A.A. (Abdullah Alhamidi) and A.Y.E.; writing—original draft preparation, A.M.P.; writing—review and editing, A.M.P., H.S., N.S.K., and A.A. (Arfat Anis); supervision, A.M.P. and S.M.A.-Z.; project administration, A.M.P. All authors have read and agreed to the published version of the manuscript.

**Funding:** This project was funded by the National Plan for Science, Technology, and Innovation (MAARIFAH), King Abdulaziz City for Science and Technology, Kingdom of Saudi Arabia, Award Number (13-ADV1044-02).

**Institutional Review Board Statement:** Not applicable.

**Informed Consent Statement:** Not Applicable.

**Data Availability Statement:** The data presented in this study are available on request from the corresponding author.

**Acknowledgments:** This project was funded by the National Plan for Science, Technology, and Innovation (MAARIFAH), King Abdulaziz City for Science and Technology, Kingdom of Saudi Arabia, Award Number (13-ADV1044-02).

**Conflicts of Interest:** The authors declare no conflict of interest.

## References

1. Hölsä, J. Persistent luminescence beats the afterglow: 400 years of persistent luminescence. *Electrochem. Soc. Interface* **2009**, *18*, 42. [CrossRef]
2. Harvey, E.N. *A History of Luminescence from the Earliest Times until 1900*; American Philosophical Society: Philadelphia, PA, USA, 1957; pp. 1–677.
3. Abbruscato, V. Optical and Electrical Properties of $SrAl_2O_4$: $Eu^{2+}$. *J. Electrochem. Soc.* **1971**, *118*, 930. [CrossRef]
4. Katsumata, T.; Sasajima, K.; Nabae, T.; Komuro, S.; Morikawa, T. Characteristics of strontium aluminate crystals used for long-duration phosphors. *J. Am. Ceram. Soc.* **1998**, *81*, 413–416. [CrossRef]
5. Katsumata, T.; Nabae, T.; Sasajima, K.; Komuro, S.; Morikawa, T. Effects of Composition on the Long Phosphorescent $SrAl_2O_4$: $Eu^{2+}$, $Dy^{3+}$ Phosphor Crystals. *J. Electrochem. Soc.* **1997**, *144*, L243. [CrossRef]
6. Nance, J.; Sparks, T.D. From streetlights to phosphors: A review on the visibility of roadway markings. *Prog. Org. Coat.* **2020**, *148*, 105749. [CrossRef]
7. Nance, J.; Sparks, T.D. Comparison of coatings for $SrAl_2O_4$: $Eu^{2+}$, $Dy^{3+}$ powder in waterborne road striping paint under wet conditions. *Prog. Org. Coat.* **2020**, *144*, 105637. [CrossRef]
8. Lin, Y.-C.; Karlsson, M.; Bettinelli, M. Inorganic phosphor materials for lighting. *Top. Curr. Chem.* **2016**, *374*, 21. [CrossRef]
9. Tan, H.; Wang, T.; Shao, Y.; Yu, C.; Hu, L. Crucial breakthrough of functional persistent luminescence materials for biomedical and information technological applications. *Front. Chem.* **2019**, *7*, 387. [CrossRef]
10. Wang, W.; Sha, A.; Lu, Z.; Yuan, D.; Jiang, W.; Liu, Z. Cement filled with phosphorescent materials for pavement: Afterglow decay mechanism and properties. *Constr. Build. Mater.* **2021**, *284*, 122798. [CrossRef]
11. Maldiney, T.; Richard, C.; Seguin, J.; Wattier, N.; Bessodes, M.; Scherman, D. Effect of Core Diameter, Surface Coating, and PEG Chain Length on the Biodistribution of Persistent Luminescence Nanoparticles in Mice. *ACS Nano* **2011**, *5*, 854–862. [CrossRef]
12. Paterson, A.S.; Raja, B.; Garvey, G.; Kolhatkar, A.; Hagström, A.E.; Kourentzi, K.; Lee, T.R.; Willson, R.C. Persistent luminescence strontium aluminate nanoparticles as reporters in lateral flow assays. *Anal. Chem.* **2014**, *86*, 9481–9488. [CrossRef]
13. Chen, W.; Wang, Y.; Zeng, W.; Han, S.; Li, G.; Guo, H.; Li, Y.; Qiang, Q. Long persistent composite phosphor $CaAl_2O_4$: $Eu^{2+}$, $Nd^{3+}/Y_3$ $Al_5$ $O_{12}$: $Ce^{3+}$: A novel strategy to tune the colors of persistent luminescence. *New J. Chem.* **2016**, *40*, 485–491. [CrossRef]
14. Zheng, R.; Xu, L.; Qin, W.; Chen, J.; Dong, B.; Zhang, L.; Song, H. Electrospinning preparation and photoluminescence properties of $SrAl_2O_4$: $Ce^{3+}$ nanowires. *J. Mater. Sci.* **2011**, *46*, 7517–7524. [CrossRef]
15. Sakirzanovas, S.; Katelnikovas, A.; Dutczak, D.; Kareiva, A.; Jüstel, T. Synthesis and $Sm^{2+}/Sm^{3+}$ doping effects on photoluminescence properties of $Sr_4Al_{14}O_{25}$. *J. Lumin.* **2011**, *131*, 2255–2262. [CrossRef]
16. Feng, X.; Feng, W.; Wang, K. Experimental and theoretical spectroscopic study of praseodymium (III) doped strontium aluminate phosphors. *J. Alloy. Compd.* **2015**, *628*, 343–346. [CrossRef]
17. Lupei, A.; Lupei, V.; Gheorghe, C.; Gheorghe, L.; Vivien, D.; Aka, G.; Antic-Fidancev, E. Disorder effects in $Nd^{3+}$-doped strontium hexa-aluminate laser crystals. *J. Phys. Condens. Matter* **2005**, *18*, 597–611. [CrossRef]
18. Lu, B.; Shi, M.; Pang, Z.; Zhu, Y.; Li, Y. Study on the optical performance of red-emitting phosphor: $SrAl_2O_4$: $Eu^{2+}$, $Dy^{3+}/Sr_2MgSi_2O_7$: $Eu^{2+}$, $Dy^{3+}$/light conversion agent for long-lasting luminous fibers. *J. Mater. Sci. Mater. Electron.* **2021**, 1–13. [CrossRef]
19. Sahu, I.P.; Bisen, D.; Sharma, R. UV excited green luminescence of $SrAl_2O_4$: $Eu^{2+}$, $Dy^{3+}$ nanophosphor. *Res. Chem. Intermed.* **2016**, *42*, 2791–2804. [CrossRef]
20. Chang, C.-C.; Yang, C.-Y.; Lu, C.-H. Preparation and photoluminescence properties of $Sr_4Al_{14}O_{25}$:$Eu^{2+}$ phosphors synthesized via the microemulsion route. *J. Mater. Sci. Mater. Electron.* **2013**, *24*, 1458–1462. [CrossRef]
21. Zhu, Y.; Zeng, J.; Li, W.; Xu, L.; Guan, Q.; Liu, Y. Encapsulation of strontium aluminate phosphors to enhance water resistance and luminescence. *Appl. Surf. Sci.* **2009**, *255*, 7580–7585. [CrossRef]
22. Lü, X.; Zhong, M.; Shu, W.; Yu, Q.; Xiong, X.; Wang, R. Alumina encapsulated $SrAl_2O_4$:$Eu^{2+}$, $Dy^{3+}$ phosphors. *Powder Technol.* **2007**, *177*, 83–86. [CrossRef]
23. Tian, S.; Wen, J.; Fan, H.; Chen, Y.; Yan, J.; Zhang, P. Sunlight-activated long persistent luminescent polyurethane incorporated with amino-functionalized $SrAl_2O_4$:$Eu^{2+}$, $Dy^{3+}$phosphor. *Polym. Int.* **2016**, *65*, 1238–1244. [CrossRef]
24. Oguzlar, S.; Ongun, M.Z.; Keskin, O.Y.; Delice, T.K.; Azem, F.A.; Birlik, I.; Ertekin, K. Investigation of Spectral Interactions between a $SrAl_2O_4$: $Eu^{2+}$, $Dy^{3+}$ Phosphor and Nano-Scale $TiO_2$. *J. Fluoresc.* **2020**, *30*, 839–847. [CrossRef] [PubMed]
25. Khattab, T.A.; Abd El-Aziz, M.; Abdelrahman, M.S.; El-Zawahry, M.; Kamel, S. Development of long-persistent photoluminescent epoxy resin immobilized with europium (II)-doped strontium aluminate. *Luminescence* **2020**, *35*, 478–485. [CrossRef] [PubMed]
26. Wan, M.; Jiang, X.; Nie, J.; Cao, Q.; Zheng, W.; Dong, X.; Fan, Z.H.; Zhou, W. Phosphor powders-incorporated polylactic acid polymeric composite used as 3D printing filaments with green luminescence properties. *J. Appl. Polym. Sci.* **2019**, *137*, 48644. [CrossRef]

27. Cheng, L.-X.; Liu, T.; Li, L.; Yang, L.; He, H.-W.; Zhang, J.-C. Self-repairing inorganic phosphors/polymer composite film for restructuring luminescent patterns. *Mater. Res. Express* **2021**, *8*, 065302. [CrossRef]
28. De Clercq, D.M.; Chan, S.V.; Hardy, J.; Price, M.B.; Davis, N.J.L.K. Reducing reabsorption in luminescent solar concentrators with a self-assembling polymer matrix. *J. Lumin.* **2021**, *236*, 118095. [CrossRef]
29. Poulose, A.M.; Anis, A.; Shaikh, H.; Alhamidi, A.; Siva Kumar, N.; Elnour, A.Y.; Al-Zahrani, S.M. Strontium Aluminate-Based Long Afterglow PP Composites: Phosphorescence, Thermal, and Mechanical Characteristics. *Polymers* **2021**, *13*, 1373. [CrossRef]
30. Ye, F.; Dong, S.; Tian, Z.; Yao, S.; Zhou, Z.; Wang, S. Fabrication and characterization of long-persistent luminescence/polymer ($Ca_2MgSi_2O_7$: $Eu^{2+}$, $Dy^{3+}$/PLA) composite fibers by electrospinning. *Opt. Mater.* **2015**, *45*, 64–68. [CrossRef]
31. Khursheed, S.; Kumar, V.; Singh, V.K.; Sharma, J.; Swart, H. Optical properties of $Sr_3B_2O_6$: $Dy^{3+}$/PMMA polymer nanocomposites. *Phys. B Condens. Matter* **2018**, *535*, 184–188. [CrossRef]
32. George, G.; Luo, Z. A Review on Electrospun Luminescent Nanofibers: Photoluminescence Characteristics and Potential Applications. *Curr. Nanosci.* **2020**, *16*, 321–362. [CrossRef]
33. Bem, D.; Swart, H.; Luyt, A.; Coetzee, E.; Dejene, F. Properties of green $SrAl_2O_4$ phosphor in LDPE and PMMA polymers. *J. Appl. Polym. Sci.* **2010**, *117*, 2635–2640. [CrossRef]
34. Ni, Z.; Fan, T.; Bai, S.; Zhou, S.; Lv, Y.; Ni, Y.; Xu, B. Effect of the Concentration of $SrAl_2O_4$: $Eu^{2+}$and $Dy^{3+}$ (SAO) on Characteristics and Properties of Environment Friendly Long-Persistent Luminescence Composites from Polylactic Acid and SAO. *Scanning* **2021**, *2021*, 6337768. [CrossRef] [PubMed]
35. Włochowicz, A.; Eder, M. Distribution of lamella thicknesses in isothermally crystallized polypropylene and polyethylene by differential scanning calorimetry. *Polymer* **1984**, *25*, 1268–1270. [CrossRef]
36. Ouchiar, S.; Stoclet, G.G.; Cabaret, C.; Gloaguen, V. Influence of the filler nature on the crystalline structure of polylactide-based nanocomposites: New insights into the nucleating effect. *Macromolecules* **2016**, *49*, 2782–2790. [CrossRef]
37. Xu, J.; Tanabe, S. Persistent luminescence instead of phosphorescence: History, mechanism, and perspective. *J. Lumin.* **2019**, *205*, 581–620. [CrossRef]
38. Clabau, F.; Rocquefelte, X.; Jobic, S.; Deniard, P.; Whangbo, M.-H.; Garcia, A.; Le Mercier, T. Mechanism of phosphorescence appropriate for the long-lasting phosphors $Eu^{2+}$-doped $SrAl_2O_4$ with codopants $Dy^{3+}$ and $B^{3+}$. *Chem. Mater.* **2005**, *17*, 3904–3912. [CrossRef]
39. Dorenbos, P. Mechanism of persistent luminescence in $Eu^{2+}$ and $Dy^{3+}$ codoped aluminate and silicate compounds. *J. Electrochem. Soc.* **2005**, *152*, H107–H110. [CrossRef]
40. Lephoto, M.A.; Ntwaeaborwa, O.M.; Pitale, S.S.; Swart, H.C.; Botha, J.R.; Mothudi, B.M. Synthesis and characterization of $BaAl_2O_4$:$Eu^{2+}$ co-doped with different rare earth ions. *Phys. B Condens. Matter* **2012**, *407*, 1603–1606. [CrossRef]
41. Kim, S.; Lee, Y.; Kim, C.; Choi, S. Analysis of Mechanical Property Degradation of Outdoor Weather-Exposed Polymers. *Polymers* **2022**, *14*, 357. [CrossRef]

 materials

*Article*

# Compressive Strength Assessment of Soil–Cement Blocks Incorporated with Waste Tire Steel Fiber

**Joaquin Humberto Aquino Rocha [1,*], Fernando Palacios Galarza [2], Nahúm Gamalier Cayo Chileno [2], Marialaura Herrera Rosas [2], Sheyla Perez Peñaranda [2], Luis Ledezma Diaz [2] and Rodrigo Pari Abasto [2]**

1 Department of Civil Engineering, COPPE, Federal University of Rio de Janeiro, Rio de Janeiro 21945970, Brazil
2 Department of Civil Engineering, Faculty of Technology, Universidad Privada del Valle, Tiquipaya Campus, Tiquipaya MQ9F+GH, Bolivia; fpalaciosg@univalle.edu (F.P.G.); ccn0025217@est.univalle.edu (N.G.C.C.); hrm2014954@est.univalle.edu (M.H.R.); pps0028721@est.univalle.edu (S.P.P.); ldl0029367@est.univalle.edu (L.L.D.); par0028483@est.univalle.edu (R.P.A.)
* Correspondence: joaquin.rocha@coc.ufrj.br; Tel.: +55-591-75983072

**Abstract:** The rapid growth in waste tire disposal has become a severe environmental concern in recent decades. Recycling rubber and steel fibers from wasted tires as construction materials helps counteract this imminent environmental crisis, mainly improving the performance of cement-based materials. Consequently, the present article aims to evaluate the potential use of waste tire steel fibers (i.e., WTSF) incorporated in the manufacture of soil–cement blocks, considering their compressive resistance as a primary output variable of comparison. The experimental methodology applied in this study comprised the elaboration of threefold mixtures of soil–cement blocks, all of them with 10% by weight in Portland cement, but with different volumetric additions of WTSF (i.e., 0%, 0.75%, and 1.5%). The assessment's outcomes revealed that the addition of 0.75% WTSF does not have a statistically significant influence on the compressive resistance of the samples. On the contrary, specimens with 1.5% WTSF displayed a 20% increase (on average) in their compressive strength. All the tested samples' results exhibited good agreement with the minimum requirements of the different standards considered. The compressive resistance was evaluated in the first place because it is the primary provision demanded by the specifications for applying soil–cement materials in building constructions. However, further research on the physical and mechanical properties of WTSF soil–cement blocks is compulsory; an assessment of the durability of soil–cement blocks with WTSF should also be carried out.

**Keywords:** scrap tire recycled steel fiber; soil–cement blocks; compressive strength; sustainability

**Citation:** Rocha, J.H.A.; Galarza, F.P.; Chileno, N.G.C.; Rosas, M.H.; Peñaranda, S.P.; Diaz, L.L.; Abasto, R.P. Compressive Strength Assessment of Soil–Cement Blocks Incorporated with Waste Tire Steel Fiber. *Materials* **2022**, *15*, 1777. https://doi.org/10.3390/ma15051777

Academic Editors: Andrea Petrella and Michele Notarnicola

Received: 31 December 2021
Accepted: 26 January 2022
Published: 26 February 2022



## 1. Introduction

Steel fiber reinforcement improves the mechanical performance of cement-based materials (e.g., concrete, mortars, blocks, and other cement-based materials). Steel fibers' usage enhances the mechanical properties of these building materials. To date, the most studied features are the compressive and flexural strength of cement-based materials reinforced with steel fibers. Moreover, their ductility and resistance for dynamic loadings have been analyzed in multiple investigations [1,2]. Past research demonstrated the technical feasibility of steel fibers' usage in construction materials, particularly incorporated in concrete composite materials [3,4]. However, using reinforcement percentages of steel fibers higher than 1% may raise manufacturing expenses [5,6]. With this in mind, the objective of sustainability in the construction industry might be achieved by utilizing waste tire steel fiber (WTSF) as a green initiative recourse [7].

In the last decade, rapid population growth and land-use changes have generated a significant demand and increase in the vehicles fleet. This situation has led to a gradual increment in tire need and tire waste, yielding a direct impact on the environment [8]. As a result, approximately more than one billion tires are discarded worldwide each year [9,10], with 4.46 million tons [11] belonging to the United States of America. On the other hand, Bolivia generates roughly 1.5 million of tire waste per year [12]. In developing countries, environmental problems are primarily related to lack of planning and inappropriate design of rubbish dumps [13]. Furthermore, these materials' natural degradation is slow and laborious [14]. Hence, tire recycling in construction materials presents a sustainable alternative, accentuating the re-utilization of rubber [7,15,16] and WTSF [2,6].

The amount of WTSF extracted can represent up to 15% of the total weight of the tires [17]. Moreover, as demonstrated by a few authors, this material possesses mechanical properties analogous to industrial steel fibers. Thus, WTSF has also been investigated as reinforcement in cement-based materials [18,19]. However, recycled steel fibers' obtention difficulty prevents manufacturing and commercialization on a large scale; the amount of WTSF collected depends primarily on the obtaining process and the tire type. Therefore, additional research is required to determine its influence on the performance of cement-based materials in the long term [9,20].

On the contrary, soil–cement composite material has been widely spread for sustainable construction. Specifically, soil–cement blocks are a low-cost alternative in construction because they require low energy consumption, no combustion process, and no transportation of materials since the blocks can be produced on-site [21]. Additionally, the production of soil–cement blocks can reutilize other solid waste or materials that negatively impact the environment but enhance the brick's mechanical properties [22]. Various investigations have improved the properties of this material with the use of different natural fibers, such as coconut [23] and palm fibers [24]. In addition, studies of granite cutting residue adjunction (GCR) [21] and recycled tire fibers [25] are also reported; the latter found that recycled steel fibers help absorb plastic energy and resist deformation.

Sukontasukkul and Jamsawang [26] demonstrated that the use of steel and polypropylene fibers improves the flexural performance of soil–cement materials. The authors used percentages less than 1%, where polypropylene fibers presented optimal results. Tajdin et al. [27] used three types of fibers: jute, polypropylene, and steel, to reinforce soil–cement materials. The results showed an increase in the material's mechanical performance, specifically for the compression, tensile and flexural strength. On the contrary, Bakam et al. [28] used cassava fibers to reinforce earth blocks. A brief literature review has shown that studies on the behavior of soil–cement blocks are still limited and even more limited regarding WTSF, unlike other cement-based materials, such as concrete and mortar, where the use of WSTF is widely studied [29–31]. Therefore, using soil–cement with WTSF as a novel construction material could positively impact the environment, thus reducing the consumption of traditional materials and the waste from used tires.

This article aims to assess the potential use of WTSF in soil–cement blocks, centralizing the study on their compressive strength behavior. The latter is considered the primary mechanical feature required by several codes to be accomplished. In addition, a classification of the type of block for use in construction can be obtained by compressive strength achieved [32–37]. Hence, an exploratory and experimental campaign was developed where the soil–cement blocks were manufactured with different percentages of WTSF (0%, 0.75%, and 1.50%), verifying their impact on the blocks' compressive strength.

## 2. Materials and Methods

Soil–cement blocks were prepared, incorporating steel fibers from waste tires (i.e., WTSF) to their matrix composition. Threefold volumetric additions of WTSF were studied: 0% (control), 0.75%, and 1.5%. All compressed earth block samples had 10% of Portland cement added to the mixture (Figure 1) since this cement quantity has demonstrated better mechanical properties outcomes [38–40].

Figure 2 presents the methodology followed in this study. One controllable input factor (% WTSF) is contemplated; the remaining factors (materials used), although they may be controllable, were kept fixed. This procedure was followed to monitor the influence of WTSF on the compressive strength of soil–cement blocks. The F-test was utilized for the hypothesis test to compare the variance of the means of different levels of factors with the individual variances. The latter was corroborated by a Tukey test, comparing the individual means from the analysis of variance (ANOVA).

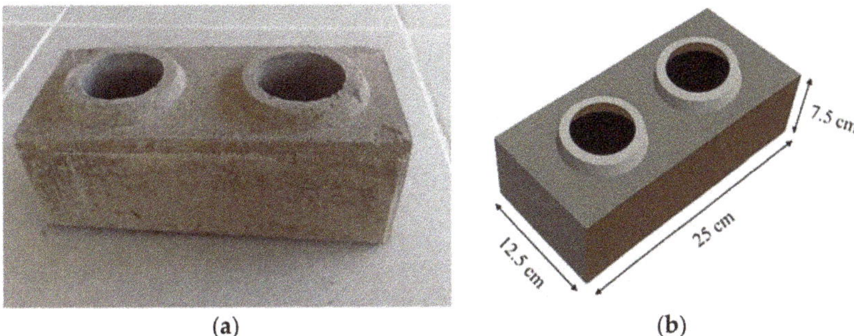


**Figure 1.** Soil–cement block sample: (**a**) 0.75% WTSF, and (**b**) dimensions.

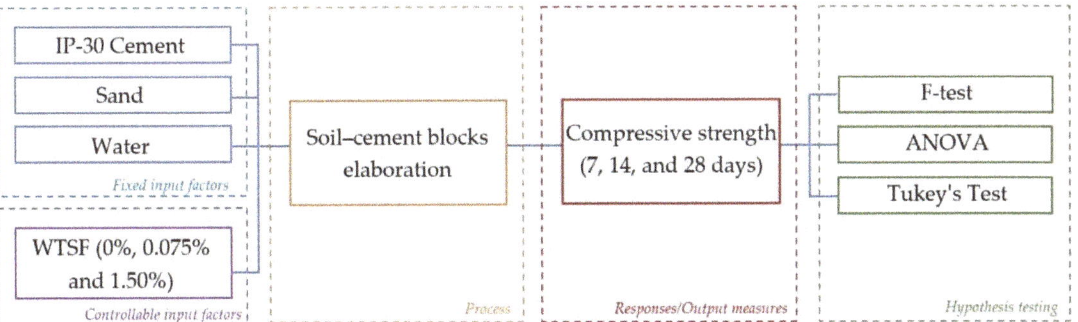


**Figure 2.** Research methodology.

Locally available IP-30 Portland cement classified as Type IP Portland Pozzolana cement as per the ASTM C595/C595M-20 [41] was employed. Chemical and physical cement features are detailed in Tables 1 and 2, respectively.

The soil material employed for the sample's preparation was extracted from the southern area of Cochabamba, Bolivia, at UTM coordinates 19K 799063.33 E 8070682.40 S. ASTM D422 [42] standard was used to perform the soil's granulometric analysis. Atterberg limits in the soil sample were determined as per the ASTM D4318 [43] specification. In addition, the soil material was classified following the AASTHO M145 standard [44] recommendations and the unified soil classification system (USCS) [45]. Finally, soil's optimal moisture content was measured employing the Proctor compaction test as per the ASTM D698 standard [46].

**Table 1.** Chemical analysis for IP-30 cement.

| Parameter | Unit | IP-30 Cement |
|---|---|---|
| Loss on ignition | % | 2.33 |
| $SiO_2$ | % | 32.83 |
| $Al_2O_3$ | % | 4.53 |
| $Fe_2O_3$ | % | 2.32 |
| CaO | % | 50.77 |
| MgO | % | 4.55 |
| $SO_3$ | % | 2.10 |

Data provided by the manufacturer.

A local tire recycling company provided the residual steel fibers for the study. The tire recycling process includes grinding using pneumatic equipment with rotating shafts; this procedure scraps the tires into smaller particles. Subsequently, the residual steel fibers are removed using magnets. Shulman [47] describes the recycling process to obtain WTSF and rubber in detail. Since WTSF were mixed with residual rubber, a previous fibers material selection was made, as shown in Figure 3. The current study employed only the selected fiber for manufacturing soil–cement blocks (Figure 3b).

**Table 2.** Physical analysis for IP-30 cement.

| Parameter | Unit | IP-30 Cement |
|---|---|---|
| Blaine | $m^2/kg$ | 448 |
| Residue T325 | % | 5.34 |
| True Density | $g/cm^3$ | 2.98 |
| Bulk Density | $g/cm^3$ | 1.05 |
| Initial Setting | h | 2.32 |
| Final Setting | h | 4.65 |
| 3-Day Strength | MPa | 19.19 |
| 7-Day Strength | MPa | 24.90 |
| 28-Day Strength | MPa | 30.63 |

Data provided by the manufacturer.

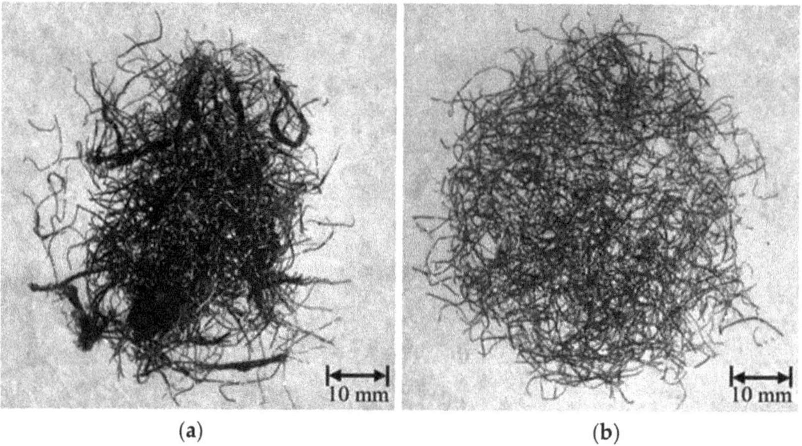


**Figure 3.** WTSF: (**a**) unclassified and (**b**) classified.

The steel fibers used had 23.22 ± 5.12 mm in length; Afroughsabet et al. [48] categorized this material as macro fibers based on its dimensions. Moreover, the diameter of the collected fibers was 0.23 ± 0.05 mm, such as those used by Pilakoutas et al. [9] and Hu et al. [49].

Soil–cement block samples were elaborated with a mechanical mixer and a manual Lego-type die press. Figure 4 shows the procedure followed to elaborate the soil–cement blocks. After the manufacturing process, the blocks were kept in the laboratory facilities for curing and drying for the next seven days. During this time, the block samples were isolated with plastic in their base to avoid rapid water loss, as recommended by Gatani [50]. Figure 5 depicts the soil–cement blocks incorporated with WTSF prior to the compression testing.

**Figure 4.** Preparation of soil–cement blocks: (**a**) preparation of materials, (**b**) placement of mixture, and (**c**) compaction.

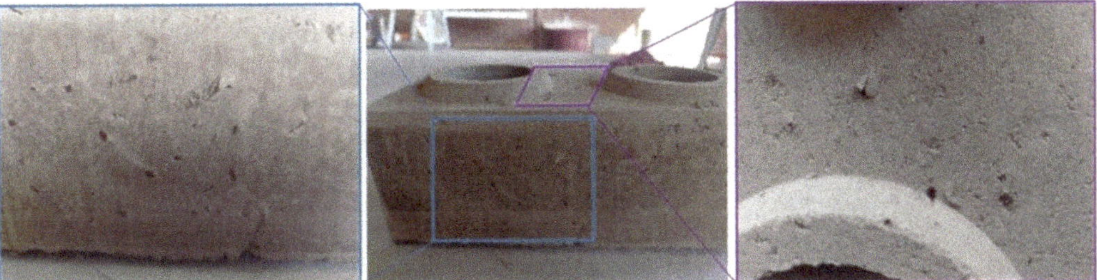

**Figure 5.** WTSF soil–cement block samples.

The blocks' compressive strength was determined for 7, 14, and 28 days, utilizing four blocks per group age and residual steel fiber percentage.

The experimental methodology procedure followed the NBR 8492 standard [51] as the basis for conducting the test.

## 3. Results and Discussion

### 3.1. Soil Properties

Figure 6 depicts the results from the granulometric analysis of the soil. It was observed that the soil material was composed of sand (55%) and slime clay (45%).

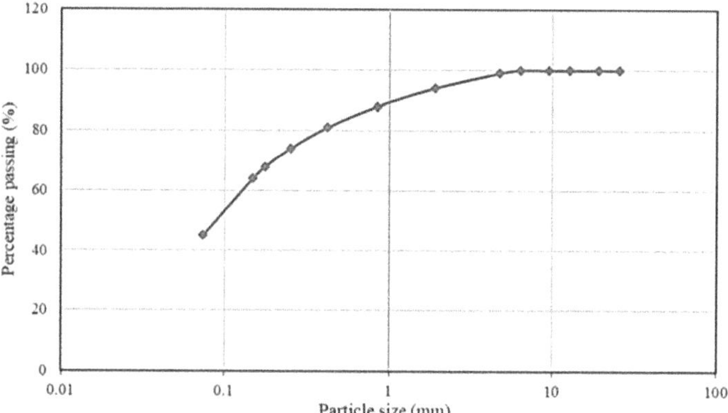


**Figure 6.** Soil's granulometric classification curve.

Atterberg's liquid limit (LL) and plasticity index (IP = LL − LP) obtained were 27 and 15, respectively. Therefore, the soil was classified as A-6 (i.e., clay soil) as per the AASTHO standard M145 [44] and clay sand (SC) according to the USCS [45] specification.

The Proctor test results are presented in Table 3 and Figure 7. Although the samples' steel fibers' addition showed independence from the optimum moisture content, the samples with steel had a higher dry density.

**Table 3.** Proctor tests results.

| Mixture | 0% WTSF | 0.75% WTSF | 1.50% WTSF |
|---|---|---|---|
| Dry density (g/cm$^3$) | 1.83 | 1.84 | 1.85 |
| Optimum moisture (%) | 13.56 | 13.99 | 13.02 |

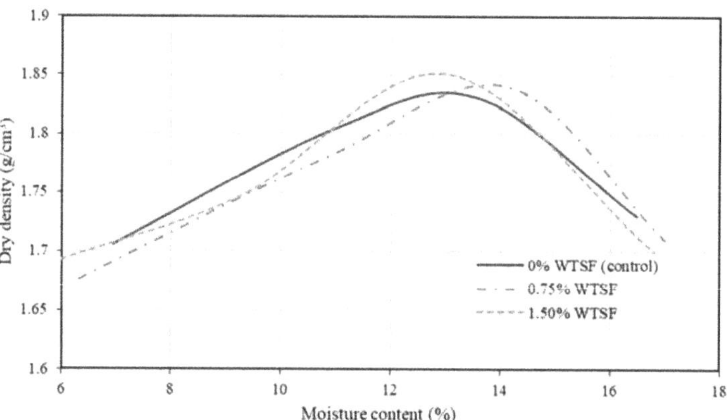


**Figure 7.** Proctor test's compaction curves.

## 3.2. *Compressive Strength*

Figure 8 depicts the compressive strength results of the tested blocks. Due to aging, an increase in compressive strength could be noticed, reaching a maximum value at 28 days. Moreover, the block samples with 1.50% WTSF presented the highest compressive strength values for all ages. However, WTSF's effect at 0.75% showed a non-defined correlation with divergent values concerning the control sample (0% WTSF).

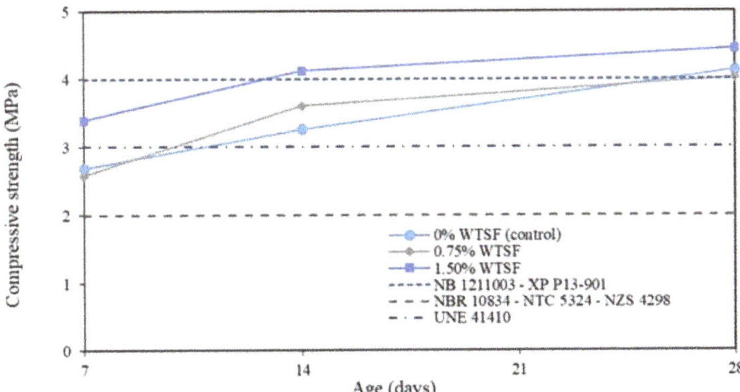


**Figure 8.** Block compression tests results.

Soil–cement blocks' compressive strengths shown in Figure 8 are satisfactory as per the Brazilian soil–cement standard [32]; they all laid above the minimum required of 2 MPa after seven days. Bolivia does not have a specific regulation for soil–cement blocks. However, the regulations of the Bolivian standard for ceramic bricks can be extended for this study [33]. Based on the latter, the specimens were typed as "Category C" with a minimum required compressive strength of 4 MPa for common materials required in building construction. Blocks with 1.50% WTSF showed good agreement with this category after 14 days of curing. The remaining blocks with 0% and 0.75% WTSF accomplished this requirement after 28 days of curing. Regarding the Colombian standard materials regulation [34], the soil–cement WTSF blocks met the minimum requirements of the "BSC-20" category (i.e., strength > 2 MPa). However, they could not reach the BSC-60 category with a minimum compressive strength required of 6 MPa.

Regarding the European regulations, the Spanish standard UNE 41410 [35] classifies compressed earth blocks according to their compressive strength: 1.3 MPa (BTC 1), 3 MPa (BTC 3), and 5 MPa (BTC 5). In this regulation, the usage of binders such as cement is optional. It was observed that all the soil–cement blocks with WTSF belonged to BTC 3 category after 14 days. Regarding the French standard XP P13-901 [36], the blocks reached the category BSC 40 (4 MPa) at 28 days. Finally, the New Zealand specification NZS 4298 [37] indicates a minimum compressive strength of 2 MPa, a value reached by all the tested samples after seven days.

These results are consistent with previous investigations on soil–cement blocks with steel fibers reinforcement that report improved mechanical properties [25,26]. Similarly, Nasir [52] and Ndyambaje [53] reported an enhancement in the compressive strength of cement-based materials reinforced with 1.5% WTSF and 1.2% WTSF, respectively; both authors employed recycled steel fibers of 40 mm in length. On the contrary, Mastali and Dalvand [54] demonstrated that the addition of lower percentages of WTSF (up to 0.75%) could also lead to positive outcomes regarding the compressive strength of cement-based materials.

Percentages used for WTSF in this study are similar to those reported in the literature [52–54]. However, these studies only consider concrete and mortar. The present study shows that these exact percentages of WTSF can also be extended for soil–cement blocks as a novel application of this tire residue. In this case, the soil–cement blocks with 1.5% WTSF have the most relevant results. Similarly, Eko et al. [25] found that 2% WSTF significantly improves the mechanical properties of unfired earth blocks. However, Eko et al. [25] used a lateritic soil and different block dimensions, which differs from the current study.

Figure 9 below shows the soil–cement WTSF blocks after the compressive strength test.

**Figure 9.** Tested soil–cement WTSF blocks: (**a**) 0.75% WTSF and (**b**) 1.50% WTSF.

In order to verify whether the compression strength average values by age are statistically significantly different, an analysis of variance (ANOVA) (Table 4) was performed, considering a significance value of 0.01.

**Table 4.** ANOVA analysis for concrete strength delimited by aging.

| Age | F | F Crit | *p* Value | Significance |
|---|---|---|---|---|
| 7 | 53.94017 | 8.02152 | 9.75561E-06 | Yes |
| 14 | 39.49235 | 8.02152 | 3.50154E-05 | Yes |
| 28 | 16.04894 | 8.02152 | 0.00108 | Yes |

To better understand Table 4, two hypotheses are proposed: null (H0), if F < F crit; there is not a significant difference between the means, and alternative (H1), if F > F crit, there is a significant difference between the means. As detailed in Table 4, F > F crit in all cases; therefore, there were statistically significant differences between the means, and all *p*-values were less than 0.01. Hence, it can be stated that there is a statistically significant difference between the block's compressive strength average values ranked by age. In order to verify the latter, a Tukey test was carried out with a similar significance value (0.01). Thus, it seemed that there were no statistically significant differences between the means of the blocks with 0% WTSF and 0.75% WTSF (Table 5). Nevertheless, there were differences between these two percentages regarding the 1.50% blocks of WTSF. Therefore, the latter indicated that percentages greater than 1.50% of WTSF addition positively impact the compressive strength.

**Table 5.** Tukey's test for concrete strength on an aging basis.

| Group (WTSF) | | *p* Value per Age (Days) | | |
|---|---|---|---|---|
| | | 7 | 14 | 28 |
| 0% | 0.75% | 0.49364751 | 0.01659152 | 0.34869971 |
| 0% | 1.50% | $4.1885 \times 10^{-5}$ | $2.6717 \times 10^{-5}$ | 0.0078263 |
| 0.75% | 1.50% | $1.424 \times 10^{-5}$ | 0.00124796 | 0.001025 |

Figure 10 shows the percentage of compressive strength variation compared to the control sample. As previously mentioned, the blocks with 0.75% of WTSF did not present statistically significant differences contrasted with the control samples. For this reason, an average variation of 1.22% was observed exclusively. Regarding the 1.50% blocks of WTSF, a positive average variation of 20.17% was detailed. However, the most significant growth in compressive strength occurred between 7 and 14 days of curing.

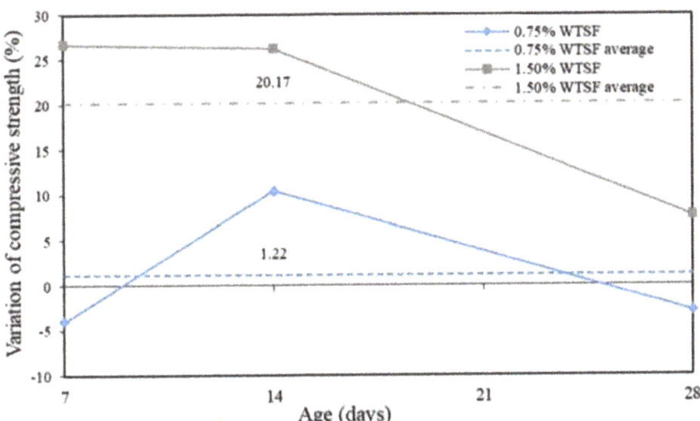


**Figure 10.** Percentage of variation in compressive strength.

## 4. Conclusions

Based on the experimental test carried out in this study, the following conclusions can be drawn:

- The Proctor test on the samples revealed that the amount of WTSF does not influence optimal humidity content in the specimens. However, block samples with WTSF showed higher dry density.
- The highest compressive strength was found for the 1.50% WTSF blocks. This latter can be extended to all the samples regarding the content of WTSF in their matrices.
- After seven days of curing, all block samples topped the minimum compressive strength requirement established in the ABNT standard [32]. Regarding the Bolivian specification for building materials [33], the blocks were classified as Category C, accomplishing the minimum strength of 4 MPa after 14 days of curing for 1.50% WTSF blocks and after 28 days for 0.75% WTSF samples. At the age of 28 days, all the soil–cement blocks beat the minimum compressive strength requirements, considering the standards of Colombia [34], Spain [35], France [36], and New Zealand [37].
- An ANOVA analysis supported by the Tukey's test on the compressive resistance's results revealed that the growth in strength for soil–cement blocks tested occurred after seven days of curing and for 1.50 % WTSF blocks only. Thus, adding higher percentages of recycled steel fibers to the blocks may positively impact their compressive strength.
- The current study was exploratory, only considering the compressive strength assessment; the main mechanical property required by the regulations for its use in construction. Therefore, subsequent research is needed to consider other mechanical properties, such as tensile strength, flexural strength, durability, plastic energy absorption, large deformation resist, and others to consolidate the soil–cement WTSF blocks usage in building construction.
- The compressive strength results are favorable for 1.5% of WTSF, reaching the minimum category for construction with ease, especially for walls with nonstructural function. Thus, soil–cement blocks with WTSF are presented as a sustainable option in developing countries due to the reuse of waste and to help reduce the housing deficit.
- The current study was limited to three mixing groups, which does not provide a deep understanding of the influence of WTSF on the compressive strength of soil–cement blocks. Future lines of research may consider higher amounts of WTSF to find an optimal percentage addition to obtain advantageous physical and mechanical properties. Additionally, WTSF with rubber can be considered for further studies since this waste is obtained directly from the recycling companies. Hence, waste tire material selection could be avoided, saving time and providing greater practicality.
- Although this exploratory study accomplished the regulations regarding the minimum number of block samples to be tested by age, it is essential to explore larger samples to have more satisfactory results and have the possibility of a broader statistical analysis.

**Author Contributions:** Conceptualization, J.H.A.R. and N.G.C.C.; methodology, J.H.A.R., N.G.C.C. and M.H.R.; formal analysis, J.H.A.R.; investigation, F.P.G., L.L.D., S.P.P. and R.P.A.; resources, F.P.G.; writing—original draft preparation, J.H.A.R., F.P.G., N.G.C.C. and M.H.R.; writing—review and editing, J.H.A.R. and F.P.G.; visualization, F.P.G., M.H.R., L.L.D., S.P.P. and R.P.A.; supervision, J.H.A.R. All authors have read and agreed to the published version of the manuscript.

**Funding:** This research received no external funding.

**Institutional Review Board Statement:** Not applicable.

**Informed Consent Statement:** Not applicable.

**Data Availability Statement:** Not applicable.

**Acknowledgments:** The contributions of the National Directorate of Research and the Academic Department of Civil Engineering of the "Universidad Privada del Valle" are acknowledged.

**Conflicts of Interest:** The authors declare no conflict of interest.

## References

1. ACI Committee 544. *State-of-the-Art Report on Fiber Reinforced Concrete*; ACI Committee 544 Report 544.1R-96; American Concrete Institute: Detroit, MI, USA, 1996.
2. Shi, X.; Brescia-Norambuena, L.; Tavares, C.; Grasley, Z. Semicircular bending fracture test to evaluate fracture properties and ductility of cement mortar reinforced by scrap tire recycled steel fiber. *Eng. Fract. Mech.* **2020**, *236*, 107228. [CrossRef]
3. Achilleos, C.; Hadjimitsis, D.; Neocleous, K.; Pilakoutas, K.; Neophytou, P.O.; Kallis, S. Proportioning of steel fibre reinforced concrete mixes for pavement construction and their impact on environment and cost. *Sustainability* **2011**, *3*, 965–983. [CrossRef]
4. Liu, Y.; Zhang, Z.; Shi, C.; Zhu, D.; Li, N.; Deng, Y. Development of ultra-high performance geopolymer concrete (UHPGC): Influence of steel fiber on mechanical properties. *Cem. Concr. Compos.* **2020**, *112*, 103670. [CrossRef]
5. Awolusi, T.F.; Oke, O.L.; Akinkurolere, O.O.; Sojobi, A.O. Application of response surface methodology: Predicting and optimizing the properties of concrete containing steel fibre extracted from waste tires with limestone powder as filler. *Case Stud. Constr. Mater.* **2019**, *10*, e00212. [CrossRef]
6. Awolusi, T.F.; Oke, O.L.; Atoyebi, O.D.; Akinkurolere, O.O.; Sojobi, A.O. Waste tires steel fiber in concrete: A review. *Innov. Infrastruct. Solut.* **2021**, *6*, 34. [CrossRef]
7. Thomas, B.S.; Gupta, R.C. A comprehensive review on the applications of waste tire rubber in cement concrete. *Renew. Sustain. Energ. Rev.* **2016**, *54*, 1323–1333. [CrossRef]
8. United Nations. *World Economic and Social Survey 2013: Sustainable Development Challenges*; United Nations: New York, NY, USA, 2013.
9. Pilakoutas, K.; Neocleous, K.; Tlemat, H. Reuse of tyre steel fibres as concrete reinforcement. *Proc. Inst. Civ. Eng. Eng. Sustain.* **2004**, *157*, 131–138. [CrossRef]
10. World Business Council for Sustainable Development. *Managing End-of-Life Tires*; Atar Roto Presse SA: Satigny, Switzerland, 2008.
11. US Tire Manufacturers Association. 2019 US Scrap Tire Management Summary. 2020. Available online: https://www.ustires.org/sites/default/files/2019%20USTMA%20Scrap%20Tire%20Management%20Summary%20Report.pdf (accessed on 30 December 2021).
12. Cámara Boliviana de Hidrocarburos. En Bolivia 1.5 Millones de Llantas se Usan al Año. 2016. Available online: http://www.cbhe.org.bo/noticias/13225-en-bolivia-1-5-millones-de-llantas-se-usan-al-ano (accessed on 30 December 2021).
13. Thomas, B.S.; Gupta, R.C.; Kalla, P.; Cseteneyi, L. Strength, abrasion and permeation characteristics of cement concrete containing discarded rubber fine aggregates. *Constr. Build. Mater.* **2014**, *59*, 204–212. [CrossRef]
14. Gupta, T.; Sharma, R.K.; Chaudhary, S. Impact resistance of concrete containing waste rubber fiber and silica fume. *Int. J. Impact Eng.* **2015**, *83*, 76–87. [CrossRef]
15. Shu, X.; Huang, B. Recycling of waste tire rubber in asphalt and portland cement concrete: An overview. *Constr. Build. Mater.* **2014**, *67*, 217–224. [CrossRef]
16. Mohajerani, A.; Burnett, L.; Smith, J.V.; Markovski, S.; Rodwell, G.; Rahman, M.T.; Kurmus, H.; Mirzababaei, M.; Arulrajah, A.; Horpibulsuk, S.; et al. Recycling waste rubber tyres in construction materials and associated environmental considerations: A review. *Resour. Conserv. Recycl.* **2020**, *155*, 104679. [CrossRef]
17. Bjegovic, D.; Baricevic, A.; Lakusic, S.; Damjanovic, D.; Duvnjak, I. Positive interaction of industrial and recycled steel fibres in fibre reinforced concrete. *J. Civ. Eng. Manag.* **2013**, *19* (Suppl. S1), S50–S60. [CrossRef]
18. Leone, M.; Centonze, G.; Colonna, D.; Micelli, F.; Aiello, M.A. Fiber-reinforced concrete with low content of recycled steel fiber: Shear behaviour. *Constr. Build. Mater.* **2018**, *161*, 141–155. [CrossRef]
19. Al-Tikrite, A.; Hadi, M.N. Mechanical properties of reactive powder concrete containing industrial and waste steel fibres at different ratios under compression. *Constr. Build. Mater.* **2017**, *154*, 1024–1034. [CrossRef]
20. Neocleous, K.; Tlemat, H.; Pilakoutas, K. Design issues for concrete reinforced with steel fibers, including fibers recovered from used tires. *J. Mater. Civ. Eng.* **2006**, *18*, 677–685. [CrossRef]

21. Nascimento, E.S.S.; de Souza, P.C.; de Oliveira, H.A.; Júnior, C.M.M.; de Oliveira Almeida, V.G.; de Melo, F.M.C. Soil-cement brick with granite cutting residue reuse. *J. Clean. Prod.* **2021**, *321*, 129002. [CrossRef]
22. da Silva Segantini, A.; Wada, P. An evaluation of the composition of soil cement bricks with construction and demolition waste. *Acta Sci. Technol.* **2011**, *33*, 179–184. [CrossRef]
23. Sivakumar Babu, G.L.; Vasudevan, A.K. Strength and stiffness response of coir fiber-reinforced tropical soil. *J. Mater. Civ. Eng.* **2008**, *20*, 571–577. [CrossRef]
24. Islam, M.S.; Iwashita, K. Performance of Composite Soil Reinforced with Jute Fiber. In *Proceedings of the Getty Seismic Adobe Project 2006 Colloquium*, 1st ed.; Hardy, M., Cancino, C., Ostergren, G., Eds.; The Getty Conservation Institute: Los Angeles, CA, USA, 2009; Volume 1, pp. 11–22.
25. Eko, R.M.; Offa, E.D.; Ngatcha, T.Y.; Minsili, L.S. Potential of salvaged steel fibers for reinforcement of unfired earth blocks. *Constr. Build. Mater.* **2012**, *35*, 340–346. [CrossRef]
26. Sukontasukkul, P.; Jamsawang, P. Use of steel and polypropylene fibers to improve flexural performance of deep soil–cement column. *Constr. Build. Mater.* **2012**, *29*, 201–205. [CrossRef]
27. Tajdini, M.; Hajialilue Bonab, M.; Golmohamadi, S. An Experimental Investigation on Effect of Adding Natural and Synthetic Fibres on Mechanical and Behavioural Parameters of Soil–Cement Materials. *Int. J. Civ. Eng.* **2018**, *16*, 353–370. [CrossRef]
28. Bakam, A.; Ndikontar, M.; Njilah, I. Essais de stabilisation de la latérite avec les fibres cellulosiques. *Afr. J. Sci. Technol.* **2004**, *5*, 22–28. [CrossRef]
29. Centonze, G.; Leone, M.; Aiello, M.A. Steel fibers from waste tires as reinforcement in concrete: A mechanical characterization. *Constr. Build. Mater.* **2012**, *36*, 46–57. [CrossRef]
30. Samarakoon, S.; Ruben, P.; Pedersen, J.; Evangelista, L. Mechanical performance of concrete made of steel fibers from tire waste. *Case Stud. Constr. Mater.* **2019**, *11*, e00259. [CrossRef]
31. Moghadam, A.; Omidinasab, F.; Abdalikia, M. The effect of initial strength of concrete wastes on the fresh and hardened properties of recycled concrete reinforced with recycled steel fibers. *Constr. Build. Mater.* **2021**, *300*, 124284. [CrossRef]
32. Associação Brasileira de Normas Técnicas. *NBR 10834: Bloco De Solo-Cimento Sem Função Estrutural—Requisitos*; Associação Brasileira de Normas Técnicas: Rio de Janeiro, Brasil, 2013.
33. Instituto Boliviano de Normalización y Calidad. *NB 1211003: Ladrillos Cerámicos—Ladrillos Macizos—Clasificación Y Requisitos (Tercera Revisión)*; Instituto Boliviano de Normalización y Calidad: La Paz, Bolivia, 2013.
34. ICONTEC. *NTC 5324: Bloques de Suelo Cemento Para Muros y Divisiones. Definiciones. Especificaciones. Métodos de Ensayo. Condiciones de Entrega*; ICONTEC: Bogota, Colombia, 2004.
35. Asociación Española de Normalización y Certificación. *UNE 41410. Compressed Earth Blocks for Walls and Partitations*; Definitions, Specifications and Test Methods; Asociación Española de Normalización y Certificación: Madrid, Spain, 2008.
36. Association française de Normalisation. *XP P13-901: Compressed Earth Blocks for Walls and Partitions: Definitions—Specifications—Test* Methods—Delivery Acceptance Conditions; Association Française de Normalisation: Paris, France, 2001.
37. Standards New Zealand. *NZS 4298: Materials and Workmanship for Earth Buildings*; Standards New Zealand: Wellington, New Zealand, 1998.
38. Carvalho, M.; Ramos, F.; Zegarra, J.; Pereira, C. Evaluation over time of the mechanical properties of soil-cement blocks used in semi-permeable pavements. *Revista Ingeniería de Construcción* **2016**, *31*, 61–70. [CrossRef]
39. Arooz, F.R.; Halwatura, R.U. Mud-concrete block (MCB): Mix design & durability characteristics. *Case Stud. Constr. Mater.* **2018**, *8*, 39–50. [CrossRef]
40. Rocha, J.H.A.; Rosas, M.H.; Chileno, N.G.C.; Tapia, G.S.C. Physical-mechanical assessment for soil-cement blocks including rice husk ash. *Case Stud. Constr. Mater.* **2021**, *14*, e00548. [CrossRef]
41. American Society for Testing and Materials. *ASTM C595/C595M-20, Standard Specification for Blended Hydraulic Cements*; American Society for Testing and Materials: West Conshohocken, PA, USA, 2020. [CrossRef]
42. American Society for Testing and Materials. *ASTM D422-463(2007)e2, Standard Test Method for Particle-Size Analysis of Soils (Withdrawn 2016)*; American Society for Testing and Materials: West Conshohocken, PA, USA, 2007. [CrossRef]
43. American Society for Testing and Materials. *ASTM D4318-17e1, Standard Test Methods for Liquid Limit, Plastic Limit, and Plasticity Index of Soils*; American Society for Testing and Materials: West Conshohocken, PA, USA, 2017. [CrossRef]
44. American Association of State Highway and Transportation Officials. *M145-91 Standard Specification for Classification of Soils and Soil-Aggregate Mixtures for Highway Construction Purposes*; American Association of State Highway and Transportation Officials: Washington, DC, USA, 1995.
45. American Society for Testing and Materials. *ASTM D2487-17e1, Standard Practice for Classification of Soils for Engineering Purposes (Unified Soil Classification System)*; American Society for Testing and Materials: West Conshohocken, PA, USA, 2017. [CrossRef]
46. American Society for Testing and Materials. *ASTM D698-12e2, Standard Test Methods for Laboratory Compaction Characteristics of Soil Using Standard Effort (12 400 ft-lbf/ft$^3$ (600 kN-m/m$^3$))*; American Society for Testing and Materials: West Conshohocken, PA, USA, 2012. [CrossRef]
47. Shulman, V. Tire Recycling. In *Waste*, 2nd ed.; Letcher, T., Vallero, D., Eds.; Elsevier: Amsterdam, The Netherlands, 2019; pp. 489–515.
48. Afroughsabet, V.; Biolzi, L.; Ozbakkaloglu, T. High-performance fiber-reinforced concrete. *J. Mater. Sci.* **2016**, *51*, 6517–6551. [CrossRef]

49. Hu, H.; Papastergiou, P.; Angelakopoulos, H.; Guadagnini, M.; Pilakoutas, K. Mechanical properties of SFRC using blended manufactured and recycled tyre steel fibres. *Constr. Build. Mater.* **2018**, *163*, 376–389. [CrossRef]
50. Gatani, M.P. Ladrillos de suelo-cemento: Mampuesto tradicional en base a un material sostenible. *Inf. Constr.* **2000**, *51*, 35–47. [CrossRef]
51. Associação Brasileira de Normas Técnicas. *NBR 8492: Tijolo De Solo Cimento—Análise Dimensional, Determinação Da Resistência à Compressão E Da Absorção De Água–Método De Ensaio*; Associação Brasileira de Normas Técnicas: Rio de Janeiro, Brasil, 2012.
52. Nasir, B. Steel Fiber Reinforced Concrete Made with Fibers Extracted from Used Tyres. Ph.D. Thesis, Addis Ababa University, Addis Ababa, Ethiopia, 2009.
53. Ndayambaje, J.C. Structural Performance and Impact Resistance of Rubberized Concrete. Ph.D. Thesis, Pan-African University for Basic Science, Technology and Inovation, Juja, Kenya, 2018.
54. Mastali, M.; Dalvand, A. Use of silica fume and recycled steel fibers in self-compacting concrete (SCC). *Constr. Build. Mater.* **2016**, *125*, 196–209. [CrossRef]

 materials

*Article*

# Development of Eco-Friendly Concrete Mix Using Recycled Aggregates: Structural Performance and Pore Feature Study Using Image Analysis

**Plaban Deb [1], Barnali Debnath [2], Murtaza Hasan [1,*], Ali S. Alqarni [3,*], Abdulaziz Alaskar [3], Abdullah H. Alsabhan [3], Mohammad Amir Khan [4], Shamshad Alam [3] and Khalid S. Hashim [5]**

1 Department of Civil Engineering, Chandigarh University, Mohali 140413, Punjab, India; plaban930@gmail.com
2 Department of Civil Engineering, North Tripura District Polytechnic, Bagbassa 799253, Tripura, India; brnali540@gmail.com
3 Department of Civil Engineering, College of Engineering, King Saud University, P.O. Box 800, Riyadh-11421, Saudi Arabia; abalaskar@ksu.edu.sa (A.A.); aalsabhan@ksu.edu.sa (A.H.A.); salam@ksu.edu.sa (S.A.)
4 Department of Civil Engineering, Galgotia College of Engineering, Knowledge Park I, Greater Noida 201310, Uttar Pradesh, India; amirmdamu@gmail.com
5 Built Environment and Sustainable Technologies (BEST) Research Institute, Liverpool John Moores University, Liverpool L3 3AF, UK; k.s.hashim@ljmu.ac.uk
* Correspondence: murtazadce@gmail.com (M.H.); aalqarni@ksu.edu.sa (A.S.A.)

**Citation:** Deb, P.; Debnath, B.; Hasan, M.; Alqarni, A.S.; Alaskar, A.; Alsabhan, A.H.; Khan, M.A.; Alam, S.; Hashim, K.S. Development of Eco-Friendly Concrete Mix Using Recycled Aggregates: Structural Performance and Pore Feature Study Using Image Analysis. *Materials* **2022**, 15, 2953. https://doi.org/10.3390/ma15082953

Academic Editors: Andrea Petrella, Michele Notarnicola and Francisco Agrela

Received: 12 March 2022
Accepted: 16 April 2022
Published: 18 April 2022



**Abstract:** The shortage of natural aggregates has compelled the developers to devote their efforts to finding alternative aggregates. On the other hand, demolition waste from old constructions creates huge land acquisition problems and environmental pollution. Both these problems can be solved by recycling waste materials. The current study aims to use recycled brick aggregates (RBA) to develop eco-friendly pervious concrete (PC) and investigate the new concrete's structural performance and pore structure distributions. Through laboratory testing and image processing techniques, the effects of replacement ratio (0%, 20%, 40%, 60%, 80%, and 100%) and particle size (4.75 mm, 9.5 mm, and 12.5 mm) on both structural performance and pore feature were analyzed. The obtained results showed that the smallest aggregate size (size = 4.75 mm) provides the best strength compared to the large sizes. The image analysis method has shown the average pore sizes of PC mixes made with smaller aggregates (size = 4.75 mm) as 1.8–2 mm, whereas the mixes prepared with an aggregate size of 9.5 mm and 12.5 mm can provide pore sizes of 2.9–3.1 mm and 3.7–4.2 mm, respectively. In summary, the results confirmed that 40–60% of the natural aggregates could be replaced with RBA without influencing both strength and pore features.

**Keywords:** recycled brick aggregate; pervious concrete; image processing; pore feature

## 1. Introduction

The growing need for natural aggregates in the construction sector is causing a scarcity of materials and creating an environmental imbalance. [1,2]. Some countries, therefore, have restricted the excess use of natural aggregates in construction [3]. The north-eastern part of India, especially Tripura, is suffering from the unavailability of natural aggregates, and these are mostly imported from the nearby country Bangladesh [4–6]. As the states do not have their own source of natural aggregates and depend on other states/countries, the cost of natural stone is much higher in such areas [7]. As the world is becoming environmentally more conscious, any project nowadays not only considers the economy of the project but also checks the environmental impact that the project would have on the livelihood of the human being. Being a nonrenewable resource, natural aggregates requires an alternative to counteract their overuse and shortages [8]. On the other hand, demolishing old structures produces a considerable amount of construction and demolition waste (CDW) that requires expensive disposal, and thereby large land areas become occupied [9,10]. Billions of tons of CDW annually are generated worldwide, causing severe environmental problems, such as water and soil pollution and infrastructural problems. For example, choking sewer systems in Chennai, India, due to the accumulation of CDW, resulted in a severe flood in 2015 [11]. Therefore, researchers are constantly trying to find innovative ways to recycle or reuse CDW to minimize their severe environmental and infrastructural problems effects, such as using recycled aggregates (RA) and recycled concrete aggregates (RCA) in construction material [7–9]. The concept of using RA as a construction material is not new, and there are several studies on the use of RA in concrete mixes or asphalt mixes [10–17]. In addition, researchers have recently tried to use RA in pervious concrete (PC) mixes [18,19].

Pervious concrete (PC) is an emerging construction concept that is a benefit to urban developers as it is the best and most sustainable way to control urban stormwater [20–22]. Currently, most developed and many developing countries are looking forward to using the PC in urban areas for its advantages from an ecological and hydrological point of view. Researchers have studied several aspects of PC as a pavement structure for managing heavy runoff in urban areas and 'Urban Heat Island' control [23–26]. PC is mainly porous concrete mixes having a certain amount of voids through which the runoff can be transferred to the lower layers and finally to the groundwater. Around 15–30% of voids are generally present in the PC mixes that can effectively produce an infiltration rate of around 0.1–3.5 cm/s [20,27]. However, the presence of these voids drastically affects the strength of PC mixes [28,29]. Hence, pervious concrete pavements cannot be implemented on any good quality service roads such as expressways, highways, or any other connecting roads. The major application of PCs can be found in parking lots, sidewalks, shoulders, etc., where low-strength pavements are also acceptable. The compressive strength of PC mixes generally ranges from 2 to 30 MPa, whereas the average strength of 'Pavement Quality Concrete (PQC)' is around 40 MPa. The foregoing studies on PCs have incorporated several governing factors such as (a) size, shape, and type of aggregate, (b) mix proportions, (c) void content, (d) water–cement ratio, (e) aggregate–binder ratio, (f) compaction type, and many others.

Zaetang et al. [30] have used lightweight aggregates (LWA) for preparing PC mixes and observed that the strength of PC mixes using LWA ranged from 2.47 MPa to 6 MPa. Considering the type of aggregates, the majority of the researchers have used natural stone aggregates, although some researchers have used lightweight aggregates, recycled aggregates, brick aggregates, steel slag aggregates, etc. [19,28,30–35]. Debnath and Sarkar [36] carried out a detailed characterization of PCs using over-burnt bricks and found that brick aggregates can be a good option for minimizing the consumption of stones. Steel slag aggregates were also used in the production of PC; the results of these studies indicated the compressive strength of the produced PC is in the range of 10–30 MPa with a permeability range of 0.2–2.8 cm/s [33,37]. Gaedicke et al. [38] found that using RA in PC could keep the compressive strength of the produced PC (PC-RA) in the range of 8–20 MPa, whereas the compressive strength of PC made of natural aggregates (PC-NA) is about 28 MPa. In addition, the permeability of the PC-RA is in the range of 0.5–1.5 cm/s, whereas the maximum permeability of PC-NA is 1.2 cm/s. Other studies on PC-RA have also mentioned that RAs are feasible enough to be used in PC mixes. However, all the previous studies focused on recycled concrete aggregates (RCA) and they have not focused particularly on recycled brick aggregates (RBAs). Thus the performance and suitability of RBAs in pervious concrete mixes are unknown. The present study addresses this research gap and aims at using recycled brick aggregates (RBA) in the production of PC. This study mostly focuses on the mechanical properties of the produced PC and the suitability of RBAs in PC mixes. Moreover, the past studies on PC-RA mainly conducted experimental tests to find out the performances, while the current study uses image analysis along with the experimental tests to investigate the performance of the produced concrete. The novelty of this research work mainly lies in the use of RBA in the PC mixes which will help to mitigate the shortage of NA as well as the nuisances created by CDW. This study primarily provides a simple overview of the structural behavior and pore feature details of PC mixes prepared with recycled brick aggregates.

## 2. Experimental Plan

### *2.1. Aggregates and Other Materials*

Demolition wastes were collected from a nearby construction site in India, and then recycled brick aggregates were collected from these, see Figure 1. The latter shows that the aggregates consist of crushed brick and mortar. The collected aggregates were sieved and graded to obtain three different aggregate gradations named RBA-12.5, RBA-9.5, and RBA-4.75, see Table 1. The general properties of RBAs are also evaluated and compared with the properties of natural stone and natural brick aggregates (NBA) (Table 2). The table provides the average values of different test results. The properties were calculated for different size fractions and then the average values are represented here for ease of comparison. Ordinary Portland Cement (OPC)-43 grade cement was used as a binder for preparing PC mixes, and sand was used as fine aggregate. The use of sand in PC mixes is generally prohibited, but some studies mentioned that sand provides better bonding and is essential when non-conventional aggregates are used in PC mixes [22]. Thus, in this study, 10% of sand by volume of coarse aggregate is used in the mix.

**Figure 1.** Recycled brick aggregate (RBA).

**Table 1.** Aggregate gradation.

| Gradation | Size of Aggregate (Percentage Passing) | | | | |
|---|---|---|---|---|---|
| | 13.2 mm | 12.5 mm | 9.5 mm | 6.3 mm | 4.75 mm |
| RBA-12.5 | 100 | 0 | 0 | 0 | 0 |
| RBA-9.5 | 100 | 100 | 0 | 0 | 0 |
| RBA-4.75 | 100 | 100 | 100 | 100 | 0 |

**Table 2.** General properties of RBA (standard specifications were followed from [39,40]).

| Properties | NA | RBA | NBA (Over-Burnt) | Standard Guidelines |
|---|---|---|---|---|
| Impact value (%) | 18.4 | 33.4 | 36.2 | IS:2386, Part IV [40] |
| Abrasion value (%) | 25.6 | 40.8 | 45.3 | IS:2386, Part IV [40] |
| Crushing value (%) | 22.5 | 36.5 | 38.7 | IS:2386, Part IV [40] |
| Specific gravity | 2.781 | 1.975 | 1.912 | IS:2386, Part III [39] |

Mix Design and Preparation of Samples

The addition of RBA in PCs was performed in different sets depending upon the quantity of RBA. Several blending proportions were used, and natural aggregate (NA) was partially/fully replaced by RBA. The blending proportions (NA:RBA) were 100:0, 80:20, 60:40, 40:60, 20:80, and 0:100. The mix design method used in this study was following the research conducted by Debnath and Sarkar [36], as there is a lack of standard specifications for mix design of PC with non-conventional aggregates. The water–cement ratio was chosen as 0.3, and the designed air void was 20%. Table 3 shows the mixed proportions of cement, aggregate, and water along with a commercial superplasticizer (BASF Master Rheobuild 1125, Master Builders Solutions India Private Limited, Navi Mumbai, Maharashtra, India). The samples were prepared in a traditional concrete mixer, and then the prepared mix was transferred into some prefabricated molds depending upon the type of tests. A curing period of 28 days was fixed for all the mixes.

**Table 3.** Mix proportion of PC mixes.

| Mix Type | NA (%) | RBA (%) | NA (kg/m$^3$) | RBA (kg/m$^3$) | Binder (kg/m$^3$) | Sand (kg/m$^3$) | Water (kg/m$^3$) | Admixture (kg/m$^3$) |
|---|---|---|---|---|---|---|---|---|
| RBA-0 | 100 | 0 | 1619.57 | 0 | 242.42 | 161.83 | 72.73 | 1.94 |
| RBA-20 | 80 | 20 | 1295.66 | 230.12 | 242.42 | 161.83 | 72.73 | 1.94 |
| RBA-40 | 60 | 40 | 971.74 | 460.24 | 242.42 | 161.83 | 72.73 | 1.94 |
| RBA-60 | 40 | 60 | 647.83 | 690.36 | 242.42 | 161.83 | 72.73 | 1.94 |
| RBA-80 | 20 | 80 | 323.91 | 920.48 | 242.42 | 161.83 | 72.73 | 1.94 |
| RBA-100 | 0 | 100 | 0 | 1150.60 | 242.42 | 161.83 | 72.73 | 1.94 |

## 3. Analysis Methods

### *3.1. Structural Behavior*

The structural performance of the PC was evaluated by checking its compressive strength and flexural strength behavior. According to IS: 516 [41], the compressive strength was performed using a cube of 150 mm × 150 mm × 150 mm. The cubes were prepared using the standard cube molds followed by IS:516 [41]. The loading rate during testing was kept as 4 tonne/mm. The flexure behavior of PC was calculated using beams of size 100 mm × 100 mm × 500 mm. A four-point loading system was used for this test, as shown in Figure 2, where the load was applied through a loading frame, and the failure load was recorded using a load cell.

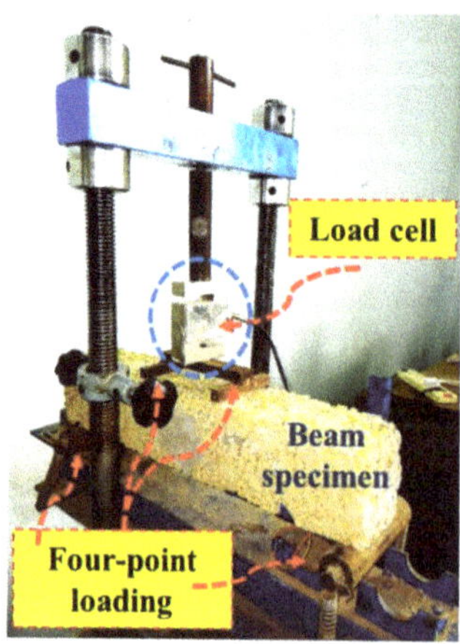


**Figure 2.** Four-point bending test.

Pore Feature and Image Analysis

The porous nature of PC was accumulated through porosity analysis in both the experimental and image analysis process. The experimental method adopted for porosity analysis was followed by ASTM C1754 [42] and other past studies [28]. Cylinders of 100 mm in diameter and 200 mm in height were chosen for performing the tests, where the basic water-displacement concept was used. In the 2nd phase, the PC mixes were analyzed using the popular 'Image Processing Technique (IPT)', where the PC images were processed and thresholded to find the number of voids present in the mixes. Image processing is a widely used method in different fields of engineering. For the past few years, this IPT is also being used for identifying several responses of structures [43]. Freely available software '*ImageJ*$^{TM}$' was used here, and the stereological analysis was conducted [44]. The method adopted for IPT is described through a flow chart in Figure 3. Initially, circular images were scanned from different locations of different PC samples. A similar cylindrical specimen was used for IPT as used for porosity calculation. Then the specimen was cut at five equal intervals to find thin slices, which means each specimen gave five slices. Three replicates were also used for each mix to quantify the variability in the pore distributions, which ultimately produced 15 slices for a particular mix type. The slices were then painted white to identify pores and solid phases properly. Once the solid and pores were identified, the slices were then scanned with a flatbed scanner, and the images' scaling was performed. After scaling and smoothening of boundaries, these were converted into binary images followed by thresholding and unnecessary noise removal. Finally, a square cross-section was cropped from each image (equal size for each type of mix), which denoted the representative area element (RAE). This RAE signifies the behavior of a whole specimen, and hence this RAE should be selected carefully. As PC mixes are porous and have variability in pore distribution, 15 RAEs were taken from each mix, which can represent the behavior of that particular mix.

**Figure 3.** Method of image processing.

# 4. Results and Discussions

## 4.1. Strength Behavior of PC-RBA

### 4.1.1. Compressive Strength

This section gives the detailed test results of compressive strength (fcs) of PC mixes produced with RBAs. The variation of fcs values for different blending proportions of RBA is shown in Figure 4, where it can be observed that the average fcs values are becoming reduced with the increased proportions of RBA. The PC mix with 100% NA has shown an fcs value of 14–16.7 MPa, whereas the fcs value of the PC mix with 100% RBA is only about 1.3–2.1 MPa. The recycled bricks have lesser strength properties such as crushing value, impact value, etc., compared to natural ones (refer to Table 2), and hence the concrete mixes produced with RBA become weaker than PC-NA. The typical lower limit of fcs for PC mixes is recommended as 3 MPa [45] and the majority of researchers have found or suggested an fcs value of at least 6 MPa for PC mixes made with single-sized aggregates [30,46]. However, it is observed in the figure that the PC mix with 100% RBA is not able to provide the lower limit of fcs, and can give an fcs value of about 1.3–2.1 MPa, irrespective of aggregate size. The PC mix with 80% RBA can produce an fcs value of about 3.2 MPa, 3.6 MPa, and 4.2 MPa for RBA-12.5, RBA-9.5, and RBA-4.75, respectively higher than the lower limit, although these do not satisfy the suggested limit (6 MPa). However, the PC mix made with 60% RBA is showing an fcs value of 6.1 MPa and 6.6 MPa for RBA-9.5 and RBA-4.75, which is very slightly higher than the suggested limit, while the mix RBA-60 with an aggregate size of 12.5 mm can reach an fcs value of 4.9 MPa. On the other hand, the PC mixes with 40% or 20% can provide fcs values much higher than the suggested limit for any aggregate size. For the mixes RBA-40 and RBA-20, the fcs values are 1.4–1.8 times and 1.7–2.4 times higher than the suggested limit. Therefore, if 6 MPa is taken as a benchmark of compressive strength for PC mixes, the natural aggregates can be replaced by 60% of RBA, when the aggregate size is 9.5 mm or less. Similarly, for an aggregate size of 12.5 mm, the natural aggregates can be replaced by 40% of RBA.

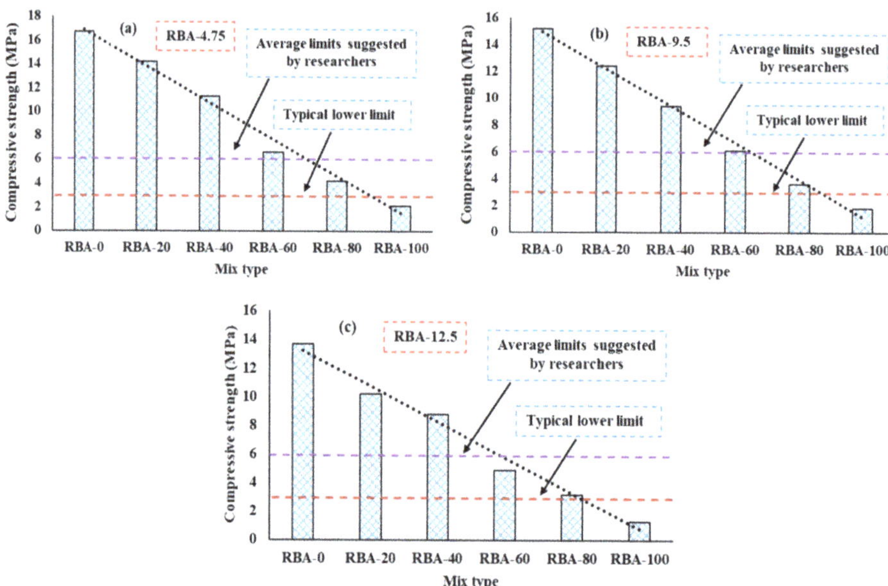


Figure 4. Compressive strength of PC mixes, (**a**) for RBA-4.75, (**b**) RBA-9.5, (**c**) RBA-12.5.

#### 4.1.2. Flexural Strength

As pervious concretes are mostly used as paved sections, the flexural response also becomes essential. The flexural strength (fr) variations for different mixes are shown in Figure 5, where it is observed that the mixes produced with a higher percentage of RBA are not suitable to withstand a high flexural load. The variations of fr are somewhat similar to that of fcs due to the poor strength of RBAs. The path of flexural load transfer mostly propagates through the aggregates, and hence the weaker RBAs cannot take more loads. The typical lower limit of fr was suggested as 1 MPa [26,39], and it is found that the PC mixes made with 100% and 80% RBA cannot provide the minimum required value of fr for RBA-4.75. When the 4.75 mm aggregate is used for preparing PC mixes, fr values are obtained as 0.43 MPa, 0.74 MPa, 1.1 MPa, 1.5 MPa, 1.8 MPa, and 2.4 MPa for the mixes RBA-100, RBA-80, RBA-60, RBA-40, RBA-20, and RBA-0, respectively. However, for RBA-9.5 and RBA-12.5, 100% and 80% and 60% RBA usage are not useful for obtaining a limit value of flexural strength. The aggregate size of 9.5 mm fr values are obtained as 0.38 MPa, 0.57 MPa, 0.82 MPa, 1.2 MPa, 1.5 MPa, and 2.1 MPa for the mixes RBA-100, RBA-80, RBA-60, RBA-40, RBA-20, and RBA-0, respectively. Similarly, the aggregate size of 12.5 mm fr values are obtained as 0.33 MPa, 0.49 MPa, 0.75 MPa, 1.1 MPa, 1.3 MPa, and 1.9 MPa for the mixes RBA-100, RBA-80, RBA-60, RBA-40, RBA-20, and RBA-0, respectively. Another important thing that can be noticed here is that the value of fr is higher for smaller aggregates compared to larger aggregates, which is primarily happening due to the presence of smaller pore sizes. The gap between two large aggregates is much higher than the gap between two small aggregates, and hence the crack propagation in the mixes with larger aggregates becomes much easier.

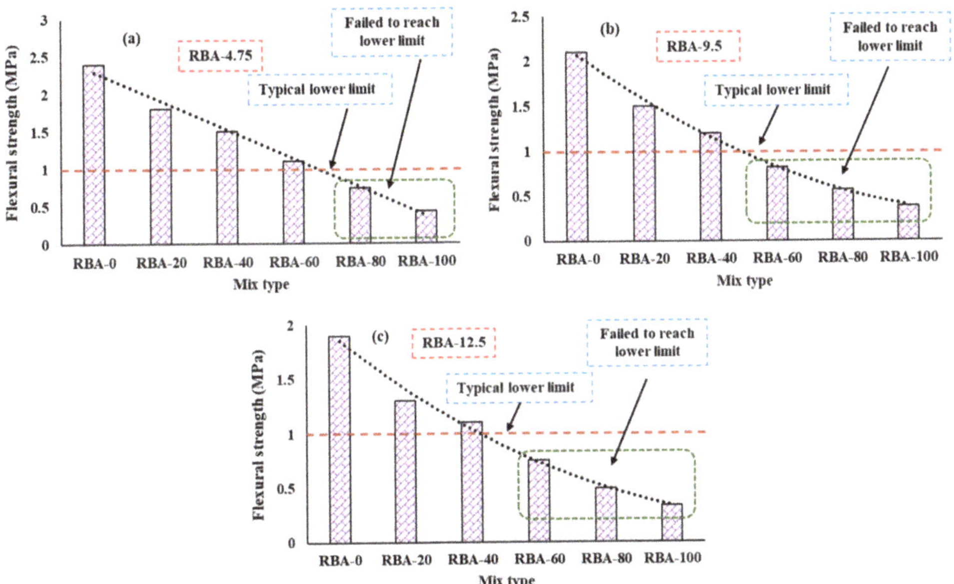


**Figure 5.** Flexural behavior of PC, (**a**) for RBA-4.75, (**b**) RBA-9.5, (**c**) RBA-12.5.

### *4.2. Pore Feature Analysis*

A suitable porous structure and sufficient amount of voids are the key factors of PCs as an adequate water percolation rate can only be achieved due to the presence of these voids. In this study, the porous nature is checked in two ways: (a) porosity measurement and (b) stereological analysis through IPT. The experimental porosity results are shown in Figure 6, where it can be found that the effect of RBAs on the change of porosity values is very little as the porosity value is slightly increasing with the increase in RBA content. It was observed from the literature that the porosity of PC mixes ranges from 15–30%, and here all the mixes show a porosity range of 20% to 30%, which is quite satisfactory. However, the difference in porosity values is much more prominent if the size of aggregates is changed. For an aggregate size of 4.75 mm, the porosity lies in the range of 20–22%, increasing up to 26–27% if a 9.5 mm size aggregate is used. Similarly, the porosity values can be achieved up to 32% if the aggregate size is kept at 12.5 mm. Such varying porosity is common because of the production of different sizes of pores in PC mixes when varying aggregates are used.

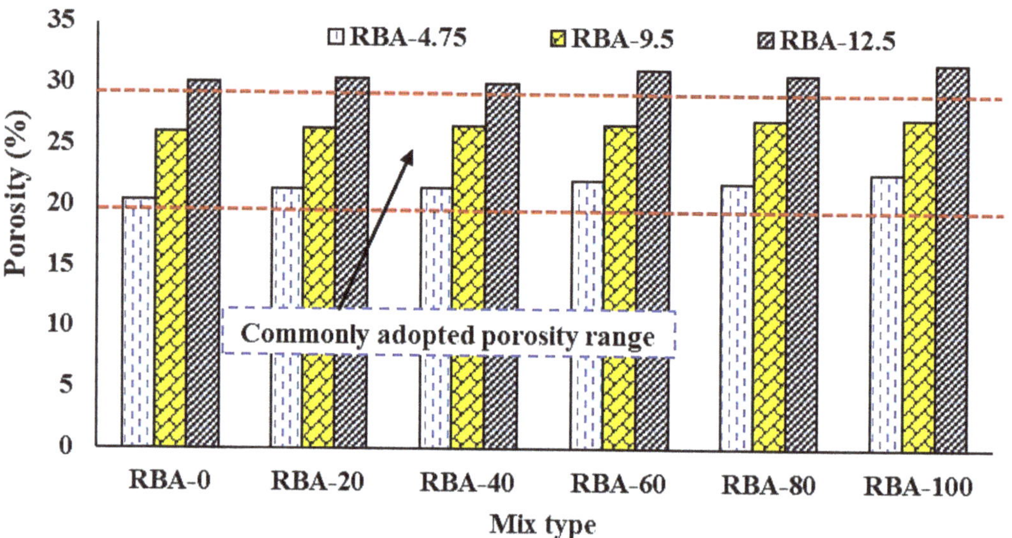


**Figure 6.** Variation of porosity in PC mixes.

This can be better understood from the RAEs collected from different mixes, reflecting the pore distribution in a particular PC mix (Figure 7). Image analysis helps to understand the size of the pores and the distribution of pores in a mix. In this study, the size of pores is identified from the stereological method, which signifies that the area fractions present in each RAE will be numerically equal to the actual porosity calculated experimentally. From this stereological image analysis, several parameters are obtained, such as the area fractions, pore histograms, and pore size present in each type of mix. The area fractions identified for each slice of each replicate of all mixes are mentioned in Table 4, along with the standard deviations and the experimental porosity values. The standard deviation values obtained for the slices S1 to S15 for each type of mix lie in the range of 0.76–1.97. This indicates that the replicates of each mix do not show any great variability and the pore distribution along the specimen is almost homogeneous. According to the stereological theory, the average values of $A_f$ need to be the same as $n_{exp}$. However, it is found that the values of $A_f$ for each mix are slightly higher than $n_{exp}$, indicating a higher void content (Figure 8). However, this variation lies in the range of 0.2% to 7%, and hence it can be assumed that the chosen RAEs can easily replicate the actual specimen.

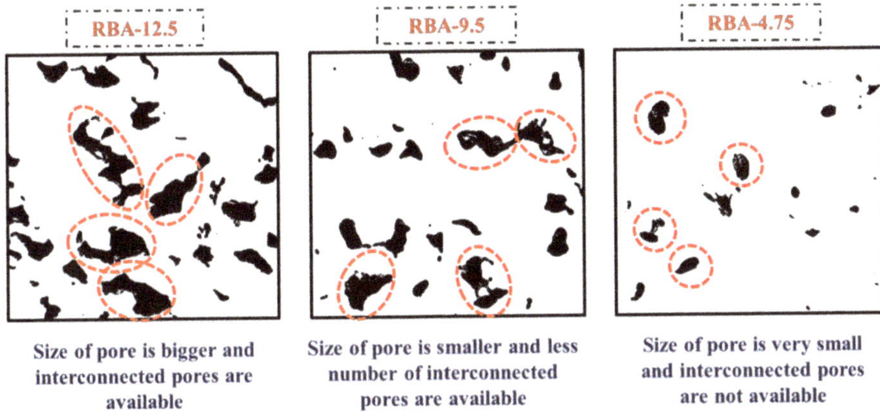


**Figure 7.** Typical RAEs for mixes with different aggregate sizes.

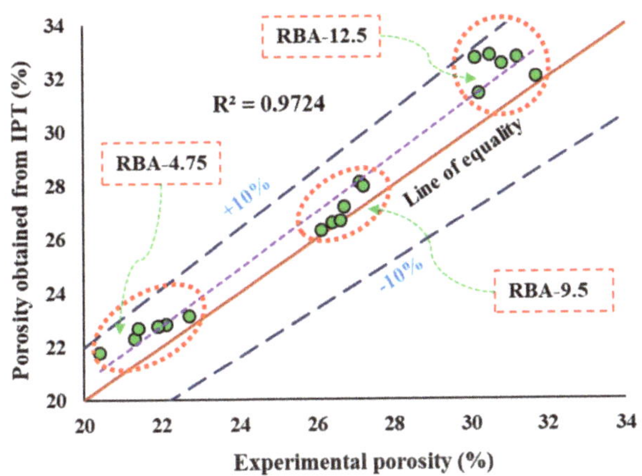


**Figure 8.** Comparison of experimental porosity with the porosity obtained from IPT.

**Table 4.** Details of area fractions obtained from RAEs.

| Mix Type | R-1 | | | | | R-2 | | | | | R-3 | | | | | St Dev | Average | Experimental Porosity ($n_{exp}$) |
|---|---|---|---|---|---|---|---|---|---|---|---|---|---|---|---|---|---|---|
| | S1 | S2 | S3 | S4 | S5 | S6 | S7 | S8 | S9 | S10 | S11 | S12 | S13 | S14 | S15 | | | |
| RBA-0 | 21.4 | 24.2 | 21.6 | 22.4 | 25.4 | 23.2 | 18.7 | 20.8 | 21.7 | 24.3 | 20.5 | 23.4 | 21.1 | 18.3 | 20.1 | 1.97 | 21.81 | 20.4 |
| RBA-20 | 22.4 | 23.6 | 20.7 | 19.7 | 21.4 | 23.7 | 26.2 | 21.5 | 20.6 | 25.4 | 23.4 | 20.2 | 19.7 | 23.6 | 22.4 | 1.94 | 22.30 | 21.3 |
| RBA-40 | 21.6 | 22.4 | 22.3 | 24.6 | 22.7 | 23.4 | 21.8 | 22.6 | 25.1 | 23.2 | 20.6 | 22.1 | 21.9 | 23.6 | 22.7 | 1.11 | 22.71 | 21.4 |
| RBA-60 | 23.6 | 25.1 | 20.4 | 22.7 | 23.4 | 21.8 | 26.7 | 21.4 | 25.3 | 24.1 | 21.3 | 20.2 | 21.4 | 23.6 | 21.4 | 1.86 | 22.83 | 22.1 |
| RBA-80 | 21.5 | 22.3 | 21.7 | 24.1 | 23.5 | 24.3 | 22.4 | 22.9 | 21.6 | 23.4 | 21.4 | 25.1 | 22.2 | 23.6 | 21.7 | 1.13 | 22.78 | 21.9 |
| RBA-100 | 22.6 | 25.4 | 23.1 | 22.3 | 21.6 | 24.8 | 22.7 | 23.3 | 21.6 | 22.9 | 24.7 | 23.5 | 24.1 | 23.5 | 21.4 | 1.17 | 23.17 | 22.7 |
| RBA-0 | 28.4 | 25.3 | 25.6 | 24.1 | 25.3 | 27.4 | 24.6 | 25.8 | 26.8 | 26.9 | 27.8 | 25.6 | 28.9 | 26.8 | 25.7 | 1.34 | 26.33 | 26.1 |
| RBA-20 | 26.4 | 25.5 | 26.5 | 27.3 | 27.7 | 26.3 | 26.9 | 25.8 | 26.1 | 25.5 | 26.2 | 27.4 | 26.7 | 26.2 | 28.4 | 0.79 | 26.59 | 26.4 |
| RBA-40 | 27.4 | 26.8 | 26.7 | 25.8 | 27.4 | 28.3 | 25.7 | 26.8 | 25.9 | 26.3 | 27.4 | 25.9 | 27.3 | 26.4 | 25.7 | 0.76 | 26.65 | 26.6 |
| RBA-60 | 26.7 | 28.6 | 25.6 | 28.4 | 27.6 | 25.7 | 28.3 | 27.4 | 26.8 | 25.8 | 26.3 | 25.8 | 29.1 | 27.4 | 28.4 | 1.15 | 27.19 | 26.7 |
| RBA-80 | 27.8 | 28.9 | 26.7 | 29.4 | 28.4 | 27.6 | 28.3 | 28.7 | 28.9 | 26.8 | 27.7 | 28.8 | 29.6 | 27.4 | 26.8 | 0.91 | 28.12 | 27.1 |
| RBA-100 | 29.4 | 27.6 | 28.7 | 28.3 | 28.4 | 27.5 | 29.5 | 27.4 | 28.5 | 26.7 | 28.7 | 26.7 | 25.9 | 27.5 | 28.8 | 1.00 | 27.97 | 27.2 |
| RBA-0 | 31.2 | 32.5 | 29.8 | 30.7 | 30.6 | 31.6 | 31.7 | 31.5 | 30.2 | 30.8 | 33.4 | 31.4 | 32.1 | 31.6 | 32.2 | 0.90 | 31.42 | 30.2 |
| RBA-20 | 32.4 | 33.6 | 32.7 | 33.4 | 31.5 | 33.8 | 31.2 | 32.4 | 32.8 | 33.6 | 34.6 | 33.4 | 32.8 | 31.8 | 32.7 | 0.89 | 32.85 | 30.5 |
| RBA-40 | 33.2 | 32.6 | 32.1 | 32.8 | 33.2 | 36.4 | 33.2 | 31.5 | 31.4 | 32.3 | 31.9 | 32.7 | 32.5 | 33.4 | 31.7 | 1.16 | 32.73 | 30.1 |
| RBA-60 | 33.5 | 31.6 | 32.7 | 32.5 | 32.4 | 33.6 | 33.8 | 31.5 | 34.2 | 31.7 | 32.6 | 31.8 | 32.7 | 33.8 | 33.4 | 0.86 | 32.79 | 31.2 |
| RBA-80 | 32.7 | 32.8 | 32.1 | 30.8 | 34.5 | 32.2 | 32.6 | 31.6 | 33.4 | 32.6 | 32.8 | 31.9 | 33.5 | 31.9 | 32.5 | 0.84 | 32.53 | 30.8 |
| RBA-100 | 34.3 | 32.1 | 31.5 | 30.6 | 32.1 | 31.6 | 30.7 | 31.8 | 31.1 | 32.5 | 32.2 | 33.1 | 30.8 | 34.6 | 31.4 | 1.16 | 32.03 | 31.9 |

Another major outcome obtained from IPT is calculating the size of the pores in each mix. For obtaining the size of pores, two assumptions are taken in this study, (a) the shapes of the pores are circular, and (b) each pore has a distinct property. Several pore sizes are obtained from a single RAE, and then the cumulative frequency distribution (CFD) curves are drawn to find the equivalent pore size. In the present study, the equivalent pore size is taken as the $d_{50}$ value, i.e., the pore size matching with 50% CFD curve is considered as $d_{50}$. The variation of CFD curves for different mixes is shown in Figure 9, and the average pore sizes for all the mixes are mentioned in Figure 10. It can be seen that the size of the pores for the mixes produced with RBA-12.5 is much larger than that of the mixes produced with RBA-4.75, which is mainly attributed to the presence of larger-sized aggregates in the mixes of RBA-12.5. It can also be understood from Figure 7, where the processed images are shown for three different sizes of aggregates. The available pore size for RBA-4.75 is much smaller as compared to RBA-12.5.

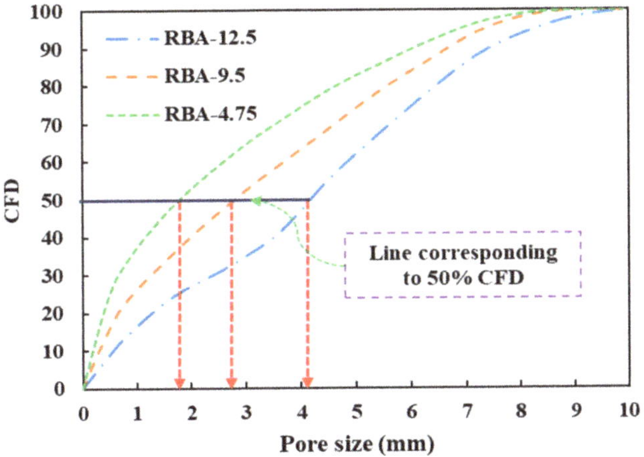


**Figure 9.** Frequency distribution and pore size calculation.

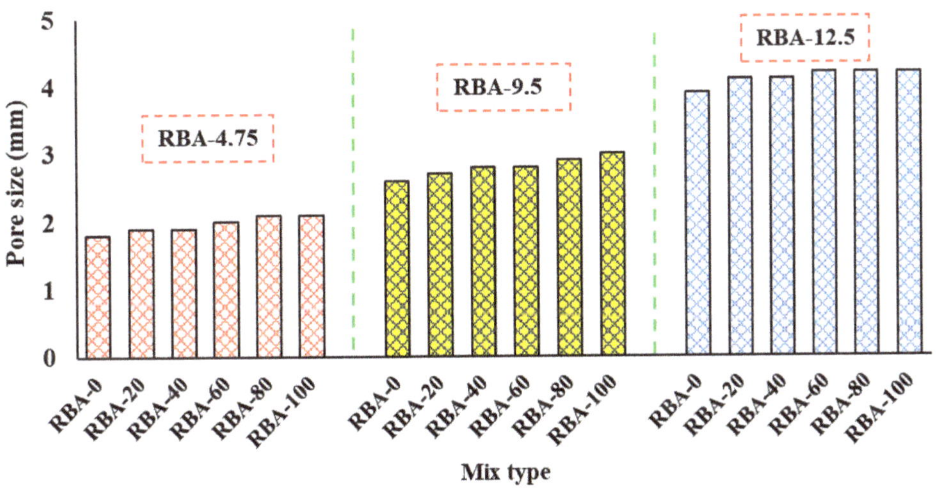


**Figure 10.** Average pore sizes for different mixes.

## 5. Summary of the Test Results

This study provides a preliminary understanding of the usage of recycled brick aggregates in the preparation of pervious concrete mixes. Several percentages of RBAs replace the natural aggregates, and the structural performance, i.e., the strength of the PC mixes, is checked through experimental tests. Being a bi-functional structure, pervious concrete also possesses internal voids for adequate water percolation. This study checks this porous nature through experiments and image processing techniques. The major outcome of the study deals with the applicability of RBAs in PC, and it is found that the full replacement of natural aggregates may not be possible as the effects of RBAs and NAs in PC mixes are not the same. If a PC mix is produced with 100% NA, the compressive strength can be obtained up to 14–16.7 MPa, which will be reduced to 1.3–2.1 MPa with the usage of 100 RBAs. The overall results show that the natural aggregates can be replaced by up to 40–60% of RBAs for ensuring a satisfactory strength. However, the aggregate size also plays an important role here as the smaller aggregates can provide better strength than large aggregates. Compressive and flexural strength values show that 60% of RBAs can be used in PCs if the aggregate size is chosen as 4.75 mm. However, if the aggregate size is larger than 4.75 mm (9.5 mm or 12.5 mm), natural aggregates can be replaced by 40% of RBAs. The porosity does not have much effect on the quantities of RBA that are being used in the mix; rather the porosity is greatly affected by the size of the RBA used in the mix. It is found that the use of 12.5 mm sized RBAs can provide a better porosity, although the use of 4.75 mm sized RBAs can also provide a satisfactory porosity. However, the image analysis reveals that the size of the pores present in PC mixes made with 12.5 mm sized aggregate is about 80% higher than that of the mixes made with 4.75 mm sized aggregates. Considering both strength and porosity, it can be decided that the size of RBA used in PC should be in the range of 4.75 mm to 9.5 mm. In short, it is clear from the test results that the RBAs can easily replace a certain percentage of NA, which will help to conserve our natural resources and also minimize the adverse environmental effects of demolition wastes.

## 6. Conclusions

The deficiency of natural aggregates for the growing demand for construction is assisting the search for alternate materials. From the perspective of natural resource conservation, other marginal materials have more environmental benefits. Moreover, the generation of substantial demolition waste creates several environmental nuisances that can only be minimized by recycling or reusing the waste materials. Both these severe issues can be curtailed by using recycled aggregates in the construction work, which will resolve the problems associated with waste disposal and reduce the consumption of natural aggregates. This study uses recycled brick aggregate in pervious concrete mixes with the intention of partially or fully replacing the natural aggregates. However, the test results disclose that the use of 100% RBAs in PC mixes may not be suitable from the structural point of view, as the use of 100% RBAs cannot provide adequate strength for the PC mixes. The overall test results and image analysis show that the smaller-sized RBAs (4.75 mm to 9.5 mm) are very useful for providing adequate pore size and sufficient strength and porosity. Although some laboratory tests and image analyses of the PC samples have been conducted in this study, further studies need to be carried out for the overall application of RBA in PC. The performance of PC under fatigue load needs further study because the pavements are often subjected to repeated loading and are more prone to fatigue failure.

## 7. Limitations and Future Scope

Although some laboratory tests and image analyses of the PC samples have been conducted in this study, further studies need to be carried out for the overall application of RBA in PC. This study is primarily focused on a particular design mix of PC for several percentages of NA and RBA. It is assumed that the mix could be resolved by optimizing the mix proportions, and several trials could be conducted to obtain an optimum mix design. This can be considered a limitation of this study and can be studied further. Moreover, the durability study and cost analysis of PC with RBA require future analysis for the practical application of PC with RBA. Although the cost analysis has not been performed in this study, the use of recycled materials will be more economical in some locations due to the high price of stones in those areas. The performance of PC under fatigue load also needs further study because the pavements are often subjected to repeated loading and are more prone to fatigue failure. The life cycle cost analysis, fatigue studies, etc., are the future research scopes of this study, which can also be conducted to understand the applicability of RBAs in PC.

**Author Contributions:** P.D.: data collection, composing, reviewing, modifying, and funding; B.D.: composing, data collection, reviewing and modifying; M.H.: modifying, preparation of the manuscript and editing; A.S.A.: drafting and preparation of the manuscript; A.A.: preparation of the manuscript; A.H.A. and S.A.: preparation of the manuscript; M.A.K.: drafting and preparation of the manuscript; K.S.H.: preparation of the manuscript. All authors have read and agreed to the published version of the manuscript.

**Funding:** The authors would like to acknowledge the support provided by Researchers Supporting Project Number (RSP-2021/284), King Saud University, Riyadh, Saudi Arabia.

**Acknowledgments:** The authors are thankful to NIT Agartala for providing the facility for lab experiments and King Saud University, Riyadh, Saudi Arabia for financial support through Project Number (RSP-2021/284).

**Conflicts of Interest:** The authors declare no conflict of interest.

## References

1. Tam, V.W.Y.; Soomro, M.; Evangelista, A.C.J. A review of recycled aggregate in concrete applications (2000–2017). *Constr. Build. Mater.* **2018**, *172*, 272–292. [CrossRef]
2. Bektas, F.; Wang, K.; Ceylan, H. Effects of crushed clay brick aggregate on mortar durability. *Constr. Build. Mater.* **2009**, *23*, 1909–1914. [CrossRef]
3. Singh, S.; Ransinchung, G.D.; Monu, K. Sustainable lean concrete mixes containing wastes originating from roads and industries. *Constr. Build. Mater.* **2019**, *209*, 619–630. [CrossRef]
4. Mazumder, A.R.; Kabir, A.; Yazdani, N. Performance of Overburnt Distorted Bricks as Aggregates in Pavement Works. *J. Mater. Civ. Eng.* **2006**, *18*, 777–785. [CrossRef]
5. Sarkar, D.; Pal, M.; Sarkar, A.K.; Mishra, U. Evaluation of the Properties of Bituminous Concrete Prepared from Brick-Stone Mix Aggregate. *Adv. Mater. Sci. Eng.* **2016**, *2016*, 2761038. [CrossRef]
6. Zachariah, J.P.; Sarkar, P.P.; Debnath, B.; Pal, M. Effect of polypropylene fibres on bituminous concrete with brick as aggregate. *Constr. Build. Mater.* **2018**, *168*, 867–876. [CrossRef]
7. Zachariah, J.P.; Sarkar, P.P.; Pal, M. A study on the moisture damage and rutting resistance of polypropylene modified bituminous mixes with crushed brick aggregate wastes. *Constr. Build. Mater.* **2021**, *269*, 121357. [CrossRef]
8. Yunchao, T.; Zheng, C.; Wanhui, F.; Yumei, N.; Cong, L.; Jieming, C. Combined effects of nano-silica and silica fume on the mechanical behavior of recycled aggregate concrete. *Nanotechnol. Rev.* **2021**, *10*, 819–838. [CrossRef]
9. Salehi, S.; Arashpour, M.; Kodikara, J.; Guppy, R. Sustainable pavement construction: A systematic literature review of environmental and economic analysis of recycled materials. *J. Clean. Prod.* **2021**, *313*, 127936. [CrossRef]
10. Prasad, D.; Singh, B.; Suman, S.K. Utilization of recycled concrete aggregate in bituminous mixtures: A comprehensive review. *Constr. Build. Mater.* **2022**, *326*, 126859. [CrossRef]
11. Ram, V.G.; Kishore, K.C.; Kalidindi, S.N. Environmental benefits of construction and demolition debris recycling: Evidence from an Indian case study using life cycle assessment. *J. Clean. Prod.* **2020**, *255*, 120258. [CrossRef]
12. Guo, H.; Shi, C.; Guan, X.; Zhu, J.; Ding, Y.; Ling, T.-C.; Zhang, H.; Wang, Y. Durability of recycled aggregate concrete—A review. *Cem. Concr. Compos.* **2018**, *89*, 251–259. [CrossRef]
13. Etxeberria, M.; Vázquez, E.; Mari, A.; Barra, M. Influence of amount of recycled coarse aggregates and production process on properties of recycled aggregate concrete. *Cem. Concr. Res.* **2007**, *37*, 735–742. [CrossRef]
14. Etxeberria, M.; Mari, A.; Vázquez, E. Recycled aggregate concrete as structural material. *Mater. Struct.* **2006**, *40*, 529–541. [CrossRef]
15. Singh, S.; Ransinchung, G.; Monu, K.; Kumar, P. Laboratory investigation of RAP aggregates for dry lean concrete mixes. *Constr. Build. Mater.* **2018**, *166*, 808–816. [CrossRef]
16. Prasad, D.; Pandey, A.; Kumar, B. Sustainable production of recycled concrete aggregates by lime treatment and mechanical abrasion for M40 grade concrete. *Constr. Build. Mater.* **2021**, *268*, 121119. [CrossRef]
17. Mohammed, T.U.; Hasnat, A.; Awal, M.A.; Bosunia, S.Z. Recycling of Brick Aggregate Concrete as Coarse Aggregate. *J. Mater. Civ. Eng.* **2015**, *27*, B4014005. [CrossRef]

18. Zhu, X.; Chen, X.; Shen, N.; Tian, H.; Fan, X.; Lu, J. Mechanical properties of pervious concrete with recycled aggregate. *Comput. Concr.* **2018**, *21*, 623–635. [CrossRef]
19. Toghroli, A.; Shariati, M.; Sajedi, F.; Ibrahim, Z.; Koting, S.; Mohamad, E.T.; Khorami, M. A review on pavement porous concrete using recycled waste materials. *Smart Struct. Syst.* **2018**, *22*, 433–440. [CrossRef]
20. Debnath, B.; Sarkar, P.P. Pervious concrete as an alternative pavement strategy: A state-of-the-art review. *Int. J. Pavement Eng.* **2018**, *21*, 1516–1531. [CrossRef]
21. Mullaney, J.; Lucke, T. Practical Review of Pervious Pavement Designs. *Clean-Soil Air Water* **2013**, *42*, 111–124. [CrossRef]
22. Debnath, B.; Sarkar, P.P. Clogging in Pervious Concrete Pavement Made with Non-conventional Aggregates: Performance Evaluation and Rehabilitation Technique. *Arab. J. Sci. Eng.* **2021**, *46*, 10381–10396. [CrossRef]
23. Sartipi, M.; Sartipi, F. Stormwater retention using pervious concrete pavement: Great Western Sydney case study. *Case Stud. Constr. Mater.* **2019**, *11*, e00274. [CrossRef]
24. Khankhaje, E.; Rafieizonooz, M.; Salim, M.R.; Khan, R.; Mirza, J.; Siong, H.C. Salmiati Sustainable clean pervious concrete pavement production incorporating palm oil fuel ash as cement replacement. *J. Clean. Prod.* **2018**, *172*, 1476–1485. [CrossRef]
25. Imran, H.; Akib, S.; Karim, M.R. Permeable pavement and stormwater management systems: A review. *Environ. Technol.* **2013**, *34*, 2649–2656. [CrossRef] [PubMed]
26. Chen, J.; Chu, R.; Wang, H.; Zhang, L.; Chen, X.; Du, Y. Alleviating urban heat island effect using high-conductivity permeable concrete pavement. *J. Clean. Prod.* **2019**, *237*, 117722. [CrossRef]
27. Chandrappa, A.K.; Biligiri, K.P. Pervious concrete as a sustainable pavement material—Research findings and future prospects: A state-of-the-art review. *Constr. Build. Mater.* **2016**, *111*, 262–274. [CrossRef]
28. Debnath, B.; Sarkar, P.P. Permeability prediction and pore structure feature of pervious concrete using brick as aggregate. *Constr. Build. Mater.* **2019**, *213*, 643–651. [CrossRef]
29. Debnath, B.; Sarkar, P.P. Prediction and model development for fatigue performance of pervious concrete made with over burnt brick aggregate. *Mater. Struct.* **2020**, *53*, 86. [CrossRef]
30. Zaetang, Y.; Wongsa, A.; Sata, V.; Chindaprasirt, P. Use of lightweight aggregates in pervious concrete. *Constr. Build. Mater.* **2013**, *48*, 585–591. [CrossRef]
31. Aliabdo, A.A.; Elmoaty, A.E.M.A.; Fawzy, A.M. Experimental investigation on permeability indices and strength of modified pervious concrete with recycled concrete aggregate. *Constr. Build. Mater.* **2018**, *193*, 105–127. [CrossRef]
32. Sata, V.; Wongsa, A.; Chindaprasirt, P. Properties of pervious geopolymer concrete using recycled aggregates. *Constr. Build. Mater.* **2013**, *42*, 33–39. [CrossRef]
33. Wang, G.; Chen, X.; Dong, Q.; Yuan, J.; Hong, Q. Mechanical performance study of pervious concrete using steel slag aggregate through laboratory tests and numerical simulation. *J. Clean. Prod.* **2020**, *262*, 121208. [CrossRef]
34. Zaetang, Y.; Wongsa, A.; Sata, V.; Chindaprasirt, P. Use of coal ash as geopolymer binder and coarse aggregate in pervious concrete. *Constr. Build. Mater.* **2015**, *96*, 289–295. [CrossRef]
35. Debnath, B.; Sarkar, P.P. Application of Nano $SiO_2$ in Pervious Concrete Pavement Using Waste Bricks as Coarse Aggregate. *Arab. J. Sci. Eng.* **2022**, 1–21. [CrossRef]
36. Debnath, B.; Sarkar, P.P. Characterization of pervious concrete using over burnt brick as coarse aggregate. *Constr. Build. Mater.* **2020**, *242*, 118154. [CrossRef]
37. Zhang, G.; Wang, S.; Wang, B.; Zhao, Y.; Kang, M.; Wang, P. Properties of pervious concrete with steel slag as aggregates and different mineral admixtures as binders. *Constr. Build. Mater.* **2020**, *257*, 119543. [CrossRef]
38. Gaedicke, C.; Marines, A.; Miankodila, F. Assessing the abrasion resistance of cores in virgin and recycled aggregate pervious concrete. *Constr. Build. Mater.* **2014**, *68*, 701–708. [CrossRef]
39. *IS 2386*; Part III, Method of Test for Aggregate for Concrete. Part III- Specific Gravity, Density, Voids, Absorption and Bulking. Bureau of Indian Standards: New Delhi, India, 1963.
40. Bureau of Indian Standards. *IS 2386*; Part IV, Methods of Test for Aggregates for Concrete, Part 4: Mechanical Properties [CED 2: Cement and Concrete]; Bureau of Indian Standards: New Delhi, India, 2002; pp. 1–37.
41. Bureau of Indian Standards. *IS:516*; Method of Tests for Strength of Concrete; Bureau of Indian Standards: New Delhi, India, 2004; pp. 516–1959. [CrossRef]
42. ASTM International. *C1754/C1754-12*; Standard Test Method for Density and Void Content of Hardened Pervious Concrete; ASTM International: West Conshohocken, PA, USA, 2012; p. 3. [CrossRef]
43. Tang, Y.; Zhu, M.; Chen, Z.; Wu, C.; Chen, B.; Li, C.; Li, L. Seismic performance evaluation of recycled aggregate concrete-filled steel tubular columns with field strain detected via a novel mark-free vision method. *Structures* **2022**, *37*, 426–441. [CrossRef]
44. Debnath, B.; Sarkar, P.P. Quantification of random pore features of porous concrete mixes prepared with brick aggregate: An application of stereology and mathematical morphology. *Constr. Build. Mater.* **2021**, *294*, 123594. [CrossRef]
45. American Concrete Institute. *ACI-522R. Report on Pervious Concrete, ACI Comm. 522*; American Concrete Institute: Indianapolis, IN, USA, 2010; pp. 1–42.
46. Joshaghani, A.; Ramezanianpour, A.A.; Ataei, O.; Golroo, A. Optimizing pervious concrete pavement mixture design by using the Taguchi method. *Constr. Build. Mater.* **2015**, *101*, 317–325. [CrossRef]

 materials

*Article*

# Comparison of Prediction Models Based on Machine Learning for the Compressive Strength Estimation of Recycled Aggregate Concrete

**Kaffayatullah Khan [1,*], Waqas Ahmad [2], Muhammad Nasir Amin [1], Fahid Aslam [3], Ayaz Ahmad [4] and Majdi Adel Al-Faiad [5]**

1 Department of Civil and Environmental Engineering, College of Engineering, King Faisal University, Al-Ahsa 31982, Saudi Arabia; mgadir@kfu.edu.sa
2 Department of Civil Engineering, COMSATS University Islamabad, Abbottabad 22060, Pakistan; waqasahmad@cuiatd.edu.pk
3 Department of Civil Engineering, College of Engineering in Al-Kharj, Prince Sattam bin Abdulaziz University, Al-Kharj 11942, Saudi Arabia; f.aslam@psau.edu.sa
4 MaREI Centre, Ryan Institute and School of Engineering, College of Science and Engineering, National University of Ireland Galway, H91 HX31 Galway, Ireland; a.ahmad8@nuigalway.ie
5 Department of Chemical Engineering, College of Engineering, King Faisal University, Al-Ahsa 31982, Saudi Arabia; malfaiad@kfu.edu.sa
* Correspondence: kkhan@kfu.edu.sa

**Abstract:** Numerous tests are used to determine the performance of concrete, but compressive strength (CS) is usually regarded as the most important. The recycled aggregate concrete (RAC) exhibits lower CS compared to natural aggregate concrete. Several variables, such as the water-cement ratio, the strength of the parent concrete, recycled aggregate replacement ratio, density, and water absorption of recycled aggregate, all impact the RAC's CS. Many studies have been carried out to ascertain the influence of each of these elements separately. However, it is difficult to investigate their combined effect on the CS of RAC experimentally. Experimental investigations entail casting, curing, and testing samples, which require considerable work, expense, and time. It is vital to adopt novel methods to the stated aim in order to conduct research quickly and efficiently. The CS of RAC was predicted in this research utilizing machine learning techniques like decision tree, gradient boosting, and bagging regressor. The data set included eight input variables, and their effect on the CS of RAC was evaluated. Coefficient correlation ($R^2$), the variance between predicted and experimental outcomes, statistical checks, and k-fold evaluations, were carried out to validate and compare the models. With an $R^2$ of 0.92, the bagging regressor technique surpassed the decision tree and gradient boosting in predicting the strength of RAC. The statistical assessments also validated the superior accuracy of the bagging regressor model, yielding lower error values like mean absolute error (MAE) and root mean square error (RMSE). MAE and RMSE values for the bagging model were 4.258 and 5.693, respectively, which were lower than the other techniques employed, i.e., gradient boosting (MAE = 4.956 and RMSE = 7.046) and decision tree (MAE = 6.389 and RMSE = 8.952). Hence, the bagging regressor is the best suitable technique to predict the CS of RAC.

**Keywords:** recycled concrete aggregate; compressive strength; green concrete; machine learning; decision tree; gradient boosting; bagging regressor

**Citation:** Khan, K.; Ahmad, W.; Amin, M.N.; Aslam, F.; Ahmad, A.; Al-Faiad, M.A. Comparison of Prediction Models Based on Machine Learning for the Compressive Strength Estimation of Recycled Aggregate Concrete. *Materials* **2022**, *15*, 3430. https://doi.org/10.3390/ma15103430

Academic Editors: Andrea Petrella and Michele Notarnicola

Received: 5 April 2022
Accepted: 27 April 2022
Published: 10 May 2022

**Publisher's Note:** MDPI stays neutral with regard to jurisdictional claims in published maps and institutional affiliations.



## 1. Introduction

Worldwide, concrete is the most utilized material in the building sector [1–6]. Its appeal originates from several characteristics, including its minimal expense, water and heat resistance, and flexibility to a variety of shapes and sizes [7–13]. Concrete might be used to build almost every sort of structure [14,15]. Concrete is composed of three fundamental components: aggregates, cement, and water [16–18]. Amongst these ingredients, aggregate is significant as it makes up around 60–75% of the overall volume of concrete [19]. Moreover, the fast growth of industrialization and urbanization has made concrete the least eco-friendly material because it uses the most natural resources. Concrete is crucial to a country's economic prosperity due to its widespread use. It utilizes around 20,000 million tons of raw materials (natural aggregates) every year [20]. Moreover, the mining and processing natural aggregates (NAs) requires considerable energy and results in increased $CO_2$ emissions [21]. Thus, increased use of concrete results in rapid depletion of natural resources and increased contamination of the environment [22–24]. Now, scholars are focusing their research on the application of alternate materials to natural ones, thereby promoting naturally responsible construction.

Modern infrastructure development necessitates extensive refurbishment of present structures, causing immense volumes of construction and demolition waste (CDW). Due to the crucial nature of CDW, it must be disposed of securely. Two concerns confront the current building sector: dwindling natural resources and a rise in CDW. Both challenges might be addressed concurrently by CDW recycling in the new building. Recycling leftover concrete from CDW has developed into a feasible alternative to NA in concrete [25,26]. Waste concrete is generated in a number of ways, including destroyed structures, abandoned precast concrete members, residual concrete in batching facilities, and concrete samples tested in laboratories [27–30]. Thus, incorporating recycled concrete aggregates (RCAs) in the building sector will be an economical and eco-friendly way to decrease CDW volume [31,32]. RAs are divided into three categories: recycled brick aggregate, RCA, and recycled mixed (bricks and concrete) aggregate. Meanwhile, RAs include a range of pollutants, including woodblocks, glass, paper fragments, and plastics [33,34]. Presently, RCA is the most often utilized in construction [35–40]. Thus, substituting RCAs from CDW for NAs in concrete will encourage sustainable development.

The process of building predictive models for concrete strength is ongoing in order to minimize needless test repetitions and material waste. There are various popular models for simulating the characteristics of concrete, including best-fit curves (based on regression analysis). However, because concrete has a nonlinear behavior [41,42], regression models developed using this approach may not adequately capture the material's underlying nature. Additionally, regression techniques may underestimate the influence of concrete constituents [43]. Artificial intelligence techniques, such as machine learning (ML), are some of the most advanced modeling approaches employed in the field of civil engineering. These methods model responses using input variables, and the output models are confirmed by experimentation. ML methods are employed to forecast concrete strength [44–48], the performance of bituminous mixtures [49], and the durability of concrete [50–52]. The majority of previous ML-based studies have focused on CS prediction for conventional concretes [53–59], using their physiochemical attributed (e.g., cement content; water content; and mass/volume of admixture and/or mineral additive); only a few articles have focused on the prediction of the characteristics of RAC. Duan et al. [60] used a nonlinear, regression-based ML model, namely an artificial neural network (ANN), to forecast the CS of RAC. Gholampour et al. [61] investigated the applicability of several regression-based ML models for predicting the mechanical properties of RAC. Deshpande et al. [62] employed ANN to predict the CS of RAC, which might possibly be used to estimate MOE when paired with semi-empirical formulae. Behnood et al. [63] predicted the properties of RAC using the M5P model tree technique—a very recent decision tree ML model [64]. Deng et al. [65] predicted the CS of RAC using a convolutional ANN-based deep learning algorithm. Nonetheless, it is critical to note that the most frequently used ML model in prior research is ANN frequently fails to accurately predict outcomes [66,67]. This is because ANN models are based on local optimization and search algorithms (e.g., the back-propagation mechanism used in several neural network-based ML models for parameter optimization) that are highly susceptible to becoming confined in (or around) local minima rather than converging to the global minimum [66]. As a result of this difficulty, when ANN models are retrained, they frequently provide inconsistently or even poorer predictions for the same set of inputs (e.g., using a larger or a different database) [68]. Recent studies have demonstrated that the bagging regressor (BR) and gradient boosting (GB) models based on a modification of the bootstrap aggregation decision tree (DT) algorithm outperforms other standalone ML models in terms of prediction accuracy of concrete CS [69–72]. These studies credit the BR and GB model's better prediction performance to its unmatched ability to handle discrete and continuous variables across monotonic and non-monotonic data domains, while simultaneously lowering variance across different subsets of the training data set. Despite the BR and GB model's benefits, an exhaustive literature analysis revealed that these models have rarely been used to forecast the CS of RAC.

The aim of this work is to determine how ML strategies might be used to anticipate the CS of RAC. One single ML algorithm, DT, and two ensemble ML approaches, GB and BR, were employed. To evaluate the performance of each method, correlation coefficients ($R^2$) and statistical tests were carried out. Furthermore, each technique's validity was confirmed using k-fold evaluation and error dispersals. This research is noteworthy because it predicts the CS of RAC utilizing both single and ensemble ML methods. The experimental explorations require substantial human effort, experimentation expenses, and time for collection, casting, curing, and testing materials. Since a variety of parameters, including waster–cement ratio (w/c), parent concrete strength, recycled aggregate replacement ratio, water absorption, and density, all influence the CS of RAC, and their combined effect is difficult to analyze experimentally. ML techniques are capable of identifying the cumulative influence of their components with minimal effort. ML methods require a data set, which may be gathered from previous research since several investigations have been conducted to determine the CS of RAC. The data collected can then be employed to train ML methods and anticipate material strength. Some previous studies also employed ML methods to estimate the properties of RAC, but with a limited number of data samples and input parameters. For example, Salimbahrami and Shakeri [73] predicted the CS of RAC using the ANN technique with 7 input variables and 124 data samples. Similarly, Duan et al. [74] predicted the CS of RAC with 6 input variables and 209 data points. This study employed different ML techniques from the previous studies and estimated the CS of RAC with 8 input parameters and 638 data points. It is expected that using a higher number of input variables and data points will result in the superior precision of ML techniques. The goal of this research is to determine the most appropriate ML approach for estimating the CS of RAC and the influence of various factors on RAC strength.

## 2. Methods

### 2.1. Data Employed for Modeling

To attain the desired outcome, ML algorithms require a diverse set of input variables [75–77]. Utilizing data gathered from the past studies (see Appendix A), the CS of RAC was calculated. To avoid bias, experimental data were picked at random from past studies. The available publications on the usage of similar materials in the CS of RAC were reviewed. While the majority of articles studied extra aspects of RAC, this analysis used CS data for modeling. The algorithms took eight variables as inputs: the RCA replacement ratio, the parent concrete strength, the aggregate–cement ratio (a/c), the water–cement ratio (w/c), the nominal maximum RCA size, the Los Angeles abrasion index of RCA, the bulk density of RCA, and the water absorption of RCA, and only CS taken as the output. The quantity of input factors and dataset size have a substantial effect on the ML method results [78–80]. A total of 638 data points were used in the current research to run ML techniques. The descriptive statistic assessment of all input factors is summarized in Table 1. The table contains the mathematical identifications for all the input factors. Figure 1 depicts the relative frequency dispersal of all variables applied in the investigation. It summarizes the number of possible interpretations for each value or combination of values.

**Table 1.** Descriptive assessment results of input factors used.

| Parameter | Water-Cement Ratio (w/c) | Aggregate-Cement Ratio (a/c) | RCA Replacement Ratio (%) | Parent Concrete Strength (MPa) | Nominal Maximum RCA Size (mm) | Bulk Density of RCA (kg/m$^3$) | Water Absorption of RCA (%) | Los Angeles Abrasion Index of RCA |
|---|---|---|---|---|---|---|---|---|
| Mean | 0.49 | 2.99 | 53.03 | 5.00 | 21.51 | 1666.16 | 3.49 | 6.75 |
| Maximum | 0.87 | 6.50 | 100.00 | 100.00 | 32.00 | 2880.00 | 11.90 | 42.00 |
| Minimum | 0.00 | 0.00 | 0.00 | 0.00 | 0.00 | 0.00 | 0.00 | 0.00 |
| Range | 0.87 | 6.50 | 100.00 | 100.00 | 32.00 | 2880.00 | 11.90 | 42.00 |
| Mode | 0.50 | 3.10 | 100.00 | 0.00 | 20.00 | 0.00 | 0.00 | 0.00 |
| Median | 0.49 | 2.90 | 50.00 | 0.00 | 20.00 | 2330.00 | 3.90 | 0.00 |
| Sum | 312 | 1913 | 33,884 | 3193 | 13,747 | 10,646.77 | 2231 | 4312 |
| Standard Deviation | 0.11 | 0.83 | 40.01 | 15.38 | 5.71 | 1115.04 | 2.94 | 13.89 |

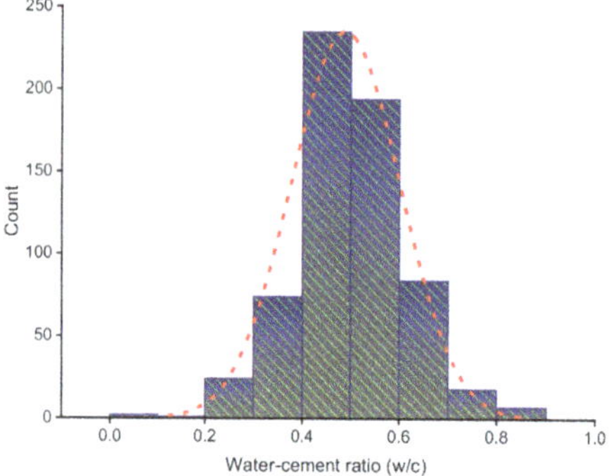


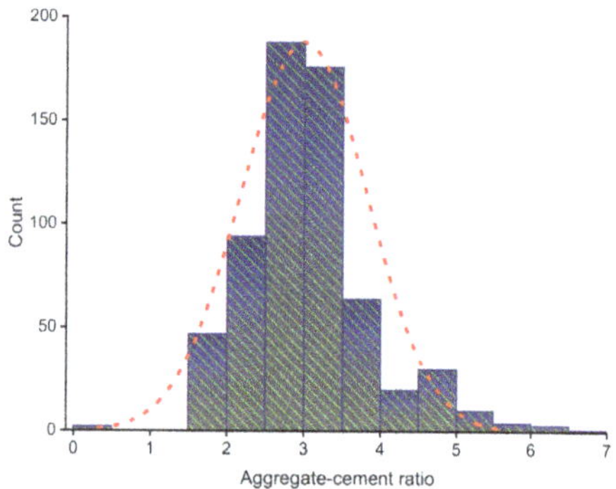


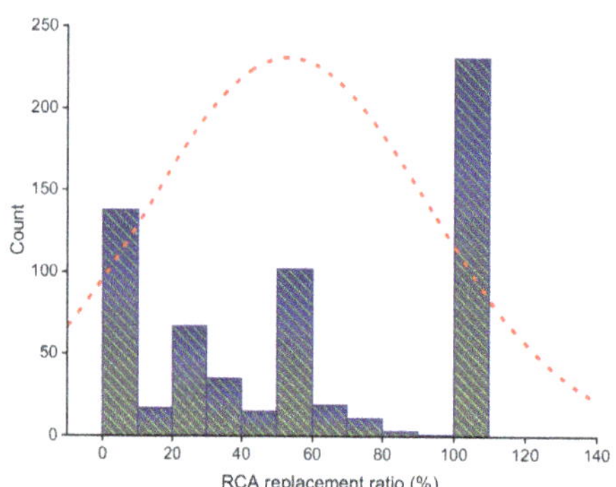


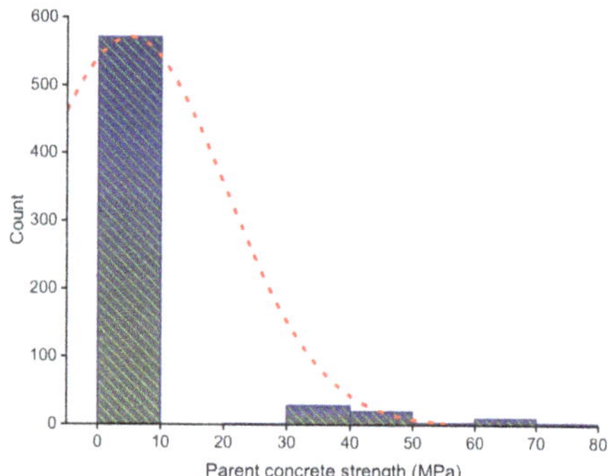


**Figure 1.** *Cont.*

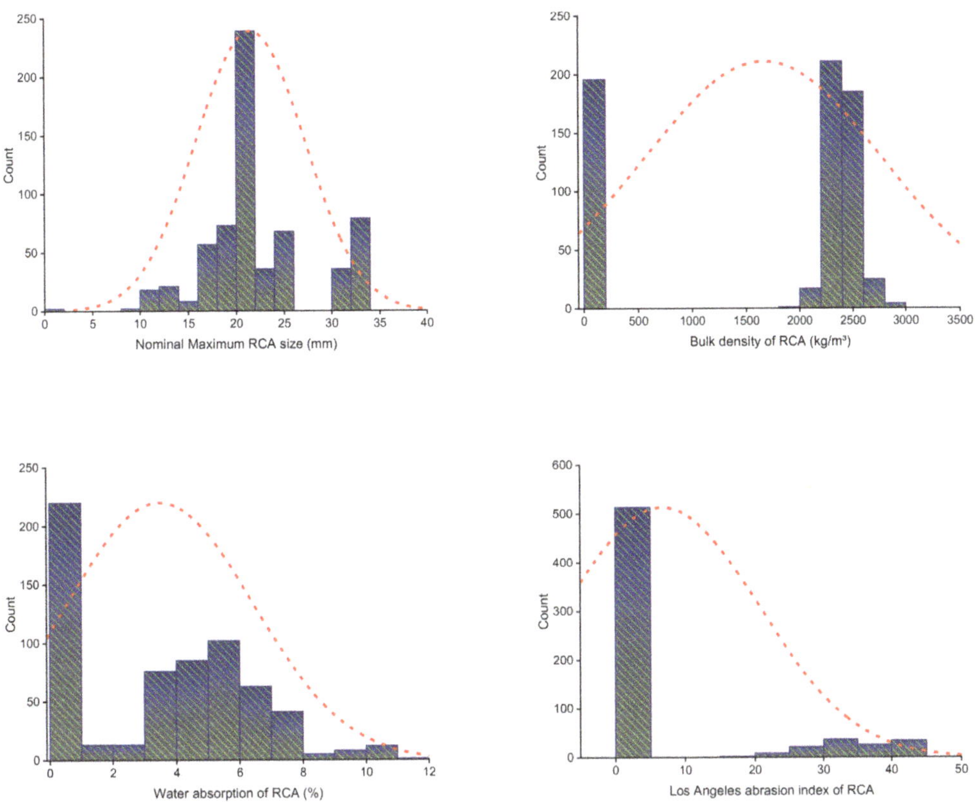


**Figure 1.** Relative frequency dispersal of input factors.

### 2.2. Machine Learning Algorithms Employed

To meet the study's aims, a single ML method (DT) and ensemble ML approaches (GB and BR) were employed with Python scripting using the Anaconda Navigator package. Spyder (Version 4.3.5) was selected to operate the DT, GB, and BR models. These ML techniques are frequently utilized to forecast required results in response to input parameters. These techniques are able to anticipate the temperature impact, the mechanical strength, and the durability of materials [81–83]. Eight input factors and one output (CS) were used throughout the modeling process. The expected result's $R^2$ score represented the accuracy of all techniques. The $R^2$ indicates the degree of deviation; a value near zero indicates greater deviation, while a value near one indicates that the data and model are virtually perfectly fit [70]. The sub-sections beneath describe the ML approaches used in this research. Moreover, statistical and k-fold analyses, as well as error evaluations, are performed on all ML methods like mean absolute error (MAE) and root mean square error (RMSE). Furthermore, sensitivity analysis (SA) is used to determine the influence of all input factors on the estimated results. The research method is depicted in Figure 2.

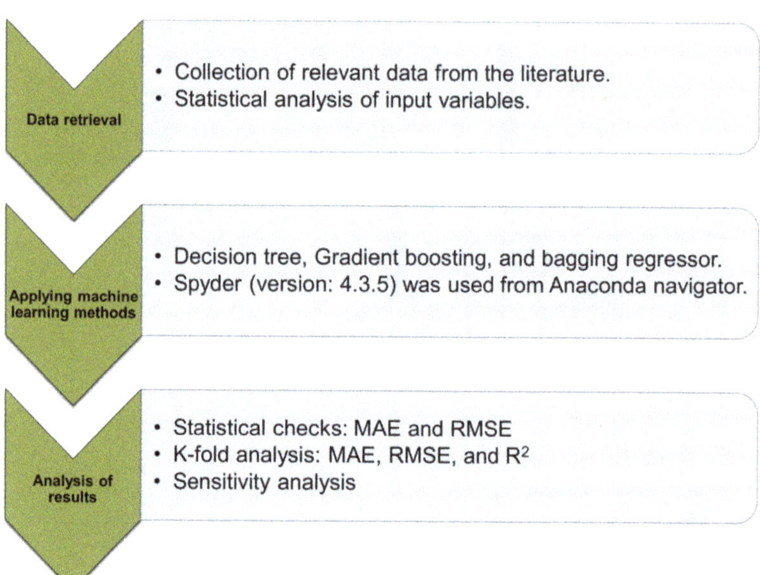


**Figure 2.** Sequence of research methods.

2.2.1. Decision Tree

DTs are formed by developing techniques for segmenting a data sample into branch-like portions. These portions unite to create an inverted tree with a root node on the upper side [84]. Figure 3 illustrates a schematic representation of the DT technique. As depicted, a DT can have both continuous and single features. Relationships between the object of assessment and the input fields are utilized to generate the branching or segmentation decision rule beneath the root node. Following the link's establishment, one or more decision rules detailing the associations among the inputs and targeted results might be generated. Decision rules approximate the values of new or undetermined interpretations accurately when they incorporate input values but not targets. At each division point, the errors are computed, and the variable with the smallest fitness function value is taken as the split position, followed by the same for the other variables.

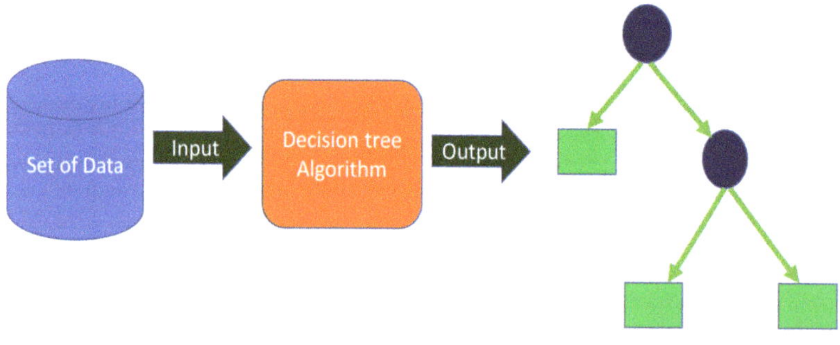


**Figure 3.** Decision tree schematic representation [85].

### 2.2.2. Gradient Boosting

Friedman [86] presented GB as an ensemble strategy for classification and regression in 1999. GB is only applicable to regression. As seen in Figure 4, the GB technique compares each iteration of the randomly chosen training set to the base model. GB for execution may be sped up and accuracy increased by randomly subsampling the training data, which also helps prevent overfitting. The lower the training data percentage, the faster the regression because the model must suit minor data with every single iteration. The GB algorithm requires tuning parameters, including n-trees and shrinkage rate, where n-trees is the number of trees to be generated; n-trees must not be kept too small, and the shrinkage factor, normally referred to as the learning rate employed to all trees in the development, should not be set too high [87].

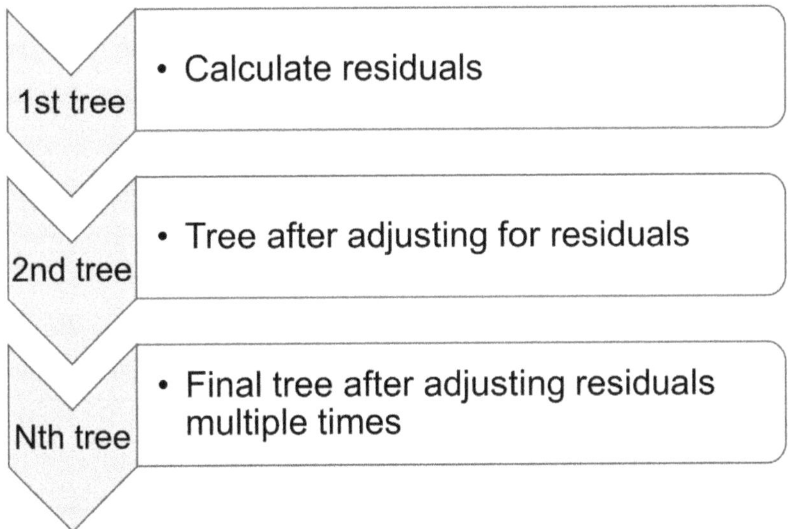


**Figure 4.** Schematic representation of gradient boosting technique [72].

### 2.2.3. Bagging Regressor

BR is a comparable SML technique that compensates for the prediction model's variance during the training stage by improving it with supplementary data. This result is established on an asymmetric selection strategy that makes use of data exchange from the original set. Utilizing sampling with the substitute, some observations may be reiterated in each new testing dataset, allowing for greater accuracy. During the BR process, each constituent has an equal probability of being included in the new dataset, regardless of its importance. There is no influence on the forecasting force of a training set that is larger in size than the training set. It is also possible to considerably reduce the variation by fine-tuning the estimate to get the desired conclusion. For subsequent model training, each of these data sets is commonly utilized to supplement the others. Using an ensemble of numerous models, the mean of all predictions from each model is used to create this ensemble. In regression, the prediction might be the average or mean of the estimates from a number of different models [88]. Twenty sub-models are employed to optimize the DT using BR to obtain an adamant output result. Figure 5 depicts the bagging algorithm's flow chart, which details the procedure until the desired output is obtained.

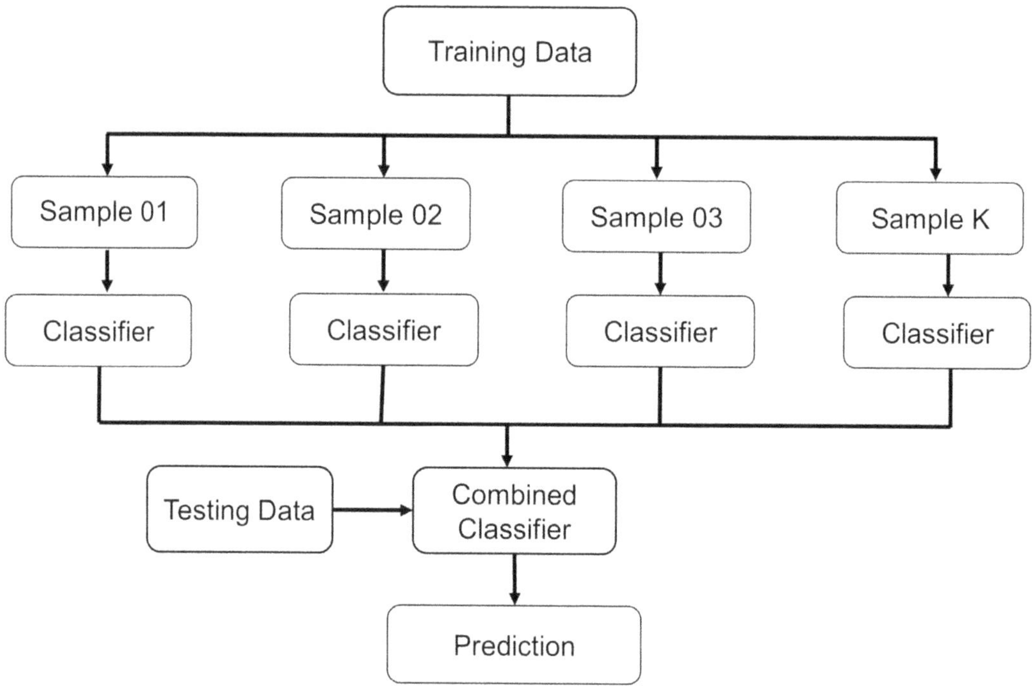


**Figure 5.** Schematic representation of the bagging regressor technique [85].

## 3. Analysis of Results

### 3.1. DT Model

Figure 6 demonstrates the DT model's results for the CS estimate of RAC. Figure 6a illustrates the relationship among experimental and anticipated results. The DT approach produced findings that were less accurate and had a moderate discrepancy between experimental and projected outcomes. The $R^2$ of 0.77 validates the DT model's lower performance in projecting the CS of RAC. Figure 6b depicts the scattering of experimental, anticipated, and error values for the DT model. The error values were evaluated, and the maximum and average values were noted to be 37.68 and 6.39 MPa, respectively. Furthermore, the dispersal of error values was found, with 11.7% of values falling below 1 MPa, 41.4% falling between 1 and 5 MPa, 28.1% falling between 5 and 10 MPa, and 18.8% falling over 10 MPa. The scattering of error numbers indicates that the DT technique works less precisely.

### 3.2. GB Model

Figure 7 shows the outcomes from the GB model's estimation of the CS of RAC. Figure 7a illustrates the relationship among experimental and estimated results. The GB method resulted in an output that was more precise and had the least degree of difference between actual and projected results. The GB model is better at forecasting the CS of RAC, with an $R^2$ of 0.85. The scattering of experimental, anticipated, and error figures for the GB model are depicted in Figure 7b. The results for the average and highest error are 4.78 and 27.96 MPa, respectively. The dispersal of errors was 20.4% lower than 1 MPa, 43.1% in the range of 1 and 5 MPa, 20.0% in the range of 5 and 10 MPa, and 16.5% larger than 10 MPa. The dispersal of errors demonstrates the GB technique's superior estimating accuracy to the DT. The GB model takes the advantage of optimized value from the twenty sub-models, resulting in the higher precision.

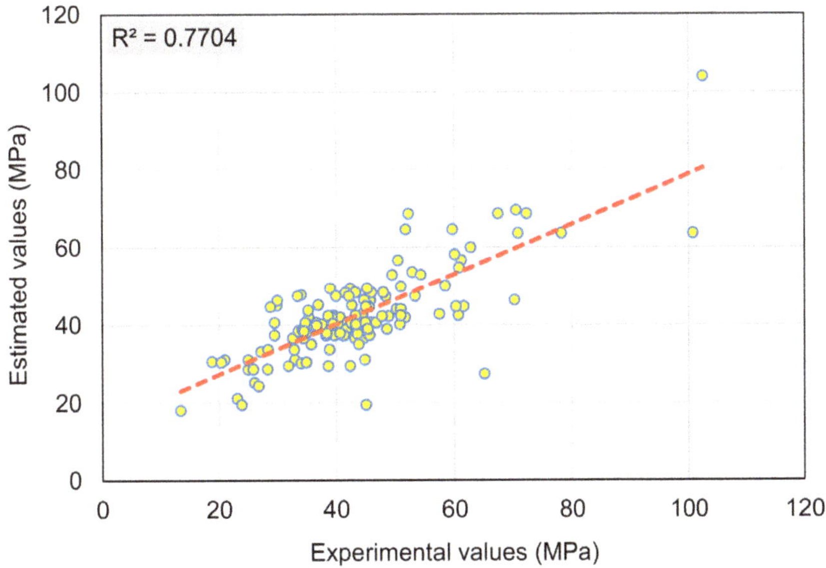


(**a**)



(**b**)

**Figure 6.** DT model: (**a**) Link among experimental and projected outcomes; (**b**) Scattering of actual and predicted results.

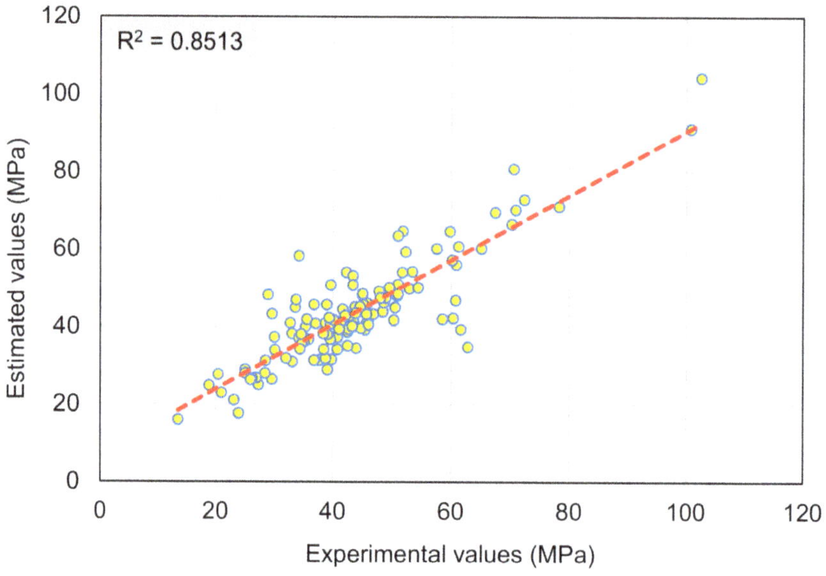


(**a**)

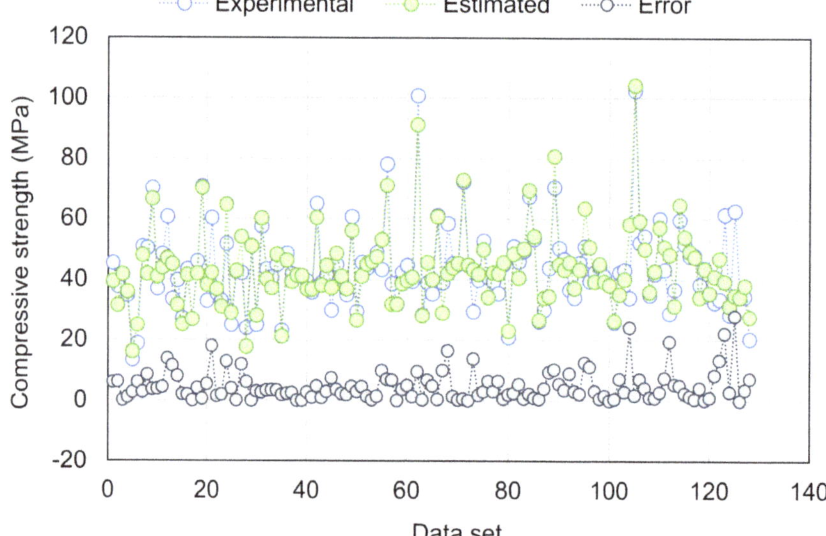


(**b**)

**Figure 7.** GB model: (**a**) Relationship among experimental and projected outcomes; (**b**) Scattering of actual and predicted results.

### *3.3. BR Model*

Figure 8a,b exemplify an evaluation of the experimental and expected findings for the BR model. Figure 8a illustrates the relationship among experimental and projected results, with an $R^2$ of 0.92 implying that the BR model is more accurate in estimating the RAC's CS than the DT and GB models. The scattering of experimental, anticipated, and error scores for the BR model are depicted in Figure 8b. The maximum and average errors were found to be 23.22 and 4.26 MPa, respectively. The dispersal of error values was 16.4% lower than 1 MPa, 54.7% in the range of 1 and 5 MPa, 21.15% in the range of 5 and 10 MPa, and only 7.8% higher than 10 MPa. These decreased error numbers suggest that the BR technique is more precise than the other models used in this investigation. Similar to the GB method, the BR method produces twenty sub-models, and the optimized sub-model based on the $R^2$ is chosen. Because the BR approach employs substitution sampling, some observations may be repeated in each new testing dataset, resulting in increased accuracy.

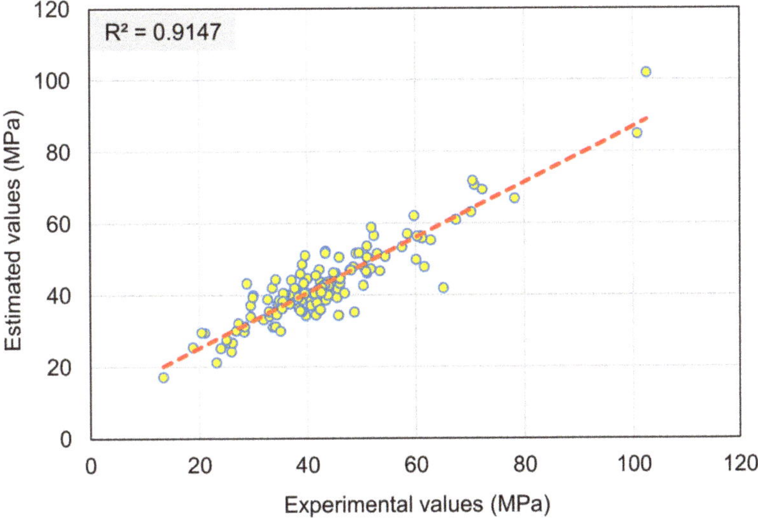


(a)

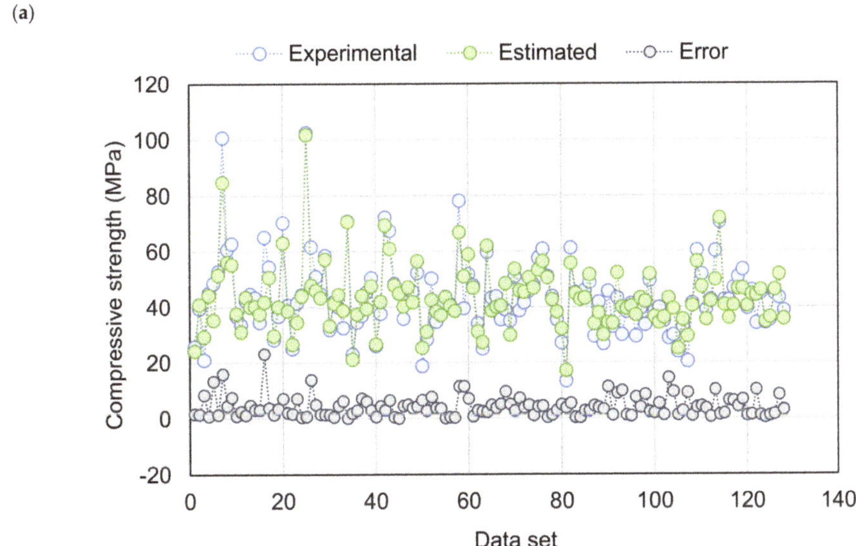


(b)

**Figure 8.** Bagging regressor model: (**a**) Relationship among experimental and forecasted results; (**b**) Scattering of actual and predicted results.

## 4. Validation of Models

The models were validated using k-fold and statistical techniques. The k-fold approach is widely used to determine the validity of a technique [89] in which the related dataset is arbitrarily distributed and classified into 10 classes. As depicted in Figure 9, nine units will be utilized for training models and one for verifying them. The model is more accurate when the errors (RMSE and MAE) are small, and the $R^2$ is greater. Moreover, the procedure should be repeated ten times to ensure that a plausible conclusion is reached. This substantial effort greatly contributes to the ML technique's exceptional correctness. Moreover, as seen in Table 2, all ML methods were statistically assessed for the inaccuracy (MAE and RMSE). These analyses also validated the BR model's superior exactness in comparison to the DT and GB models, owing to their lower error values. The approaches' predictive performance was assessed statistically using Equations (1) and (2), which were obtained from earlier work [90,91].

$$\text{MAE} = \frac{1}{n}\sum_{i=1}^{n}|x_i - x| \tag{1}$$

$$\text{RMSE} = \sqrt{\sum \frac{\left(y_{pred} - y_{ref}\right)^2}{n}} \tag{2}$$

where $n$ = total quantity of data points, $x$, $y_{ref}$ = experimental values in the data set, and $x_i$, $y_{pred}$ = projected values from techniques

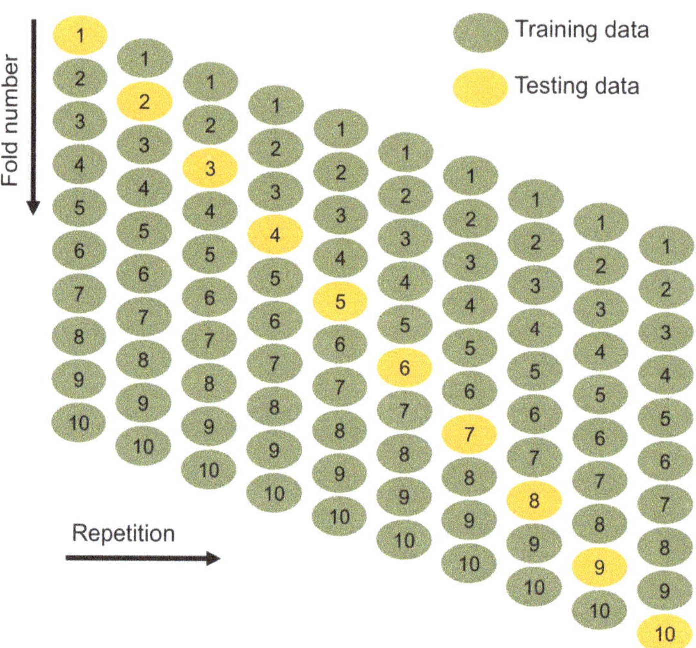


**Figure 9.** Schematic depiction of k-fold assessment [92].

**Table 2.** Statistical assessments of the models.

| Model | MAE | RMSE |
|---|---|---|
| Decision tree | 6.389 | 8.952 |
| Gradient boosting | 4.956 | 7.046 |
| Bagging | 4.258 | 5.693 |

MAE, RMSE, and $R^2$ were computed to determine the effectiveness of the k-fold process, and their values are shown in Table 3. Figures 10–12 illustrate the comparison of k-fold analysis for all of the methods used. The MAE for the DT model was in the range of 6.39 and 14.68 MPa, having an average of 11.83 MPa. When compared to the GB method, the MAE varied from 4.78 to 14.60 MPa, having an average of 10.27 MPa. MAE for the BR model ranged from 4.26 to 10.82 MPa, having an average of 8.10 MPa (Figure 10). The average RMSE for the DT, GB, and BR methods were 13.81, 11.05, and 10.69 MPa, respectively (Figure 11). Moreover, the average $R^2$ for the DT, GB, and BR models were 0.53, 0.67, and 0.71, respectively (Figure 12). In comparison with the GB and DT methods, the BR method with smaller errors (MAE and RMSE) and superior $R^2$ is more exact in estimating the CS of RAC.

**Table 3.** Results obtained from the k-fold assessment.

| K-Fold | Decision Tree | | | Gradient Boosting | | | Bagging Regressor | | |
|---|---|---|---|---|---|---|---|---|---|
| | MAE | RMSE | $R^2$ | MAE | RMSE | $R^2$ | MAE | RMSE | $R^2$ |
| 1 | 12.37 | 16.63 | 0.66 | 12.70 | 10.73 | 0.74 | 10.82 | 14.06 | 0.87 |
| 2 | 14.53 | 13.55 | 0.27 | 8.03 | 9.78 | 0.53 | 7.65 | 8.80 | 0.57 |
| 3 | 13.90 | 8.95 | 0.43 | 12.14 | 8.58 | 0.85 | 8.87 | 10.78 | 0.75 |
| 4 | 10.63 | 14.62 | 0.73 | 8.97 | 14.06 | 0.84 | 7.90 | 11.97 | 0.40 |
| 5 | 14.68 | 17.60 | 0.77 | 11.76 | 12.92 | 0.81 | 7.35 | 9.80 | 0.82 |
| 6 | 9.80 | 12.57 | 0.32 | 13.60 | 7.99 | 0.83 | 7.97 | 10.12 | 0.85 |
| 7 | 12.10 | 11.57 | 0.39 | 4.96 | 7.05 | 0.37 | 8.22 | 10.58 | 0.78 |
| 8 | 12.11 | 17.63 | 0.77 | 13.14 | 9.16 | 0.74 | 10.31 | 15.05 | 0.76 |
| 9 | 11.76 | 12.56 | 0.48 | 9.11 | 8.18 | 0.61 | 4.26 | 5.69 | 0.77 |
| 10 | 6.39 | 12.44 | 0.45 | 11.44 | 17.06 | 0.27 | 7.62 | 9.99 | 0.54 |

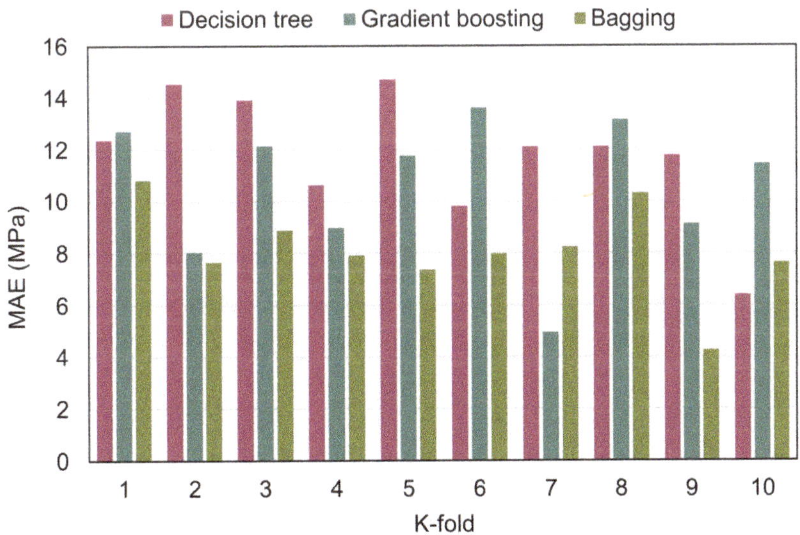


**Figure 10.** Mean absolute error distribution from k-fold analysis.

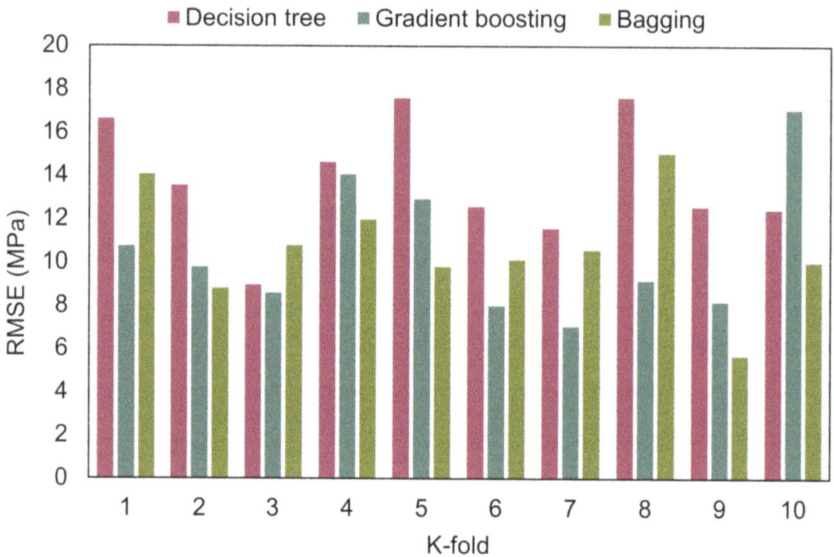


**Figure 11.** Root mean square error distribution from k-fold analysis.

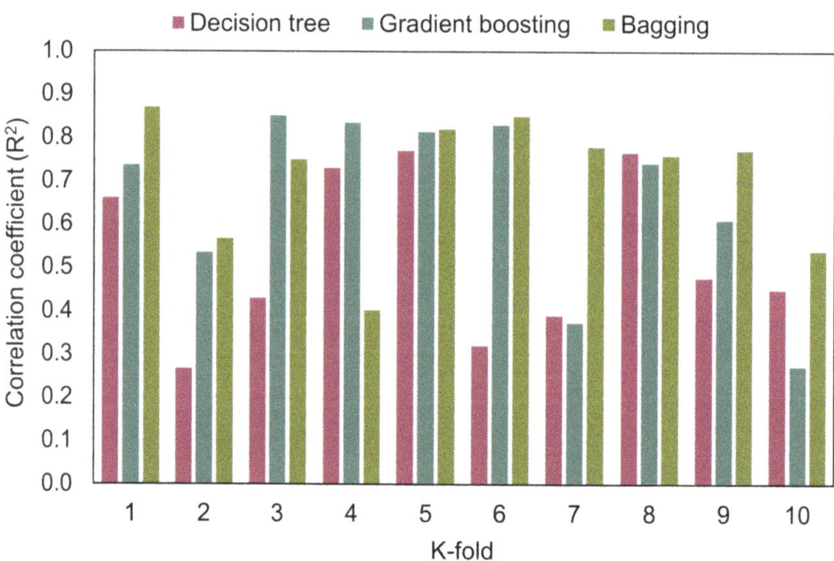


**Figure 12.** Correlation coefficient ($R^2$) distribution from the k-fold analysis.

## 5. Sensitivity Analysis

This evaluation intends to find out the influence of input factors on RAC's CS prediction. The input factors have a major influence on the anticipated result [93]. The effect of the input factors on the CS forecast of RAC is seen in Figure 13. The analysis found that the essential ingredient was the RCA replacement ratio, accounting for around 21% of the total, followed by parent concrete strength at approximately 18% and w/c at approximately 17%. The remaining input factors had a smaller effect on the forecast of RAC's CS, with the Los Angeles abrasion index of RCA, water absorption of RCA, a/c, nominal maximum RCA size, and bulk density of RCA contributing to about 13%, 9%, 9%, 7%, and 6%, respectively. SA produced results that were related to the quantity of inputs and the data sample used to create the models. Equations (3) and (4) were used to determine the effect of an input parameter on the technique's output.

$$N_i = f_{max}(x_i) - f_{min}(x_i), \tag{3}$$

$$S_i = \frac{N_i}{\sum_{j-i}^{n} N_j}, \tag{4}$$

where, $f_{max}(x_i)$ is the highest anticipated result over the *ith* output, $f_{min}(x_i)$ is the least anticipated results over the *ith* output, and $S_i$ is the percentage contribution of a specific input factor.

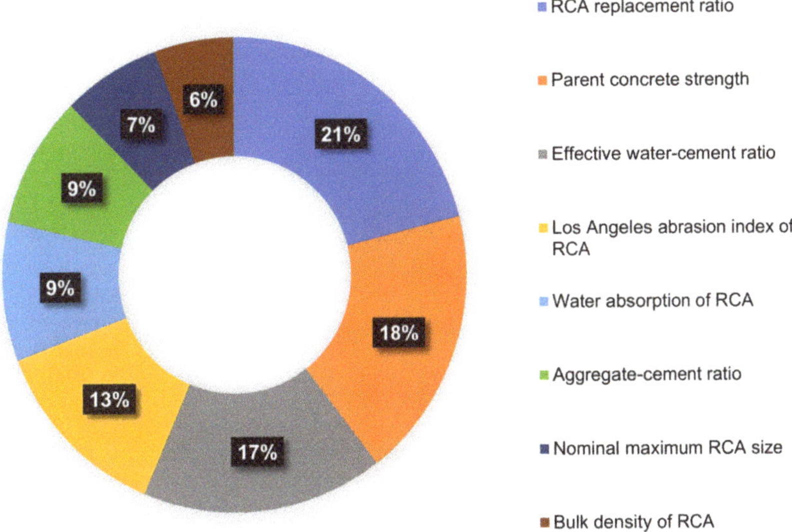


**Figure 13.** Input variables contribution to predicting outcomes.

## 6. Discussion

The goal of this work was to add to the body of knowledge concerning the application of modern strategies for evaluating the CS of RAC. This sort of study will benefit the building sector by facilitating the advancement of fast and cost-efficient material property prediction tools. Furthermore, by encouraging eco-friendly strategies through these measures, the approval and usage of RAC in the building sector will be hastened. Figure 14 illustrates the benefits of RAC in the construction industry. Urbanization and industrialization need considerable infrastructure renewal, resulting in enormous CDW volumes. As a result, landfill area is becoming increasingly scarce as necessary areas are turned into garbage ditches, estate and waste dumping costs continue to rise. As a result, waste management has become a priority in emerging countries and is a worldwide concern that requires a long-term solution. Furthermore, extracting and managing NAs for concrete uses a considerable amount of energy and produces $CO_2$ [21]. Thus, including RCA in the manufacturing process of concrete may result in increased energy savings, resource conservation, building sustainability, cost savings, and a large reduction in CDW.

This analysis illustrates how ML strategies might be used to foretell the CS of RAC. Three ML methods, including DT, GB, and BR, were employed. DT is a single ML method, while GB and BR are ensemble ML methods. Each approach was evaluated for exactness to determine the most effective prediction. The BR model, with an $R^2$ of 0.92, gave more precise findings than the GB and DT models, which had an $R^2$ of 0.85 and 0.77, respectively. Moreover, the accuracy of all techniques was tested by the statistical k-fold analysis techniques. The model's precision increases as the number of error values decreases. However, defining and suggesting the ideal ML model for forecasting outcomes across several domains is challenging since a model's precision is highly reliant on the input factors and size of the data set employed during modeling. Ensembled ML methods frequently take advantage of the weak learner by producing sub-models that may be trained on data and tweaked to improve the $R^2$. Figure 15 illustrates the dispersion of $R^2$ for the GB and BR sub-models. The $R^2$ for the GB sub-models were 0.818, 0.844, and 0.869, respectively. Similarly, the $R^2$ values for the lowest, average, and maximum BR sub-models were 0.899, 0.907, and 0.915, respectively. These findings indicated that BR sub-models had better $R^2$ values than GB sub-models, indicating that the BR model was more precise in estimating RAC's CS. In addition, an SA was carried out to find out the influence of all input factors on the RAC's projected CS. The execution of a model might be impacted by the model's input factors and the quantity of data points. SA was used to find out the contribution of each of the eight input factors to the anticipated output. The three most significant input factors were discovered to be the RCA replacement ratio, parent concrete strength, and w/c.

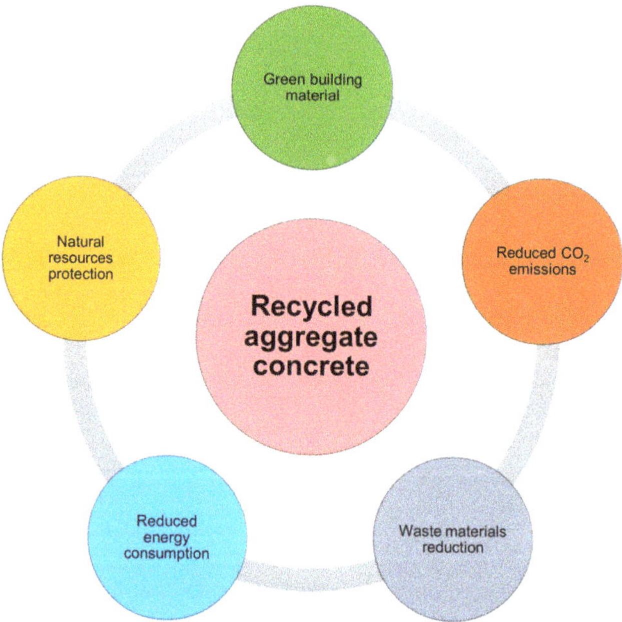


**Figure 14.** Benefits related to the adoption and application of recycled aggregate concrete.

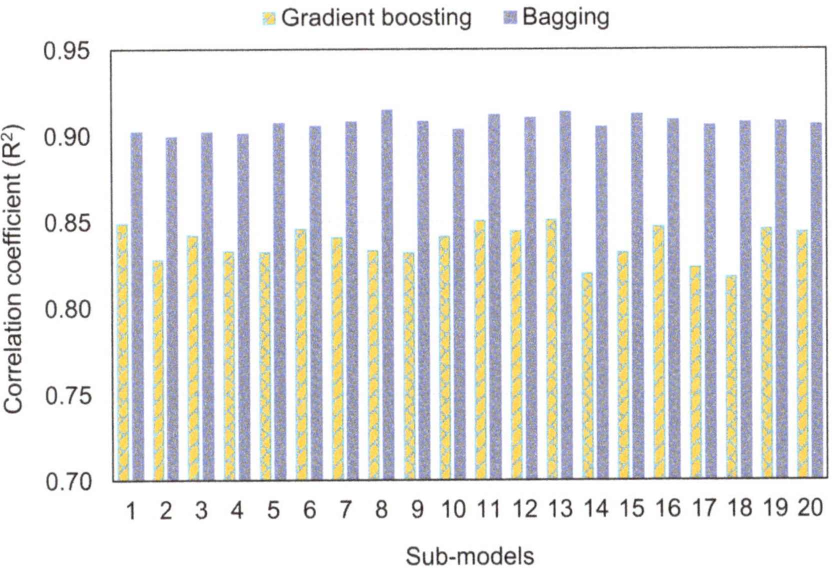


**Figure 15.** Correlation coefficient of sub-models.

## 7. Conclusions

The goal of this research was to estimate the compressive strength (CS) of recycled aggregate concrete (RAC) with the application of both single and ensemble machine learning (ML) algorithms. To predict outcomes, a decision tree (DT) and two ensemble approaches—gradient boosting (GB) and bagging regressor BR—were used. As a result of this analysis, the following findings have been drawn:

- Ensemble ML approaches outperformed the single ML approach in estimating the CS of RAC, with the BR model achieving the greatest accuracy. Correlation coefficients ($R^2$) were 0.92, 0.85, and 0.77 for the BR, GB, and DT models, respectively. Ensemble ML models (BR and GB) produced findings that were within a reasonable range and did not significantly diverge from experimental results. In comparison, the single ML model (DT) had a lower accuracy and was not suggested for estimating RAC strength.
- The model's performance was confirmed by statistical tests and k-fold analysis. These evaluations also validated the BR model's maximum accuracy, as seen by its reduced error values when compared to the GB and DT models.
- Sensitivity analysis indicated that the RCA replacement ratio was the most influential factor determining the model's outcome, accounting for around 21% of the total, followed by parent concrete strength at around 18% and the water–cement ratio at 16%. However, the other input parameters contributed less to the estimation of RAC's CS, with Los Angeles abrasion index of RCA, water absorption of RCA, aggregate–cement ratio, nominal maximum RCA size, bulk density of RCA accounting for around 13%, 9%, 9%, 7%, and 6%, respectively.
- This sort of study will benefit the construction industry by allowing for the advancement of rapid and cost-efficient approaches for estimating the strength of materials. Furthermore, by supporting eco-friendly construction through these measures, the adoption and application of RAC in construction will be promoted.

This study suggests that future research should use experimental procedures, mixed proportions, field trials, and other numerical evaluation techniques in order to enhance the number of data points and input parameters. In addition, to improve the models' responsiveness, environmental factors like temperature and humidity and a complete description of the raw materials may be incorporated as input variables. Furthermore, edge detection methods might be employed to detect cracks in concrete [94,95]. Nevertheless, algorithms for exact product identification and categorization are not confined to edge detection methods. This is a significant restriction of the proposal's objectives, and its limits should be assumed with greater rigor and realism in developing the arguments for future research.

**Author Contributions:** M.N.A.: Conceptualization, Resources, Project Administration, Supervision, Funding acquisition, Writing, Reviewing and Editing. W.A.: Conceptualization, Data curation, Software, Methodology, Investigation, Validation, Writing—Original draft. K.K.: Methodology, Investigation, Writing, Reviewing and Editing. F.A.: Resources, Visualization, Writing, Reviewing and Editing. A.A.: Data curation, Software, Validation, Writing, Reviewing and Editing. M.A.A.-F.: Formal analysis, Visualization, Writing, Reviewing and Editing. All authors have read and agreed to the published version of the manuscript.

**Funding:** The work was supported by the Deanship of Scientific Research, Vice Presidency for Graduate Studies and Scientific Research, King Faisal University, Al-Ahsa, Saudi Arabia. (Project No. GRANT505).

**Institutional Review Board Statement:** Not applicable.

**Informed Consent Statement:** Not applicable.

**Data Availability Statement:** The data used in this research has been properly cited and reported in the main text.

**Acknowledgments:** The authors acknowledge the Deanship of Scientific Research, Vice Presidency for Graduate Studies and Scientific Research at King Faisal University, Al-Ahsa, Saudi Arabia, for the financial support under the Annual Research Grants (Project No. GRANT505).

**Conflicts of Interest:** The authors declare no conflict of interest.

## Appendix A

**Table A1.** Data set retrieved from the literature and used for the analysis.

| Refs. | Water-Cement Ratio ($w_{eff}/c$) | Aggregate-Cement Ratio (a/c) | RCA Replacement Ratio (RCA %) | Parent Concrete Strength (MPa) | Nominal Maximum RCA Size (mm) | Bulk Density of RCA (kg/m$^3$) | Water Absorption of RCA (%) | Los Angeles Abrasion of RCA | Compressive Strength (MPa) |
|---|---|---|---|---|---|---|---|---|---|
| [96] | 0.5 | 2.6 | - | - | 20 | - | - | - | 42.8 |
| | 0.5 | 2.5 | 20 | - | 20 | - | - | - | 42.7 |
| | 0.5 | 2.5 | 50 | - | 20 | - | - | - | 41.3 |
| | 0.5 | 2.3 | 100 | - | 20 | - | - | - | 41.8 |
| [97] | 0.45 | 3.3 | - | - | 20 | - | - | - | 51.2 |
| | 0.45 | 3.3 | 30 | - | 20 | 2400 | 4.9 | - | 50.6 |
| | 0.45 | 3.3 | 50 | - | 20 | 2400 | 4.9 | - | 50.8 |
| | 0.45 | 3.3 | 100 | - | 20 | 2400 | 4.9 | - | 50.2 |
| | 0.39 | 2.6 | - | - | 20 | - | - | - | 60.3 |
| | 0.39 | 2.6 | 30 | - | 20 | 2400 | 4.9 | - | 60.8 |
| | 0.39 | 2.6 | 50 | - | 20 | 2400 | 4.9 | - | 61.2 |
| | 0.39 | 2.6 | 100 | - | 20 | 2400 | 4.9 | - | 60.2 |
| | 0.29 | 2.2 | - | - | 20 | - | - | - | 70.5 |
| | 0.29 | 2.2 | 30 | - | 20 | 2400 | 4.9 | - | 70.2 |
| | 0.29 | 2.2 | 50 | - | 20 | 2400 | 4.9 | - | 70.8 |
| | 0.29 | 2.2 | 100 | - | 20 | 2400 | 4.9 | - | 70 |
| [98] | 0.36 | 2.4 | - | 41.6 | 16 | - | - | - | 48.4 |
| | 0.36 | 2.3 | 100 | 41.6 | 16 | - | - | - | 44.5 |
| | 0.36 | 2.2 | 100 | 41.6 | 16 | - | - | - | 38.7 |

Table A1. *Cont.*

| Refs. | Water-Cement Ratio ($w_{eff}/c$) | Aggregate-Cement Ratio (a/c) | RCA Replacement Ratio (RCA %) | Parent Concrete Strength (MPa) | Nominal Maximum RCA Size (mm) | Bulk Density of RCA (kg/m$^3$) | Water Absorption of RCA (%) | Los Angeles Abrasion of RCA | Compressive Strength (MPa) |
|---|---|---|---|---|---|---|---|---|---|
| | 0.36 | 2.4 | - | 50.6 | 16 | - | - | - | 48.9 |
| | 0.36 | 2.3 | 100 | 50.6 | 16 | - | - | - | 46.1 |
| | 0.36 | 2.2 | 100 | 50.6 | 16 | - | - | - | 42.4 |
| | 0.36 | 2.4 | - | 63.2 | 16 | - | - | - | 48.9 |
| | 0.36 | 2.3 | 100 | 63.2 | 16 | - | - | - | 52.5 |
| | 0.36 | 2.2 | 100 | 63.2 | 16 | - | - | - | 50.7 |
| | 0.36 | 2.4 | - | 35.6 | 16 | - | - | - | 48.9 |
| | 0.36 | 2.3 | 100 | 35.6 | 16 | - | - | - | 45.2 |
| | 0.36 | 2.2 | 100 | 35.6 | 16 | - | - | - | 42 |
| | 0.36 | 2.4 | - | 66 | 16 | - | - | - | 48.9 |
| | 0.36 | 2.3 | 100 | 66 | 16 | - | - | - | 49.6 |
| | 0.36 | 2.2 | 100 | 66 | 16 | - | - | - | 45.1 |
| | 0.36 | 2.7 | - | 72.3 | 16 | - | - | - | 52.3 |
| | 0.36 | 2.4 | 100 | 72.3 | 16 | - | - | - | 54.4 |
| | 0.36 | 2.3 | 100 | 72.3 | 16 | - | - | - | 48.2 |
| [99] | 0.47 | 2.5 | - | 38.4 | 20 | - | - | - | 39 |
| | 0.47 | 2.5 | 15 | 38.4 | 20 | 2410 | 5.8 | - | 38.1 |
| | 0.45 | 2.5 | 30 | 38.4 | 20 | 2410 | 5.8 | - | 37 |
| | 0.42 | 2.4 | 60 | 38.4 | 20 | 2410 | 5.8 | - | 35.8 |
| | 0.38 | 2.3 | 100 | 38.4 | 20 | 2410 | 5.8 | - | 34.5 |
| [100] | 0.6 | 4.6 | - | - | 15 | - | - | - | 43.5 |
| | 0.6 | 4.1 | 100 | - | 15 | 2450 | 5.6 | - | 38.2 |
| | 0.45 | 3.3 | - | - | 15 | - | - | - | 61.7 |
| | 0.45 | 2.9 | 100 | - | 15 | 2450 | 5.6 | - | 52.8 |
| | 0.35 | 2.6 | - | - | 15 | - | - | - | 74.4 |
| | 0.35 | 2.3 | 100 | - | 15 | 2450 | 5.6 | - | 62.8 |
| | 0.45 | 3.2 | 25 | - | 15 | - | - | - | 60.7 |
| | 0.45 | 3.1 | 50 | - | 15 | 2450 | 5.6 | - | 59.4 |
| [101] | 0.57 | 3.1 | - | - | 20 | - | - | - | 48.3 |
| | 0.57 | 3.1 | 20 | - | 20 | 2330 | 6.3 | - | 44.9 |
| | 0.57 | 3.1 | 50 | - | 20 | 2330 | 6.3 | - | 44.7 |
| | 0.57 | 3 | 100 | - | 20 | 2330 | 6.3 | - | 46.8 |
| | 0.57 | 3 | - | - | 20 | - | - | - | 40.2 |
| | 0.57 | 3.1 | 20 | - | 20 | 2330 | 6.3 | - | 43.2 |
| | 0.57 | 2.9 | 50 | - | 20 | 2330 | 6.3 | - | 39.7 |
| | 0.57 | 2.9 | 100 | - | 20 | 2330 | 6.3 | - | 43.3 |
| | 0.57 | 3 | - | - | 20 | - | - | - | 46 |
| | 0.57 | 2.8 | 20 | - | 20 | 2330 | 6.3 | - | 43 |
| | 0.57 | 2.7 | 50 | - | 20 | 2330 | 6.3 | - | 38.1 |
| | 0.57 | 2.9 | 100 | - | 20 | 2330 | 6.3 | - | 39.1 |
| [102] | 0.5 | 2.4 | 100 | - | 25 | - | - | - | 30.2 |
| | 0.5 | 2.3 | 100 | - | 25 | - | - | - | 36.2 |
| | 0.7 | 3.3 | 100 | - | 25 | - | - | - | 27.7 |
| | 0.7 | 3.2 | 100 | - | 25 | - | - | - | 20.4 |
| [103] | 0.43 | 3 | - | - | 32 | - | - | - | 35.9 |
| | 0.43 | 2.9 | 33 | - | 32 | 2520 | 9.3 | - | 34.1 |
| | 0.43 | 2.8 | 53 | - | 32 | 2520 | 9.3 | - | 29.6 |
| | 0.43 | 2.8 | 72 | - | 32 | 2520 | 9.3 | - | 30.3 |
| | 0.43 | 2.7 | 100 | - | 32 | 2520 | 9.3 | - | 26.7 |
| [104] | 0.42 | 3 | - | - | 32 | - | - | - | 36.8 |
| | 0.37 | 2.9 | 30 | - | 32 | 2442 | 6 | - | 37.2 |

Table A1. *Cont.*

| Refs. | Water-Cement Ratio ($w_{eff}/c$) | Aggregate-Cement Ratio (a/c) | RCA Replacement Ratio (RCA %) | Parent Concrete Strength (MPa) | Nominal Maximum RCA Size (mm) | Bulk Density of RCA (kg/m$^3$) | Water Absorption of RCA (%) | Los Angeles Abrasion of RCA | Compressive Strength (MPa) |
|---|---|---|---|---|---|---|---|---|---|
| | 0.34 | 2.8 | 50 | - | 32 | 2442 | 6 | - | 37.8 |
| | 0.38 | 2 | 70 | - | 32 | 2442 | 6 | - | 36.7 |
| | 0.27 | 2.7 | 100 | - | 32 | 2442 | 6 | - | 35.2 |
| [105] | 0.55 | 4 | - | - | 25 | - | - | - | 42 |
| | 0.55 | 3.9 | 25 | - | 25 | 2430 | 4.4 | - | 42 |
| | 0.52 | 3.6 | 50 | - | 25 | 2430 | 4.4 | - | 41 |
| | 0.5 | 3.5 | 100 | - | 25 | 2430 | 4.4 | - | 40 |
| [106] | 0.55 | 4 | - | - | 25 | - | - | - | 35.5 |
| | 0.55 | 3.9 | 25 | - | 25 | 2430 | 4.5 | - | 38.8 |
| | 0.52 | 3.6 | 50 | - | 25 | 2430 | 4.5 | - | 39.4 |
| | 0.5 | 3.5 | 100 | - | 25 | 2430 | 4.5 | - | 38.3 |
| [107] | 0.41 | 3.1 | - | - | 20 | - | - | - | 59.4 |
| | 0.42 | 3.2 | 10 | - | 20 | 2165 | 6.8 | - | 62.2 |
| | 0.43 | 3.4 | 20 | - | 20 | 2165 | 6.8 | - | 58.4 |
| | 0.44 | 3.5 | 30 | - | 20 | 2165 | 6.8 | - | 61.3 |
| | 0.45 | 3.7 | 50 | - | 20 | 2165 | 6.8 | - | 60.8 |
| | 0.45 | 4.4 | 100 | - | 20 | 2165 | 6.8 | - | 61 |
| [108] | 0.51 | 2.6 | - | - | 20 | - | - | - | 48.6 |
| | 0.49 | 2.5 | 20 | - | 20 | 2570 | 3.5 | - | 45.3 |
| | 0.48 | 2.5 | 50 | - | 20 | 2570 | 3.5 | - | 42.5 |
| | 0.46 | 2.5 | 80 | - | 20 | 2570 | 3.5 | - | 39.2 |
| | 0.45 | 2.5 | 100 | - | 20 | 2570 | 3.5 | - | 37.1 |
| [109] | 0.49 | 4.7 | - | - | 16 | 2270 | - | - | 37.7 |
| | 0.49 | 3.9 | 100 | - | 16 | 2270 | - | - | 34.6 |
| | 0.36 | 2.4 | - | - | 16 | 2270 | - | - | 57.9 |
| | 0.36 | 2.2 | 100 | - | 16 | 2270 | - | - | 56.4 |
| | 0.49 | 3.7 | - | - | 16 | 2780 | - | - | 39.8 |
| | 0.49 | 4.4 | 100 | - | 16 | 2780 | - | - | 40.1 |
| | 0.36 | 2.4 | - | - | 16 | 2780 | - | - | 58.3 |
| | 0.36 | 2.3 | 100 | - | 16 | 2780 | - | - | 60.2 |
| | 0.49 | 5.1 | - | - | 16 | 2565 | - | - | 40.1 |
| | 0.49 | 4.2 | 100 | - | 16 | 2565 | - | - | 35.3 |
| | 0.36 | 2.7 | - | - | 16 | 2565 | - | - | 61.8 |
| | 0.36 | 2.4 | 100 | - | 16 | 2565 | - | - | 57.5 |
| [110] | 0.47 | 3.3 | - | - | 32 | - | - | - | 31.2 |
| | 0.41 | 3.3 | 30 | - | 32 | 2449 | 6 | - | 31 |
| | 0.38 | 3.2 | 50 | - | 32 | 2449 | 6 | - | 29.3 |
| | 0.36 | 3.1 | 70 | - | 32 | 2449 | 6 | - | 28.4 |
| | 0.32 | 3 | 100 | - | 32 | 2449 | 6 | - | 27.2 |
| [111] | 0.45 | 2.8 | - | - | 20 | - | - | - | 66.8 |
| | 0.45 | 2.8 | 20 | - | 20 | 2570 | 3.5 | - | 62.4 |
| | 0.45 | 2.7 | 50 | - | 20 | 2570 | 3.5 | - | 55.8 |
| | 0.45 | 2.7 | 100 | - | 20 | 2570 | 3.5 | - | 42 |
| | 0.55 | 2.6 | - | - | 20 | - | - | - | 48.6 |
| | 0.55 | 2.5 | 20 | - | 20 | 2570 | 3.5 | - | 45.3 |
| | 0.55 | 2.5 | 50 | - | 20 | 2570 | 3.5 | - | 42.5 |
| | 0.55 | 2.5 | 100 | - | 20 | 2570 | 3.5 | - | 38.1 |
| [112] | 0.65 | 3.1 | - | - | 19 | - | - | - | 21.8 |
| | 0.65 | 3.1 | 100 | - | 19 | 2390 | 4.4 | - | 22.1 |
| | 0.5 | 2.9 | - | - | 19 | - | - | - | 26.7 |

**Table A1.** *Cont.*

| Refs. | Water-Cement Ratio ($w_{eff}$/c) | Aggregate-Cement Ratio (a/c) | RCA Replacement Ratio (RCA %) | Parent Concrete Strength (MPa) | Nominal Maximum RCA Size (mm) | Bulk Density of RCA (kg/m$^3$) | Water Absorption of RCA (%) | Los Angeles Abrasion of RCA | Compressive Strength (MPa) |
|---|---|---|---|---|---|---|---|---|---|
| | 0.5 | 2.9 | 100 | - | 19 | 2390 | 4.4 | - | 25.1 |
| | 0.48 | 2.8 | - | - | 19 | - | - | - | 28.9 |
| | 0.48 | 2.8 | 100 | - | 19 | 2390 | 4.4 | - | 27.2 |
| | 0.43 | 2.6 | - | - | 19 | - | - | - | 31.1 |
| | 0.43 | 2.6 | 100 | - | 19 | 2390 | 4.4 | - | 28.7 |
| | 0.4 | 2.4 | - | - | 19 | - | - | - | 33.7 |
| | 0.4 | 2.4 | 100 | - | 19 | 2390 | 4.4 | - | 29.5 |
| [113] | 0.54 | 3.1 | - | - | 32 | - | - | - | 26.8 |
| | 0.35 | 3.1 | 100 | - | 32 | 2512 | 6.3 | - | 24.6 |
| | 0.49 | 3.1 | 100 | - | 32 | 2670 | 1.8 | - | 26.9 |
| | 0.46 | 2.7 | - | - | 32 | - | - | - | 34.3 |
| | 0.31 | 2.7 | 100 | - | 32 | 2512 | 6.3 | - | 30.2 |
| | 0.43 | 2.7 | 100 | - | 32 | 2670 | 1.8 | - | 34.2 |
| | 0.42 | 2.4 | - | - | 32 | - | - | - | 38.6 |
| | 0.28 | 2.4 | 100 | - | 32 | 2512 | 6.3 | - | 35.5 |
| | 0.39 | 2.4 | 100 | - | 32 | 2670 | 1.8 | - | 38.4 |
| [114] | 0.7 | 4.1 | - | - | 30 | - | - | - | 18.1 |
| | 0.67 | 3.9 | 100 | - | 30 | 2520 | 3.8 | 34 | 18 |
| | 0.67 | 3.9 | 100 | - | 30 | 2510 | 3.9 | 39 | 15.4 |
| | 0.35 | 3.1 | - | - | 30 | - | - | - | 37.5 |
| | 0.35 | 4.3 | 100 | - | 30 | 2520 | 3.8 | 34 | 36.4 |
| | 0.36 | 2.1 | 100 | - | 30 | 2510 | 3.9 | 39 | 35.7 |
| | 0.34 | 2.1 | - | - | 30 | - | - | - | 48.4 |
| | 0.34 | 1.9 | 100 | - | 30 | 2520 | 3.8 | 34 | 44.4 |
| | 0.34 | 2.2 | 100 | - | 30 | 2510 | 3.9 | 39 | 43.8 |
| [115] | 0.47 | 3.3 | - | - | 32 | - | - | - | 31.2 |
| | 0.41 | 3.3 | 30 | - | 32 | 2449 | 6 | - | 31 |
| | 0.38 | 3.2 | 50 | - | 32 | 2449 | 6 | - | 29.3 |
| | 0.36 | 3.1 | 70 | - | 32 | 2449 | 6 | - | 28.4 |
| | 0.32 | 3 | 100 | - | 32 | 2449 | 6 | - | 27.2 |
| [116] | 0.55 | 2.6 | - | - | 20 | - | - | - | 48.6 |
| | 0.55 | 2.6 | 20 | - | 20 | 2580 | 3.5 | - | 45.3 |
| | 0.55 | 2.5 | 50 | - | 20 | 2580 | 3.5 | - | 42.5 |
| | 0.55 | 2.5 | 100 | - | 20 | 2580 | 3.5 | - | 38.1 |
| | 0.5 | 2.6 | - | - | 20 | - | - | - | 54.1 |
| | 0.5 | 2.6 | 20 | - | 20 | 2580 | 3.5 | - | 51.7 |
| | 0.5 | 2.6 | 50 | - | 20 | 2580 | 3.5 | - | 47.1 |
| | 0.5 | 2.6 | 100 | - | 20 | 2580 | 3.5 | - | 43.4 |
| | 0.45 | 2.8 | - | - | 20 | - | - | - | 66.8 |
| | 0.45 | 2.8 | 20 | - | 20 | 2580 | 3.5 | - | 62.4 |
| | 0.45 | 2.7 | 50 | - | 20 | 2580 | 3.5 | - | 56.8 |
| | 0.45 | 2.5 | 100 | - | 20 | 2580 | 3.5 | - | 52.1 |
| | 0.4 | 2.9 | - | - | 20 | - | - | - | 72.3 |
| | 0.4 | 2.8 | 20 | - | 20 | 2580 | 3.5 | - | 69.6 |
| | 0.4 | 2.8 | 50 | - | 20 | 2580 | 3.5 | - | 65.3 |
| | 0.4 | 2.8 | 100 | - | 20 | 2580 | 3.5 | - | 58.5 |
| [117] | 0.5 | 2.9 | - | - | 25 | - | - | - | 39.5 |
| | 0.5 | 2.9 | 30 | - | 25 | 2530 | 1.9 | - | 36.7 |
| | 0.5 | 2.9 | 50 | - | 25 | 2530 | 1.9 | - | 38 |
| | 0.5 | 2.8 | 100 | - | 25 | 2530 | 1.9 | - | 36 |
| | 0.5 | 2.8 | 30 | - | 25 | - | - | - | 32.6 |
| | 0.5 | 2.8 | 50 | - | 25 | 2400 | 6.2 | - | 30.4 |
| | 0.5 | 2.7 | 100 | - | 25 | 2400 | 6.2 | - | 29.5 |

Table A1. *Cont.*

| Refs. | Water-Cement Ratio ($w_{eff}$/c) | Aggregate-Cement Ratio (a/c) | RCA Replacement Ratio (RCA %) | Parent Concrete Strength (MPa) | Nominal Maximum RCA Size (mm) | Bulk Density of RCA (kg/m$^3$) | Water Absorption of RCA (%) | Los Angeles Abrasion of RCA | Compressive Strength (MPa) |
|---|---|---|---|---|---|---|---|---|---|
| [118] | 0.58 | 3.2 | - | - | 32 | - | - | - | 44.6 |
| | 0.52 | 3.2 | 50 | - | 32 | 2720 | 4.8 | - | 41.4 |
| | 0.45 | 3.2 | 100 | - | 32 | 2720 | 4.8 | - | 40.7 |
| | 0.52 | 3.2 | 50 | - | 32 | 2650 | 4.6 | - | 38.3 |
| | 0.46 | 3.2 | 100 | - | 32 | 2650 | 4.6 | - | 36.6 |
| | 0.52 | 3.2 | 50 | - | 32 | 2880 | 4.4 | - | 41.2 |
| | 0.47 | 3.2 | 100 | - | 32 | 2880 | 4.4 | - | 40.3 |
| [119] | 0.41 | 2.6 | - | - | 20 | - | - | - | 42.3 |
| | 0.39 | 2.5 | 20 | - | 20 | 2338 | 5.2 | 40.2 | 47.4 |
| | 0.36 | 2.5 | 50 | - | 20 | 2338 | 5.2 | 40.2 | 47.3 |
| | 0.32 | 2.3 | 100 | - | 20 | 2338 | 5.2 | 40.2 | 54.8 |
| [120] | 0.52 | 2.9 | - | - | 22 | - | - | - | 48 |
| | 0.52 | 2.9 | 10 | - | 22 | - | - | - | 46.9 |
| | 0.52 | 2.8 | 20 | - | 22 | - | - | - | 47.7 |
| | 0.52 | 2.8 | 30 | - | 22 | - | - | - | 50.8 |
| | 0.52 | 2.8 | 40 | - | 22 | - | - | - | 48 |
| | 0.52 | 2.8 | 50 | - | 22 | - | - | - | 49.5 |
| | 0.52 | 2.7 | 100 | - | 22 | - | - | - | 50.3 |
| | 0.54 | 3.2 | - | - | 22 | - | - | - | 23.5 |
| | 0.54 | 3.1 | 25 | - | 22 | - | - | - | 21.6 |
| | 0.54 | 3.1 | 100 | - | 22 | - | - | - | 20.5 |
| [121] | 0.76 | 4.5 | 100 | - | 30 | - | - | - | 21.1 |
| | 0.76 | 4.5 | 100 | - | 30 | - | - | - | 22 |
| | 0.76 | 4.5 | 100 | - | 30 | - | - | - | 23.1 |
| | 0.76 | 4.5 | 100 | - | 30 | - | - | - | 23.5 |
| | 0.76 | 4.5 | 100 | - | 30 | - | - | - | 20.4 |
| | 0.76 | 4.5 | 100 | - | 30 | - | - | - | 18.9 |
| | 0.76 | 4.5 | 100 | - | 30 | - | - | - | 21.2 |
| | 0.66 | 3.9 | 100 | - | 30 | - | - | - | 25.7 |
| | 0.66 | 3.9 | 100 | - | 30 | - | - | - | 28 |
| | 0.66 | 3.9 | 100 | - | 30 | - | - | - | 25.1 |
| | 0.66 | 3.9 | 100 | - | 30 | - | - | - | 27.5 |
| | 0.66 | 3.9 | 100 | - | 30 | - | - | - | 26.1 |
| | 0.66 | 3.9 | 100 | - | 30 | - | - | - | 27.4 |
| | 0.66 | 3.9 | 100 | - | 30 | - | - | - | 27.7 |
| | 0.66 | 3.9 | 100 | - | 30 | - | - | - | 25 |
| | 0.57 | 3.3 | 100 | - | 30 | - | - | - | 30.5 |
| | 0.57 | 3.3 | 100 | - | 30 | - | - | - | 32.7 |
| | 0.57 | 3.3 | 100 | - | 30 | - | - | - | 32.8 |
| | 0.57 | 3.3 | 100 | - | 30 | - | - | - | 33.1 |
| | 0.48 | 2.7 | 100 | - | 30 | - | - | - | 35.3 |
| | 0.48 | 2.7 | 100 | - | 30 | - | - | - | 35.2 |
| | 0.48 | 2.7 | 100 | - | 30 | - | - | - | 32.5 |
| | 0.48 | 2.7 | 100 | - | 30 | - | - | - | 33.8 |
| | 0.41 | 2.2 | 100 | - | 30 | - | - | - | 41.9 |
| | 0.41 | 2.2 | 100 | - | 30 | - | - | - | 38.4 |
| | 0.41 | 2.2 | 100 | - | 30 | - | - | - | 38.7 |
| | 0.41 | 2.2 | 100 | - | 30 | - | - | - | 41.2 |
| [122] | 0.54 | 3.1 | - | - | 32 | - | - | - | 26.8 |
| | 0.35 | 3.1 | 100 | - | 32 | 2512 | 6.3 | - | 24.6 |
| | 0.49 | 3.1 | 100 | - | 32 | 2670 | 1.8 | - | 26.9 |
| | 0.46 | 3.3 | - | - | 32 | - | - | - | 34.3 |

Table A1. *Cont.*

| Refs. | Water-Cement Ratio ($w_{eff}/c$) | Aggregate-Cement Ratio (a/c) | RCA Replacement Ratio (RCA %) | Parent Concrete Strength (MPa) | Nominal Maximum RCA Size (mm) | Bulk Density of RCA (kg/m$^3$) | Water Absorption of RCA (%) | Los Angeles Abrasion of RCA | Compressive Strength (MPa) |
|---|---|---|---|---|---|---|---|---|---|
| | 0.27 | 3.3 | 100 | - | 32 | 2512 | 6.3 | - | 30.2 |
| | 0.41 | 3.3 | 100 | - | 32 | 2670 | 1.8 | - | 34.2 |
| | 0.42 | 3 | - | - | 32 | - | - | - | 38.6 |
| | 0.24 | 3 | 100 | - | 32 | 2512 | 6.3 | - | 35.5 |
| | 0.38 | 3 | 100 | - | 32 | 2670 | 1.8 | - | 38.4 |
| [123] | 0.4 | 3.1 | 50 | - | 12 | 2420 | 6.8 | - | 43.3 |
| | 0.45 | 3.1 | 50 | - | 12 | 2400 | 6.8 | - | 39.6 |
| | 0.5 | 3.2 | 50 | - | 12 | 2400 | 6.8 | - | 38.1 |
| | 0.55 | 3.2 | 50 | - | 12 | 2400 | 6.8 | - | 34.5 |
| | 0.6 | 3.3 | 50 | - | 12 | 2400 | 6.8 | - | 31.6 |
| | 0.4 | 3.1 | 50 | - | 22 | 2420 | 8.8 | - | 46.1 |
| | 0.45 | 3.1 | 50 | - | 22 | 2420 | 8.8 | - | 45.8 |
| | 0.5 | 3.2 | 50 | - | 22 | 2420 | 8.8 | - | 39.9 |
| | 0.55 | 3.3 | 50 | - | 22 | 2420 | 8.8 | - | 36.3 |
| | 0.6 | 3.3 | 50 | - | 22 | 2420 | 8.8 | - | 34.7 |
| [124] | 0.5 | 3.5 | - | - | 20 | - | - | - | 28.3 |
| | 0.5 | 3.5 | 20 | - | 20 | 2400 | - | - | 27.2 |
| | 0.5 | 3.5 | 40 | - | 20 | 2400 | - | - | 26.5 |
| | 0.5 | 3.5 | 60 | - | 20 | 2400 | - | - | 25.4 |
| | 0.5 | 3.5 | 80 | - | 20 | 2400 | - | - | 25.1 |
| | 0.5 | 3.5 | 100 | - | 20 | 2400 | - | - | 20.4 |
| | 0.5 | 3.8 | 20 | - | 20 | 2630 | - | - | 26.4 |
| | 0.5 | 4.1 | 40 | - | 20 | 2630 | - | - | 25.9 |
| | 0.5 | 4.5 | 60 | - | 20 | 2630 | - | - | 23.5 |
| | 0.5 | 4.8 | 80 | - | 20 | 2630 | - | - | 15.4 |
| [125] | 0.51 | 3.6 | - | - | 32 | - | - | - | 43.4 |
| | 0.57 | 3.6 | 50 | - | 32 | 2489 | 2.4 | 34 | 45.2 |
| | 0.62 | 3.6 | 100 | - | 32 | 2489 | 2.4 | 34 | 45.7 |
| [126] | 0.65 | 3.3 | - | - | 19 | - | - | - | 20.2 |
| | 0.65 | 3.2 | 25 | - | 19 | 2440 | 5.8 | 33.6 | 18.5 |
| | 0.65 | 3.1 | 50 | - | 19 | 2440 | 5.8 | 33.6 | 18 |
| | 0.65 | 3.1 | 75 | - | 19 | 2440 | 5.8 | - | 16.5 |
| | 0.42 | 2.7 | - | - | 19 | - | - | 33.6 | 40 |
| | 0.42 | 2.7 | 25 | - | 19 | 2440 | 5.8 | 33.6 | 33 |
| | 0.42 | 2.6 | 50 | - | 19 | 2440 | 5.8 | 33.6 | 34.5 |
| | 0.42 | 2.5 | 75 | - | 19 | 2440 | 5.8 | 33.6 | 34 |
| [127] | 0.65 | 3.4 | - | - | 16 | - | - | - | 31.9 |
| | 0.66 | 3.3 | 20 | - | 16 | 2400 | 5 | - | 31.7 |
| | 0.68 | 3.1 | 50 | - | 16 | 2400 | 5 | - | 32.4 |
| | 0.68 | 2.8 | 100 | - | 16 | 2400 | 5 | - | 30.1 |
| | 0.5 | 2.6 | - | - | 16 | - | - | - | 44.8 |
| | 0.51 | 2.5 | 20 | - | 16 | 2400 | 5 | - | 43.7 |
| | 0.53 | 2.3 | 50 | - | 16 | 2400 | 5 | - | 37.5 |
| | 0.56 | 2.1 | 100 | - | 16 | 2400 | 5 | - | 40.5 |
| [128] | 0.45 | 1.9 | - | - | 19 | 2420 | 5.4 | - | 35.2 |
| | 0.45 | 3.4 | 64 | - | 19 | 2420 | 5.4 | - | 41.4 |
| | 0.45 | 2.3 | 100 | - | 19 | 2420 | 5.4 | - | 43.9 |
| | 0.45 | 2.1 | - | - | 19 | 2500 | 3.3 | - | 34.1 |
| | 0.45 | 3.1 | 64 | - | 19 | 2500 | 3.3 | - | 44.8 |
| | 0.45 | 2.5 | 100 | - | 19 | 2500 | 3.3 | - | 45.9 |

**Table A1.** *Cont.*

| Refs. | Water-Cement Ratio ($w_{eff}/c$) | Aggregate-Cement Ratio (a/c) | RCA Replacement Ratio (RCA %) | Parent Concrete Strength (MPa) | Nominal Maximum RCA Size (mm) | Bulk Density of RCA (kg/m$^3$) | Water Absorption of RCA (%) | Los Angeles Abrasion of RCA | Compressive Strength (MPa) |
|---|---|---|---|---|---|---|---|---|---|
| [129] | 0.65 | 3.4 | - | - | 16 | - | - | - | 31.9 |
| | 0.65 | 3.3 | 20 | - | 16 | 2400 | 5 | 34 | 31.7 |
| | 0.65 | 3.1 | 50 | - | 16 | 2400 | 5 | 34 | 32.4 |
| | 0.65 | 2.8 | 100 | - | 16 | 2400 | 5 | 34 | 30.1 |
| | 0.5 | 2.6 | - | - | 16 | - | - | - | 44.8 |
| | 0.5 | 2.5 | 20 | - | 16 | 2400 | 5 | 34 | 43.7 |
| | 0.5 | 2.8 | 50 | - | 16 | 2400 | 5 | 34 | 37.5 |
| | 0.5 | 2.1 | 100 | - | 16 | 2400 | 5 | 34 | 40.5 |
| [130] | 0.43 | 3.1 | - | - | 20 | - | - | - | 51.8 |
| | 0.43 | 3 | 25 | - | 20 | 2661 | 1.9 | - | 47 |
| | 0.43 | 2.9 | 50 | - | 20 | 2602 | 2.6 | - | 46 |
| | 0.43 | 2.8 | 100 | - | 20 | 2510 | 3.9 | 38.8 | 42.5 |
| [131] | 0.45 | 2.3 | - | - | 19 | - | - | - | 44.4 |
| | 0.45 | 2.3 | 100 | - | 19 | 2490 | 4.8 | 37 | 41 |
| | 0.55 | 2.9 | - | - | 19 | - | - | - | 36.7 |
| | 0.55 | 2.9 | 100 | - | 19 | 2490 | 4.8 | 37 | 33.3 |
| | 0.65 | 3.5 | - | - | 19 | - | - | - | 30.4 |
| | 0.65 | 3.5 | 100 | - | 19 | 2490 | 4.8 | 37 | 24.8 |
| [132] | 0.6 | 4.6 | - | - | 19 | - | - | - | 25 |
| | 0.6 | 4.6 | 25 | - | 19 | - | - | - | 26.7 |
| | 0.6 | 4.5 | 50 | - | 19 | - | - | - | 21.5 |
| | 0.6 | 4.5 | 75 | - | 19 | - | - | - | 21.4 |
| | 0.6 | 4.4 | 100 | - | 19 | - | - | - | 20 |
| | 0.45 | 2.6 | - | - | 19 | - | - | - | 39.5 |
| | 0.45 | 2.6 | 25 | - | 19 | - | - | - | 38.3 |
| | 0.45 | 2.5 | 50 | - | 19 | - | - | - | 37 |
| | 0.45 | 2.5 | 75 | - | 19 | - | - | - | 35 |
| | 0.45 | 2.5 | 100 | - | 19 | - | - | - | 33.3 |
| [133] | 0.49 | 3.1 | - | - | 25 | - | - | - | 44.3 |
| | 0.37 | 3 | 100 | 26.3 | 25 | 2490 | 2.9 | - | 37.6 |
| | 0.43 | 3 | 100 | 42.7 | 25 | 2570 | 2.9 | - | 43.3 |
| | 0.36 | 2.9 | 100 | 42.7 | 25 | 2440 | 5.6 | - | 42.6 |
| | 0.36 | 2.9 | 100 | 65.3 | 25 | 2470 | 5.3 | - | 44.7 |
| [134] | 0.53 | 6.5 | - | - | 32 | - | - | - | 39.3 |
| | 0.43 | 5.4 | 100 | - | 32 | 2263 | 6 | - | 33.2 |
| | 0.49 | 5.1 | 100 | - | 32 | 2283 | 4.2 | - | 35.6 |
| | 0.53 | 5.1 | 100 | - | 32 | 2292 | 4.3 | - | 34.6 |
| | 0.6 | 5.3 | 100 | - | 32 | 2301 | 5 | - | 37.3 |
| | 0.54 | 6.4 | 90 | - | 32 | 2609 | 1.5 | - | 45.4 |
| | 0.46 | 5.9 | 60 | - | 32 | 2518 | 2.7 | - | 54.3 |
| | 0.44 | 5.8 | 60 | - | 32 | 2584 | 1.6 | - | 54.4 |
| | 0.45 | 6.4 | 25 | - | 32 | 2594 | 1.6 | - | 53.4 |
| [135] | 0.43 | 3 | - | - | 32 | - | - | - | 34.8 |
| | 0.47 | 2.9 | 30 | - | 32 | - | - | - | 31.9 |
| | 0.49 | 2.8 | 50 | - | 32 | - | - | - | 30.6 |
| | 0.54 | 2.7 | 100 | - | 32 | - | - | - | 29.7 |

**Table A1.** *Cont.*

| Refs. | Water-Cement Ratio ($w_{eff}$/c) | Aggregate-Cement Ratio (a/c) | RCA Replacement Ratio (RCA %) | Parent Concrete Strength (MPa) | Nominal Maximum RCA Size (mm) | Bulk Density of RCA (kg/m$^3$) | Water Absorption of RCA (%) | Los Angeles Abrasion of RCA | Compressive Strength (MPa) |
|---|---|---|---|---|---|---|---|---|---|
| [136] | 0.66 | 4.6 | - | - | 20 | - | - | - | 21 |
| | 0.66 | 4.6 | 30 | - | 20 | 2340 | 5.3 | - | 20 |
| | 0.61 | 4.3 | 50 | - | 20 | 2340 | 5.3 | - | 19 |
| | 0.58 | 4 | 100 | - | 20 | 2340 | 5.3 | - | 18 |
| | 0.55 | 3.8 | - | - | 20 | - | - | - | 21 |
| | 0.55 | 3.8 | 30 | - | 20 | 2340 | 5.3 | - | 23 |
| | 0.51 | 3.5 | 50 | - | 20 | 2340 | 5.3 | - | 24 |
| | 0.48 | 3.4 | 100 | - | 20 | 2340 | 5.3 | - | 21 |
| | 0.5 | 3.5 | - | - | 20 | - | - | - | 31 |
| | 0.5 | 3.5 | 30 | - | 20 | 2340 | 5.3 | - | 25 |
| | 0.47 | 3.2 | 50 | - | 20 | 2340 | 5.3 | - | 29 |
| | 0.44 | 3 | 100 | - | 20 | 2340 | 5.3 | - | 30 |
| | 0.48 | 3.3 | - | - | 20 | - | - | - | 33 |
| | 0.48 | 3.3 | 30 | - | 20 | 2340 | 5.3 | - | 39 |
| | 0.44 | 3.1 | 50 | - | 20 | 2340 | 5.3 | - | 31 |
| | 0.42 | 2.9 | 100 | - | 20 | 2340 | 5.3 | - | 34 |
| [137] | 0.6 | 4.3 | - | - | 32 | - | - | - | 36.6 |
| | 0.6 | 3.8 | 100 | - | 32 | 2264 | 2 | - | 33.6 |
| | 0.52 | 3.6 | - | - | 32 | - | - | - | 41.8 |
| | 0.52 | 3.2 | 100 | - | 32 | 2276 | 2 | - | 41.1 |
| | 0.47 | 3 | - | - | 32 | - | - | - | 48.6 |
| | 0.47 | 2.7 | 100 | - | 32 | 2273 | 2 | - | 48.1 |
| [138] | 0.6 | 3 | - | 37.3 | 12 | - | - | - | 39.5 |
| | 0.59 | 3.2 | 10 | 37.3 | 12 | 2010 | 10.9 | - | 40 |
| | 0.57 | 3.5 | 30 | 37.3 | 12 | 2010 | 10.9 | - | 38.6 |
| | 0.54 | 3.8 | 50 | 37.3 | 12 | 2010 | 10.9 | - | 37.6 |
| | 0.46 | 4.6 | 100 | 37.3 | 12 | 2010 | 10.9 | - | 38.6 |
| | 0.45 | 3.2 | - | 37.3 | 12 | - | - | - | 53.3 |
| | 0.44 | 3.3 | 10 | 37.3 | 12 | 2010 | 10.9 | - | 53.7 |
| | 0.42 | 3.7 | 30 | 37.3 | 12 | 2010 | 10.9 | - | 51 |
| | 0.67 | 4 | 50 | 37.3 | 12 | 2010 | 10.9 | - | 47.8 |
| | 0.68 | 4.8 | 100 | 37.3 | 12 | 2010 | 10.9 | - | 45.1 |
| | 0.67 | 3.3 | - | 37.3 | 12 | - | - | - | 65.2 |
| | 0.7 | 3.4 | 10 | 37.3 | 12 | 2010 | 10.9 | - | 64.6 |
| | 0.53 | 3.8 | 30 | 37.3 | 12 | 2010 | 10.9 | - | 65.4 |
| | 0.53 | 4.1 | 50 | 37.3 | 12 | 2010 | 10.9 | - | 63.2 |
| | 0.53 | 5 | 100 | 37.3 | 12 | 2010 | 10.9 | - | 63 |
| [139] | 0.54 | 3 | - | 41.4 | 20 | - | - | - | 49.8 |
| | 0.54 | 3 | 20 | 41.4 | 20 | 2451 | 7.3 | 40 | 50.5 |
| | 0.54 | 3 | 50 | 41.4 | 20 | 2451 | 7.3 | 40 | 48.1 |
| | 0.54 | 2.9 | 100 | 41.4 | 20 | 2451 | 7.3 | 40 | 45.2 |
| | 0.45 | 3.1 | - | 41.4 | 20 | - | - | - | 59.7 |
| | 0.45 | 3.1 | 20 | 41.4 | 20 | 2451 | 7.3 | 40 | 64.7 |
| | 0.45 | 3.1 | 50 | 41.4 | 20 | 2451 | 7.3 | 40 | 55 |
| | 0.45 | 3.1 | 100 | 41.4 | 20 | 2451 | 7.3 | 40 | 53.9 |
| | 0.4 | 3.2 | - | 41.4 | 20 | - | - | - | 78.7 |
| | 0.4 | 3.2 | 20 | 41.4 | 20 | 2451 | 7.3 | 40 | 69.9 |
| | 0.4 | 3.2 | 50 | 41.4 | 20 | 2451 | 7.3 | 40 | 63.8 |
| | 0.4 | 3.1 | 100 | 41.4 | 20 | 2451 | 7.3 | 40 | 62.8 |
| [140] | 0.48 | 4.1 | - | - | 10 | - | - | - | 38.9 |
| | 0.48 | 3.5 | 100 | - | 10 | 2360 | 4.7 | 15.1 | 38.6 |
| | 0.39 | 3.1 | 100 | - | 10 | 2280 | 6.2 | 22.1 | 38.1 |
| | 0.29 | 2.6 | 100 | - | 10 | 2220 | 7.8 | 25 | 39.3 |

**Table A1.** *Cont.*

| Refs. | Water-Cement Ratio ($w_{eff}/c$) | Aggregate-Cement Ratio (a/c) | RCA Replacement Ratio (RCA %) | Parent Concrete Strength (MPa) | Nominal Maximum RCA Size (mm) | Bulk Density of RCA (kg/m$^3$) | Water Absorption of RCA (%) | Los Angeles Abrasion of RCA | Compressive Strength (MPa) |
|---|---|---|---|---|---|---|---|---|---|
| | 0.34 | 2.3 | - | - | 10 | - | - | - | 61.9 |
| | 0.31 | 2.1 | 100 | - | 10 | 2360 | 4.7 | 15.1 | 60.1 |
| | 0.27 | 1.8 | 100 | - | 10 | 2280 | 6.2 | 22.1 | 60.2 |
| | 0.19 | 1.5 | 100 | - | 10 | 2220 | 7.8 | 25 | 62.8 |
| [141] | 0.52 | 3 | 100 | - | 25 | 2490 | 4.9 | - | 37.6 |
| | 0.52 | 3 | 100 | - | 25 | 2570 | 2.9 | - | 43.3 |
| | 0.52 | 2.9 | 100 | - | 25 | 2440 | 5.6 | - | 42.6 |
| | 0.52 | 2.9 | 100 | - | 25 | 2470 | 5.3 | - | 44.7 |
| | 0.52 | 3.1 | - | - | 25 | - | - | - | 44.3 |
| | 0.58 | 3.2 | - | - | 32 | - | - | - | 44.6 |
| | 0.52 | 3.2 | 53 | - | 32 | 2720 | 4.8 | - | 41.4 |
| [142] | 0.58 | 3.2 | 100 | - | 32 | 2720 | 4.8 | - | 40.7 |
| | 0.52 | 3.2 | 54 | - | 32 | 2650 | 4.6 | - | 38.3 |
| | 0.58 | 3.2 | 100 | - | 32 | 2650 | 4.6 | - | 36.6 |
| | 0.52 | 3.2 | 53 | - | 32 | 2880 | 4.4 | - | 41.2 |
| | 0.58 | 3.2 | 100 | - | 32 | 2880 | 4.4 | - | 40.3 |
| [143] | 0.41 | 1.7 | 15 | - | 20 | 2330 | 4.4 | - | 50.8 |
| | 0.41 | 1.7 | 30 | - | 20 | 2330 | 4.4 | - | 44.9 |
| | 0.41 | 1.7 | 45 | - | 20 | 2330 | 4.4 | - | 44.6 |
| | 0.41 | 1.7 | 60 | - | 20 | 2330 | 4.4 | - | 42.4 |
| | 0.41 | 1.7 | 15 | - | 20 | 2370 | 4 | - | 54 |
| | 0.41 | 1.7 | 30 | - | 20 | 2370 | 4 | - | 56 |
| | 0.41 | 1.7 | 45 | - | 20 | 2370 | 4 | - | 54.4 |
| | 0.41 | 1.7 | 60 | - | 20 | 2370 | 4 | - | 40.6 |
| | 0.41 | 1.7 | 15 | - | 20 | 2390 | 3.6 | - | 55.2 |
| | 0.41 | 1.7 | 30 | - | 20 | 2390 | 3.6 | - | 53.5 |
| | 0.41 | 1.7 | 45 | - | 20 | 2390 | 3.6 | - | 56.9 |
| | 0.41 | 1.7 | 60 | - | 20 | 2390 | 3.6 | - | 54.7 |
| | 0.41 | 1.7 | 15 | - | 20 | 2320 | 4.6 | - | 50.5 |
| | 0.41 | 1.7 | 30 | - | 20 | 2320 | 4.6 | - | 48.9 |
| | 0.41 | 1.7 | 45 | - | 20 | 2320 | 4.6 | - | 45.8 |
| | 0.41 | 1.7 | 60 | - | 20 | 2320 | 4.6 | - | 40 |
| | 0.41 | 1.7 | 15 | - | 20 | 2390 | 3.7 | - | 54.4 |
| | 0.41 | 1.7 | 30 | - | 20 | 2390 | 3.7 | - | 50.2 |
| | 0.41 | 1.7 | 45 | - | 20 | 2390 | 3.7 | - | 49.5 |
| | 0.41 | 1.7 | 60 | - | 20 | 2390 | 3.7 | - | 40.4 |
| | 0.41 | 1.7 | 15 | - | 20 | 2390 | 3.5 | - | 45 |
| | 0.41 | 1.7 | 30 | - | 20 | 2390 | 3.5 | - | 46.9 |
| | 0.41 | 1.7 | 45 | - | 20 | 2390 | 3.5 | - | 51.4 |
| | 0.41 | 1.7 | 60 | - | 20 | 2390 | 3.5 | - | 53.2 |
| | 0.41 | 1.7 | 15 | - | 20 | 2380 | 3.8 | - | 55.3 |
| | 0.41 | 1.7 | 30 | - | 20 | 2380 | 3.8 | - | 55.9 |
| | 0.41 | 1.7 | 45 | - | 20 | 2380 | 3.8 | - | 52.6 |
| | 0.41 | 1.7 | 60 | - | 20 | 2380 | 3.8 | - | 48 |
| | 0.41 | 1.7 | 15 | - | 20 | 2380 | 3.8 | - | 49.1 |
| | 0.41 | 1.7 | 30 | - | 20 | 2380 | 3.8 | - | 49.9 |
| | 0.41 | 1.7 | 45 | - | 20 | 2380 | 3.8 | - | 50.3 |
| | 0.41 | 1.7 | 60 | - | 20 | 2380 | 3.8 | - | 47.5 |
| | 0.41 | 1.7 | 15 | - | 20 | 2400 | 3.5 | - | 43.2 |
| | 0.41 | 1.7 | 30 | - | 20 | 2400 | 3.5 | - | 53.7 |
| | 0.41 | 1.7 | 45 | - | 20 | 2400 | 3.5 | - | 50 |
| | 0.41 | 1.7 | 60 | - | 20 | 2400 | 3.5 | - | 43.3 |

**Table A1.** *Cont.*

| Refs. | Water-Cement Ratio ($w_{eff}/c$) | Aggregate-Cement Ratio (a/c) | RCA Replacement Ratio (RCA %) | Parent Concrete Strength (MPa) | Nominal Maximum RCA Size (mm) | Bulk Density of RCA (kg/m$^3$) | Water Absorption of RCA (%) | Los Angeles Abrasion of RCA | Compressive Strength (MPa) |
|---|---|---|---|---|---|---|---|---|---|
| | 0.41 | 1.7 | 15 | - | 20 | 2370 | 4 | - | 52.9 |
| | 0.41 | 1.7 | 30 | - | 20 | 2370 | 4 | - | 49.9 |
| | 0.41 | 1.7 | 45 | - | 20 | 2370 | 4 | - | 53.7 |
| | 0.41 | 1.7 | 60 | - | 20 | 2370 | 4 | - | 46 |
| [144] | 0.48 | 5.1 | - | 36 | 25 | - | - | - | 41.3 |
| | 0.48 | 5 | 27 | 36 | 25 | 2250 | 7 | - | 51.4 |
| | 0.48 | 4.9 | 64 | 36 | 25 | 2250 | 7 | - | 45.6 |
| | 0.48 | 5 | 37 | 36 | 25 | 2250 | 7 | - | 44.7 |
| | 0.48 | 5 | 37 | 36 | 25 | 2250 | 7 | - | 41.9 |
| [145] | 0.5 | 2.4 | 100 | - | 25 | 2452 | 4.1 | - | 51 |
| | 0.5 | 2.4 | 100 | - | 25 | 2452 | 4.1 | - | 49 |
| | 0.5 | 2.4 | 100 | - | 25 | 2452 | 4.1 | - | 48 |
| | 0.5 | 2.6 | - | - | 25 | 2452 | 4.1 | - | 52 |
| | 0.5 | 2.5 | 50 | - | 25 | 2452 | 4.1 | - | 51 |
| | 0.5 | 2.5 | 50 | - | 25 | 2452 | 4.1 | - | 51 |
| | 0.5 | 2.5 | 50 | - | 25 | 2452 | 4.1 | - | 51 |
| | 0.5 | 2.5 | 25 | - | 25 | 2452 | 4.1 | - | 52 |
| | 0.5 | 2.5 | 25 | - | 25 | 2452 | 4.1 | - | 50 |
| | 0.5 | 2.5 | 25 | - | 25 | 2452 | 4.1 | - | 49 |
| [146] | 0.38 | 2 | - | - | 25 | - | - | - | 54.1 |
| | 0.28 | 2 | 100 | - | 25 | 2260 | 7.5 | - | 38.3 |
| | 0.28 | 2 | 100 | - | 25 | 2260 | 7.5 | - | 32.9 |
| | 0.23 | 2 | 100 | - | 25 | 2260 | 7.5 | - | 33.2 |
| | 0.46 | 2.4 | - | - | 25 | - | - | - | 42.2 |
| | 0.34 | 2.4 | 100 | - | 25 | 2260 | 7.5 | - | 31.3 |
| | 0.34 | 2.4 | 100 | - | 25 | 2260 | 7.5 | - | 28.4 |
| | 0.28 | 2.4 | 100 | - | 25 | 2260 | 7.5 | - | 28 |
| | 0.58 | 3.1 | - | - | 25 | - | - | - | 28.8 |
| | 0.43 | 3.1 | 100 | - | 25 | 2260 | 7.5 | - | 26.5 |
| | 0.43 | 3.1 | 100 | - | 25 | 2260 | 7.5 | - | 23.3 |
| | 0.35 | 3.1 | 100 | - | 25 | 2260 | 7.5 | - | 21.6 |
| | 0.67 | 3.5 | - | - | 25 | - | - | - | 23.6 |
| | 0.49 | 3.5 | 100 | - | 25 | 2260 | 7.5 | - | 21.6 |
| | 0.49 | 3.5 | 100 | - | 25 | 2260 | 7.5 | - | 18 |
| | 0.4 | 3.5 | 100 | - | 25 | 2260 | 7.5 | - | 18.8 |
| | 0.8 | 4.2 | - | - | 25 | - | - | - | 17.3 |
| | 0.59 | 4.2 | 100 | - | 25 | 2260 | 7.5 | - | 16.1 |
| | 0.59 | 4.2 | 100 | - | 25 | 2260 | 7.5 | - | 13.4 |
| | 0.48 | 4.2 | 100 | - | 25 | 2260 | 7.5 | - | 13.9 |
| [147] | 0.6 | 3.6 | - | - | 20 | - | - | - | 38 |
| | - | - | - | - | - | - | - | - | - |
| | 0.59 | 3.3 | 20 | - | 20 | 2320 | 5.3 | 42 | 41 |
| | 0.57 | 3.3 | 50 | - | 20 | 2320 | 5.3 | 42 | 44 |
| | 0.54 | 3 | 100 | - | 20 | 2320 | 5.3 | 42 | 45 |
| | 0.46 | 2.6 | - | - | 20 | - | - | 42 | 51.5 |
| | 0.45 | 2.5 | 20 | - | 20 | 2320 | 5.3 | - | 50.5 |
| | 0.44 | 2.5 | 50 | - | 20 | 2320 | 5.3 | 42 | 45 |
| | 0.42 | 2.3 | 100 | - | 20 | 2320 | 5.3 | 42 | 56 |
| | 0.67 | 3.6 | - | - | 20 | - | - | - | 37 |
| | 0.68 | 3.4 | 20 | - | 20 | 2320 | 5.3 | 42 | 33.5 |
| | 0.67 | 3 | 50 | - | 20 | 2320 | 5.3 | 42 | 32 |

Table A1. *Cont.*

| Refs. | Water-Cement Ratio ($w_{eff}/c$) | Aggregate-Cement Ratio (a/c) | RCA Replacement Ratio (RCA %) | Parent Concrete Strength (MPa) | Nominal Maximum RCA Size (mm) | Bulk Density of RCA (kg/m³) | Water Absorption of RCA (%) | Los Angeles Abrasion of RCA | Compressive Strength (MPa) |
|---|---|---|---|---|---|---|---|---|---|
| | 0.7 | 2.3 | 100 | - | 20 | 2320 | 5.3 | 42 | 32 |
| | 0.53 | 2.7 | - | - | 20 | - | - | - | 45 |
| | 0.53 | 2.5 | 20 | - | 20 | 2320 | 5.3 | 42 | 44 |
| | 0.53 | 2.2 | 50 | - | 20 | 2320 | 5.3 | 42 | 41 |
| | 0.52 | 1.8 | 100 | - | 20 | 2320 | 5.3 | 42 | 41.5 |
| | 0.51 | 3.1 | - | - | 20 | - | - | - | 46.5 |
| | 0.52 | 3.2 | 20 | - | 20 | 2320 | 5.3 | 42 | 44 |
| | 0.54 | 3 | 50 | - | 20 | 2320 | 5.3 | 42 | 41 |
| | 0.58 | 2.8 | 100 | - | 20 | 2320 | 5.3 | 42 | 33.5 |
| | 0.42 | 2.7 | - | - | 20 | - | - | - | 58 |
| | 0.42 | 2.9 | 20 | - | 20 | 2320 | 5.3 | 42 | 53.5 |
| | 0.44 | 2.7 | 50 | - | 20 | 2320 | 5.3 | 42 | 54 |
| | 0.49 | 2.5 | 100 | - | 20 | 2320 | 5.3 | 42 | 40 |
| [148] | 0.42 | 2.6 | 50 | - | 20 | 2330 | 6.1 | 34.6 | 41.6 |
| | 0.51 | 2.3 | 100 | - | 20 | 2330 | 6.1 | 34.6 | 31.4 |
| | 0.52 | 2.6 | 50 | - | 20 | 2330 | 6.1 | 34.6 | 35.5 |
| | 0.61 | 2.3 | 100 | - | 20 | 2330 | 6.1 | 34.6 | 26 |
| | 0.44 | 2.6 | 50 | - | 20 | 2320 | 5.8 | 32.2 | 44.6 |
| | 0.51 | 2.3 | 100 | - | 20 | 2320 | 5.8 | 32.2 | 36.7 |
| | 0.62 | 2.3 | 100 | - | 20 | 2320 | 5.8 | 32.2 | 29.5 |
| | 0.41 | 2.8 | 20 | - | 20 | 2360 | 3.9 | 30.8 | 46.1 |
| | 0.42 | 2.6 | 50 | - | 20 | 2360 | 3.9 | 30.8 | 45.1 |
| | 0.45 | 2.3 | 100 | - | 20 | 2360 | 3.9 | 30.8 | 42.9 |
| | 0.5 | 2.8 | 20 | - | 20 | 2360 | 3.9 | 30.8 | 39.3 |
| | 0.52 | 2.6 | 50 | - | 20 | 2360 | 3.9 | 30.8 | 39.5 |
| | 0.54 | 2.3 | 100 | - | 20 | 2360 | 3.9 | 30.8 | 37.7 |
| | 0.42 | 2.8 | 20 | - | 20 | 2350 | 4.5 | 28.5 | 48.1 |
| | 0.43 | 2.6 | 50 | - | 20 | 2350 | 4.5 | 28.5 | 41 |
| | 0.4 | 2.3 | 100 | - | 20 | 2350 | 4.5 | 28.5 | 38.7 |
| | 0.51 | 2.8 | 20 | - | 20 | 2350 | 4.5 | 28.5 | 42.7 |
| | 0.52 | 2.6 | 50 | - | 20 | 2350 | 4.5 | 28.5 | 35.4 |
| | 0.5 | 2.3 | 100 | - | 20 | 2350 | 4.5 | 28.5 | 31.4 |
| | 0.42 | 2.8 | 20 | - | 20 | 2350 | 4.7 | 30.1 | 48.5 |
| | 0.42 | 2.6 | 50 | - | 20 | 2350 | 4.7 | 30.1 | 45.4 |
| | 0.43 | 2.3 | 100 | - | 20 | 2350 | 4.7 | 30.1 | 37 |
| | 0.52 | 2.8 | 20 | - | 20 | 2350 | 4.7 | 30.1 | 41.3 |
| | 0.52 | 2.6 | 50 | - | 20 | 2350 | 4.7 | 30.1 | 36.8 |
| | 0.56 | 2.3 | 100 | - | 20 | 2350 | 4.7 | 30.1 | 31.2 |
| [149] | 0.41 | 2.6 | - | - | 32 | - | - | - | 47.2 |
| | 0.38 | 2.6 | 33 | - | 32 | 2578 | 9.3 | - | 42.4 |
| | 0.36 | 2.6 | 53 | - | 32 | 2578 | 9.3 | - | 45.7 |
| | 0.34 | 2.6 | 72 | - | 32 | 2578 | 9.3 | - | 36.7 |
| | 0.31 | 2.6 | 100 | - | 32 | 2578 | 9.3 | - | 38.9 |
| [150] | 0.47 | 3.8 | - | - | 20 | - | - | - | 53.1 |
| | 0.47 | 3.7 | 20 | - | 20 | 2336 | 3.6 | - | 50 |
| | 0.47 | 3.6 | 50 | - | 20 | 2315 | 3.6 | - | 45.3 |
| | 0.47 | 3.6 | 75 | - | 20 | 2295 | 3.6 | - | 44 |
| | 0.47 | 3.5 | 100 | - | 20 | 2273 | 3.6 | - | 41.6 |
| [151] | 0.29 | 2.9 | - | - | 10 | - | - | - | 102.1 |
| | 0.29 | 2.8 | 20 | 100 | 10 | 2470 | 3.7 | 24 | 108 |
| | 0.29 | 2.8 | 50 | 100 | 10 | 2470 | 3.7 | 24 | 104.8 |
| | 0.29 | 2.7 | 100 | 100 | 10 | 2470 | 3.7 | 24 | 108.5 |
| | 0.29 | 2.8 | 20 | 60 | 10 | 2390 | 4.9 | 25.2 | 102.5 |

**Table A1.** *Cont.*

| Refs. | Water-Cement Ratio ($w_{eff}$/c) | Aggregate-Cement Ratio (a/c) | RCA Replacement Ratio (RCA %) | Parent Concrete Strength (MPa) | Nominal Maximum RCA Size (mm) | Bulk Density of RCA (kg/m$^3$) | Water Absorption of RCA (%) | Los Angeles Abrasion of RCA | Compressive Strength (MPa) |
|---|---|---|---|---|---|---|---|---|---|
| | 0.29 | 2.7 | 50 | 60 | 10 | 2390 | 4.9 | 25.2 | 103.1 |
| | 0.29 | 2.6 | 100 | 60 | 10 | 2390 | 4.9 | 25.2 | 100.8 |
| | 0.29 | 2.8 | 20 | 40 | 10 | 2300 | 5.9 | 24.3 | 104.3 |
| | 0.29 | 2.7 | 50 | 40 | 10 | 2300 | 5.9 | 24.3 | 96.8 |
| | 0.29 | 2.5 | 100 | 40 | 10 | 2300 | 5.9 | 24.3 | 91.2 |
| [152] | 0.65 | 4.6 | - | - | 20 | - | - | - | 18 |
| | 0.65 | 4.7 | 25 | - | 20 | 2380 | 6.94 | 29 | 14.7 |
| | 0.65 | 4.8 | 50 | - | 20 | 2380 | 6.94 | 29 | 14.6 |
| | 0.65 | 4.8 | 75 | - | 20 | 2380 | 6.94 | 29 | 14.2 |
| | 0.72 | 5.8 | - | - | 20 | - | - | - | 30.8 |
| | 0.72 | 5.9 | 20 | - | 20 | 2380 | 6.94 | 29 | 26.8 |
| | 0.72 | 6 | 40 | - | 20 | 2380 | 6.94 | 29 | 26.6 |
| | 0.45 | 1.9 | - | - | 16 | - | - | - | 66.9 |
| | 0.45 | 2.3 | 20 | - | 16 | 2380 | 6.94 | 29 | 49.3 |
| | 0.45 | 2.5 | 40 | - | 16 | 2380 | 6.94 | 29 | 40.9 |
| [153] | 0.6 | 3.5 | - | - | 16 | - | - | - | 42 |
| | 0.6 | 3.4 | 20 | - | 16 | 2380 | 6.9 | - | 42.9 |
| | 0.6 | 3.4 | 50 | - | 16 | 2380 | 6.9 | - | 42.5 |
| | 0.6 | 3.2 | 100 | - | 16 | 2380 | 6.9 | - | 40.9 |
| | 0.5 | 2.7 | - | - | 16 | - | - | - | 50.2 |
| | 0.5 | 2.6 | 20 | - | 16 | 2380 | 6.9 | - | 51.6 |
| | 0.5 | 2.5 | 50 | - | 16 | 2380 | 6.9 | - | 51.6 |
| | 0.5 | 2.4 | 100 | - | 16 | 2380 | 6.9 | - | 50.3 |
| [154] | 0.5 | 3.4 | - | - | 22 | - | - | - | 46.7 |
| | 0.5 | 3.4 | 50 | - | 12 | 2380 | - | - | 46.9 |
| | 0.5 | 3.4 | 50 | - | 22 | 2380 | - | - | 46.4 |
| | 0.5 | 3.4 | 100 | - | 22 | 2380 | - | - | 48.6 |
| [155] | 0.52 | 2.2 | - | - | 19 | - | - | - | 29.9 |
| | 0.49 | 2.1 | 25 | - | 19 | 2500 | 6.6 | - | 32.6 |
| [156] | 0.5 | 3.5 | 50 | - | 8 | 2330 | 3.8 | 41.4 | 33 |
| | - | - | - | - | - | - | - | - | - |
| [157] | 0.5 | 3.2 | 50 | - | 8 | 2280 | 4.1 | - | 29.1 |
| | 0.68 | 3.8 | - | - | 20 | - | - | - | 34.5 |
| | 0.68 | 3.6 | 100 | - | 20 | 2450 | 3.1 | - | 35 |
| | 0.68 | 3.4 | 100 | - | 20 | 2370 | 7.1 | - | 29.2 |
| | 0.68 | 3.4 | 100 | - | 20 | 2360 | 7.8 | - | 27.7 |
| | 0.51 | 3.3 | - | - | 20 | - | - | - | 48.3 |
| | 0.51 | 3.1 | 100 | - | 20 | 2450 | 3.1 | - | 47.6 |
| | 0.51 | 3 | 100 | - | 20 | 2370 | 7.1 | - | 42 |
| | 0.51 | 3 | 100 | - | 20 | 2360 | 7.8 | - | 42.9 |
| | 0.44 | 2.5 | - | - | 20 | - | - | - | 61.6 |
| | 0.44 | 2.4 | 100 | - | 20 | 2450 | 3.1 | - | 60 |
| | 0.44 | 2.3 | 100 | - | 20 | 2370 | 7.1 | - | 53.7 |
| | 0.44 | 2.3 | 100 | - | 20 | 2360 | 7.8 | - | 53.2 |
| | 0.34 | 2.2 | - | - | 20 | - | - | - | 80.8 |
| | 0.34 | 2.1 | 100 | - | 20 | 2450 | 3.1 | - | 78.2 |
| | 0.34 | 2 | 100 | - | 20 | 2370 | 7.1 | - | 71.2 |
| [158] | 0.34 | 2 | 100 | - | 20 | 2360 | 7.8 | - | 65.4 |
| | 0.5 | 3.1 | - | - | 19 | - | - | - | 36.5 |
| | 0.5 | 3 | 30 | - | 19 | 2570 | 2.7 | - | 33.6 |
| | 0.5 | 3 | 60 | - | 19 | 2570 | 2.7 | - | 30.4 |
| [159] | 0.5 | 2.8 | 100 | - | 19 | 2570 | 2.7 | - | 29.1 |
| | 0.65 | 3.1 | - | - | 20 | - | - | - | 40.5 |
| | 0.65 | 3.2 | 20 | - | 20 | 2300 | 5.2 | 40.2 | 39.5 |

**Table A1.** *Cont.*

| Refs. | Water-Cement Ratio ($w_{eff}/c$) | Aggregate-Cement Ratio (a/c) | RCA Replacement Ratio (RCA %) | Parent Concrete Strength (MPa) | Nominal Maximum RCA Size (mm) | Bulk Density of RCA (kg/m$^3$) | Water Absorption of RCA (%) | Los Angeles Abrasion of RCA | Compressive Strength (MPa) |
|---|---|---|---|---|---|---|---|---|---|
| | 0.65 | 3.2 | 50 | - | 20 | 2300 | 5.2 | 40.2 | 40.8 |
| | 0.65 | 3.2 | 100 | - | 20 | 2300 | 5.2 | 40.2 | 43.7 |
| | 0.65 | 3.1 | - | - | 20 | - | - | - | 40.5 |
| | 0.65 | 3.1 | 20 | - | 20 | 2300 | 5.5 | 28.6 | 41 |
| | 0.65 | 3.1 | 50 | - | 20 | 2300 | 5.5 | 28.6 | 38.8 |
| [160] | 0.65 | 3.2 | 100 | - | 20 | 2300 | 5.5 | 28.6 | 39.9 |
| | 0.42 | 2.7 | - | - | 25 | - | - | - | 38.6 |
| | 0.4 | 2.7 | 16 | - | 25 | 2200 | 5.4 | - | 32.7 |
| | 0.39 | 2.2 | 37 | - | 25 | 2200 | 5.4 | - | 31.7 |
| [161] | 0.36 | 2.7 | 52 | - | 25 | 2200 | 5.4 | - | 29 |
| | 0.86 | 4.6 | - | - | 22 | - | - | - | 23.9 |
| | 0.65 | 3.4 | - | - | 22 | - | - | - | 38.7 |
| | 0.41 | 2.9 | - | - | 22 | - | - | - | 71.1 |
| | 0.87 | 4.6 | 100 | - | 22 | 2451 | 7.8 | - | 19.7 |
| | 0.66 | 3.4 | 100 | - | 22 | 2387 | 6.9 | - | 35.7 |
| | 0.42 | 2.8 | 100 | - | 22 | 2362 | 4.2 | - | 66.8 |
| | 0.86 | 4.6 | 100 | - | 22 | 2456 | 7.5 | - | 21.8 |
| | 0.65 | 3.5 | 100 | - | 22 | 2455 | 6.4 | - | 36.1 |
| | 0.42 | 2.9 | 100 | - | 22 | 2496 | 4.2 | - | 68.5 |
| | 0.81 | 4.9 | - | - | 22 | - | - | - | 27.5 |
| | 0.63 | 3.6 | - | - | 22 | - | - | - | 42.4 |
| | 0.4 | 3 | - | - | 22 | - | - | - | 72.3 |
| | 0.84 | 4.5 | 100 | - | 22 | 2401 | 7.6 | - | 21 |
| | 0.63 | 3.5 | 100 | - | 22 | 2484 | 5.4 | - | 41.1 |
| | 0.4 | 2.8 | 100 | - | 22 | 2363 | 3.6 | - | 70.2 |
| | 0.82 | 4.7 | 100 | - | 22 | 2447 | 6.9 | - | 23.6 |
| | 0.64 | 3.4 | 100 | - | 22 | 2458 | 5.8 | - | 39.7 |
| [162] | 0.42 | 2.9 | 100 | - | 22 | 2464 | 3.9 | - | 66.5 |
| | 0.64 | 3 | - | - | 19 | - | - | - | 33 |
| | 0.77 | 3.1 | 100 | - | 19 | 2268 | 4.9 | - | 27.5 |
| [163] | 0.7 | 3.4 | 100 | - | 19 | 1946 | 11.9 | - | 29.9 |
| | 0.6 | 3.6 | - | - | 19 | - | - | - | 47.8 |
| | 0.59 | 3.3 | 20 | - | 19 | 2320 | 5.3 | 37 | 49.3 |
| | 0.57 | 3.3 | 50 | - | 19 | 2320 | 5.3 | 37 | 47.5 |
| | 0.54 | 3 | 100 | - | 19 | 2320 | 5.3 | 37 | 53.7 |
| | 0.46 | 2.6 | - | - | 19 | - | - | - | 62 |
| | 0.45 | 2.5 | 20 | - | 19 | 2320 | 5.3 | 37 | 64.8 |
| | 0.44 | 2.5 | 50 | - | 19 | 2320 | 5.3 | 37 | 63.5 |
| | 0.42 | 2.3 | 100 | - | 19 | 2320 | 5.3 | 37 | 65.1 |
| | 0.67 | 3.6 | - | - | 19 | - | - | - | 62 |
| | 0.68 | 3.4 | 20 | - | 19 | 2320 | 5.3 | 37 | 64.8 |
| | 0.67 | 3 | 50 | - | 19 | 2320 | 5.3 | 37 | 63.5 |
| | 0.7 | 2.3 | 100 | - | 19 | 2320 | 5.3 | 37 | 65.1 |
| | 0.53 | 2.7 | - | - | 19 | - | - | - | 57.3 |
| | 0.53 | 2.5 | 20 | - | 19 | 2320 | 5.3 | 37 | 54.9 |
| | 0.53 | 2.2 | 50 | - | 19 | 2320 | 5.3 | 37 | 51.5 |
| | 0.52 | 1.8 | 100 | - | 19 | 2320 | 5.3 | 37 | 50.3 |
| | 0.51 | 3.1 | - | - | 19 | - | - | - | 60.1 |
| | 0.52 | 3.2 | 20 | - | 19 | 2320 | 5.3 | 37 | 56.5 |
| | 0.54 | 3 | 50 | - | 19 | 2320 | 5.3 | 37 | 48.9 |
| | 0.58 | 2.8 | 100 | - | 19 | 2320 | 5.3 | 37 | 43.1 |
| | 0.42 | 2.7 | - | - | 19 | - | - | - | 72.9 |
| | 0.42 | 2.9 | 20 | - | 19 | 2320 | 5.3 | 37 | 67.4 |
| | 0.44 | 2.7 | 50 | - | 19 | 2320 | 5.3 | 37 | 61.2 |

## References


1. Dong, W.; Li, W.; Tao, Z. A comprehensive review on performance of cementitious and geopolymeric concretes with recycled waste glass as powder, sand or cullet. *Resour. Conserv. Recycl.* **2021**, *172*, 105664. [CrossRef]
2. Baturkin, D.; Hisseine, O.A.; Masmoudi, R.; Tagnit-Hamou, A.; Massicotte, L. Valorization of recycled FRP materials from wind turbine blades in concrete. *Resour. Conserv. Recycl.* **2021**, *174*, 105807. [CrossRef]
3. Yu, J.; Mishra, D.K.; Hu, C.; Leung, C.K.Y.; Shah, S.P. Mechanical, environmental and economic performance of sustainable Grade 45 concrete with ultrahigh-volume Limestone-Calcined Clay (LCC). *Resour. Conserv. Recycl.* **2021**, *175*, 105846. [CrossRef]
4. Hu, H.-B.; He, Z.-H.; Shi, J.-Y.; Liang, C.-F.; Shibro, T.-G.; Liu, B.-J.; Yang, S.-Y. Mechanical properties, drying shrinkage, and nano-scale characteristics of concrete prepared with zeolite powder pre-coated recycled aggregate. *J. Clean. Prod.* **2021**, *319*, 128710. [CrossRef]
5. Golafshani, E.M.; Behnood, A.; Hosseinikebria, S.S.; Arashpour, M. Novel metaheuristic-based type-2 fuzzy inference system for predicting the compressive strength of recycled aggregate concrete. *J. Clean. Prod.* **2021**, *320*, 128771. [CrossRef]
6. Mikhailenko, P.; Piao, Z.; Kakar, M.R.; Bueno, M.; Poulikakos, L.D. Durability and surface properties of low-noise pavements with recycled concrete aggregates. *J. Clean. Prod.* **2021**, *319*, 128788. [CrossRef]
7. Kazmi, S.M.S.; Munir, M.J.; Wu, Y.-F. Application of waste tire rubber and recycled aggregates in concrete products: A new compression casting approach. *Resour. Conserv. Recycl.* **2021**, *167*, 105353. [CrossRef]
8. Mistri, A.; Dhami, N.; Bhattacharyya, S.K.; Barai, S.V.; Mukherjee, A.; Biswas, W.K. Environmental implications of the use of bio-cement treated recycled aggregate in concrete. *Resour. Conserv. Recycl.* **2021**, *167*, 105436. [CrossRef]
9. Ahmad, W.; Farooq, S.H.; Usman, M.; Khan, M.; Ahmad, A.; Aslam, F.; Yousef, R.A.; Abduljabbar, H.A.; Sufian, M. Effect of coconut fiber length and content on properties of high strength concrete. *Materials* **2020**, *13*, 1075. [CrossRef]
10. Khan, M.; Ali, M. Effectiveness of hair and wave polypropylene fibers for concrete roads. *Constr. Build. Mater.* **2018**, *166*, 581–591. [CrossRef]
11. Khan, M.; Cao, M.; Ali, M. Cracking behaviour and constitutive modelling of hybrid fibre reinforced concrete. *J. Build. Eng.* **2020**, *30*, 101272. [CrossRef]
12. Khan, M.; Cao, M.; Xie, C.; Ali, M. Efficiency of basalt fiber length and content on mechanical and microstructural properties of hybrid fiber concrete. *Fatigue Fract. Eng. Mater. Struct.* **2021**, *44*, 2135–2152. [CrossRef]
13. Khan, M.; Cao, M.; Chaopeng, X.; Ali, M. Experimental and analytical study of hybrid fiber reinforced concrete prepared with basalt fiber under high temperature. *Fire Mater.* **2022**, *46*, 205–226. [CrossRef]
14. Hosseini, P.; Booshehrian, A.; Delkash, M.; Ghavami, S.; Zanjani, M.K. Use of nano-$SiO_2$ to improve microstructure and compressive strength of recycled aggregate concretes. In *Nanotechnology in Construction 3*; Springer: Berlin/Heidelberg, Germany, 2009; pp. 215–221.
15. Xiong, Z.; Lin, L.; Qiao, S.; Li, L.; Li, Y.; He, S.; Li, Z.; Liu, F.; Chen, Y. Axial performance of seawater sea-sand concrete columns reinforced with basalt fibre-reinforced polymer bars under concentric compressive load. *J. Build. Eng.* **2022**, *47*, 103828. [CrossRef]
16. Priya, K.L.; Ragupathy, R. Effect of sugarcane bagasse ash on strength properties of concrete. *Int. J. Res. Eng. Technol.* **2016**, *5*, 159–164.
17. Khan, M.; Cao, M.; Xie, C.; Ali, M. Effectiveness of hybrid steel-basalt fiber reinforced concrete under compression. *Case Stud. Constr. Mater.* **2022**, *16*, e00941. [CrossRef]
18. Khan, M.; Cao, M.; Ai, H.; Hussain, A. Basalt Fibers in Modified Whisker Reinforced Cementitious Composites. *Period. Polytech. Civ. Eng.* **2022**. [CrossRef]
19. Kosmatka, S.H.; Kerkhoff, B.; Panarese, W.C. *Design and Control of Concrete Mixtures*; Portland Cement Association: Skokie, IL, USA, 2002; Volume 5420.
20. Behera, M.; Bhattacharyya, S.K.; Minocha, A.K.; Deoliya, R.; Maiti, S. Recycled aggregate from C&D waste & its use in concrete—A breakthrough towards sustainability in construction sector: A review. *Constr. Build. Mater.* **2014**, *68*, 501–516.
21. Limbachiya, M.; Meddah, M.S.; Ouchagour, Y. Use of recycled concrete aggregate in fly-ash concrete. *Constr. Build. Mater.* **2012**, *27*, 439–449. [CrossRef]
22. Yang, H.; Liu, L.; Yang, W.; Liu, H.; Ahmad, W.; Ahmad, A.; Aslam, F.; Joyklad, P. A comprehensive overview of geopolymer composites: A bibliometric analysis and literature review. *Case Stud. Constr. Mater.* **2022**, *16*, e00830. [CrossRef]
23. Zhang, B.; Ahmad, W.; Ahmad, A.; Aslam, F.; Joyklad, P. A scientometric analysis approach to analyze the present research on recycled aggregate concrete. *J. Build. Eng.* **2022**, *46*, 103679. [CrossRef]
24. Alyousef, R.; Ahmad, W.; Ahmad, A.; Aslam, F.; Joyklad, P.; Alabduljabbar, H. Potential use of recycled plastic and rubber aggregate in cementitious materials for sustainable construction: A review. *J. Clean. Prod.* **2021**, *329*, 129736. [CrossRef]
25. Kisku, N.; Joshi, H.; Ansari, M.; Panda, S.; Nayak, S.; Dutta, S.C. A critical review and assessment for usage of recycled aggregate as sustainable construction material. *Constr. Build. Mater.* **2017**, *131*, 721–740. [CrossRef]
26. Albayati, A.; Wang, Y.; Wang, Y.; Haynes, J. A sustainable pavement concrete using warm mix asphalt and hydrated lime treated recycled concrete aggregates. *Sustain. Mater. Technol.* **2018**, *18*, e00081. [CrossRef]
27. Pedro, D.; De Brito, J.; Evangelista, L. Influence of the use of recycled concrete aggregates from different sources on structural concrete. *Constr. Build. Mater.* **2014**, *71*, 141–151. [CrossRef]
28. Ossa, A.; García, J.; Botero, E. Use of recycled construction and demolition waste (CDW) aggregates: A sustainable alternative for the pavement construction industry. *J. Clean. Prod.* **2016**, *135*, 379–386. [CrossRef]

29. Rao, A.; Jha, K.N.; Misra, S. Use of aggregates from recycled construction and demolition waste in concrete. *Resour. Conserv. Recycl.* **2007**, *50*, 71–81. [CrossRef]
30. Thomas, C.; De Brito, J.; Gil, V.; Sainz-Aja, J.; Cimentada, A. Multiple recycled aggregate properties analysed by X-ray microtomography. *Constr. Build. Mater.* **2018**, *166*, 171–180. [CrossRef]
31. Tang, Y.; Zhu, M.; Chen, Z.; Wu, C.; Chen, B.; Li, C.; Li, L. Seismic performance evaluation of recycled aggregate concrete-filled steel tubular columns with field strain detected via a novel mark-free vision method. *Structures* **2022**, *37*, 426–441. [CrossRef]
32. Yunchao, T.; Zheng, C.; Wanhui, F.; Yumei, N.; Cong, L.; Jieming, C. Combined effects of nano-silica and silica fume on the mechanical behavior of recycled aggregate concrete. *Nanotechnol. Rev.* **2021**, *10*, 819–838. [CrossRef]
33. Vegas, I.; Broos, K.; Nielsen, P.; Lambertz, O.; Lisbona, A. Upgrading the quality of mixed recycled aggregates from construction and demolition waste by using near-infrared sorting technology. *Constr. Build. Mater.* **2015**, *75*, 121–128. [CrossRef]
34. Xie, J.-H.; Guo, Y.-C.; Liu, L.-S.; Xie, Z.-H. Compressive and flexural behaviours of a new steel-fibre-reinforced recycled aggregate concrete with crumb rubber. *Constr. Build. Mater.* **2015**, *79*, 263–272. [CrossRef]
35. Azúa, G.; González, M.; Arroyo, P.; Kurama, Y. Recycled coarse aggregates from precast plant and building demolitions: Environmental and economic modeling through stochastic simulations. *J. Clean. Prod.* **2019**, *210*, 1425–1434. [CrossRef]
36. Lu, B.; Shi, C.; Cao, Z.; Guo, M.; Zheng, J. Effect of carbonated coarse recycled concrete aggregate on the properties and microstructure of recycled concrete. *J. Clean. Prod.* **2019**, *233*, 421–428. [CrossRef]
37. Muduli, R.; Mukharjee, B.B. Effect of incorporation of metakaolin and recycled coarse aggregate on properties of concrete. *J. Clean. Prod.* **2019**, *209*, 398–414. [CrossRef]
38. Kim, Y.; Hanif, A.; Kazmi, S.M.S.; Munir, M.J.; Park, C. Properties enhancement of recycled aggregate concrete through pretreatment of coarse aggregates–Comparative assessment of assorted techniques. *J. Clean. Prod.* **2018**, *191*, 339–349. [CrossRef]
39. Mohseni, E.; Saadati, R.; Kordbacheh, N.; Parpinchi, Z.S.; Tang, W. Engineering and microstructural assessment of fibre-reinforced self-compacting concrete containing recycled coarse aggregate. *J. Clean. Prod.* **2017**, *168*, 605–613. [CrossRef]
40. Bai, G.; Zhu, C.; Liu, C.; Liu, B. An evaluation of the recycled aggregate characteristics and the recycled aggregate concrete mechanical properties. *Constr. Build. Mater.* **2020**, *240*, 117978. [CrossRef]
41. Awoyera, P.O. Nonlinear finite element analysis of steel fibre-reinforced concrete beam under static loading. *J. Eng. Sci. Technol.* **2016**, *11*, 1669–1677.
42. Cao, M.; Mao, Y.; Khan, M.; Si, W.; Shen, S. Different testing methods for assessing the synthetic fiber distribution in cement-based composites. *Constr. Build. Mater.* **2018**, *184*, 128–142. [CrossRef]
43. Sadrmomtazi, A.; Sobhani, J.; Mirgozar, M.A. Modeling compressive strength of EPS lightweight concrete using regression, neural network and ANFIS. *Constr. Build. Mater.* **2013**, *42*, 205–216. [CrossRef]
44. Öztaş, A.; Pala, M.; Özbay, E.a.; Kanca, E.; Caglar, N.; Bhatti, M.A. Predicting the compressive strength and slump of high strength concrete using neural network. *Constr. Build. Mater.* **2006**, *20*, 769–775. [CrossRef]
45. Sarıdemir, M. Predicting the compressive strength of mortars containing metakaolin by artificial neural networks and fuzzy logic. *Adv. Eng. Softw.* **2009**, *40*, 920–927. [CrossRef]
46. Ni, H.-G.; Wang, J.-Z. Prediction of compressive strength of concrete by neural networks. *Cem. Concr. Res.* **2000**, *30*, 1245–1250. [CrossRef]
47. Sobhani, J.; Najimi, M.; Pourkhorshidi, A.R.; Parhizkar, T. Prediction of the compressive strength of no-slump concrete: A comparative study of regression, neural network and ANFIS models. *Constr. Build. Mater.* **2010**, *24*, 709–718. [CrossRef]
48. Awoyera, P.O.; Kirgiz, M.S.; Viloria, A.; Ovallos-Gazabon, D. Estimating strength properties of geopolymer self-compacting concrete using machine learning techniques. *J. Mater. Res. Technol.* **2020**, *9*, 9016–9028. [CrossRef]
49. Shafabakhsh, G.H.; Ani, O.J.; Talebsafa, M. Artificial neural network modeling (ANN) for predicting rutting performance of nano-modified hot-mix asphalt mixtures containing steel slag aggregates. *Constr. Build. Mater.* **2015**, *85*, 136–143. [CrossRef]
50. Hodhod, O.A.; Ahmed, H.I. Modeling the corrosion initiation time of slag concrete using the artificial neural network. *HBRC J.* **2014**, *10*, 231–234. [CrossRef]
51. Carmichael, R.P. *Relationships between Young's Modulus, Compressive Strength, Poisson's Ratio, and Time for Early Age Concrete*; ENGR 082 Project Final Report; Swarthmore College: Swarthmore, PA, USA, 2009.
52. Bal, L.; Buyle-Bodin, F. Artificial neural network for predicting drying shrinkage of concrete. *Constr. Build. Mater.* **2013**, *38*, 248–254. [CrossRef]
53. Chou, J.-S.; Tsai, C.-F.; Pham, A.-D.; Lu, Y.-H. Machine learning in concrete strength simulations: Multi-nation data analytics. *Constr. Build. Mater.* **2014**, *73*, 771–780. [CrossRef]
54. Young, B.A.; Hall, A.; Pilon, L.; Gupta, P.; Sant, G. Can the compressive strength of concrete be estimated from knowledge of the mixture proportions?: New insights from statistical analysis and machine learning methods. *Cem. Concr. Res.* **2019**, *115*, 379–388. [CrossRef]
55. Akande, K.O.; Owolabi, T.O.; Twaha, S.; Olatunji, S.O. Performance comparison of SVM and ANN in predicting compressive strength of concrete. *IOSR J. Comput. Eng.* **2014**, *16*, 88–94. [CrossRef]
56. Behnood, A.; Behnood, V.; Gharehveran, M.M.; Alyamac, K.E. Prediction of the compressive strength of normal and high-performance concretes using M5P model tree algorithm. *Constr. Build. Mater.* **2017**, *142*, 199–207. [CrossRef]
57. Duan, Z.-H.; Kou, S.-C.; Poon, C.-S. Prediction of compressive strength of recycled aggregate concrete using artificial neural networks. *Constr. Build. Mater.* **2013**, *40*, 1200–1206. [CrossRef]

58. de Melo, V.V.; Banzhaf, W. Improving the prediction of material properties of concrete using Kaizen Programming with Simulated Annealing. *Neurocomputing* **2017**, *246*, 25–44. [CrossRef]
59. Yeh, I.C.; Lien, L.-C. Knowledge discovery of concrete material using genetic operation trees. *Expert Syst. Appl.* **2009**, *36*, 5807–5812. [CrossRef]
60. Duan, Z.-H.; Kou, S.-C.; Poon, C.-S. Using artificial neural networks for predicting the elastic modulus of recycled aggregate concrete. *Constr. Build. Mater.* **2013**, *44*, 524–532. [CrossRef]
61. Gholampour, A.; Mansouri, I.; Kisi, O.; Ozbakkaloglu, T. Evaluation of mechanical properties of concretes containing coarse recycled concrete aggregates using multivariate adaptive regression splines (MARS), M5 model tree (M5Tree), and least squares support vector regression (LSSVR) models. *Neural Comput. Appl.* **2020**, *32*, 295–308. [CrossRef]
62. Deshpande, N.; Londhe, S.; Kulkarni, S. Modeling compressive strength of recycled aggregate concrete by Artificial Neural Network, Model Tree and Non-linear Regression. *Int. J. Sustain. Built Environ.* **2014**, *3*, 187–198. [CrossRef]
63. Behnood, A.; Olek, J.; Glinicki, M.A. Predicting modulus elasticity of recycled aggregate concrete using M5′ model tree algorithm. *Constr. Build. Mater.* **2015**, *94*, 137–147. [CrossRef]
64. Frank, E.; Hall, M.; Trigg, L.; Holmes, G.; Witten, I.H. Data mining in bioinformatics using Weka. *Bioinformatics* **2004**, *20*, 2479–2481. [CrossRef] [PubMed]
65. Deng, F.; He, Y.; Zhou, S.; Yu, Y.; Cheng, H.; Wu, X. Compressive strength prediction of recycled concrete based on deep learning. *Constr. Build. Mater.* **2018**, *175*, 562–569. [CrossRef]
66. Cunningham, P.; Carney, J.; Jacob, S. Stability problems with artificial neural networks and the ensemble solution. *Artif. Intell. Med.* **2000**, *20*, 217–225. [CrossRef]
67. Cook, R.; Lapeyre, J.; Ma, H.; Kumar, A. Prediction of compressive strength of concrete: Critical comparison of performance of a hybrid machine learning model with standalone models. *J. Mater. Civ. Eng.* **2019**, *31*, 04019255. [CrossRef]
68. Han, T.; Siddique, A.; Khayat, K.; Huang, J.; Kumar, A. An ensemble machine learning approach for prediction and optimization of modulus of elasticity of recycled aggregate concrete. *Constr. Build. Mater.* **2020**, *244*, 118271. [CrossRef]
69. Zhu, Y.; Ahmad, A.; Ahmad, W.; Vatin, N.I.; Mohamed, A.M.; Fathi, D. Predicting the Splitting Tensile Strength of Recycled Aggregate Concrete Using Individual and Ensemble Machine Learning Approaches. *Crystals* **2022**, *12*, 569. [CrossRef]
70. Ahmad, A.; Ahmad, W.; Aslam, F.; Joyklad, P. Compressive strength prediction of fly ash-based geopolymer concrete via advanced machine learning techniques. *Case Stud. Constr. Mater.* **2022**, *16*, e00840. [CrossRef]
71. Ahmad, W.; Ahmad, A.; Ostrowski, K.A.; Aslam, F.; Joyklad, P.; Zajdel, P. Application of Advanced Machine Learning Approaches to Predict the Compressive Strength of Concrete Containing Supplementary Cementitious Materials. *Materials* **2021**, *14*, 5762. [CrossRef]
72. Yuan, X.; Tian, Y.; Ahmad, W.; Ahmad, A.; Usanova, K.I.; Mohamed, A.M.; Khallaf, R. Machine Learning Prediction Models to Evaluate the Strength of Recycled Aggregate Concrete. *Materials* **2022**, *15*, 2823. [CrossRef]
73. Salimbahrami, S.R.; Shakeri, R. Experimental investigation and comparative machine-learning prediction of compressive strength of recycled aggregate concrete. *Soft Comput.* **2021**, *25*, 919–932. [CrossRef]
74. Duan, J.; Asteris, P.G.; Nguyen, H.; Bui, X.-N.; Moayedi, H. A novel artificial intelligence technique to predict compressive strength of recycled aggregate concrete using ICA-XGBoost model. *Eng. Comput.* **2021**, *37*, 3329–3346. [CrossRef]
75. Sufian, M.; Ullah, S.; Ostrowski, K.A.; Ahmad, A.; Zia, A.; Śliwa-Wieczorek, K.; Siddiq, M.; Awan, A.A. An Experimental and Empirical Study on the Use of Waste Marble Powder in Construction Material. *Materials* **2021**, *14*, 3829. [CrossRef] [PubMed]
76. Shah, M.I.; Memon, S.A.; Khan Niazi, M.S.; Amin, M.N.; Aslam, F.; Javed, M.F. Machine Learning-Based Modeling with Optimization Algorithm for Predicting Mechanical Properties of Sustainable Concrete. *Adv. Civ. Eng.* **2021**, *2021*, 6682283. [CrossRef]
77. Ziolkowski, P.; Niedostatkiewicz, M. Machine learning techniques in concrete mix design. *Materials* **2019**, *12*, 1256. [CrossRef] [PubMed]
78. Ahmad, A.; Farooq, F.; Niewiadomski, P.; Ostrowski, K.; Akbar, A.; Aslam, F.; Alyousef, R. Prediction of compressive strength of fly ash based concrete using individual and ensemble algorithm. *Materials* **2021**, *14*, 794. [CrossRef] [PubMed]
79. Olalusi, O.B.; Awoyera, P.O. Shear capacity prediction of slender reinforced concrete structures with steel fibers using machine learning. *Eng. Struct.* **2021**, *227*, 111470. [CrossRef]
80. Dutta, S.; Samui, P.; Kim, D. Comparison of machine learning techniques to predict compressive strength of concrete. *Comput. Concr.* **2018**, *21*, 463–470.
81. Song, Y.-Y.; Ying, L.U. Decision tree methods: Applications for classification and prediction. *Shanghai Arch. Psychiatry* **2015**, *27*, 130.
82. Hillebrand, E.; Medeiros, M.C. The benefits of bagging for forecast models of realized volatility. *Econom. Rev.* **2010**, *29*, 571–593. [CrossRef]
83. Feng, W.; Wang, Y.; Sun, J.; Tang, Y.; Wu, D.; Jiang, Z.; Wang, J.; Wang, X. Prediction of thermo-mechanical properties of rubber-modified recycled aggregate concrete. *Constr. Build. Mater.* **2022**, *318*, 125970. [CrossRef]
84. Karbassi, A.; Mohebi, B.; Rezaee, S.; Lestuzzi, P. Damage prediction for regular reinforced concrete buildings using the decision tree algorithm. *Comput. Struct.* **2014**, *130*, 46–56. [CrossRef]
85. Zou, Y.; Zheng, C.; Alzahrani, A.M.; Ahmad, W.; Ahmad, A.; Mohamed, A.M.; Khallaf, R.; Elattar, S. Evaluation of Artificial Intelligence Methods to Estimate the Compressive Strength of Geopolymers. *Gels* **2022**, *8*, 271. [CrossRef]

86. Friedman, J.H. Greedy function approximation: A gradient boosting machine. *Ann. Stat.* **2001**, *29*, 1189–1232. [CrossRef]
87. Dahiya, N.; Saini, B.; Chalak, H.D. Gradient boosting-based regression modelling for estimating the time period of the irregular precast concrete structural system with cross bracing. *J. King Saud Univ.-Eng. Sci.* 2021. [CrossRef]
88. Huang, J.; Sun, Y.; Zhang, J. Reduction of computational error by optimizing SVR kernel coefficients to simulate concrete compressive strength through the use of a human learning optimization algorithm. *Eng. Comput.* **2021**, 1–18. [CrossRef]
89. Ahmad, A.; Chaiyasarn, K.; Farooq, F.; Ahmad, W.; Suparp, S.; Aslam, F. Compressive Strength Prediction via Gene Expression Programming (GEP) and Artificial Neural Network (ANN) for Concrete Containing RCA. *Buildings* **2021**, *11*, 324. [CrossRef]
90. Farooq, F.; Ahmed, W.; Akbar, A.; Aslam, F.; Alyousef, R. Predictive modeling for sustainable high-performance concrete from industrial wastes: A comparison and optimization of models using ensemble learners. *J. Clean. Prod.* **2021**, *292*, 126032. [CrossRef]
91. Aslam, F.; Farooq, F.; Amin, M.N.; Khan, K.; Waheed, A.; Akbar, A.; Javed, M.F.; Alyousef, R.; Alabdulijabbar, H. Applications of gene expression programming for estimating compressive strength of high-strength concrete. *Adv. Civ. Eng.* **2020**, *2020*, 8850535. [CrossRef]
92. Wang, Q.; Ahmad, W.; Ahmad, A.; Aslam, F.; Mohamed, A.; Vatin, N.I. Application of Soft Computing Techniques to Predict the Strength of Geopolymer Composites. *Polymers* **2022**, *14*, 1074. [CrossRef]
93. Ahmad, A.; Ostrowski, K.A.; Maślak, M.; Farooq, F.; Mehmood, I.; Nafees, A. Comparative Study of Supervised Machine Learning Algorithms for Predicting the Compressive Strength of Concrete at High Temperature. *Materials* **2021**, *14*, 4222. [CrossRef]
94. Dorafshan, S.; Thomas, R.J.; Maguire, M. Comparison of deep convolutional neural networks and edge detectors for image-based crack detection in concrete. *Constr. Build. Mater.* **2018**, *186*, 1031–1045. [CrossRef]
95. Hoang, N.-D.; Nguyen, Q.-L. Metaheuristic Optimized Edge Detection for Recognition of Concrete Wall Cracks: A Comparative Study on the Performances of Roberts, Prewitt, Canny, and Sobel Algorithms. *Adv. Civ. Eng.* **2018**, *2018*, 7163580. [CrossRef]
96. Yoda, K.; Yoshikane, T.; Nakashima, Y.; Soshiroda, T. Recycled cement and recycled concrete in Japan. In Proceedings of the Second International RILEM Symposium on Demolition and Reuse of Concrete and Masonry, Tokyo, Japan, 7–11 November 1988; pp. 527–536.
97. Limbachiya, M.C.; Leelawat, T.; Dhir, R.K. Use of recycled concrete aggregate in high-strength concrete. *Mater. Struct.* **2000**, *33*, 574–580. [CrossRef]
98. Ajdukiewicz, A.; Kliszczewicz, A. Influence of recycled aggregates on mechanical properties of HS/HPC. *Cem. Concr. Compos.* **2002**, *24*, 269–279. [CrossRef]
99. Gómez-Soberón, J.M.V. Porosity of recycled concrete with substitution of recycled concrete aggregate: An experimental study. *Cem. Concr. Res.* **2002**, *32*, 1301–1311. [CrossRef]
100. Vázquez, E.; Hendriks, C.F.; Janssen, G.M.T. Influence of recycled concrete aggregates on concrete durability. In Proceedings of the International RILEM Conference on the Use of Recycled Materials in Building and Structures, Barcelona, Spain, 8–11 November 2004; pp. 554–562.
101. Poon, C.S.; Shui, Z.H.; Lam, L.; Fok, H.; Kou, S.C. Influence of moisture states of natural and recycled aggregates on the slump and compressive strength of concrete. *Cem. Concr. Res.* **2004**, *34*, 31–36. [CrossRef]
102. Lin, Y.-H.; Tyan, Y.-Y.; Chang, T.-P.; Chang, C.-Y. An assessment of optimal mixture for concrete made with recycled concrete aggregates. *Cem. Concr. Res.* **2004**, *34*, 1373–1380. [CrossRef]
103. Xiao, J.Z.; Li, J.B.; Zhang, C. On relationships between the mechanical properties of recycled aggregate concrete: An overview. *Mater. Struct.* **2006**, *39*, 655–664. [CrossRef]
104. Wei, X.U. Experimental study on influence of recycled coarse aggregates contents on properties of recycled aggregate concrete. *Concrete* **2006**, *10*, 45–47.
105. Etxeberria, M.; Marí, A.R.; Vázquez, E. Recycled aggregate concrete as structural material. *Mater. Struct.* **2007**, *40*, 529–541. [CrossRef]
106. Etxeberria, M.; Vázquez, E.; Marí, A.; Barra, M. Influence of amount of recycled coarse aggregates and production process on properties of recycled aggregate concrete. *Cem. Concr. Res.* **2007**, *37*, 735–742. [CrossRef]
107. Evangelista, L.; de Brito, J. Mechanical behaviour of concrete made with fine recycled concrete aggregates. *Cem. Concr. Compos.* **2007**, *29*, 397–401. [CrossRef]
108. Poon, C.S.; Kou, S.C.; Lam, L. Influence of recycled aggregate on slump and bleeding of fresh concrete. *Mater. Struct.* **2007**, *40*, 981–988. [CrossRef]
109. Ajdukiewicz, A.B.; Kliszczewicz, A.T. Comparative tests of beams and columns made of recycled aggregate concrete and natural aggregate concrete. *J. Adv. Concr. Technol.* **2007**, *5*, 259–273. [CrossRef]
110. Min-ping, H.U. Mechanical properties of concrete prepared with different recycled coarse aggregates replacement rate. *Concrete* **2007**, *2*, 16.
111. Kou, S.C.; Poon, C.S.; Chan, D. Influence of fly ash as cement replacement on the properties of recycled aggregate concrete. *J. Mater. Civ. Eng.* **2007**, *19*, 709–717. [CrossRef]
112. Rahal, K. Mechanical properties of concrete with recycled coarse aggregate. *Build. Environ.* **2007**, *42*, 407–415. [CrossRef]
113. Wang, Z.W. Production and properties of high quality recycled aggregates. *Concrete* **2007**, *3*, 74–77.
114. Casuccio, M.; Torrijos, M.C.; Giaccio, G.; Zerbino, R. Failure mechanism of recycled aggregate concrete. *Constr. Build. Mater.* **2008**, *22*, 1500–1506. [CrossRef]
115. Min-ping, H. Mechanical properties of recycled aggregate concrete at early ages. *Concrete* **2008**, *223*, 37–41.

116. Kou, S.C.; Poon, C.S.; Chan, D. Influence of fly ash as a cement addition on the hardened properties of recycled aggregate concrete. *Mater. Struct.* **2008**, *41*, 1191–1201. [CrossRef]
117. Yang, K.-H.; Chung, H.-S.; Ashour, A.F. Influence of Type and Replacement Level of Recycled Aggregates on Concrete Properties. *ACI Mater. J.* **2008**, *105*, 289–296.
118. Zhou, H.; Liu, B.K.; Lu, G. Experimental research on the basic mechanical properties of recycled aggregate concrete. *J. Anhui Inst. Architect. Indust.* **2008**, *16*, 4.
119. Domingo-Cabo, A.; Lázaro, C.; López-Gayarre, F.; Serrano-López, M.A.; Serna, P.; Castaño-Tabares, J.O. Creep and shrinkage of recycled aggregate concrete. *Constr. Build. Mater.* **2009**, *23*, 2545–2553. [CrossRef]
120. Padmini, A.K.; Ramamurthy, K.; Mathews, M.S. Influence of parent concrete on the properties of recycled aggregate concrete. *Constr. Build. Mater.* **2009**, *23*, 829–836. [CrossRef]
121. Yang, X.; Wu, J.; Liang, J.G. Experimental study on relationship between tensile strength and compressive strength of recycled aggregate concrete. *Sichuan Build. Sci.* **2009**, *35*, 190–192.
122. Ye, H. Experimental study on mechanical properties of concrete made with high quality recycled aggregates. *Sichuan Build Sci* **2009**, *35*, 195–199.
123. Corinaldesi, V. Mechanical and elastic behaviour of concretes made of recycled-concrete coarse aggregates. *Constr. Build. Mater.* **2010**, *24*, 1616–1620. [CrossRef]
124. Kumutha, R.; Vijai, K. Strength of concrete incorporating aggregates recycled from demolition waste. *ARPN J. Eng. Appl. Sci.* **2010**, *5*, 64–71.
125. Malešev, M.; Radonjanin, V.; Marinković, S. Recycled concrete as aggregate for structural concrete production. *Sustainability* **2010**, *2*, 1204–1225. [CrossRef]
126. Zega, C.J.; Di Maio, A.A. Recycled concretes made with waste ready-mix concrete as coarse aggregate. *J. Mater. Civ. Eng.* **2011**, *23*, 281–286. [CrossRef]
127. Belén, G.-F.; Fernando, M.-A.; Diego, C.L.; Sindy, S.-P. Stress–strain relationship in axial compression for concrete using recycled saturated coarse aggregate. *Constr. Build. Mater.* **2011**, *25*, 2335–2342. [CrossRef]
128. Fathifazl, G.; Razaqpur, A.G.; Isgor, O.B.; Abbas, A.; Fournier, B.; Foo, S. Creep and drying shrinkage characteristics of concrete produced with coarse recycled concrete aggregate. *Cem. Concr. Compos.* **2011**, *33*, 1026–1037. [CrossRef]
129. Gonzalez-Fonteboa, B.; Martinez-Abella, F.; Eiras-Lopez, J.; Seara-Paz, S. Effect of recycled coarse aggregate on damage of recycled concrete. *Mater. Struct.* **2011**, *44*, 1759–1771. [CrossRef]
130. Chakradhara Rao, M.; Bhattacharyya, S.K.; Barai, S.V. Influence of field recycled coarse aggregate on properties of concrete. *Mater. Struct.* **2011**, *44*, 205–220. [CrossRef]
131. Somna, R.; Jaturapitakkul, C.; Chalee, W.; Rattanachu, P. Effect of the water to binder ratio and ground fly ash on properties of recycled aggregate concrete. *J. Mater. Civ. Eng.* **2012**, *24*, 16–22. [CrossRef]
132. Abd Elhakam, A.; Awad, E. Influence of self-healing, mixing method and adding silica fume on mechanical properties of recycled aggregates concrete. *Constr. Build. Mater.* **2012**, *35*, 421–427. [CrossRef]
133. Cui, Z.L.; Lu, S.S.; Wang, Z.S. Influence of recycled aggregate on strength and anti-carbonation properties of recycled aggregate concrete. *J. Build. Mater* **2012**, *15*, 264–267.
134. Hoffmann, C.; Schubert, S.; Leemann, A.; Motavalli, M. Recycled concrete and mixed rubble as aggregates: Influence of variations in composition on the concrete properties and their use as structural material. *Constr. Build. Mater.* **2012**, *35*, 701–709. [CrossRef]
135. Li, H.; Xiao, J.Z. On fatigue strength of recycled aggregate concrete based on its elastic modulus. *J. Build. Mater* **2012**, *15*, 260–263.
136. Limbachiya, M.; Meddah, M.S.; Ouchagour, Y. Performance of Portland/Silica Fume Cement Concrete Produced with Recycled Concrete Aggregate. *ACI Mater. J.* **2012**, *109*, 91–100.
137. Marinković, S.; Radonjanin, V.; Malešev, M.; Ignjatović, I. Comparative environmental assessment of natural and recycled aggregate concrete. *Waste Manag.* **2010**, *30*, 2255–2264. [CrossRef] [PubMed]
138. Pereira, P.; Evangelista, L.; De Brito, J. The effect of superplasticizers on the mechanical performance of concrete made with fine recycled concrete aggregates. *Cem. Concr. Compos.* **2012**, *34*, 1044–1052. [CrossRef]
139. Barbudo, A.; De Brito, J.; Evangelista, L.; Bravo, M.; Agrela, F. Influence of water-reducing admixtures on the mechanical performance of recycled concrete. *J. Clean. Prod.* **2013**, *59*, 93–98. [CrossRef]
140. Butler, L.; West, J.S.; Tighe, S.L. Effect of recycled concrete coarse aggregate from multiple sources on the hardened properties of concrete with equivalent compressive strength. *Constr. Build. Mater.* **2013**, *47*, 1292–1301. [CrossRef]
141. Chen, Z.P.; Xu, J.J.; Zheng, H.H.; Su, Y.S.; Xue, J.Y.; Li, J.T. Basic mechanical properties test and stress-strain constitutive relations of recycled coarse aggregate concrete. *J. Build. Mater.* **2013**, *16*, 24–32.
142. Hou, Y.L.; Zheng, G. Mechanical properties of recycled aggregate concrete in different age. *J. Build. Mater.* **2013**, *16*, 683–687.
143. Ismail, S.; Ramli, M. Engineering properties of treated recycled concrete aggregate (RCA) for structural applications. *Constr. Build. Mater.* **2013**, *44*, 464–476. [CrossRef]
144. Manzi, S.; Mazzotti, C.; Bignozzi, M.C. Short and long-term behavior of structural concrete with recycled concrete aggregate. *Cem. Concr. Compos.* **2013**, *37*, 312–318. [CrossRef]
145. Matias, D.; De Brito, J.; Rosa, A.; Pedro, D. Mechanical properties of concrete produced with recycled coarse aggregates–Influence of the use of superplasticizers. *Constr. Build. Mater.* **2013**, *44*, 101–109. [CrossRef]

146. Sheen, Y.-N.; Wang, H.-Y.; Juang, Y.-P.; Le, D.-H. Assessment on the engineering properties of ready-mixed concrete using recycled aggregates. *Constr. Build. Mater.* **2013**, *45*, 298–305. [CrossRef]
147. Thomas, C.; Setién, J.; Polanco, J.; Alaejos, P.; De Juan, M.S. Durability of recycled aggregate concrete. *Constr. Build. Mater.* **2013**, *40*, 1054–1065. [CrossRef]
148. Ulloa, V.A.; García-Taengua, E.; Pelufo, M.-J.; Domingo, A.; Serna, P. New views on effect of recycled aggregates on concrete compressive strength. *ACI Mater. J.* **2013**, *110*, 1–10.
149. Xiao, J.; Li, H.; Yang, Z. Fatigue behavior of recycled aggregate concrete under compression and bending cyclic loadings. *Constr. Build. Mater.* **2013**, *38*, 681–688. [CrossRef]
150. Younis, K.H.; Pilakoutas, K. Strength prediction model and methods for improving recycled aggregate concrete. *Constr. Build. Mater.* **2013**, *49*, 688–701. [CrossRef]
151. Andreu, G.; Miren, E. Experimental analysis of properties of high performance recycled aggregate concrete. *Constr. Build. Mater.* **2014**, *52*, 227–235. [CrossRef]
152. Beltrán, M.G.; Agrela, F.; Barbudo, A.; Ayuso, J.; Ramírez, A. Mechanical and durability properties of concretes manufactured with biomass bottom ash and recycled coarse aggregates. *Constr. Build. Mater.* **2014**, *72*, 231–238. [CrossRef]
153. Beltrán, M.G.; Barbudo, A.; Agrela, F.; Galvín, A.P.; Jiménez, J.R. Effect of cement addition on the properties of recycled concretes to reach control concretes strengths. *J. Clean. Prod.* **2014**, *79*, 124–133. [CrossRef]
154. Çakır, Ö.; Sofyanlı, Ö.Ö. Influence of silica fume on mechanical and physical properties of recycled aggregate concrete. *HBRC J.* **2015**, *11*, 157–166. [CrossRef]
155. Carneiro, J.A.; Lima, P.R.L.; Leite, M.B.; Toledo Filho, R.D. Compressive stress–strain behavior of steel fiber reinforced-recycled aggregate concrete. *Cem. Concr. Compos.* **2014**, *46*, 65–72. [CrossRef]
156. Dilbas, H.; Şimşek, M.; Çakır, Ö. An investigation on mechanical and physical properties of recycled aggregate concrete (RAC) with and without silica fume. *Constr. Build. Mater.* **2014**, *61*, 50–59. [CrossRef]
157. Duan, Z.H.; Poon, C.S. Properties of recycled aggregate concrete made with recycled aggregates with different amounts of old adhered mortars. *Mater. Des.* **2014**, *58*, 19–29. [CrossRef]
158. Folino, P.; Xargay, H. Recycled aggregate concrete–Mechanical behavior under uniaxial and triaxial compression. *Constr. Build. Mater.* **2014**, *56*, 21–31. [CrossRef]
159. Gayarre, F.L.; Perez, C.L.-C.; Lopez, M.A.S.; Cabo, A.D. The effect of curing conditions on the compressive strength of recycled aggregate concrete. *Constr. Build. Mater.* **2014**, *53*, 260–266. [CrossRef]
160. Kang, T.H.K.; Kim, W.; Kwak, Y.-K.; Hong, S.-G. Flexural Testing of Reinforced Concrete Beams with Recycled Concrete Aggregates. *ACI Struct. J.* **2014**, *111*, 607–616. [CrossRef]
161. Pedro, D.; De Brito, J.; Evangelista, L. Performance of concrete made with aggregates recycled from precasting industry waste: Influence of the crushing process. *Mater. Struct.* **2015**, *48*, 3965–3978. [CrossRef]
162. Pepe, M.; Toledo Filho, R.D.; Koenders, E.A.B.; Martinelli, E. Alternative processing procedures for recycled aggregates in structural concrete. *Constr. Build. Mater.* **2014**, *69*, 124–132. [CrossRef]
163. Thomas, C.; Sosa, I.; Setién, J.; Polanco, J.A.; Cimentada, A.I. Evaluation of the fatigue behavior of recycled aggregate concrete. *J. Clean. Prod.* **2014**, *65*, 397–405. [CrossRef]

*Article*

# Investigating the Effectiveness of Recycled Agricultural and Cement Manufacturing Waste Materials Used in Oil Sorption

**Marina Valentukeviciene *** and **Ramune Zurauskiene**

Environmental Engineering Faculty, Vilnius Gediminas Technical University, 10223 Vilnius, Lithuania; ramune.zurauskiene@vilniustech.lt
* Correspondence: marina.valentukeviciene@vilniustech.lt; Tel.: +370-616-53746

**Abstract:** This research investigates how sorbents made from recycled waste materials affect the properties of water used to remove residues flushed from oil tanks transported by rail. The mineral sorbent was added to water following the flushing process. Water temperatures were maintained at 21 °C and 70 °C for a contact period of 30 min. The experiments demonstrated that: when the sorbent is active, turbidity removal efficiency was about 64%; color removal efficiency of 56% was obtained; and total iron concentration removal was approximately 68%. The effect of the characteristics of the materials on the adsorption capacity was evaluated using the removed amount of oil per one gram of every sorbent. It was found that straw sorbent oil adsorption capacity was up to 33 mg/g, peat sorbent 37 mg/g, and mineral sorbent 1.83 mg/g. The following were also measured during the experiment: temperature, pH, chemical oxygen usage, total iron concentrations, suspended matter, and oil concentrations. The findings show that recycled sorbents obtained from waste materials are environmentally sustainable and can be reused to treat water that has been used to flush oil transported in rail tanks.

**Keywords:** recycled materials; environmental engineering; water reuse; oil removal

**Citation:** Valentukeviciene, M.; Zurauskiene, R. Investigating the Effectiveness of Recycled Agricultural and Cement Manufacturing Waste Materials Used in Oil Sorption. *Materials* **2022**, *15*, 218. https://doi.org/10.3390/ma15010218

Academic Editors: Andrea Petrella and Michele Notarnicola

Received: 24 November 2021
Accepted: 26 December 2021
Published: 28 December 2021

**Publisher's Note:** MDPI stays neutral with regard to jurisdictional claims in published maps and institutional affiliations.



## 1. Introduction

The most significant sources of oil pollution include those produced by factories, combustion engines, and/or leaking vehicle parts [1]. The most common pollutants are gasoline, lead, zinc, chrome, iron, arsenic, pesticides, nitrates, and toxic compounds [2].

Typically, oil residue is cleaned from rail tanks that have been used to transport fluids, although the process of removal can be highly problematic [3]. This process comprises the following steps. Firstly, emptying the tank by removing carbohydrate until the level of the oil product is sufficiently lowered so that cleaning equipment can be used.

Secondly, introducing air to flush out harmful gases, but only to the point that steam generated by the oil product does not exceed the highest allowed concentration. Then, the third step is to use special cleaning equipment to cleanse the tank wall and floor bed. This is achieved through oil flow created by the cleaning equipment used to measure both the concentration of carbohydrates in the gasses and static electricity energy. Note that both of these procedures should be performed at the same time. Finally, there are a few more steps: flushing the dissolved oil particles and dispersive sludge into a special container or via a pipeline; flushing the emulsified water from the tank; conducting sorption during degassing, forced ventilation and introduction of air; by the end, controlling the quality of the sludge that has been removed.

Currently, several methods are used to remove sludge from the bed of the tank floor. One of these is the hydraulic method, which is energy efficient and economically efficient when compared to other hydraulic methods used to clean tanks. In addition, the absence of complicated parts in this form of cleaning makes for a highly reliable system which requires little oversight, thereby saving time [3]. Sludge needs to be reduced during the above mentioned procedures, although this usually extends the time between cleaning, and increases the amount of wastewater in the capture vessel, as well as the oil product transportation systems [3].

Cleaning technologies used to purify water include reverse osmosis with nanofiltration and ultrafiltration, ultrafiltration with adsorption by activated charcoal, electrochemical oxidation, biological treatment, and flocculation [4]. These are discussed in numerous published articles that review various transport cleaning wastewater management methods used throughout the world [5]. Such methods include membrane filtration, adsorption, floatation and chemical coagulation, cleaning using dispersion resulting from ultrasound, nano-technologies, catalysis processes, and other hybrid technologies.

Electrochemical methods can be used to destabilize oil product emulsion in wastewater, and the most commonly used electrochemical methods for cleaning are electrocoagulation and enhanced flocculation [5]. Hydrocarbons and suspended materials can be removed from wastewater by a physical-chemical method using coagulants and flocculants. During the coagulation process various particles (suspended, colloidal particles, oil) are destabilized causing them to attach to other particles. With this method, ever-larger particle flakes form, so that when flake density becomes higher than that of water, they sink to the bottom of the tank and are removed using the sedimentation method [5]. The aforementioned authors [6] compared membrane bioreactor processes (MBP) with chemical coagulation and ozonation, and concluded that better results could be achieved using MBP instead of coagulation.

When sorption is used to treat wastewater, oil products are removed using an adsorbent, for example: polypropylene, activated carbon, polyacrylamide, the base of which is made of chitosan [5]. Of all the methods used to remove pollution produced by wastewater treatment technologies, sorption proved to be the best due to its low price, simplicity, and effectiveness [7].

Oil products in wastewater take different forms, for example unstable, dispersed oil products as well as stable emulsified oil products that float on the surface [8]. A number of researchers [9] have demonstrated that electrocoagulation using iron and aluminum electrodes remove oil products from rail tanks. Their experiments showed a reduction by 90.18% of Chemical Oxygen Demand indicators in the flushed wastewater. Other researchers [10] have investigated the use of natural fibers such as rice husks to treat water polluted with oil products. One clear advantage of this cleaning method is its ability to absorb different compounds from multicomponent solutions, thereby achieving a high level of reused water quality, especially from highly polluted sources. Notwithstanding, sorbent made from rice husks has limitations due to the prior removal of most remaining pollutants. This is due to the fact that the rice peel and straw contain high levels of silicon dioxide.

Skimming, precipitation, centrifugation, and biological methods are common techniques used to separate oil pollution from contaminated wastewater. Yet, most of these are unable to remove emulsified oil products, because the size of the compound particles is small. Moreover, proposed new technologies are expensive and do not effectively remove quantities of residual oil particles [11]. Since the membrane method is complicated, sorption methods are more acceptable when treating reused wastewater.

Previous research [11] shows that sorbents able to adsorb oil products can be divided into three main groups:

(1) Inorganic mineral products
(2) Organic synthetic products
(3) Organic natural products.

Accordingly, researchers have explored the removal of oil products from wastewater using barley straw, their aim being to evaluate the possibility of removing oil products using naturally disintegrating sorbents [11]. Various types of oil products were used in the experiments: sea Balaiem, Western Desert crude oil, Russian oil, and Arabian heavy crude oil. It was found that by increasing the temperature, sorption capacity rises, whereas at low temperatures oil particles tend to clog the pores of a sorbent, thereby preventing the absorption of oil. An attempt was made to remove adsorbed oil from the sorbent by mechanically pressing it several times and using it repeatedly for oil particle sorption. After three cycles, sawdust sorption capacity was measured which was equal to 50% from initial sorption capacity.

Thus, the main aim of our research was to identify the most effective, recyclable sorbents that can remove oil and other pollutants from reused wastewater obtained from the cleaning of railway tanks.

## 2. Materials and Methods

Storm-water that has been captured is commonly used to remove oil products residuals and other pollutants from railway transport tanks. The contaminated wastewater in the rail tank is eliminated from the washing area using special pumps and then stored in a reservoir. Oil, has a chemical composition of lipids, which, due to their structure, liquefy at higher temperatures and solidify at lower temperatures. During washing, hydraulic pumps inject the water into a heat exchange unit where it is heated to 70 °C. Hot water is then fed to the cleaning equipment, close to which are 14 cisterns, each with a capacity of 700 $m^3$. Each of the cleaning devices has four steam supply stands, capable of processing four tanks at once. The inside of the tank is washed with the heated water, thereafter the wastewater is treated in a separator that removes the oil products which allows the treated wastewater to be reused.

The operator of the cleaning equipment first places the nozzle of the steaming device into the tank's interior, then opens the steam valve to eject hot water at pressure. The tank is steamed for 20 min until the residue liquefies. Upon completion, the inner valve at the top of the rail tank is opened and steaming of the cistern resumes. This expels the oil products and voids the tank, with the result that the liquefied oil product residue drains into a collecting repository. Once the residue has been drained, the steam is turned off and the steaming equipment is withdrawn.

Water is fed into the nozzles of the washing apparatus impeller which spins on horizontal and vertical axles. For railway transport, most cleaning technologies are performed using high pressure and spinning nozzles, the cleaning being performed through an aperture in the center of the cistern top.

The contaminated wastewater from the cistern drains out through the pipelines to the primary treatment equipment. Sand and other sinking materials are separated in two sand and grease precipitators which have a depth of two meters. Oil products in the cleaning equipment rise to the surface due to the difference in density and are collected with the brush skimmer. Wastewater collected from the cleaning equipment is stored in reservoirs and re-used for cleaning processes. The optic method is used to identify if the wastewater is no longer recyclable due to its increased color and turbidity, in which case it is then returned for additional treatment.

Oily wastewater was obtained directly after washing processes were conducted with water at 70 °C. After standing still at the reservoirs, the temperature was lowered to 21 °C. All experiments were conducted with both temperatures to obtain the best reliable results.

Three different sorbents were used for experimental wastewater treatment:

- Gypsum granules, produced by absorbing granule limestone. According to the manufacturer's specification it is recyclable sorbent for oil compounds removal
- Barley straw sorbent, obtained from local farmers as a waste product due to its low quality
- Hydrophobic peat sorbent, which can be used to absorb large amounts of oil products.

Two natural sorbents were used in this research (waste straw from a local Lithuanian farming area, and waste peat from an excavation area (JSC Durpeta)), and one mineral sorbent from recycled wasted materials produced by JSC Akmenes Cementas (Naujoji Akmene, Lithuania). All sorbents were dried in laboratory conditions with an approximate air humidity of 70%, and room temperature equal to 18 °C, before being mixed with wastewater. Natural sorbents were mechanically cut and sieved into useful fractions with the size varying in a range of approximately 2.0–3.0 cm. Mineral sorbent was obtained directly from the producing factory in fractions approximately between 4–6 mm. The straw sorbent has an approximate bulk density—0.04 g/cm$^3$, peat sorbent has an approximate bulk density of 0.03 g/ cm$^3$ and the highest density was measured for the mineral sorbent, equal to 0.75 g/ cm$^3$ from cement production recycling processes.

Before the experiments were conducted, five 1000 mL volume glass containers were filled with 500 mL of investigated wastewater. Sorbent was added to four of the containers using a different quantity of sorbent in each sample. According to the highest porosity of natural sorbents, 1 g, 2 g, 3 g, or 4 g of straw or hydrophobic peat sorbent was used in each sample. Following experiments with mineral sorbent (high density compared with natural sorbents) were carried out using 5 g, 10 g, 20 g, 30 g, and 40 g respectively.

With the experimental equipment, these sorbents were mixed with the investigated wastewater by rotating at 60 r/min rate. The separation method is sedimentary process by which mineral sorbent falls to the bottom of sedimentation tank. This was carried out for the sedimentation time of 30 min between solids and liquids, before the water quality analysis. When hydrophobic peat and straw sorbents were added to water, the separation method was carried out in fluidized beds (30 min) where both sorbents were separated from the treated water on the top of used reservoirs. Both separation methods ensured that the samples taken for water quality analysis included the treated water only.

These samples were analyzed using the ICP emission instrument on Perkin–Elmer ICP-400 (The Perkin–Elmer Plasma 400 ICP Emission Spectrometer, PerkinElmer Inc., Waltham, MA, USA). All measurements investigated in this research were within the specified acceptance criterion of $\pm$10% of the known value, and the typical deviation for most concentrations estimated less than 3%.

The quality of the water obtained was determined using indicators considered to be the most important: concentration of oil products, total suspended solids, total iron concentration, turbidity, color, conductivity, pH, and Chemical Oxygen Demand (COD). A portable conductivity measurement device (Cond315i, WTW, Weilheim, Germany) and pH-meter instrument (WTW pH 323, WTW, Weilheim, Germany) were used to measure individual samples of wastewater, conductivity, pH, and temperature in the tank. The infrared spectrophotometric analysis (IR) method was used to analyze the oil concentration. In the course of the experiment, a spectrophotometer (Spectroquant Nova 60 MERCK, Merck Group, Darmstadt, Germany) was used to measure turbidity and the color of wastewater (before and after sorption). Chemical Oxygen Demand (COD) is a chemical test, used as a chemical oxidation of the wastewater samples. The dichromate COD method measures close to the maximum possible oxygen demand of the wastewater sample using the standard method ISO 15705:2002—a specific version of the sealed tube (Micro) method [12]. Total suspended solids were measured after filtering of the obtained sample through a 0.45 µm filter membrane and all filters were dried at 105 °C and weighed. Iron concentrations in wastewater were determined using atomic absorption spectrometry at the laboratory and by colorimetric methods directly on site of wastewater treatment plants. The data were evaluated using "MathCad" created software with a type I error (a) of 0.05.

## 3. Results and Discussion

See Table 1 for wastewater quality data of the sample investigated from tank washing processes.

**Table 1.** The wastewater quality data of investigated samples from tank washing processes.

| Nr. | Parameter | Measuring Unit | Value |
|---|---|---|---|
| 1. | Total suspended solids | mg/L | 119.80 |
| 2. | Total iron concentration | mg/L | 0.53 |
| 3. | Turbidity | FTU | 35.00 |
| 4. | Color | mL/L (Pt scale) | 659.02 |
| 5. | Conductivity | μS/cm | 287 |
| 6. | pH | pH | 7.32 |
| 7. | COD | mg/L | 1370.0 |
| 8. | Oil concentration | mg/L | 42.00 |

When 3 g of straw sorbent (the smallest amount) was added to the investigated wastewater at a temperature of 21 °C, the total iron concentration decreased to 0.4 mg/L. However, when the same amount is added to wastewater at a temperature of 70 °C, the pollution concentration increased (Figure 1).

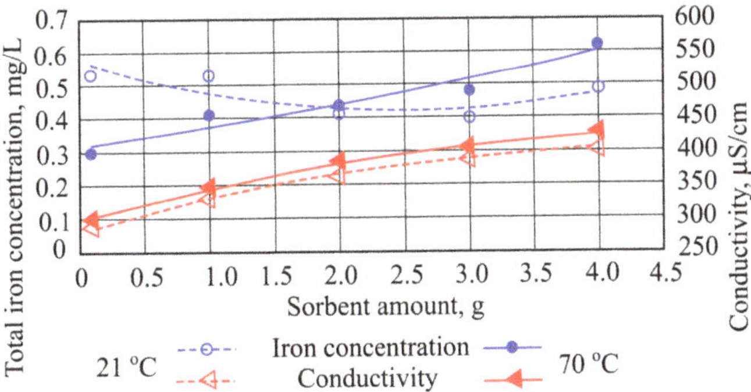


**Figure 1.** Total iron concentration and conductivity when different quantities of straw sorbent are added to the wastewater.

When the wastewater was treated with 3 g of sorbent at 21 °C, the highest measured turbidity value 42.5 FTU was obtained. When the same amount of sorbent (3 g) is added to a water temperature of 70 °C, turbidity value decreases to 62.0 FTU (Figure 2).

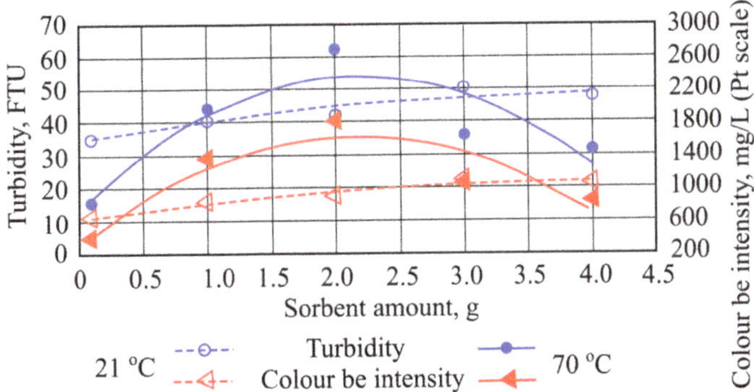


**Figure 2.** Turbidity values and color intensity when different quantities of straw sorbent are added to the wastewater.

Adding different quantities of straw sorbent to the wastewater at a temperature of 21 °C increases the color parameter from 659.02 mg/L to 1101.69 mg/L. However, when sorbent is added at 70 °C temperature, color increased to 1799.59 mg/L (Figure 2). Obtained total suspended solids interacted with used sorbents and were significantly affected by numerous factors. The wastewater from oil reservoirs washing processes was affected by rainfall and catchment runoff; reservoir corrosion; sediment disturbance, for example, by pumps; waste discharge; wastewater flow; and transported oil types. Under the pH conditions prevalent in wastewater treatment plants, iron salts are unstable and react with transported compounds to form insoluble hydroxides, which settle out as rust-colored silt. Wastewater in which this occurs can stain oil transportation reservoirs and plumbing fixtures. In the pumping system, iron compounds can settle out in the pipelines and gradually reduce the flow rate through the pipes.

When measuring COD at a water temperature of 21 °C using only 2 g of sorbent, most of the pollutants were removed (Figure 3). In other cases, this COD indicator was found to be higher than the initial value. When adding amounts of sorbent into 70 °C temperature water, the COD indicator increased until it reached its highest value of 3868.8 $mgO_2/L$ after adding 4 g of straw sorbent (Figure 3).

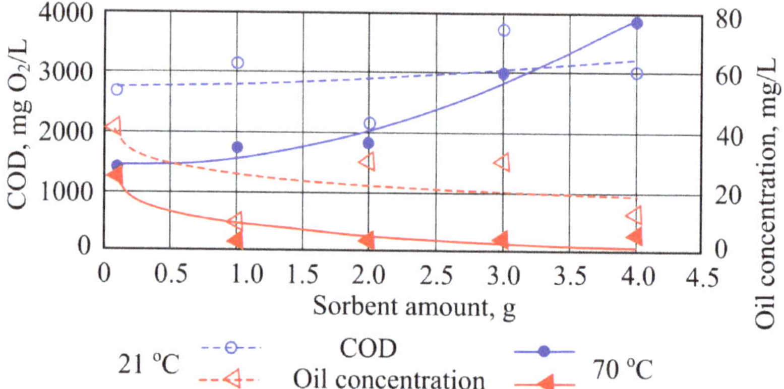


**Figure 3.** COD and oil concentrations when different quantities of straw sorbent are added to the wastewater.

When straw sorbent was introduced to the investigated wastewater at the controlled temperatures, oil product concentration values decreased. The best result was obtained when the temperature of the wastewater was 21 °C and 1 g of straw sorbent was used. In this case the sorbent absorbed 33 mg of oil pollutant. The most effective oil product sorption at 70 °C occurs when the straw sorbent adsorbed 22.5 mg of oil products (Figure 3).

It was found that a decrease in sorption capacity [13] occurs when suspended solids are measured at a wastewater temperature of 21 °C, and when straw sorbent is added in quantities of 1 g and 2 g, during which 41 mg/L of suspended solids was removed. However, adding a higher quantity of straw sorbent results some increases of the concentration of suspended solids from 42 mg/L to 161.8 mg/L (Figure 4).

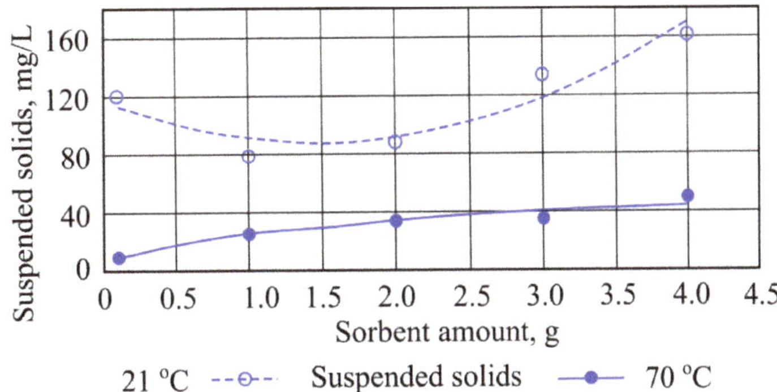


**Figure 4.** Suspended solids concentrations when straw sorbent is added to the wastewater.

Experimental results obtained from peat sorbent at a wastewater temperature 21 °C are presented below in Figures 5–8.

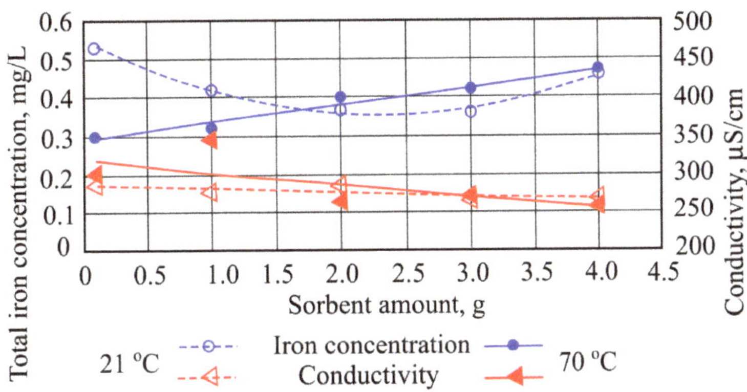


**Figure 5.** Total iron concentration and conductivity when different quantities of hydrophobic peat sorbent is added to the wastewater.

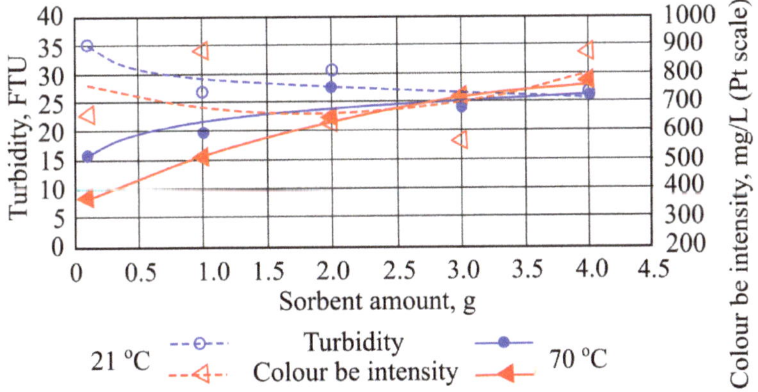


**Figure 6.** Turbidity and color be intensity values when different quantities of hydrophobic peat sorbent are added to the wastewater.

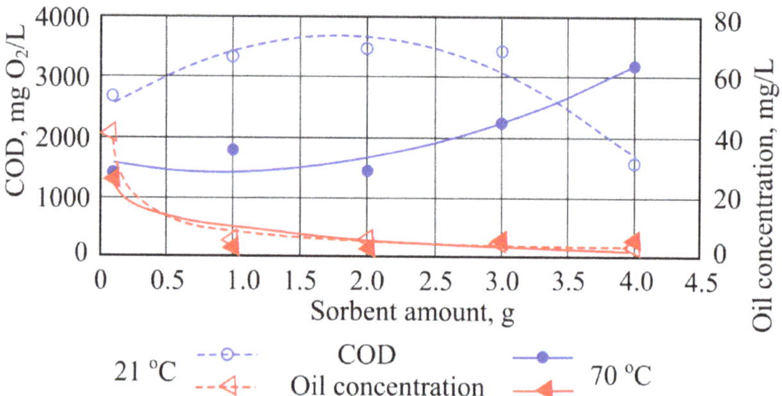


**Figure 7.** Chemical oxygen usage and oil concentration when different quantities of hydrophobic peat sorbent are added to the wastewater.

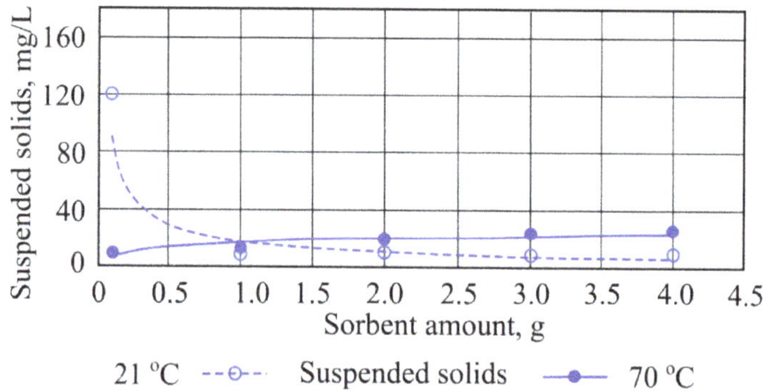


**Figure 8.** Suspended solids concentration when different quantities of hydrophobic peat sorbent are added to the wastewater.

The biggest reductions are noted in total iron concentration [14], by 0.17 mg/L from the starting value when the investigated water is at 21 °C and 3 g of hydrophobic peat sorbent has been added. When quantities of hydrophobic peat sorbent (4 g) are added at a water temperature of 70 °C, the total iron concentration captured is higher, and pollutant concentration is higher by 0.17 mg/L from the starting value (Figure 5). After adding the highest sorbent amount (4 g) into both temperatures of investigated waters, total iron concentration value was measured to be very similar—0.46–0.47 mg/L.

Figure 5 shows that in both temperatures (low and high) of the investigated water (and with the addition of hydrophobic peat sorbent), the indicator of conductivity falls at most by 21 µS/cm from initial value at 21 °C water, and by 38 µS/cm from initial value at 70 °C.

When measuring the turbidity indicator during the experiment using different investigated water temperatures, it was noted that at 21 °C water turbidity falls the most by 9.8 FTU from its initial value. However, at 70 °C when adding quantities of the sorbent, the turbidity indicator does not fall, while adding higher masses, its turbidity increases up to 11.8 FTU from a starting value (Figure 6).

Adding different hydrophobic peat sorbent masses into the water with a temperature of 21 °C produced inconsistent results. For example, by adding quantities of the smallest and highest sorbent masses (1 g and 4 g), color intensity indicator increases up to 218.34 mg/L (Pt scale) from the starting value. When 2 g and 3 g of this sorbent is added into water of the same temperature, color intensity falls to 96.71 mg/L (Pt scale) from the starting value. By adding different masses of hydrophobic peat sorbent to 70 °C temperature water, higher color intensity values were captured. The highest value (396.81 mg/L (Pt scale) rise from initial value) was determined when using 4 g of the sorbent (Figure 6).

The experiments found that chemical oxygen demand removal at 21 °C temperature occurs only when the highest mass of hydrophobic peat sorbent is used. In these conditions the decrease in chemical oxygen usage indicator was 1094.4 mg/L. In other conditions the concentration of this pollutant rises from 2668.8 mg/L to 3494.4 mg/L. When this indicator was investigated using water at a temperature of 70 °C, chemical oxygen usage sorption was recorded only when 2 g quantities of this sorbent were added. In other conditions the rise in pollutant was measured up to 1795.2 from the starting value (Figure 7).

Analysis of the results obtained where hydrophobic peat sorbent has been used showed that at both 21 °C and 70 °C the sorbent is highly efficient in adsorbing oil products. The best result is obtained by adding 4 g of sorbent into the water at 21 °C, and 2 g into water that is 70 °C. Water at room temperature (Figure 7) increases the capacity of sorbent sorption to remove oil products.

During experiments to investigate the concentration of suspended solids, we noted that at 70 °C the pollutant concentration is significantly lower when compared to water at room temperature. When adding different hydrophobic peat sorbent masses into 70 °C temperature water that contains a high suspended solids concentration, an increase is observed that reaches 18 mg/L at most from the starting concentration. When hydrophobic peat sorbents were added to water at 21 °C, there was a notable decrease in the concentration of suspended solids (Figure 8).

Total iron is effectively adsorbed when granulated mineral sorbent granules were added in water at 21 °C and 70 °C. At 21 °C the most effective sorption takes place at 30 g of sorbent when 0.36 mg/L of total iron is adsorbed. At 70 °C the most effective sorption takes place when the highest quantities of sorbent are added, and during the sorption, the highest amount of pollutant is adsorbed (Figure 9).

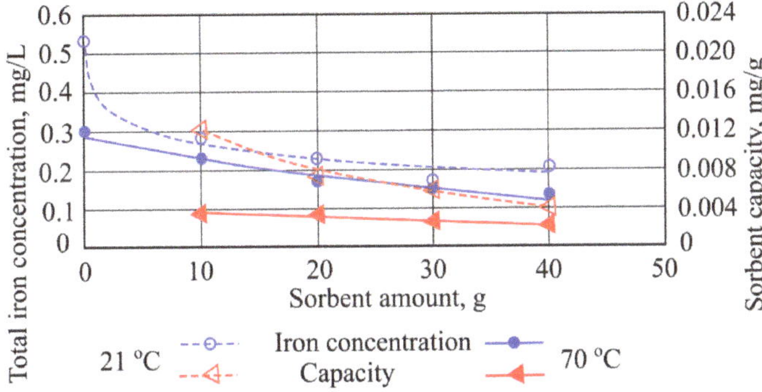


**Figure 9.** Total iron concentration and sorption capacity of mineral granulated sorbent.

Figure 9 shows the highest sorption capacity (0.0035 mg/g) differs only by a factor of 1.6 from the lowest measurement (0.002).

Figure 10 shows results of electrical conductivity change when different quantities of mineral granulated sorbent were added. The highest values of this indicator in 21 °C and 70 °C water were obtained when the highest amount of sorbent was added. From a starting value they differ by 8.4–8.6 times.

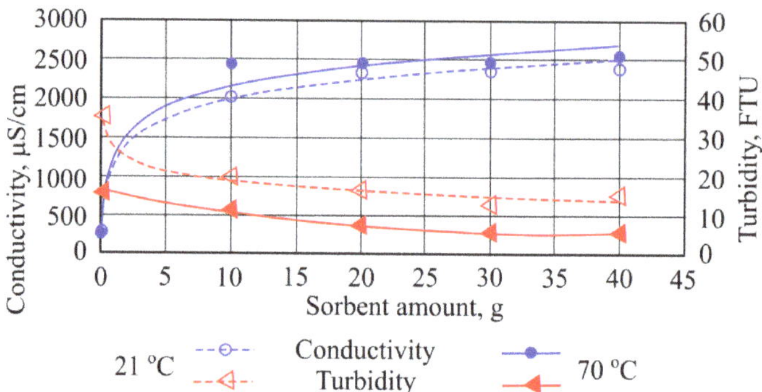


**Figure 10.** Electrical conductivity and turbidity using different quantities of mineral granulated sorbent.

When investigating different quantities of mineral granule sorbent and their influence on the water under investigation (i.e., measuring turbidity indicators), it was noted that at a room temperature of 21 °C the best results were obtained after adding 30 g of granulated sorbent. High turbidity removal from water was noted in hot water (70 °C) when adding larger quantities of mineral granule sorbent, so that lower turbidity indicator values are obtained. In this case, the highest turbidity removal efficiency was measured using 40 g of granulated sorbent (Figure 10).

When adding different quantities of mineral granule sorbent to the investigated water, color intensity was notably decreased. Based on the results obtained, the more effective color removal takes place at room temperatures of the investigated water (Figure 11).

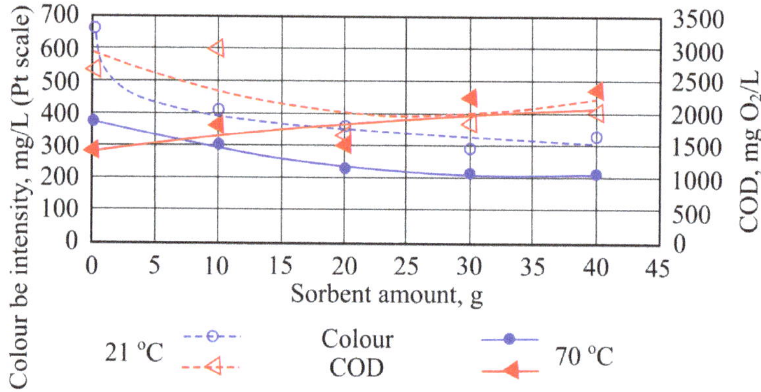


**Figure 11.** Color and chemical oxygen demand, when different quantities of mineral granulated sorbent are added to the water.

The best result is obtained by adding 20 g of mineral granulated sorbent, and in these conditions the removal efficiency of chemical oxygen demand is 37.4%. At the higher temperature, water chemical oxygen demand is not removed. When adding 30 g of mineral granule sorbent COD values rise by 835.2 mg/L from starting value (Figure 11).

When investigating the influence of granulated mineral sorbent on oil product concentration in different temperature water, it was found that in both cases oil product concentration decreased when the smallest quantity of mineral sorbent was added. (Figure 12).

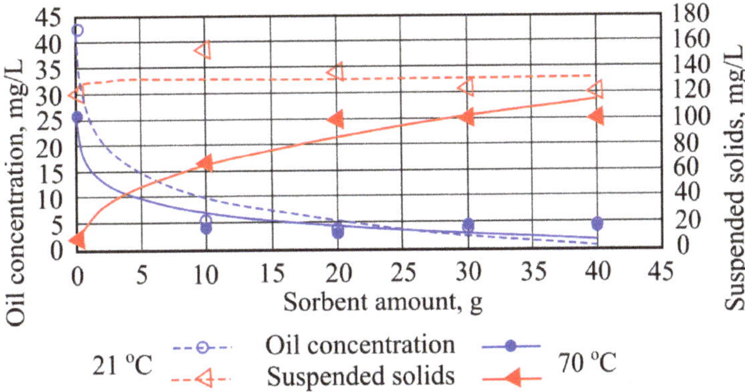


**Figure 12.** Oil concentration and suspended solids when different quantities of mineral granulated sorbent are added to the water.

When investigating the concentration of suspended solids at 70 °C when using different sorbent amounts, we noted that this indicator rises as the quantity of sorbent added is increased. However, when using 20 g, 30 g, and 40 g of sorbent masses, results changed insignificantly—only by 1 mg/L (Figure 12).

The results obtained were used to compose Freundlich isotherms. Figure 13 shows the results for granulated mineral sorbent. The Freundlich isotherms are at 21 °C (a) and 70 °C (b) of investigated water.

$$q_s = a_F C_e^{1/n},$$

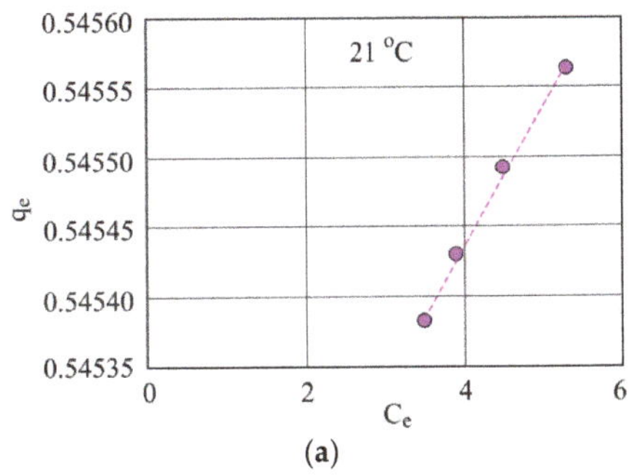


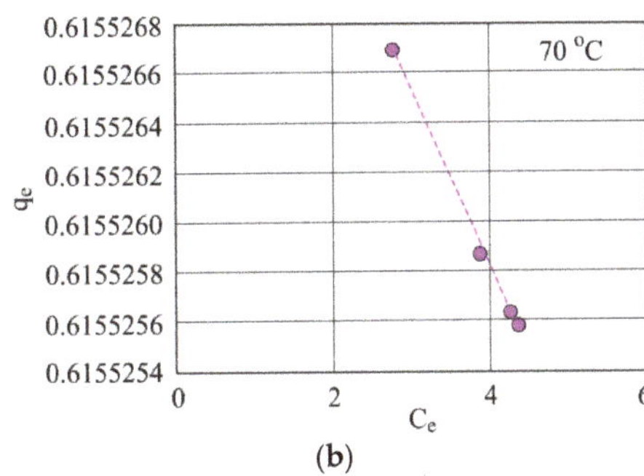


**Figure 13.** Freundlich isotherm when investigated water is at (**a**) 21 °C and (**b**) 70 °C using granulated mineral sorbent.

Here:

$a_F$ and $n$ are Freundlich constants;

$C_e$ is equilibrium adsorbed compounds concentration in water, mg/L.

Plotting Freundlich isotherm on a logarithmic scale, $a_F$ and $n$ are Freundlich constants estimated from the intercept and the slope, accordingly. Using numerical values, $a_F$ expresses the adsorbent capacity; the higher its value, the larger the capacity; $n$ is the heterogeneity constant. The more heterogeneous the sorbent, the closer $n$ value is to zero. Freundlich isotherm gives more exact results than the Langmuir isotherm for a large selection of heterogeneous sorbents and complex sorption admixtures systems.

The Freundlich isotherm shown above demonstrates that when investigated water is at 21 °C and oil concentration rises, sorbent sorption capacity decreases. However, at the higher temperature of 70 °C when oil product concentration is higher, sorbent sorption capacity falls (Figure 14).

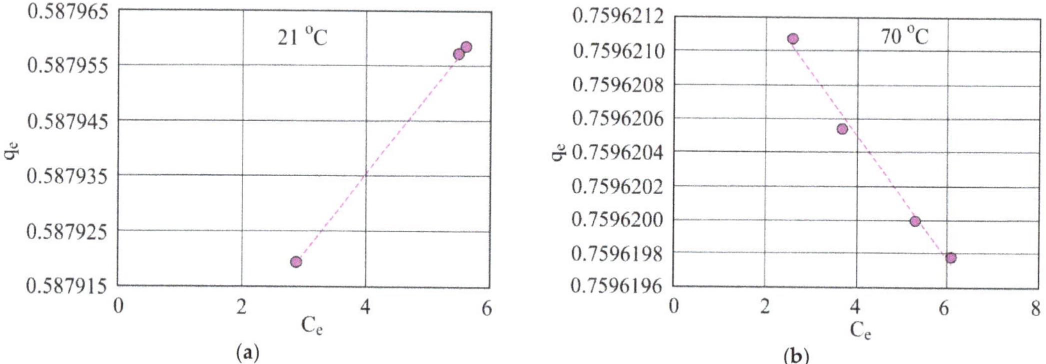


**Figure 14.** Freundlich isotherm when investigated water temperature is at (**a**) 21 °C and (**b**) 70 °C using hydrophobic peat sorbent.

In the Freundlich isotherms (Figure 14) adsorbing oil products by using hydrophobic peat sorbent, it was determined that oil concentration rises in the investigated water at room temperature, and sorption capacity also rises. This is consistent with the previously shown isotherms. For the investigated water at 70 °C it is anticipated that oil concentration will rise and the sorbent sorption capacity lower accordingly.

When comparing the results of the experiment there is a noticeable connection between certain indicator results according to some references [15,16]. The linked indicator graph, Figure 15, shows that when color intensity value is at its highest, common iron concentration is also at its highest, while the color intensity value at its lowest matches with the lowest value of the common iron concentration. This trend is noticeable both at temperatures of 21 °C and at 70 °C.

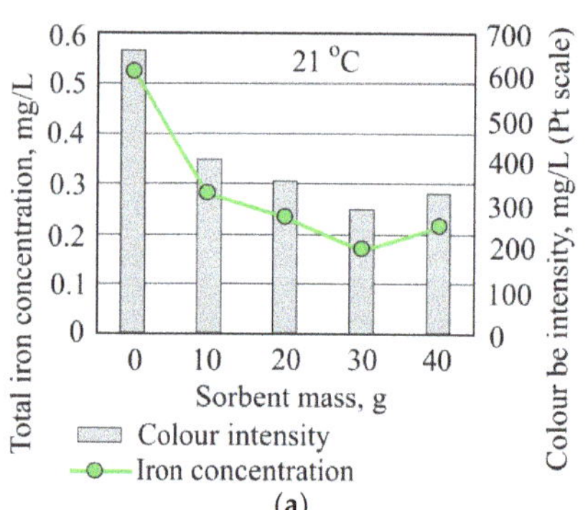


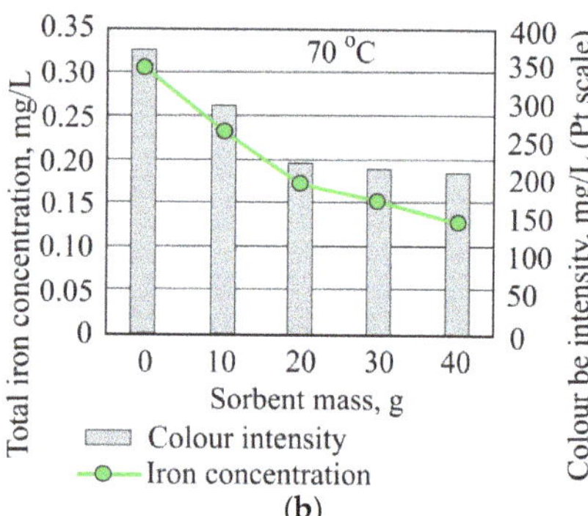


**Figure 15.** Total iron concentration and color intensity when investigated water is at (**a**) 21 °C and (**b**) 70 °C using granulated mineral sorbent.

It was also noted that when using hydrophobic peat sorbent for experiments with sufficiently high chemical oxygen demand values, higher measures of oil product concentration are found, while measuring a lower value the chemical oxygen demand is also determined as being lower (Figure 16).

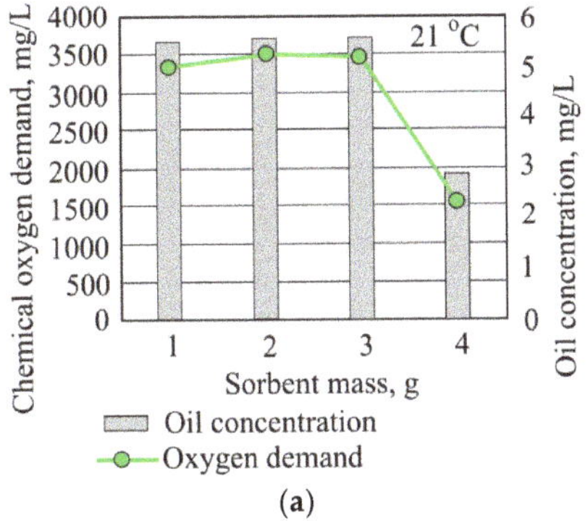


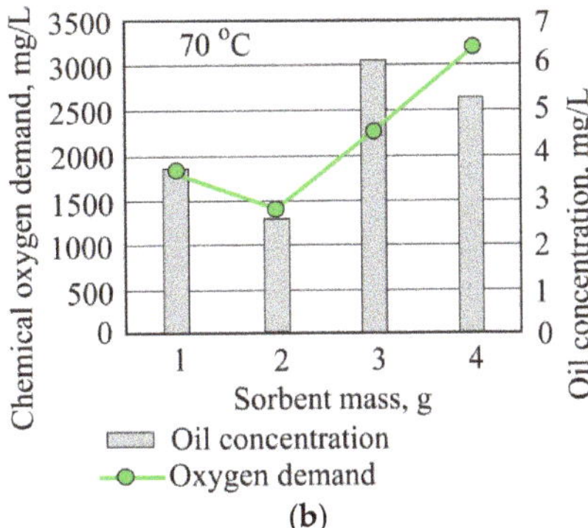


**Figure 16.** Chemical oxygen demand and oil concentration when investigated water is at (**a**) 21 °C and (**b**) 70 °C using hydrophobic peat sorbent.

The use of mineral granular sorbent for research produces a noticeable pattern (Figure 17) between chemical oxygen demand and oil concentration in water, similar to what is found when using hydrophobic peat sorbent.

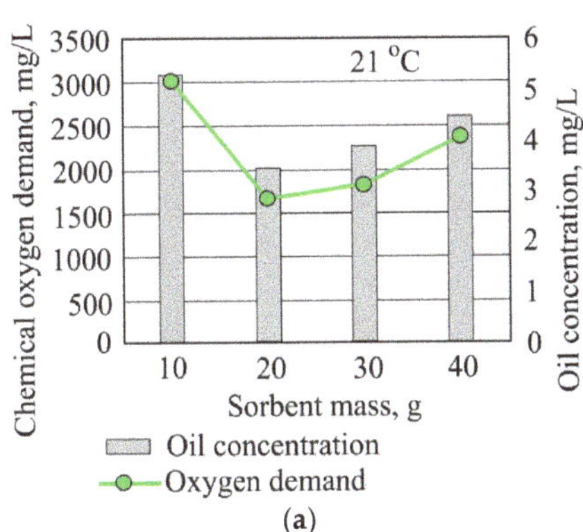


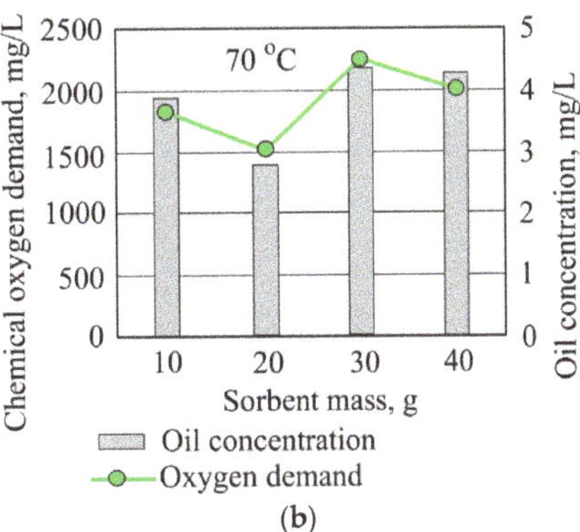


**Figure 17.** Chemical oxygen demand and oil concentration when investigated water is at (**a**) 21 °C and (**b**) 70 °C using granulated mineral sorbent.

The dosage effect is the capacity of selected sorbent, or the quantity required to remove an amount of targeted pollutant. The mechanism could be explained according to the total amount of investigated materials, e.g., for the biggest dosage, the highest obtained removal was 37 mg of oil per one gram of hydrophobic peat sorbent. In some cases, overdosing phenomena occurred with undesirable effects, i.e., increased concentrations values on the investigated materials. This was illustrated by increased turbidity values and color intensity when larger quantities of straw sorbent were added to the wastewater. The dosage effect on sorption performance or capacity is developed by equilibrium data of the model designed. Industrial technologists were evaluating color intensity and total iron concentration in preferably lowest values, thereafter the dosage effect was evaluated as efficient.

The authors of this article were looking for the most efficient sorbent and which mineral granulated sorbent is appropriate due to it being a wasted material from production sites, while building construction sectors are only interested in limited size fractions. Carbon based exhausted sorbents are very expensive to reuse or to manage in the environment [17,18]. Peat sorbent was investigated in the wastewater treatment plant of oil reservoir washing pollutant removal and it only provided some useful results. Some researchers were working with oil polluted water treatment from a related recycling and reuse of car wash water [19–21] and were interested in the efficient sorbent usage and they continue research with wasted material and similar sorbents. According to the Environmental Law of European Union we proposed to incorporate exhausted mineral sorbent in cement production materials for local market use.

The influence of the concentration on oil retention can be explained by the phenomenon of concentration polarization [22]. Generally, chemical sorption and catalytic processes allow us to have partial and selective removal in addition to an important production flow, which makes the sorption technique competitive when compared to other processes [23,24].

The combination of filtration technique with sorption process can lead us to an optimal treatment of oil and other pollutants from the reused water of oily reservoirs washing processes, with a lower cost than when using synthetic sorbents and without harming the environment [25,26]. Whereas many agricultural residue-derived sorbents have been reported [26,27], the breakthrough was explored by using porous Fe/C bio-char adsorbent and a novel phenolic foam-derived magnetic carbon foam.

## 4. Conclusions

1. The investigated sorbents can be used for the treatment of target wastewater. The selection of sorbents handling method will depend on the quantity of sorbents to be obtained from local sources: straw sorbent seasonal availability, peat excavating frequency, and wasted mineral sorbent accumulating at the cement production factory. Depending on the nature of sorbents, however, after the oily wastewater treatment, regenerating or additional methods of disposal may be necessary. The promising results were obtained with the mineral sorbent and future research will be focused on batch-type processes.
2. At 70 °C, the sorption capability of iron in the investigated water using mineral sorbent was reduced by 2.5 times when compared to water at 21 °C.
3. Raising the temperature of the water from 21 °C to 70 °C, lowers the effectiveness of all pollutant removal under investigation by 2.0–15.0% on average.
4. Removing total iron from water weakens the color of the investigated water due to trivalent iron ions, giving the water a pink or brown tinge.
5. The effect of the characteristics of the materials on the adsorption capacity was evaluated using removed amount of oil per one gram of every sorbent. Straw sorbent oil adsorption capacity was up to 33 mg/g; peat sorbent 37 mg/g, and mineral sorbent 1.83 mg/g.

**Author Contributions:** Conceptualization, R.Z. and M.V.; methodology, R.Z.; software, R.Z.; validation, M.V. and R.Z.; formal analysis, R.Z.; investigation, R.Z.; resources, M.V.; data curation, R.Z.; writing—original draft preparation, M.V.; writing—review and editing, R.Z. and M.V.; visualization, R.Z.; supervision, R.Z. and M.V.; project administration, R.Z.; funding acquisition, R.Z. All authors have read and agreed to the published version of the manuscript.

**Funding:** This research received no external funding.

**Data Availability Statement:** The data presented in this study are available on request from the corresponding author.

**Acknowledgments:** The authors express their appreciation for the significant editing contribution of the NGO "Materials users and producers association", a professional team and authors of numerous publications at Vilnius (Lithuania) and individually to Ronald Ringer.

**Conflicts of Interest:** The authors declare no conflict of interest.

## References


1. Hashim, N.H.; Zayadi, N. Pollutants Characterization of Car Wash Wastewater. *MATEC Web Conf.* **2016**, *47*, 5008. [CrossRef]
2. Lau, W.J.; Ismail, A.F.; Firdaus, S. Car wash industry in Malaysia: Treatment of car wash effluent using ultrafiltration and nanofiltration membranes. *Sep. Purif. Technol.* **2013**, *104*, 26–31. [CrossRef]
3. Nekrasov, V. Modern efficient methods of steel vertical oil tanks clean-up. *MATEC Web Conf.* **2016**, *86*, 4050. [CrossRef]
4. Ucar, D. Membrane processes for the reuse of car washing wastewater. *J. Water Reuse Desalination* **2018**, *8*, 169. [CrossRef]
5. Jamaly, S.; Giwa, A.; Hasan, S.W. Recent improvements in oily wastewater treatment: Progress, challenges and future opportunities. *J. Environ. Sci.* **2015**, *37*, 15–30. [CrossRef]
6. Boluarte, I.A.R.; Andersen, M.; Pramanik, B.K.; Chang, C.; Bagshaw, S.; Farago, L.; Jegatheesan, V.; Shu, L. Reuse of car wash wastewater by chemical coagulation and membrane bioreactor treatment processes. *Int. Biodeterior. Biodegrad.* **2016**, *113*, 44–48. [CrossRef]
7. Medyńska-Juraszek, A.; Álvarez, M.L.; Białowiec, A.; Jerzykiewicz, M. Characterization and Sodium Cations Sorption Capacity of Chemically Modified Biochars Produced from Agricultural and Forestry Wastes. *Materials* **2021**, *14*, 4714. [CrossRef]
8. Huang, S.; Ras RH, A.; Tian, X. Antifouling membranes for oily wastewater treatment: Interplay between wetting and membrane fouling. *Colloid Interface Sci.* **2018**, *36*, 90–109. [CrossRef]
9. Mohammadi, M.J.; Takdastan, A.; Jorfi, S.; Neisi, A.; Farhadi, M.; Yari, A.R.; Dobaradaran, S.; Khaniabadi, Y.O. Electrocoagulation process to Chemical and Biological Oxygen Demand treatment from carwash grey water in Ahvaz megacity, Iran. *Data Brief* **2017**, *11*, 634–639. [CrossRef]
10. Baiseitov, D.A.; Tulepov, M.I.; Sassykova, L.R.; Gabdrashova, S.E.; Gul'dana, E.; Zhumbai, D.A.; Kudaibergenov, K.K.; Mansurov, Z.A. The Sorbents for Collection of Oil and Petroleum of the Phytogenesis. *Int. J. Chem. Sci.* **2015**, *13*, 1027–1033.
11. Husseien, M.; Amer, A.A.; El-Maghraby, A.; Taha, N.A. Availability of barley straw application on oil spill clean up. *Int. J. Environ. Sci. Technol.* **2009**, *6*, 123–130. [CrossRef]
12. ISO 15705:2002*Water quality—Determination of the Chemical Oxygen Demand Index (ST-COD)—Small-Scale Sealed-Tube Method*, Technical Committee ISO/TC 147/SC 2–Secretariat: Berlin, Germany, 2002.
13. Sobik-Szołtysek, J.; Wystalska, K.; Malińska, K.; Meers, E. Influence of Pyrolysis Temperature on the Heavy Metal Sorption Capacity of Biochar from Poultry Manure. *Materials* **2021**, *14*, 6566. [CrossRef] [PubMed]
14. Akpoveta, O.V.; Osakwe, S.A. Determination of Heavy Metal Content in Refined Petroleum Products. *J. Appl. Chem.* **2014**, *7*, 01–02.
15. Banerjee, S.S.; Joshi, M.V.; Jayaram, R.V. Treatment of oil spill by sorbtion technique using fatty acid grafted sawdust. *Chemosphere* **2006**, *64*, 1026–1031. [CrossRef] [PubMed]
16. Bucurioiu, R.; Petrache, M.; Vlasceanu, V.; Petrescu, M.G. Study on oil wastewater treatment with polymeric reagents. *Sci. Study Res.* **2016**, *17*, 562.
17. Fadali, O.A.; Ebrahiem, E.E.; Farrag, T.E.; Mahmoud, M.S.; El-Gamil, A. Treatment of oily wastewater produced from refinery processes using adsorbtion technique. *Minia J. Eng. Technol. MJET* **2013**, *32*, 88–101.
18. Yavuz, Y.; Koparal, A.S.; Ogutveren, U.B. Treatment of petroleum refinery wastewater by elektrochemical methods. *Desalination* **2010**, *258*, 201–205. [CrossRef]
19. Marcus, A.C.; Ekpete, O.A. Impact of Discharged process Wastewater from an Oil Refinery on the Physicochemical Quality of a Receiving Waterbody in River State, Nigeria. *J. Appl. Chem.* **2014**, *7*, 01–08.
20. Murari, H.V. *Recycling and Reuse of Car Wash Water*; University Of Mumbai: Mumbai, India, 2014; pp. 1–45.
21. Ngamlerdpokin, K.; Kumjadpai, S.; Chatanon, P.; Tungmanee, U.; Chuenchunchom, S.; Jaruwat, P.; Lertsathitphongs, P.; Hunsom, M. Remediation of biodiesel wastewater by chemical- and elektro-coagulation: A comparative study. *J. Environ. Manag.* **2011**, *92*, 2454–2460. [CrossRef]
22. Sose, A.T.; Kulkarni, S.J.; Sose, M.T. Oil Industry–Analysis, Effects and Removal of Heavy Metals. *Int. J. Eng. Sci. Res. Technol.* **2017**, *6*, 254–257.
23. Tony, M.A.; Purcell, P.J.; Zhao, Y.Q.; Tayeb, A.M.; El-Sherbiny, M.F. Photo-catalytic degradation of an oil-water emulsion using the photo-fenton treatment process: Effects and statistical optimization. *J. Environ. Sci. Health Part A* **2009**, *44*, 179–187. [CrossRef] [PubMed]
24. Valentukevičienė, M.; Bagdžiūnaitė-Litvinaitienė, L.; Chadyšas, V.; Litvinaitis, A. Evaluating the impacts of integrated pollution on water quality of the trans-boundary Neris (Viliya) river. *Sustainability* **2018**, *10*, 4239. [CrossRef]
25. Wokoma, O.; Edori, O. Heavy metals content of an oily wastewater effluent from an oil firm at the point of discharge. *Int. J. Chem. Pharm. Technol.* **2017**, *2*, 154–161.
26. Zhang, Y.; Wang, Q.; Li, R.; Lou, Z.; Li, Y. A Novel Phenolic Foam-Derived Magnetic Carbon Foam Treated as Adsorbent for Rhodamine B: Characterization and Adsorption Kinetics. *Crystals* **2020**, *10*, 159. [CrossRef]
27. Zhang, Y.; Lou, Z.; Wang, C.; Wang, W.; Cai, J. Synthesis of Porous Fe/C Bio-Char Adsorbent for Rhodamine B from Waste Wood: Characterization, Kinetics and Thermodynamics. *Processes* **2019**, *7*, 150. [CrossRef]

 materials

*Review*

# Recycling of Carbon Fiber-Reinforced Composites—Difficulties and Future Perspectives

**Dragana Borjan [1], Željko Knez [1,2] and Maša Knez [1,*]**

[1] Laboratory for Separation Processes and Product Design, Faculty of Chemistry and Chemical Engineering, University of Maribor, Smetanova ulica 17, 2000 Maribor, Slovenia; dragana.borjan@um.si (D.B.); zeljko.knez@um.si (Ž.K.)

[2] Laboratory for Chemistry, Faculty of Medicine, University of Maribor, Taborska ulica 8, 2000 Maribor, Slovenia

* Correspondence: masa.knez@um.si

**Abstract:** Carbon fiber-reinforced composites present an exciting combination of properties and offer clear advantages that make them a perfect replacement for a spread of materials. Consequently, in recent years, their production has dramatically increased as well as the quantity of waste materials. As future legislations are likely to prevent the use of landfills and incineration to dispose of composite waste, alternative solutions such as recycling are considered as one of the urgent problems to be settled. This study presents the leading technologies for recycling carbon fiber-reinforced composites, focusing on chemical recycling using sub- and supercritical fluids. These new reaction media have been demonstrated to be more manageable and efficient in recovering clean fibers with good mechanical properties. The conventional technologies of carbon fibers recycling have also been reviewed and described with both advantages and drawbacks.

**Keywords:** composite materials; carbon fiber-reinforced composites; recycling techniques; chemical recycling; subcritical fluids; supercritical fluids; supercritical alcohols

**Citation:** Borjan, D.; Knez, Ž.; Knez, M. Recycling of Carbon Fiber-Reinforced Composites—Difficulties and Future Perspectives. *Materials* **2021**, *14*, 4191. https://doi.org/10.3390/ma14154191

Academic Editors: Andrea Petrella and Michele Notarnicola

Received: 30 June 2021
Accepted: 23 July 2021
Published: 27 July 2021



## 1. Introduction

As carbon fiber-reinforced composites present an exciting combination of properties such as corrosion resistance, durability, low thermal expansion, high strength-to-weight ratios, and strength, their demand has increased in many industrial fields over the past few decades, from architecture to infrastructure and automotive [1,2]. They offer clear advantages that make them a perfect replacement for a spread of materials, including aluminum, granite, steel, and wood. Consequently, composites are fast becoming the material of choice [3]. In 2019, the European composites market size was approximately USD 17.88 billion. Expansion at a compound growth rate of 7.5% per year from 2019 to 2025 is anticipated. Therefore, reaching USD 27.54 billion by 2025 is expected [4]. However, each sector does not manifest the same interest in composites [1]. To illustrate, in aerospace and aircraft, the selection of materials is motivated by their performance and fuel efficiency [5]. This makes the high stiffness and relatively low weight of carbon fibers a desirable replacement. Conversely, in general engineering and surface transportation, the employment of carbon fibers is determined by cost constraints, usually less critical performance need, and high production rate requirements.

The crucial drawback of composites is that they are challenging to recycle, especially carbon fiber-reinforced composites due to their hardness and chemical stability [6]. However, the significant commercial value of carbon fiber-reinforced composites recycling lies in recovering long high modulus fibers with a high intrinsic value for their reuse in high-grade applications. Turning carbon fiber-reinforced composites waste into a valuable resource and shutting the loop in their life cycle is vital for the continued use of the material in some applications [7,8].

All of this motivates new research and developments to enhance the recyclability of composite materials. Some review papers on composite recycling are available in the literature [9–14], but composite recycling remains a relatively new area. Currently, chemical, mechanical, and thermal methods have been used to recycle composites [15,16]. A chemical process is used to recover both the clean fibers and fillers and depolymerized matrix in monomers or petrochemical feedstock. Mechanical recycling is mainly based on crushing, grinding, milling, and shredding the composite part into smaller pieces, which can then be further ground into powder. Thermal recycling involves heat breaking the scrap composite and combusting the resin matrix, thereby recovering the carbon fibers. This paper seeks to examine the implementation of engineering optimization techniques in composite recycling focused on novel techniques that use sub- and supercritical fluids.

## 2. Difficulties with Composite Disposal

The utilization of composite materials has simplified modern life. Still, the extensive use in every industrial segment has caused a rapidly increasing amount of waste and high issues globally despite all its advantages. Additionally, most carbon fiber-reinforced composites produced thus far are still operational. Still, familiar sources of waste include end-of-life components, manufacturing cut-offs, out-of-date prepregs, production tools, as well as testing materials.

The conventional way to handle composite waste is incineration or disposal in landfills. However, new European waste directives on landfills and incineration will pressure these traditional disposal routes for composite materials. To preserve the environment, legislation must be instigated, usually combined with economic instruments, such as taxes, to enforce recycling [17].

Environmental, as well as economic, are the crucial aspects for recycling directions today [18]. Thermoplastics have a possible environmental concern since they are mainly non-biodegradable and challenging to obliterate naturally [19]. On the other hand, materials reinforced with virgin carbon fiber-reinforced composite are not likely to be used in many products because of the high cost. Carbon fibers represent an expensive raw material; their price as of 2020 is 5–20 USD/square meters [20]. These days, researchers concentrate on looking for new composite materials and developing better, optimized fabrication processes to lower the fabrication costs and upgrade the material properties [21,22]. However, excluding future issues with composite disposal, it is also economically favorable to implement composite recycling. Recycling activities supply the potentiality to use low-priced carbon fibers for applications that do not demand high strength and eventually open sustainable and secure sources of carbon fiber material. Concern for the environment, limiting the utilization of finite resources, and the need to manage waste disposal, have led to increasing pressure to recycle materials at the end of their useful lives [23]. To ease the continued use of the composite material in some applications, such as wind turbine blades and the automotive industry (e.g., the BMW Group uses recycled carbon fibers for the manufacturing of the reinforcement of the C-pillar with sheet molding compound, and Toyota uses Mitsubishi Rayon's sheet molding compound material for the manufacturing of the hatch door frame), it is crucial to modify composite waste into a valuable resource and close the loop within the composite life cycle [7,24,25]. Certainly, the enormous use of composite materials, thanks to their outstanding characteristics, leads to a rise in the quantity of waste produced.

Even though there are many valuable utilizations for thermoset composite materials, recycling at the end of the life cycle could also be a more complex issue [26]. Nonetheless, the perceived insufficiency of recyclability is increasingly essential and seen as a critical barrier for developing or even using composite materials in some markets. In addition to their specific issues, there are other problems related to recycling any material from end-of-life components, such as the necessity to handle contamination and collecting, identifying, sorting, and separating the scrap material (Figure 1). Furthermore, in their publication,

Pompidou et al. (2012) discussed the relationship between recyclers and customers linked by an exchange of carbon fibers at the end-of-life stage [2].

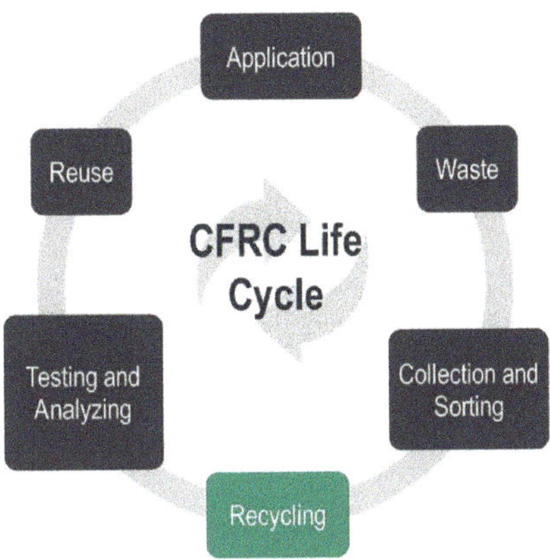


**Figure 1.** A scheme of carbon fiber-reinforced composite (CFRC) life cycle.

As a result of composites' complex composition (fibers, matrix, and fillers), it is tough to fractionate them into elemental components. Composite waste is mainly disposed of in landfills or incinerated with none recycling approaches. Albeit composite waste is relatively inert in regard to other waste (they produce no leachate and methane gas), replacements should be taken to reduce the quantity of waste disposed of and thus bring down the effect on the environment.

## 3. Usual Recycling Procedures

To comply with the legislation, manufacturers of composite products should have products for real recycling solutions. Several methods have been tried and classified into three categories: mechanical grinding to thermal and chemical degradation of resins. Some of the most usual recycling procedures are shown in Figure 2 and described below.

**Figure 2.** The most usual procedures for recycling carbon fiber-reinforced composites.

### 3.1. Mechanical Recycling

Mechanical recycling involves the use of crushing, grinding, milling, and/or shredding techniques. All the constituents of the primary composite are minimized in size to particles with a length from 50 μm to 10 mm [27]. The resulting scrap pieces are mixtures of fiber, polymer, and filler and can be segregated by sieving into powdered products (rich in resin) and fibrous products (rich in fibers) [28,29].

Mechanical recycling can be used as a charge, or partial reinforcement in other products leads to limited incorporation in new materials [30]. It is suitable for scrap composite material that is relatively clean and uncontaminated and from the known origin. It presents few advantages as it recovers both fibers and resins without using or producing hazardous materials. Moreover, this technique is more suitable for glass fiber-reinforced composites. Even though the technologies developed give powder and fibrous recyclates, which can be reused, the powder recyclates have the meager potential for reuse back into the original thermoset compounds. The fibrous recyclates have some prospective reinforcement materials, but they are not as good as virgin reinforcement. Furthermore, there are problems related to the bonding of the recyclate with polymers, and usually, they do not fit the thermosetting polymers.

Finally, the degradation of mechanical properties of the recovered fibers is significant; their architectures are coarse, non-consistent, and unstructured, and hence, the possibilities for their re-manufacturing are limited [11].

### 3.2. Thermal Recycling

Thermal recycling processes involve heat to break the scrap composite down and combust the resin matrix, thereby recovering the carbon fibers. The most used thermal recycling techniques are pyrolysis and fluidized bed procedure.

A fluidized bed operates by balancing the downward gravity forces of the weight of the particles in the charge with the upward details created by the high gas flow. This process implies thermal decomposition of the polymer matrix followed by the release and collection of discrete carbon fibers. The operating temperature of the fluidized bed is chosen to be adequate to cause the polymer to decompose, leaving clean fibers but not too high that it degrades the carbon fiber substantially [31].

Pyrolysis is a thermal decomposition method for polymers at high temperatures from 350 to 700 °C in the absence of oxygen and an inert atmosphere, e.g., $N_2$ [32–34]. It allows the recovery of long, high modules fibers, and because of that, it is one of the most widespread recycling processes. Many factors affect the pyrolysis procedure, including the composition of the waste feedstock, reactor type, and process parameters (heating rate, pressure, residence time, temperature) [35,36]. The operating temperatures have a remarkable effect on the fiber's characteristics (mechanical, electrical, and surface properties) and thus, they need to be adjusted to the type of composites to be treated. A higher temperature can be used, but this will result in some severe degradation of recyclates. Meyer et al. (2009) successfully performed the technical viability of the optimized process parameters using pyrolysis to obtain the reclaimed carbon fibers with properties similar to virgin carbon fibers [37].

Giorgini et al. (2014) used pyrolysis to recover carbon fibers and reported almost total degradation of the polymeric epoxy matrix phase with obtaining solid residue, in which substantially unharmed carbon fibers can be separated, together with oil and gas fractions that can be abused as energy and chemical feedstock [38].

On the other hand, the fluid bed process gives short, recycled carbon fibers in a fluffy form but can treat contaminated wastes with metals. This process produces an excellent fiber product, but it is not in the same form as an existing virgin fiber product.

Although the fluidized bed process is generally more straightforward in theory than pyrolysis, pyrolysis can produce potentially useful organic products from the polymer. There would need to be further processing to isolate them from the mixture of products made, and it seems likely that this would only be cost-effective on an industrial scale.

Even though the thermal recycling processes have the advantage of being able to tolerate more contaminated scrap materials, development work is therefore needed to identify how the material can be reprocessed into cost-effective new products. These may have varying degrees of char on the recycled fibers, limiting the reuse options or requiring further processing to remove them.

The dominances of the pyrolysis process are that there is no utilization of chemical solvents and that all the outcomes can be recovered and reused in one form or another. In general, the gases are reused to obtain demanded energy for the process. Recovered carbon fibers have mechanical properties similar to virgin fibers (from 4 to 20% loss in tensile strength) and can be re-manufactured [11,39]. However, the properties of recycled fibers are susceptible to processing parameters. Finally, pyrolysis is the only process used commercially, but the recycled carbon fibers are chopped or milled.

### 3.3. Chemical Recycling

Two principal demands stimulate the progress of chemical recycling technologies, the demand to safely and efficiently process materials challenging to treat with mechanical recycling, and the request to produce high-quality recycled materials [40]. Solvolysis is a chemical treatment using a solvent to degrade the resin. The solvolysis process can recover both the clean fibers and fillers and depolymerized matrix in the form of monomers or petrochemical feedstock [41]. However, it has a low contamination tolerance (e.g., no metals or painting pieces). Solvolysis can offer many possibilities based on catalysts, pressure, temperature, and a wide range of solvents. Solvolysis can be classified according to higher pressure and temperature (temperature > 200 °C) and lower pressure and temperature (temperature < 200 °C), but generally, lower temperatures are requested to degrade the polymer compared to pyrolysis. In chemical recycling, resin degradation is reached using solvents (solvolysis) or water (hydrolysis). The employment of dangerous and concentrated chemicals results in environmental impact, so water or alcohol, which are relatively environmentally friendly, usually replace harmful chemicals [42]. However, it is essential to emphasize that water can be effectively used only in sub- or supercritical conditions.

To achieve a higher dissolution efficiency and faster dissolution rate, catalysts are typically used, and the solvent characteristics are also tuned and optimized. In addition, the dissolution reagents are used to depolymerize the matrix of composites. The reclaimed fiber retains most of its mechanical properties.

Xu et al. (2013) published the tensile strength of the recovered carbon fibers in more than 95% of the virgin ones according to the single fiber tensile test. A mixed solution of hydrogen peroxide and N,N-dimethylformamide was used in this research. According to reported results observed using a scanning electron microscope, the surface of the carbon fibers was smooth with few residues of epoxy resin [43]. Li et al. (2012) used an efficient solution of acetone and hydrogen peroxide for the chemical recycling of carbon fiber/epoxy composites by oxidative degradation. Clean carbon fibers with tensile strength higher than 95% of their original strength can be secured after reacting at 60 °C for 30 min [44].

Sub- and supercritical fluids have also been considered solvents. Semi-long or long recycled carbon fibers can be produced using sub- and supercritical fluids or catalysts such as benzyl alcohol, which dissolves the resin rapidly during the recycling process [45]. However, the utilization of sub- and supercritical fluids can lead to high conditions. Composite recycling by solvolysis has been researched at ambient conditions but at high temperatures and pressures above 300 °C and 50 bar generally for carbon fiber-reinforced composites. Expensive reactors that can withstand high temperatures and pressures and corrosions are requested when sub- and supercritical conditions are used. As a result, there is a trade-off problem between the cost of facilities and the solvent valuables that needs to be solved before applying the solvolysis at an industrial scale.

On the other hand, catalysts are necessary to enhance the reaction at lower temperatures. Nevertheless, this is not always efficient enough to justify the use of catalysts or strong alkaline or acidic conditions that necessarily lead to fiber damages. At ambient conditions, the recycling treatment was performed in extreme conditions, mostly acidic with solutions that can be dangerous in terms of safety and environment.

## 4. Utilization of Sub- and Supercritical Fluids

Sub- and supercritical fluids have distinctive characteristics and valuable potential that may enhance various chemical process operations of multiple materials [46]. The utilization of sub- and supercritical fluids may replace many environmentally harmful solvents currently used in industry, such as organic solvents. Therefore, high-pressure technologies involving sub- and supercritical fluids serve the opportunity to obtain new products with specific properties or to design new processes that are environmentally friendly and sustainable [47,48]. Figure 3 shows a phase diagram of water with marked sub- and supercritical regions [49]. Sub- and supercritical fluids have been reported to achieve efficient and fast resin degradations thanks to enhanced characteristics: liquid-like high mass transport coefficient and pressure-dependent solvent power and gas-like low viscosity and high diffusivity. A clear sub- or supercritical fluid density is easily changed by relatively small variations in pressure and temperature; the viscosity is relatively low but may increase with temperature, whereas the surface tension is essentially nonexistent. Diffusivity is high, which induces interesting transport phenomena in condensed phases [50]. Sub- and supercritical fluids have recently been used in technological areas to develop green processes and can be classed as green reaction media since they are cost-effective and readily available, and they have low potential toxicity [51–53]. In addition, they can be recycled afterward by distillation and can dissolve many organic and inorganic compounds. In addition to the dissolving of organic materials, sub- and supercritical fluids can penetrate porous solids, which are still relatively innocuous under atmospheric conditions. Sub- and supercritical fluids are great reaction media for the depolymerization or decomposition of polymers as the reaction is fast and selective. Composite materials such as carbon fiber-reinforced composites can be decomposed into smaller molecular components and fiber materials. Even though it is not necessary, the use of catalysts can remarkably upgrade the decomposition reaction [54,55].

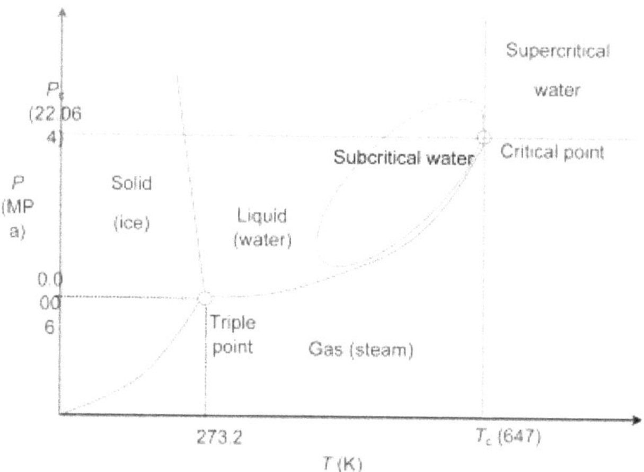


**Figure 3.** Phase diagram of water with marked sub- and supercritical region, where Pc represents critical pressure and $T_C$ critical temperature. Reprinted with permission from ref. [49]. Copyright 2016 Elsevier.

The solvolysis process for carbon fiber-reinforced composites recycling using sub- and supercritical fluids will be reviewed in the following paragraphs. This will manifest the benefits of using this type of solvent regarding its efficiency and reactivity for the decomposition of the polymer matrix and the recovery of the fibers with proper surface and mechanical properties.

Several types of sub- and supercritical fluids (Table 1) have been used for carbon fiber recycling, such as:

- Water [56–58];
- Methanol [59,60];
- Ethanol [61];
- Propanol [62–64];
- Acetone [65].

Under supercritical conditions, water has high diffusivity as well as a high mass transfer coefficient. On the other hand, the viscosity of supercritical water is low, and the dielectric constant is significantly reduced, so the hydrogen bonding essentially disappears. These characteristics allow supercritical water to effectively break down the polymer matrix while causing no damage to fibers [66]. Yuyan et al. (2009) reported that the average tensile strength of the fibers reclaimed using subcritical water was about 98.2% that of the virgin fibers. The scanning electron microscopy and atomic force microscope measurements were employed to observe the fibers' surface, and no cracks or defects were found. In addition, fibers were clean without resin residues [56].

On the other hand, Piñero-Hernanz et al. (2008) studied the potential for recycling carbon fiber-reinforced epoxy composites in the water at supercritical or near-critical conditions. Experiments were done in a batch-type reactor, and the tensile strength of the reclaimed carbon fibers varied between 90% and 98% compared to the virgin fibers [57]. Additionally, supercritical water was used to solvolysis carbon fiber/thermoset matrix and demonstrate the environmental feasibility of composite recycling [58].

In addition, supercritical methanol was reported to decompose epoxy resin used for matrix resin and recycling of the carbon fiber-reinforced composite. Conditions were 250–350 °C and 100 bar for 5–120 min. When the composite was treated in a semiflow reactor, the recovered carbon fiber had no heat damage and maintained the plain fabric shape [59]. Okajima et al. (2014) recovered carbon fibers with a solid close to virgin fibers using supercritical methanol at 270 °C and 80 bar for 90 min [60].

Hyde et al. (2006) used supercritical propanol to extract and eliminate the epoxy resin from the surface of a carbon fiber composite material. The process was effective at a temperature above 450 °C and pressure above 50 bar. In terms of tensile strength, the recycled fibers were almost as strong as the virgin. It indicated that their structural integrity was minimally damaged [62]. Yan et al. (2016) investigated the effects of degradation temperature on the efficiency of recovering carbon fibers from their epoxy resin composites by supercritical 1-propanol. The obtained results indicated that the mechanical properties of the recycled fibers decreased slightly with temperature [63]. Jiang et al. (2009) researched carbon fiber/epoxy resin composites using supercritical n-propanol and obtained a tensile strength of the recycled carbon fiber close to the corresponding as-received carbon fibers [64].

Okajima and Sako (2019) investigated the chemical recycling of carbon fiber-reinforced composites using supercritical acetone. They reported that the decomposition efficiency increased with increasing reaction pressure and acetone density, to a maximum value of 95.6% at 350 °C, 140 bar, and 60 min [65]. Furthermore, Okajima et al. (2019) used subcritical acetone and various supercritical solvents (methanol, ethanol, 1-propanol, 1-butanol, 2-butanol, tert-butanol, acetone, and methyl ethyl ketone) as solvents in a composite recycling procedure at 320 °C in the reaction time range of 6–120 min. The decomposition rate depended on the solvent, but sub- and supercritical acetone were selected as optimal for fast degradation of the matrix resin. The recovered carbon fibers saved the shape of the fabric sheets, and their tensile strength reduction was insignificant [67].

Sokoli et al. (2017) compared the degradation of composites using near-critical water and supercritical acetone and varied parameters such as temperature (in a range from 260 to

300 °C), pressure (from 60 to 300 bar), and composite/solvent ratio (from 0.29 to 2.1 g/mL). They determined the exact conditions for achieving nearly complete degradation of the resin using supercritical acetone. The tensile strength of recovered carbon fibers was retained using solvents, water, and acetone [68].

**Table 1.** Most used sub- and supercritical fluids as a solvent in carbon fiber-reinforced composite recycling.

| Fluid | Critical Data * | Reference |
|---|---|---|
| water | Tc = 647.1 K; Pc = 220.6 bar | [69] |
| methanol | Tc = 512.6 K; Pc = 81.0 bar | [70] |
| ethanol | Tc = 514.0 K; Pc = 61.4 bar | [71] |
| propanol | Tc = 536.8 K; Pc = 52.0 bar | [71] |
| acetone | Tc = 508.0 K; Pc = 48.0 bar | [72] |

* Pc—critical pressure; Tc—critical temperature.

Chemical recycling with sub- and supercritical fluids is a more recent approach; nevertheless, it is already recognized for producing recycled carbon fibers with virtually no mechanical degradation, especially when using sub- and supercritical alcohols and enabling recovering valuable chemicals from the matrix [73]. They are inexpensive and easy to handle, exhibiting low critical pressures (typically 20–60 atm) but high critical temperatures (usually 200–300 °C). However, the extraction process using sub- and supercritical alcohols does not significantly alter the thermal stability of the carbon fibers. Piñero-Hernanz et al. (2008) obtained carbon fibers that retain 85–99% of the strength of the virgin fibers using batch and semi-continuous-type reactors. Subcritical and supercritical alcohols (methanol, ethanol, 1-propanol, and acetone) were employed as reactive-extraction media in their study [61].

Generally, recycling using sub- and supercritical fluids is a quick and simple method and can be operated semi-continuously [74]. Furthermore, the decomposition of polymers proceeds rapidly and selectively, and the fibers compare favorably to the virgin fibers, with only a slight loss of tensile strength [75,76]. In addition, compared to thermal treatments, preservation of brittleness, length, and orientation is better [77]. Total removal and subsequent recovery of the epoxy resin residue are possible. Resin is immediately removed by dissolution and can be recovered ex situ from the solution by evaporation. The quality of the recovered fibers in terms of surface texture is high; the fibers appear to be utterly free from epoxy resin polymer and have a clean and smooth surface [78]. Furthermore, there is no apparent reduction in diameter or surface scratches on the fibers (the extraction process does not appear to damage the fibers).

## 5. Table Review

Recycling methods and required conditions for their realization, such as temperature, pressure, time, heating rate, atmosphere, and solvent, and obtained tensile strength of recovered fibers compared to virgin fibers, have been reviewed in Table 2.

**Table 2.** Literature summary on most used composite recycling methods.

| Recycling Method | Parameters | Tensile Strength of Recovered Fiber | Reference |
|---|---|---|---|
| Thermal (pyrolysis) | 400–600 °C<br>10 °C/min<br>120 min<br>air and $N_2$ atmosphere | undefined | [37] |
| Thermal (pyrolysis) | 500–600 °C<br>150 min | undefined | [38] |
| Chemical (solvolysis) | 90 °C<br>30 min<br>in a solution of $H_2O_2$/DMF (1:1, $v/v$) | $\tau > 95\%$ | [43] |

Table 2. *Cont.*

| Recycling Method | Parameters | Tensile Strength of Recovered Fiber | Reference |
|---|---|---|---|
| Chemical | 60 °C<br>30 min<br>Ac and $H_2O_2$ | $\tau > 95\%$ | [44] |
| Chemical<br>(subcritical) | 260–290 °C<br>10–400 bar<br>75–105 min<br>$H_2O$ | $\tau = 98.2\%$ | [56] |
| Chemical<br>(near- and supercritical) | 250–400 °C<br>40–270 bar<br>30 min<br>$H_2O$ | $90\% < \tau < 98\%$ | [57] |
| Chemical<br>(supercritical) | 400 °C<br>250 bar<br>30 min<br>$H_2O$ | undefined | [58] |
| Chemical<br>(supercritical) | 270 °C<br>80 bar<br>90 min<br>MeOH | $\tau = 91\%$ | [60] |
| Chemical<br>(supercritical) | above 450 °C<br>above 50 bar<br>several minutes<br>PrOH | $\tau \approx 95.4\%$ | [62] |
| Chemical<br>(supercritical) | 260–340 °C<br>120 min<br>1-PrOH | $94.6\% < \tau < 95.2\%$ | [63] |
| Chemical<br>(supercritical) | 310 °C<br>52 bar<br>20 min<br>n-PrOH | $88.6\% < \tau < 99.1\%$ | [64] |
| Chemical<br>(supercritical) | PrOH<br>20–140 bar<br>60 min<br>Ac | undefined | [65] |
| Chemical<br>(sub- and supercritical) | 320 °C<br>10 bar<br>20 min<br>MeOH, EtOH, 1-PrOH, 1-BuOH,<br>2-BuOH, tert-BuOH, Ac, MEK | negligible reduction | [67] |
| Chemical<br>(sub- and supercritical) | 300–450 °C<br>47–153 bar<br>15.5 min<br>MeOH, EtOH, 1-PrOH, Ac | $85\% < \tau < 99\%$ | [61] |

## 6. Conclusions and Future Perspectives

Up-to-date changes to waste management legislation and possible future directions mean that recycling directions need to be in place. The composite industry is under intense pressure to provide viable recycling scenarios for their materials if they want to continue to have a place in the market. A review of different existing recycling technologies for carbon fiber-reinforced composites has been presented in this paper. For each recycling technology, specific advantages and drawbacks have been given.

Chemical recycling gives the recovery of long fibers with good mechanical properties in regard to mechanical recycling. Accordingly, fibers resulting from this recycling technique can be reused in the re-manufacturing of new composites. The loop in the carbon fiber-reinforced composite life cycle can then be closed properly.

However, chemical processes may employ solvents that can have adverse effects on the environment and human health. Thus, to reduce these impacts, sub- and supercritical fluids have been used as green reaction media to solve carbon fiber-reinforced composites. Compared to traditional recycling techniques, solvolysis is an excellent alternative to recover fibers. It provides high mechanical properties and fiber length and a high potential for material recovery from resin. Even though this process has shown excellent results for recycling carbon fiber-reinforced composites at the laboratory scale, further investigations need to be managed to design a cost-effective pilot plant to be used at an industrial scale.

Further research studies have been undertaken as this material has been used more frequently in new applications. Research and developments on emerging green technologies using sub- and supercritical fluids are anticipated to expand soon. Overall, the recycling of carbon fiber-reinforced composites using solvolysis in sub- and supercritical conditions has been proven to be a great alternative, and it has been highlighted in this review.

The topic that is discussed is innovative and sustainable. The major limitation on the way to its (possible) industrial application is the scaling-up. This challenge is related to the knowledge of the basic process properties such as thermodynamic and mass transfer data. This should be determined for each system experimentally at the conditions pertaining to the process condition. Therefore, this process is challenging and requires fundamental research of thermodynamic and mass transfer data.

**Author Contributions:** D.B. prepared the concept of the work, performed most of an extensive literature search, and wrote most of the paper. M.K. devised the content of the article and, with Ž.K. supervised the writing. Ž.K. arranged the financial management of the project. All authors have read and agreed to the published version of the manuscript.

**Funding:** The authors would like to acknowledge the Slovenian Research Agency (ARRS) for financing this research in the frame of ARRS Program P2-0046 and Laboratory for Separation Processes and Product Design 0794-007.

**Institutional Review Board Statement:** Not applicable.

**Informed Consent Statement:** Not applicable.

**Data Availability Statement:** Data sharing is not applicable.

**Acknowledgments:** The authors would like to acknowledge the Slovenian Research Agency (ARRS) for financing this research in the frame of ARRS Program P2-0046 and Laboratory for Separation Processes and Product Design 0794-007.

**Conflicts of Interest:** The authors declare no conflict of interest.

## Abbreviations

| | |
|---|---|
| Ac | acetone |
| BuOH | butanol |
| DMF | N,N-dimethylformamide |
| EtOH | ethanol |
| $H_2O$ | water |
| $H_2O_2$ | hydrogen peroxide |
| MEK | methyl ethyl ketone |
| MeOH | methanol |
| $N_2$ | nitrogen |
| PrOH | propanol |
| $\tau$ | tensile strength of the recovered carbon fiber |

## References


1. Chalaye, H. Chalaye Composite Materials: Drive and Innovation 2002. Available online: http://www.epsilon.insee.fr/jspui/bitstream/1/56266/1/4p158-anglais.pdf (accessed on 27 January 2021).
2. Pompidou, S.; Prinçaud, M.; Perry, N.; Leray, D. Recycling of carbon fiber: Identification of bases for a synergy between recyclers and designers. In *Advanced Composite Materials and Processing; Robotics; Information Management and PLM; Design Engineering*; American Society of Mechanical Engineers: Nantes, France, 2012; Volume 3, pp. 551–560.
3. Pickering, S.J. Recycling thermoset composite materials. In *Wiley Encyclopedia of Composites*; John Wiley & Sons, Inc.: Hoboken, NJ, USA, 2012; p. weoc214. ISBN 978-1-118-09729-8.
4. Europe Composites Market Size; Industry Growth Report, 2019–2025. Available online: https://www.grandviewresearch.com/industry-analysis/europe-composites-market (accessed on 27 January 2021).
5. McConnell, V.P. Launching the carbon fibre recycling industry. *Reinf. Plast.* **2010**, *54*, 33–37. [CrossRef]
6. Navarro, C.A.; Giffin, C.R.; Zhang, B.; Yu, Z.; Nutt, S.R.; Williams, T.J. A structural chemistry look at composites recycling. *Mater. Horiz.* **2020**, *7*, 2479–2486. [CrossRef]
7. Chen, J.; Wang, J.; Ni, A. Recycling and reuse of composite materials for wind turbine blades: An overview. *J. Reinf. Plast. Compos.* **2019**, *38*, 567–577. [CrossRef]
8. Lee, C.-K.; Kim, Y.-K.; Pruitichaiwiboon, P.; Kim, J.-S.; Lee, K.-M.; Ju, C.-S. Assessing Environmentally friendly recycling methods for composite bodies of railway rolling stock using life-cycle analysis. *Transp. Res. Part D Transp. Environ.* **2010**, *15*, 197–203. [CrossRef]
9. Liu, Y.; Farnsworth, M.; Tiwari, A. A review of optimisation techniques used in the composite recycling area: State-of-the-art and steps towards a research agenda. *J. Clean. Prod.* **2017**, *140*, 1775–1781. [CrossRef]
10. Oliveux, G.; Dandy, L.O.; Leeke, G.A. Current status of recycling of fibre reinforced polymers: Review of technologies, reuse and resulting properties. *Prog. Mater. Sci.* **2015**, *72*, 61–99. [CrossRef]
11. Pimenta, S.; Pinho, S.T. Recycling carbon fibre reinforced polymers for structural applications: Technology review and market outlook. *Waste Manag.* **2011**, *31*, 378–392. [CrossRef] [PubMed]
12. Verma, S.; Balasubramaniam, B.; Gupta, R.K. Recycling, Reclamation and re-manufacturing of carbon fibres. *Curr. Opin. Green Sustain. Chem.* **2018**, *13*, 86–90. [CrossRef]
13. Zhang, J.; Chevali, V.S.; Wang, H.; Wang, C.-H. Current status of carbon fibre and carbon fibre composites recycling. *Compos. B Eng.* **2020**, *193*, 108053. [CrossRef]
14. Anane-Fenin, K.; Akinlabi, E.T. Recycling of Fibre Reinforced Composites: A Review of Current Technologies. In Proceedings of the 4th International Conference on Development and Investment in Infrastructure—Strategies for Africa, Livingstone, Zambia, 30 August–1 September 2017; p. 12.
15. Pakdel, E.; Kashi, S.; Varley, R.; Wang, X. Recent progress in recycling carbon fibre reinforced composites and dry carbon fibre wastes. *Resour. Conserv. Recycl.* **2021**, *166*, 105340. [CrossRef]
16. A Review on the Recycling of Waste Carbon Fibre/Glass Fibre-Reinforced Composites: Fibre Recovery, Properties and Life-Cycle Analysis. Available online: https://link.springer.com/article/10.1007/s42452-020-2195-4 (accessed on 18 May 2021).
17. Pickering, S.J. Recycling technologies for thermoset composite materials—Current status. *Compos. A Appl. Sci. Manuf.* **2006**, *37*, 1206–1215. [CrossRef]
18. Meng, F.; Olivetti, E.A.; Zhao, Y.; Chang, J.C.; Pickering, S.J.; McKechnie, J. Comparing Life cycle energy and global warming potential of carbon fiber composite recycling technologies and waste management options. *ACS Sustain. Chem. Eng.* **2018**, *6*, 9854–9865. [CrossRef]
19. La Rosa, A.D.; Banatao, D.R.; Pastine, S.J.; Latteri, A.; Cicala, G. Recycling treatment of carbon fibre/epoxy composites: Materials recovery and characterization and environmental impacts through life cycle assessment. *Compos. B Eng.* **2016**, *104*, 17–25. [CrossRef]
20. TradeWheel. Carbon Fiber Price per Kg. Available online: https://www.tradewheel.com/p/carbon-fiber-price-per-kg-665053/ (accessed on 8 September 2020).
21. Kim, Y.N.; Kim, Y.-O.; Kim, S.Y.; Park, M.; Yang, B.; Kim, J.; Jung, Y.C. Application of supercritical water for green recycling of epoxy-based carbon fiber reinforced plastic. *Compos. Sci. Technol.* **2019**, *173*, 66–72. [CrossRef]
22. Perry, N.; Bernard, A.; Laroche, F.; Pompidou, S. Improving design for recycling—Application to composites. *CIRP Ann.* **2012**, *61*, 151–154. [CrossRef]
23. Utekar, S.; Suriya, V.K.; More, N.; Rao, A. Comprehensive study of recycling of thermosetting polymer composites—driving force, challenges and methods. *Compos. B Eng.* **2021**, *207*, 108596. [CrossRef]
24. Pillain, B.; Loubet, P.; Pestalozzi, F.; Woidasky, J.; Erriguible, A.; Aymonier, C.; Sonnemann, G. Positioning supercritical solvolysis among innovative recycling and current waste management scenarios for carbon fiber reinforced plastics thanks to comparative life cycle assessment. *J. Supercrit. Fluids* **2019**, *154*, 104607. [CrossRef]
25. Bledzki, A.K.; Seidlitz, H.; Krenz, J.; Goracy, K.; Urbaniak, M.; Rösch, J.J. Recycling of carbon fiber reinforced composite polymers—Review—Part 2: Recovery and application of recycled carbon fibers. *Polymers* **2020**, *12*, 3003. [CrossRef] [PubMed]
26. Morales Ibarra, R. Recycling of thermosets and their composites. In *Thermosets*; Elsevier: Amsterdam, The Netherlands, 2018; pp. 639–666. ISBN 978-0-08-101021-1.

27. Yang, Y.; Boom, R.; Irion, B.; van Heerden, D.-J.; Kuiper, P.; de Wit, H. Recycling of composite materials. *Chem. Eng. Process. Process. Intensif.* **2012**, *51*, 53–68. [CrossRef]
28. Thomas, R.; Vijayan, P.; Thomas, S. Recycling of thermosetting polymers: Their blends and composites. In *Recent Developments in Polymer Recycling*; Fainleib, A., Grigoryeva, O., Eds.; Transworld Research Network: Trivandrum, India, 2012; Chapter 4; p. 33.
29. Palmer, J.; Ghita, O.R.; Savage, L.; Evans, K.E. Successful closed-loop recycling of thermoset composites. *Compos. A Appl. Sci. Manuf.* **2009**, *40*, 490–498. [CrossRef]
30. Howarth, J.; Mareddy, S.S.R.; Mativenga, P.T. Energy intensity and environmental analysis of mechanical recycling of carbon fibre composite. *J. Clean. Prod.* **2014**, *81*, 46–50. [CrossRef]
31. Pickering, S.J.; Yip, H.; Kennerley, J.R.; Kelly, R.; Rudd, C.D. The recycling of carbon fibre composites using a fluidised bed process. In *FRC 2000—Composites for the Millennium*; Elsevier: Amsterdam, The Netherlands, 2000; pp. 565–572. ISBN 978-1-85573-550-7.
32. Torres, A.; de Marco, I.; Caballero, B.M.; Laresgoiti, M.F.; Legarreta, J.A.; Cabrero, M.A.; González, A.; Chomón, M.J.; Gondra, K. Recycling by pyrolysis of thermoset composites: Characteristics of the liquid and gaseous fuels obtained. *Fuel* **2000**, *79*, 897–902. [CrossRef]
33. Cheung, K.-Y.; Lee, K.-L.; Lam, K.-L.; Chan, T.-Y.; Lee, C.-W.; Hui, C.-W. Operation strategy for multi-stage pyrolysis. *J. Anal. Appl. Pyrolysis* **2011**, *91*, 165–182. [CrossRef]
34. Naqvi, S.R.; Prabhakara, H.M.; Bramer, E.A.; Dierkes, W.; Akkerman, R.; Brem, G. A Critical review on recycling of end-of-life carbon fibre/glass fibre reinforced composites waste using pyrolysis towards a circular economy. *Resour. Conserv. Recycl.* **2018**, *136*, 118–129. [CrossRef]
35. Pohjakallio, M.; Vuorinen, T.; Oasmaa, A. Chemical routes for recycling—Dissolving, catalytic, and thermochemical technologies. In *Plastic Waste and Recycling*; Elsevier: Amsterdam, The Netherlands, 2020; pp. 359–384. ISBN 978-0-12-817880-5.
36. Lam, K.-L.; Oyedun, A.O.; Cheung, K.-Y.; Lee, K.-L.; Hui, C.-W. Modelling pyrolysis with dynamic heating. *Chem. Eng. Sci.* **2011**, *66*, 6505–6514. [CrossRef]
37. Meyer, L.O.; Schulte, K.; Grove-Nielsen, E. CFRP-recycling following a pyrolysis route: Process optimization and potentials. *J. Compos. Mater.* **2009**, *43*, 1121–1132. [CrossRef]
38. Giorgini, L.; Benelli, T.; Mazzocchetti, L.; Leonardi, C.; Zattini, G.; Minak, G.; Dolcini, E.; Tosi, C.; Montanari, I. Pyrolysis as a way to close a CFRC life cycle: Carbon fibers recovery and their use as feedstock for a new composite production. In Proceedings of the 7th International Conference on Times of Polymers and Composites (TOP), Ischia, Italy, 22–26 June 2014; pp. 354–357.
39. Morin, C.; Loppinet-Serani, A.; Cansell, F.; Aymonier, C. Near- and supercritical solvolysis of Carbon Fibre Reinforced Polymers (CFRPs) for recycling carbon fibers as a valuable resource: State of the art. *J. Supercrit. Fluids* **2012**, *66*, 232–240. [CrossRef]
40. Jiang, J.; Deng, G.; Chen, X.; Gao, X.; Guo, Q.; Xu, C.; Zhou, L. On the successful chemical recycling of carbon fiber/epoxy resin composites under the mild condition. *Compos. Sci. Technol.* **2017**, *151*, 243–251. [CrossRef]
41. Yildirir, E.; Onwudili, J.A.; Williams, P.T. Recovery of carbon fibres and production of high quality fuel gas from the chemical recycling of carbon fibre reinforced plastic wastes. *J. Supercrit. Fluids* **2014**, *92*, 107–114. [CrossRef]
42. Oliveux, G.; Dandy, L.; Leeke, G.A. A Step Change in the Recycling of Composite Materials. In Proceedings of the 16th European Conference on Composite Materials, Seville, Spain, 22–26 June 2014; p. 8.
43. Xu, P.; Li, J.; Ding, J. Chemical Recycling of carbon fibre/epoxy composites in a mixed solution of peroxide hydrogen and N,N-dimethylformamide. *Compos. Sci. Technol.* **2013**, *82*, 54–59. [CrossRef]
44. Li, J.; Xu, P.-L.; Zhu, Y.-K.; Ding, J.-P.; Xue, L.-X.; Wang, Y.-Z. A promising strategy for chemical recycling of carbon fiber/thermoset composites: Self-accelerating decomposition in a mild oxidative system. *Green Chem.* **2012**, *14*, 3260. [CrossRef]
45. Morales Ibarra, R.; Sasaki, M.; Goto, M.; Quitain, A.T.; García Montes, S.M.; Aguilar-Garib, J.A. Carbon fiber recovery using water and benzyl alcohol in subcritical and supercritical conditions for chemical recycling of thermoset composite materials. *J. Mater. Cycles Waste Manag.* **2015**, *17*, 369–379. [CrossRef]
46. Fleming, O.S.; Kazarian, S.G. Polymer processing with supercritical fluids. In *Supercritical Carbon Dioxide*; Kemmere, M.F., Meyer, T., Eds.; Wiley-VCH Verlag GmbH & Co. KGaA: Weinheim, Germany, 2006; pp. 205–238. ISBN 978-3-527-60672-6.
47. Knez, Ž.; Pantić, M.; Cör, D.; Novak, Z.; Knez Hrnčič, M. Are supercritical fluids solvents for the future? *Chem. Eng. Process. Process Intensif.* **2019**, *141*, 107532. [CrossRef]
48. Loppinet-Serani, A.; Aymonier, C.; Cansell, F. Supercritical water for environmental technologies. *J. Chem. Technol. Biotechnol.* **2010**, *85*, 583–589. [CrossRef]
49. Hrnčič, M.K.; Kravanja, G.; Knez, Ž. Hydrothermal treatment of biomass for energy and chemicals. *Energy* **2016**, *116*, 1312–1322. [CrossRef]
50. Brunner, G. Applications of supercritical fluids. *Annu. Rev. Chem. Biomol. Eng.* **2010**, *1*, 321–342. [CrossRef] [PubMed]
51. Bai, Y.; Wang, Z.; Feng, L. Chemical recycling of carbon fibers reinforced epoxy resin composites in oxygen in supercritical water. *Mater. Des.* **2010**, *31*, 999–1002. [CrossRef]
52. Khalil, Y.F. Sustainability assessment of solvolysis using supercritical fluids for carbon fiber reinforced polymers waste management. *Sustain. Prod. Consum.* **2019**, *17*, 74–84. [CrossRef]
53. Machida, H.; Takesue, M.; Smith, R.L. Green chemical processes with supercritical fluids: Properties, materials, separations and energy. *J. Supercrit. Fluids* **2011**, *60*, 2–15. [CrossRef]
54. Huang, H.; Liu, W.; Liu, Z. An additive manufacturing-based approach for carbon fiber reinforced polymer recycling. *CIRP Ann.* **2020**, *69*, 33–36. [CrossRef]

55. Liu, Y.; Liu, J.; Jiang, Z.; Tang, T. Chemical recycling of carbon fibre reinforced epoxy resin composites in subcritical water: Synergistic effect of phenol and KOH on the decomposition efficiency. *Polym. Degrad. Stab.* **2012**, *97*, 214–220. [CrossRef]
56. Yuyan, L.; Guohua, S.; Linghui, M. Recycling of carbon fibre reinforced composites using water in subcritical conditions. *Mater. Sci. Eng. A* **2009**, *520*, 179–183. [CrossRef]
57. Piñero-Hernanz, R.; Dodds, C.; Hyde, J.; García-Serna, J.; Poliakoff, M.; Lester, E.; Cocero, M.J.; Kingman, S.; Pickering, S.; Wong, K.H. Chemical recycling of carbon fibre reinforced composites in nearcritical and supercritical water. *Compos. A Appl. Sci. Manuf.* **2008**, *39*, 454–461. [CrossRef]
58. Prinçaud, M.; Aymonier, C.; Loppinet-Serani, A.; Perry, N.; Sonnemann, G. Environmental Feasibility of the recycling of carbon fibers from CFRPs by solvolysis using supercritical water. *ACS Sustain. Chem. Eng.* **2014**, *2*, 1498–1502. [CrossRef]
59. Okajima, I.; Watanabe, K.; Shimamura, Y.; Awaya, T.; Sako, T. Chemical Recycling of Carbon Fiber Reinforced Plastic with Supercritical Alcohol. Available online: https://citeseerx.ist.psu.edu/viewdoc/download?doi=10.1.1.1040.1841&rep=rep1&type=pdf (accessed on 26 February 2021).
60. Okajima, I.; Hiramatsu, M.; Shimamura, Y.; Awaya, T.; Sako, T. Chemical recycling of carbon fiber reinforced plastic using supercritical methanol. *J. Supercrit. Fluids* **2014**, *91*, 68–76. [CrossRef]
61. Piñero-Hernanz, R.; García-Serna, J.; Dodds, C.; Hyde, J.; Poliakoff, M.; Cocero, M.J.; Kingman, S.; Pickering, S.; Lester, E. Chemical recycling of carbon fibre composites using alcohols under subcritical and supercritical conditions. *J. Supercrit. Fluids* **2008**, *46*, 83–92. [CrossRef]
62. Hyde, J.R.; Lester, E.; Kingman, S.; Pickering, S.; Wong, K.H. Supercritical Propanol, a possible route to composite carbon fibre recovery: A viability study. *Compos. A Appl. Sci. Manuf.* **2006**, *37*, 2171–2175. [CrossRef]
63. Yan, H.; Lu, C.; Jing, D.; Chang, C.; Liu, N.; Hou, X. Recycling of carbon fibers in epoxy resin composites using supercritical 1-propanol. *New Carbon Mater.* **2016**, *31*, 46–54. [CrossRef]
64. Jiang, G.; Pickering, S.; Lester, E.; Turner, T.; Wong, K.; Warrior, N. Characterisation of carbon fibres recycled from carbon fibre/epoxy resin composites using supercritical n-propanol. *Compos. Sci. Technol.* **2009**, *69*, 192–198. [CrossRef]
65. Okajima, I.; Sako, T. Recycling fiber-reinforced plastic using supercritical acetone. *Polym. Degrad. Stab.* **2019**, *163*, 1–6. [CrossRef]
66. Knight, C.C.; Zeng, C.; Zhang, C.; Liang, R. Fabrication and properties of composites utilizing reclaimed woven carbon fiber by sub-critical and supercritical water recycling. *Mater. Chem. Phys.* **2015**, *149–150*, 317–323. [CrossRef]
67. Okajima, I.; Watanabe, K.; Haramiishi, S.; Nakamura, M.; Shimamura, Y.; Sako, T. Recycling of carbon fiber reinforced plastic containing amine-cured epoxy resin using supercritical and subcritical fluids. *J. Supercrit. Fluids* **2017**, *119*, 44–51. [CrossRef]
68. Sokoli, H.U.; Beauson, J.; Simonsen, M.E.; Fraisse, A.; Brøndsted, P.; Søgaard, E.G. Optimized process for recovery of glass- and carbon fibers with retained mechanical properties by means of near- and supercritical fluids. *J. Supercrit. Fluids* **2017**, *124*, 80–89. [CrossRef]
69. Kestin, J.; Sengers, J.V.; Kamgar-Parsi, B.; Levelt Sengers, J.M.H. Thermophysical properties of fluid H2O. *J. Phys. Chem. Ref. Data* **1984**, *13*, 175–183. [CrossRef]
70. Isothermal Properties for Methanol. Available online: https://webbook.nist.gov/cgi/fluid.cgi?T=313.15&PLow=1&PHigh=100&PInc=1&Applet=on&Digits=5&ID=C67561&Action=Load&Type=IsoTherm&TUnit=K&PUnit=bar&DUnit=kg%2Fm3&HUnit=kJ%2Fmol&WUnit=m%2Fs&VisUnit=uPa*s&STUnit=N%2Fm&RefState=DEF (accessed on 21 April 2021).
71. Gude, M.; Teja, A.S. Vapor-liquid critical properties of elements and compounds. 4. Aliphatic alkanols. *J. Chem. Eng. Data* **1995**, *40*, 1025–1036. [CrossRef]
72. Informatics, N.O.; NIST Standard Reference Database. Chemistry WebBook. Available online: https://webbook.nist.gov/chemistry/ (accessed on 22 April 2021).
73. Jiang, G.; Pickering, S.J.; Lester, E.; Blood, P.; Warrior, N.; Pickering, S.J. Recycling carbon fibre/epoxy resin composites using supercritical propanol. In Proceedings of the 16th International Conference on Composite Materials, Kyoto, Japan, 8–13 July 2007; p. 5.
74. Knight, C.C.; Zeng, C.; Zhang, C.; Wang, B. Recycling of woven carbon-fibre-reinforced polymer composites using supercritical water. *Environ. Technol.* **2012**, *33*, 639–644. [CrossRef]
75. Goto, M.; Jin, F.; Zhou, Q.; Wu, B. Supercritical water process for the chemical recycling of waste plastics. In *AIP Conference Proceedings, Proceedings of the AIP 2nd International Symposium on Aqua Science, Water Resource and Low Carbon Energy, Sanya Hainan, China, 10 December 2009*; AIP Publishing: Melville, NY, USA, 2010; pp. 169–172.
76. Reaction Kinetics of CFRP Degradation in Supercritical Fluids. Available online: https://link.springer.com/article/10.1007/s10924-017-1114-2 (accessed on 1 June 2021).
77. Dauguet, M.; Mantaux, O.; Perry, N.; Zhao, Y.F. Recycling of CFRP for high value applications: Effect of sizing removal and environmental analysis of the supercritical fluid solvolysis. *Procedia CIRP* **2015**, *29*, 734–739. [CrossRef]
78. Kumar, S.; Krishnan, S. Recycling of carbon fiber with epoxy composites by chemical recycling for future perspective: A review. *Chem. Pap.* **2020**, *74*, 3785–3807. [CrossRef]

*materials*

MDPI

*Article*

# Characterization and Sodium Cations Sorption Capacity of Chemically Modified Biochars Produced from Agricultural and Forestry Wastes

**Agnieszka Medyńska-Juraszek [1,*], María Luisa Álvarez [2], Andrzej Białowiec [3] and Maria Jerzykiewicz [4]**

1 Institute of Soil Sciences and Environmental Protection, Wroclaw University of Environmental and Life Sciences, 53 Grunwaldzka Str., 50-357 Wrocław, Poland
2 Department of Geological and Mining Engineering, Universidad Politécnica de Madrid, 28003 Madrid, Spain; marialuisa.alvarez@upm.es
3 Department of Applied Bioeconomy, Wrocław University of Environmental and Life Sciences, 37a Chełmońskiego Str., 51-630 Wrocław, Poland; andrzej.bialowiec@upwr.edu.pl
4 Faculty of Chemistry, Wroclaw University, 14 Joliot-Curie St., 50-383 Wrocław, Poland; maria.jerzykiewicz@chem.uni.wroc.pl
* Correspondence: agnieszka.medynska-juraszek@upwr.edu.pl

**Abstract:** Excessive amounts of sodium cations ($Na^+$) in water is an important limiting factor to reuse poor quality water in agriculture or industry, and recently, much attention has been paid to developing cost-effective and easily available water desalination technology that is not limited to natural resources. Biochar seems to be a promising solution for reducing high loads of inorganic contaminant from water and soil solution, and due to the high availability of biomass in agriculture and forestry, its production for these purposes may become beneficial. In the present research, wheat straw, sunflower husk, and pine-chip biochars produced at 250, 450 and 550 °C under simple torrefaction/pyrolysis conditions were chemically modified with ethanol or HCl to determine the effect of these activations on Na sorption capacity from aqueous solution. Biochar sorption property measurements, such as specific surface area, cation exchange capacity, content of base cations in exchangeable forms, and structural changes of biochar surface, were performed by FTIR and EPR spectrometry to study the effect of material chemical activation. The sorption capacity of biochars and activated carbons was investigated by performing batch sorption experiments, and adsorption isotherms were tested with Langmuir's and Freundlich's models. The results showed that biochar activation had significant effects on the sorption characteristics of $Na^+$, increasing its capacity (even 10-folds) and inducing the mechanism of ion exchange between biochar and saline solution, especially when ethanol activation was applied. The findings of this study show that biochar produced through torrefaction with ethanol activation requires lower energy demand and carbon footprint and, therefore, is a promising method for studying material applications for environmental and industrial purposes.

**Keywords:** biochar; modification; sodium; sorbent; saline water; soil

**Citation:** Medyńska-Juraszek, A.; Álvarez, M.L.; Białowiec, A.; Jerzykiewicz, M. Characterization and Sodium Cations Sorption Capacity of Chemically Modified Biochars Produced from Agricultural and Forestry Wastes. *Materials* **2021**, *14*, 4714. https://doi.org/10.3390/ma14164714

Academic Editor: Carlos Javier Duran-Valle

Received: 27 July 2021
Accepted: 17 August 2021
Published: 20 August 2021



## 1. Introduction

Due to the global climate change and intensive irrigation of arable soil with saline waters, the problem of soil salinization has increased rapidly in the last decade. An extensive are of arable soils in many regions of the world has been affected by salinization, and it is estimated that approximately 33% of irrigated lands and 20% of total cultivated lands have been affected [1], thus causing a decline in soil productivity [2]. The problem is mostly observed in the arid and semi-arid climatic zones; however, due to the water shortage and human activities such as soil mineral fertilization, the use of sewage sludge, or application of salts for road de-icing during the winter season [3], the problem of soil salinization occurs locally even in more humid regions. Snow melts and rainfall in temperate climate zone are beneficial for diluting soil solution and minimizing the salt stress problem in plants. In arid climates, the excess of salts is removed from soil by irrigation; however, this technique requires ample availability of quality water resources, which is often limited in high salinity-affected areas, and it creates a practical constraint for their implementation [4]. One of the most important contaminants limiting the reuse of saline water is the excessive amount of sodium ($Na^+$) and various techniques has been developed for saline water remediation [5]. The use of calcium-containing minerals, such as gypsum and calcium chloride or sulfuric acid [6,7], is very common. However, these methods are considered very expensive, environmentally unfriendly, and energy demanding; therefore, the pre-treatment of saline water before applying these methods can be deployed to reduce total desalination costs [8]. Finding an efficient and inexpensive sorbent for removing ions causing salinity from water resources is challenging. Recently, much attention has been paid to bio-waste-based materials, which is considered a green technique, offering twin benefits of waste load reduction and land and water reclamation [4]. The agricultural residues contribute as an easily available and important source of materials to be used as an environmental catalyst for the removal of contaminants [9]. Biochar obtained through the torrefaction/pyrolysis of agricultural wastes has been studied because of its high efficiency for adsorption of different inorganic pollutants [10] in soil and aqueous solutions [11,12]. Biochar seems to be a very promising material for water desalination, promoting the removal of excessive amounts of $Na^+$ from solution by exchanging it with calcium, magnesium and potassium cations present on biochar surface in large amounts [4]. In general, various mechanisms of the biochar-$Na^+$ sorption have been described, suggesting that physical sorption on biochar porous structure [5], preferential sorption, and ion sorption or $Na^+$ exchange with biochar-borne Ca and Mg ions can all be responsible for the sorption process on biochar material [13]. In previous studies, it has been reported that the incorporation of biochar into salt-affected soil alleviated salinity stress in crop plants because of its high salt ($Na^+$) potential [14]. From the practical point of view, when choosing a proper biochar based on its properties, e.g., specific surface area, cation exchange capacity, functional groups or mineral content, it is very important to match the proper selective material for cation sorption. Biochar produced from non-woody feedstock, such as manures and plant residues, is richer in nutrients, has a higher pH, and has less stable carbon than biochar produced from lignocellulose feedstock, such as wood [15,16]. Surface sorption mainly by oxygen functional groups or $Na^+$ exchange with biochar-derived Ca and Mg ions are dominant mechanisms of sodium immobilization on biochar. These properties can be optimized through the appropriate designing of the torrefaction/pyrolysis process and careful selection of the biomass for thermochemical conversion. The presence of mineral (ash) components (e.g. carbonates, phosphates, or oxides) can pose both positive and negative effects on the supportive removal of contaminants where different sorption properties can be obtained during biochar deashing [17,18]. In terms of biochars derived from crop residues, e.g., straw or husk, the chemical pre-treatment might be necessary to remove excess amounts of $Na^+$ [19], as part of the Na that exists in chlorides, carboxylates, or carbonates in biochar can be easily leachable, contributing to the salinization process [20]. Since water leaching is not efficient in the removal of organic compounds from biochar, acid or alcohol washing can be a more successful alternative [21]. Biochar, produced through the pyrolysis of inexpensive agricultural and forest residues, has been widely used as an alternative low-cost adsorbent to treat environmental pollution [22]. Many different human activities can increase salt pollution in surface water and drinking water resources and considering the growing problem with soil and water salinization, the prospect of biochar application for removal of sodium salts necessitates further and detailed studies.

The purpose of the study was to investigate the efficiency of $Na^+$ removal from aqueous solution by three biochars derived from agricultural and forestry waste (sunflower husks, wheat straw and pinewood) with (ethanol, HCl) and without chemical modification (ethanol, HCl) and to recommend the most effective procedure of chemical modification of biochar with beneficial physiochemical characteristics to reduce $Na^+$ from the solution. Recently, more attention has been paid to saline soils and different benefits of biochar in minimizing plant salt stress, explained mainly by indirect mechanisms of soil property improvements, e.g., increased content of water in soil. The knowledge about the direct mechanisms of salt sorption on biochar surfaces is less recognized. This paper describes possible mechanisms of sodium sorption on biochar, studying the efficiency of $Na^+$ sorption from solution and showing possible new applications of the material.

## 2. Materials and Methods

### 2.1. Biochar Preparation and Activation

Three types of biochar were used: wheat straw biochar (WSBC), sunflower biochar (SBC) and pinewood chips biochar (PBC), obtained as waste from agricultural and forestry activities, produced under pyrolytic conditions, respectively, at 550, 450 and 250 °C with a heating rate of (10–25 $°C \cdot min^{-1}$) and residence time of 1 h. All biochar samples were air-dried, ground, and sieved through a 2 mm mesh prior to all the experiments. Three chemical treatments were tested: non-modified biochar, HCl-washed biochar, and ethanol-washed biochar. To produce ethanol- and HCl-modified biochars, the procedure described by Dietrich et al. [21] was performed. Briefly, 5 g of each biochar was washed with either 0.1 M ethanol or 0.1 M HCl at a ratio of 1:9. The HCl-char and ethanol-char mixture was shaken for 24 h at 30 r.p.m. using the rotary shaker Multi RS-60 (Biosan, Saratoga Springs, NY, USA). Both types of biochar were then vacuum-extracted from solution and dried at 80 °C until constant weight. To remove residual HCl solution, the HCl-washed biochar was treated with pH-adjusted ($NaHCO_3$, pH of 7.5) deionized water before drying. All determinations were performed in triplicate.

### 2.2. Biochar Analysis

The Brunauer–Emmett–Teller (BET) surface area, cation exchange capacity (CEC), pH in deionized water, CNHSO elemental composition, ash, the total contents and exchangeable cations content of Ca, Mg, K and Na were determined to describe properties of the materials. The Brunauer–Emmett–Teller (BET) surface area was determined using a specific surface area analyzer Gemini VII 2390 Series (Micrometrics, Norcross, GA, USA). Exchangeable cations ($Ca^{2+}$, $Mg^{2+}$, $K^+$, $Na^+$) were determined according to the modified method described by Munera-Echeverri et al. [23] and analyzed on a Microwave Plasma-Atomic Emission Spectrometer MP-AES 4200 (Agilent Technologies, Santa Clara, CA, USA). Cation exchange capacity was calculated as a sum of base cations. The pH values were measured at a ratio of 1:5 (*w/v*) in deionized water after the sample was shaken for 1h at 130 rpm with a calibration check pH meter (Mettler-Toledo, Graifensee, Switzerland). The ash content was determined by weight loss after combustion at 750 °C for 6 h in a muffle furnace according to ASTM D7348-13 [24]. The elemental composition was analyzed on the CHNS Vario EL Cube analyzer (Elementar, Langenselbold, Germany). To determine changes of chemical composition and biochar structure after activation with agents, Fourier Transform Infrared Spectra (FT-IR) and Electron Paramagnetic Resonance spectroscope (EPR) analyses were performed. Biochar samples were dried in an oven drier at 60 °C for 6 h to prepare pellets for FT-IR analysis. FT-IR analysis of biochar samples were recorded using a Bruker Vertex 70 FT-IR spectrometer (Bruker, Karlsruhe, Germany) using standard KBr pellets–about 1 mg sample for 400 mg of KBr. Electron Paramagnetic Resonance (EPR) spectra were obtained with Bruker Elexsys E500 spectrometer (Bruker, Billerica, MA, USA) equipped with NMR teslameter (ER036TM) (Bruker, Karlsruhe, Germany) and frequency counter (ER 41 FC) (Bruker, Karlsruhe, Germany) at room temperature. For quantitative measurements double rectangular cavity resonator (ER 4105DR) (Bruker, Karlsruhe, Germany) operating at the mode TD104 was applied. In one cavity standards of radical concentration were placed while the analysed sample in the second cavity. After tuning, the spectra were recorded separately for both cavities without changing any of the measurement parameters. X-band spectra were measured at microwave power of

20 mW, modulation amplitude of 1 G. For measurements the solid samples (about 20.0 mg) were placed in quartz tubes of 5 mm outer diameter. The Li/ LiF standard was used for g-parameter calibration (g = 2.00223). As standards of spin concentration Pahokee peat standard humic acid (1S103H) and Leonardite standard humic acid (1S104H) extracted and distributed by International Humic Substances Society (IHSS) [25], and additionally, Bruker alanine pills were used as standards of spin concentration analysis. For quantitative measurements, a double rectangular cavity resonator–(ER 4105DR) (Bruker, Karlsruhe, Germany) operating at the mode TD104 was used. The standards of the radical species concentration were placed in the first cavity and the analyzed sample in the second cavity. After tuning, the spectra were recorded for both cavities separately without changing any of the measurement parameters.

### 2.3. Sodium Adsorption on Biochar Experiment

For the adsorption experiment, sodium chloride (NaCl), ACS reagent, ≥99.0% (Sigma Aldrich, Saint Louis, MO, USA) was diluted in Milli-Q® ultrapure water (Merckmilipore, Burlington, MA, USA) to obtain 1%; 0.2%; 0.5% and 1.0% NaCl solutions. For that, 1 g of each biochar (WSBC, SBC and PBC) was weighed in a falcon tube for every 10 mL of NaCl solution. The scheme of the experiment and tested variants of prepared samples were listed in Appendix A, Table A1. The concentrations of NaCl in the solution imitated the amounts of sodium delivered to soil during different activities, mainly soil fertilization, field irrigations with low-quality waters, road run offs and saline waters deposited in the environment from mining activities, where concentrations of sodium can reach up to 5%. The content of potential exchangeable $Na^{+}$, $Ca^{2+}$, $Mg^{2+}$ and $K^{+}$ in NaCl solutions used in the sorption experiment is presented in Appendix A Table A2. Biochar samples were shaken for 24 h, and approx. 1 mL of solution was collected after each treatment. Sodium concentrations were analyzed on Microwave Plasma-Atomic Emission Spectrometer MP-AES 4200 (Agilent Technologies, Santa Clara, CA, USA), after sample filtration on Munktell No. 2 filters (Ahlstrom Munksjö, Helsinki, Finland) in dilutions of 1:10 or 1:100 for torch protection.

### 2.4. Calculations and Statistical Analysis

Sodium sorption batch experiments were performed in triplicate. The data are presented as the mean values with the relative standard deviation (RSD). Student's *t*-tests were used to test for significant differences in element content between non-modified biochars and biochars modified with EtOH and HCl ($p < 0.05$). The obtained data were compiled using Microsoft Excel 2016 and Statistica Statsoft 13.3. FT-IR spectra were performed for absorbance to simplify the interpretation of intensity ratios. Characteristic areas of the spectra were integrated using data analysis and graphing software OriginPro2019 (OriginLab Corporation, Northampton, MA, USA.) The percentages were calculated according to the whole spectrum area. The isotherms of Na adsorption on each type of biochar were determined. Langmuir's (Equation (1)) and Freundlich's (Equation (2)) models were tested.

$$qe = \left(\frac{(qm \cdot b \cdot Ce)}{(1 + b \cdot Ce)}\right) \tag{1}$$

where:

*qe*—the equilibrium (instantaneous) amount of adsorbed Na ions on units of biochar mass, mg/g,
*qm*—the maximum amount of adsorbed Na ions on units of biochar mass, mg/g, to form a complete monolayer on the surface,
*Ce*—equilibrium concentration of Na ions, mg/L,
*B*—Langmuir adsorption constant related to the energy of adsorption, L/mg.

$$qe = Kf \cdot Ce^{\left(\frac{1}{n}\right)} \tag{2}$$

where:

*qe*—the equilibrium (instantaneous) amount of adsorbed Na ions on units of biochar mass, mg/g,
*Kf*—Freundlich adsorption constant, $mg \cdot g^{-1} \cdot (L \cdot mg^{-1})^{\wedge}(1/n)$,
$1/n$—empirical constant: heterogenicity coefficient.

For the determination of isotherms parameters nonlinear regression analysis was done. The regression analysis was done using the Statistica 13 software (StatSoft, Inc., TIBCO Software Inc., Palo Alto, CA, USA). For the validation of model parameters, the determination coefficient ($R^2$) was calculated at the statistical significance ($p < 0.05$). After that both models were compared with application of the Akaike Information Criterion (AIC) to indicate the simplest model matching to raw data similarly. AIC was evaluated according to the following Equation (3):

$$AIC = n \cdot ln\left(\sum_{i=1}^{n} e_i^2\right) + 2 \cdot K \quad (3)$$

where:

*AIC*—value of Akaike analysis,
*n*—the number of measurements,
*e*—the value of the rest of the model for a particular measurement point,
*K*—the number of regression coefficients, including an intercept in the model.

Generally, models with a larger number of predictors are more accurate but tend to over-fitting. The AIC approach can be used to preserve good accuracy and a low number of predictors in the compared models. When models for a particular variable are compared, a model with a lower AIC is better.

## 3. Results and Discussion

### 3.1. Biochar Sorption Properties

Previous studies have shown that the adsorption of $Na^+$ is related to cation exchange capacity (CEC) of biochar as the CEC value is closely related to the concentration of oxygenated functional groups contributing to the sorption process of different cations [5]. In terms of studied biochars, CEC was strongly related to feedstock type used during the pyrolysis process, and wheat straw biochar presented the best potential for cation sorption, having the highest specific surface area, cation exchange capacity and pH (Tables 1 and 2). However, very high content of exchangeable $Na^+$–27.01 cmol(+)/kg in WSBC, compared with 1.13 cmol(+)/kg in SBC and 1.99 cmol(+)/kg PBC–suggested that WSBC can become an $Na^+$ donor to the solution. The contribution of biochar as a donor of $Na^+$ was investigated by Sarpong et al. [19], studying the effect of halophyte *Atriplex*-derived biochar application to saline soil and the process of $Na^+$ leaching from biochar. The results of the study showed that biochar derived from biomass grown on saline soils contained higher amounts of $Na^+$ readily exchangeable with cations in soil solution, contributing to the salinization process. The highest content of potentially exchangeable $Na^+$ cations ($K^+$, $Ca^{2+}$ and $Mg^{2+}$) was determined in PBC (Table 1), suggesting that hard-wood derived biochar will have the best abilities for salt removal via an ion exchange mechanism.

Sample treatment with ethanol caused an almost twofold decrease in ash content and CEC in WSBC, increasing the biochar acidity and removing almost half of the exchangeable $Na^+$ during this modification (Table 1). In SBC and PBC, cation exchange capacity was improved as the content of base cations ($Ca^{2+}$, $Mg^{2+}$ and $K^+$) in exchangeable forms increased, probably due to the partial dissolution of low molecular weight organic compounds and rinsing of the excess of ${NH_4}^{3+}$ cations and surface changes [17,23]. HCl pre-treatment had very adverse effects on the tested biochars. In terms of WSBC, it did not cause significant sample de-ashing, while in PBC, almost 6% of ash was removed by acid. In SBC and PBC, the sample treatment with HCl caused an increase in CEC and acidity; however, this increase was balanced by the release of $Ca^{2+}$ in exchangeable forms and probable decomposition of carbonates present on the biochar surface after treatment with HCl. pH is a very important property regarding the surface charges on biochar. When biochar is applied to aqueous solution for metal removal, the solution pH strongly influences its surface charge. At a solution pH of 3–7, biochars become negatively charged, which favors positively charged ion sorption [21]. At higher temperatures, the number of negatively charged groups on biochars is reduced, and those acidic modifications may have a positive effect on surface charges and occurrence of negatively charged functional groups attracting $Na^+$. The changes of the surface charges were not investigated in this study; however, the results from the FTIR analysis and SSA determination showed that chemical modification caused changes on biochar surface, and depending on a feedstock type and initial pH, the material improved or had no significant effect on $Na^+$ sorption on tested biochars. An increase in SSA after tested chemical-pretreatments was observed in all three biochars; however, significant differences were only determined for ethanol-treated samples. The best effect of the SSA increase was observed after ethanol treatment of WSBC, and the SSA increased from 265 to 387 g/m$^2$. The increase was also significant for EtOH-SBC (from 80.5 to 93.2 g/m$^2$), while no significant changes were observed for EtOH-PBC (Table 2). Comparing both methods of biochar modifications, ethanol pre-treatment was more efficient for SSA improvement compared with HCl.

**Table 1.** Changes of cation sorption properties after biochar modification.

| Biochar | pH | Ash | CEC | $Na^+$ | $K^+$ | $Ca^{2+}$ | $Mg^{2+}$ |
|---|---|---|---|---|---|---|---|
| | - | % DM | | | cmol(+)/kg | | |
| Non-WSBC | 9.98 [1] ± 0.03 [2] | 36.4 ± 0.6 | 30.72 | 27.01 ± 0.2 | 1.42 ± 0.02 | 1.43 ± 0.04 | 0.19 ± 0.02 |
| EtOH-WSBC | 9.43 [1] ± 0.03 [2] | 19.5 | 19.67 | 16.02 ± 0.02 | 1.11 ± 0.02 | 1.98 ± 0.02 | 0.56 ± 0.02 |
| HCl-WSBC | 8.89 ± 0.02 | 33.8 | 27.71 | 23.24 ± 0.02 | 1.15 ± 0.02 | 2.34 ± 0.02 | 0.98 ± 0.02 |
| Non-SBC | 8.83 ± 0.02 | 20.3 ± 0.3 | 3.08 | 1.13 ± 0.1 | 1.01 ± 0.01 | 0.93 ± 0.02 | 0.11 ± 0.03 |
| EtOH-SBC | 8.67 ± 0.02 | 19.7 | 3.69 | 0.98 ± 0.02 | 1.03 ± 0.02 | 1.34 ± 0.02 | 0.34 ± 0.02 |
| HCl-SBC | 8.02 ± 0.02 | 21.0 | 4.96 | 1.11 ± 0.02 | 0.98 ± 0.02 | 1.98 ± 0.02 | 0.89 ± 0.02 |
| Non-PBC | 6.81 ± 0.02 | 10.5 ± 0.9 | 6.64 | 1.99 ± 0.7 | 1.45 ± 0.03 | 2.93 ± 0.01 | 0.26 ± 0.02 |
| EtOH-PBC | 6.63 ± 0.02 | 7.6 | 7.31 | 1.48 ± 0.02 | 1.94 ± 0.02 | 3.65 ± 0.02 | 0.24 ± 0.02 |
| HCl-PBC | 6.43 ± 0.02 | 4.08 | 8.13 | 1.78 ± 0.02 | 1.34 ± 0.02 | 3.89 ± 0.02 | 1.12 ± 0.02 |

[1] mean values ($n$ = 3) [2] RSD values ($n$ = 3).

**Table 2.** Specific surface area of non-modified and modified biochars.

| Biochar | SSA (g/m$^2$) |
|---|---|
| Non-WSBC | 265 [1] ± 2.1 [2] $_a$ |
| HCl-WSBC | 278 ± 3.1 $_b$ |
| EtOH-WSBC | 387 ± 4.5 $_b$ |
| Non-SBC | 80.5 ± 2.1 $_a$ |
| HCl-SBC | 85.4 ± 1.4 $_a$ |
| EtOH-SBC | 93.2 ± 1.9 $_b$ |
| Non-PBC | 16.5 ± 1.8 $_a$ |
| HCl-PBC | 14.5 ± 1.3 $_a$ |
| EtOH-PBC | 17.8 ± 0.9 $_a$ |

[1] mean values ($n$ = 3) [2] RSD values ($n$ = 3); $_{a,b}$ Different lowercase letters (a and b) indicate significant differences between non-modified and modified biochars ($p$ < 0.05).

### 3.2. Surface and Structural Changes of Biochar after Modifications

Generally, the content of carbon was strongly related to the temperature of the pyrolysis, and the lowest values were found for WSBC produced at 550 °C, while in the case of PBC, the temperature characteristic for torrefaction at 250 °C preserved the C content. All biochars had low nitrogen content (<0.9 %), and as expected, the lowest was in PBC (0.42%), and the highest was in SBC (0.95%). Hydrogen content was similar for all non-modified biochars, from 2.2% in non-WSBC to 2.57% in non-SBC. Surprisingly the content of sulfur in all non-modified biochars was very low (0.001–0.003%) and increased significantly (even up to 0.0425%) after biochar treatment with agents (Table 3). Biochar modifications changed the elemental composition of biochars, especially when EtOH was used as an agent. The effect of chemical activation depended also on the biochar type. EtOH treatment significantly increased content of C, N, H and S in WSBC and SBC, decreasing O content in both materials. In contrast, in PBC, EtOH treatment caused material oxidation, increasing O content from 7.58% to 23.69% and decreasing C from 89.55% to 72.36%, which could be related to the elution of low molecular weight products of torrefaction. HCl treatment was less effective, and even in the case of H where protonation was expected after HCl treatment, the elemental composition did not change significantly. The obtained BC molar ratios emphasize the presence of aromatic structural features and reduced content of O-containing polar functional groups on the biochar surface (low molar O/C ratio and polarity index) after WSBC and SBC treatment with EtOH and HCl, and opposite results and increase in polar oxygen-containing surface functional groups on PBC.

**Table 3.** Elemental composition of non-modified and activated biochars and the values of their molar ratios.

| Biochar | C | N | H | O | S | C/N | H/C | O/C |
|---|---|---|---|---|---|---|---|---|
| | | | (% w/dw) | | | | (molar) | |
| Non-WSBC | 63.61 [1] ± 0.2 [2] $_a$ | 0.74 ± 0.1 $_a$ | 2.2 ± 0.01 $_a$ | 33.42 ± 0.9 $_a$ | 0.001 ± 0.0 $_a$ | 85.98 | 0.03 | 0.52 |
| HCl-WSBC | 66.12 ± 0.1 $_a$ | 0.64 ± 0.05 $_b$ | 1.98 ± 0.02 $_a$ | 30.92 ± 1.1 $_a$ | 0.0355 ± 0.01 $_b$ | 104.09 | 0.02 | 0.46 |
| EtOH-WSBC | 89.42 ± 3.2 $_b$ | 0.63 ± 0.04 $_b$ | 1.81 ± 0.01 $_b$ | 8.11 ± 0.4 $_b$ | 0.0425 ± 0.01 $_b$ | 143.07 | 0.02 | 0.09 |
| Non-SBC | 79.74 ± 2.1 $_a$ | 0.95 ± 0.1 $_a$ | 2.57 ± 0.03 $_a$ | 16.73 ± 0.3 $_a$ | 0.001 ± 0.03 $_a$ | 83.49 | 0.03 | 0.20 |
| HCl-SBC | 78.95 ± 1.2 $_a$ | 0.92 ± 0.11 $_a$ | 1.80 ± 0.03 $_b$ | 18.28 ± 1.1 $_a$ | 0.047 ± 0.04 $_b$ | 85.81 | 0.02 | 0.23 |
| EtOH-SBC | 80.33 ± 1.3 $_a$ | 0.87 ± 0.12 $_b$ | 2.82 ± 0.04 $_b$ | 15.92 ± 0.9 $_a$ | 0.048 ± 0.01 $_b$ | 91.80 | 0.03 | 0.19 |
| Non-PBC | 89.55 ± 2.7 $_a$ | 0.42 ± 0.03 $_a$ | 2.44 ± 0.01 $_a$ | 7.58 ± 0.3 $_a$ | 0.003 ± 0.01 $_a$ | 213.21 | 0.03 | 0.08 |
| HCl-PBC | 75.92 ± 2.1 $_b$ | 0.53 ± 0.04 $_b$ | 3.10 ± 0.05 $_b$ | 20.43 ± 0.7 $_b$ | 0.013 ± 0.02 $_b$ | 143.24 | 0.04 | 0.26 |
| EtOH-PBC | 72.36 ± 3.1 $_b$ | 0.69 ± 0.06 $_b$ | 3.21 ± 0.03 $_b$ | 23.69 ± 0.9 $_b$ | 0.034 ± 0.03 $_b$ | 104.11 | 0.04 | 0.32 |

[1] mean values ($n$ = 3) [2] RSD values ($n$ = 3); $_{a,b}$ Different lowercase letters (a and b) indicate significant differences between non-modified and modified biochars ($p < 0.05$).

Obtained FT-IR spectra were typical for biochar material, showing bonds related to –OH (hydrogen groups) at 3400 cm$^{-1}$, C–H (aliphatic groups) at 2950–2850 cm$^{-1}$ and C–C (aromatic groups) at 1630 cm$^{-1}$ and C=O (carboxylic groups) at 1624 cm$^{-1}$ (Figure 1). The strongest peaks were observed for all three biochars at 3400 cm$^{-1}$, showing that hydrogen groups were dominant on their surface, followed by the presence of carboxylic groups bands and weak peak of aliphatic groups (the highest in SBC). Similar FTIR spectra were obtained in a study by Rostamian et al. [8], describing rice husk biochar properties and Na sorption capacity. WSBC was also more visible compared with other biochar peaks located at 469, 803 and 1098 cm$^{-1}$ ascribed to bending vibration, symmetric stretching, and asymmetric stretching of Si–O bonds, typically found in biochars derived from monocot plants such as grasses and cereals due to the presence of silica. In WSBC and PBC, an increase in the aliphatic C–H and aromatic C–C vibration bands and decrease in –OH vibration bands were observed after a sample treatment with EtOH and HCl; however, the effect depended on the agent type. For WSBC, treatment with EtOH significantly increased the number of carboxyl and =C=O of amides, ketones and quinones compared to non-WSBC and HCl-WSBC. Sharp peak located at 1098 cm$^{-1}$ was increased after WSBC treatment with ethanol, suggesting that an alcohol agent could increase silica content of WSBC. HCl sample treatment had no significant effect on tested biochar surfaces, with the exception of PBC, where a strong carboxyl peak was indicated after acid activation. The findings of our study are in contradiction to the results of Li et al. 2014 [17], who presented that wheat straw-derived BC modified with different concentrations of HCl (1.0 mol/L and 6.0 mol/L) developed a more heterogeneous porous structure compared with untreated samples. Surprisingly, from three tested biochars, sunflower husk biochar was the most "resistant" material to chemical activation, and very similar spectra were obtained for raw and activated SBC. Functional groups present at the biochar surface are one of the crucial factors determining the physical adsorption process of $Na^+$ [26,27].

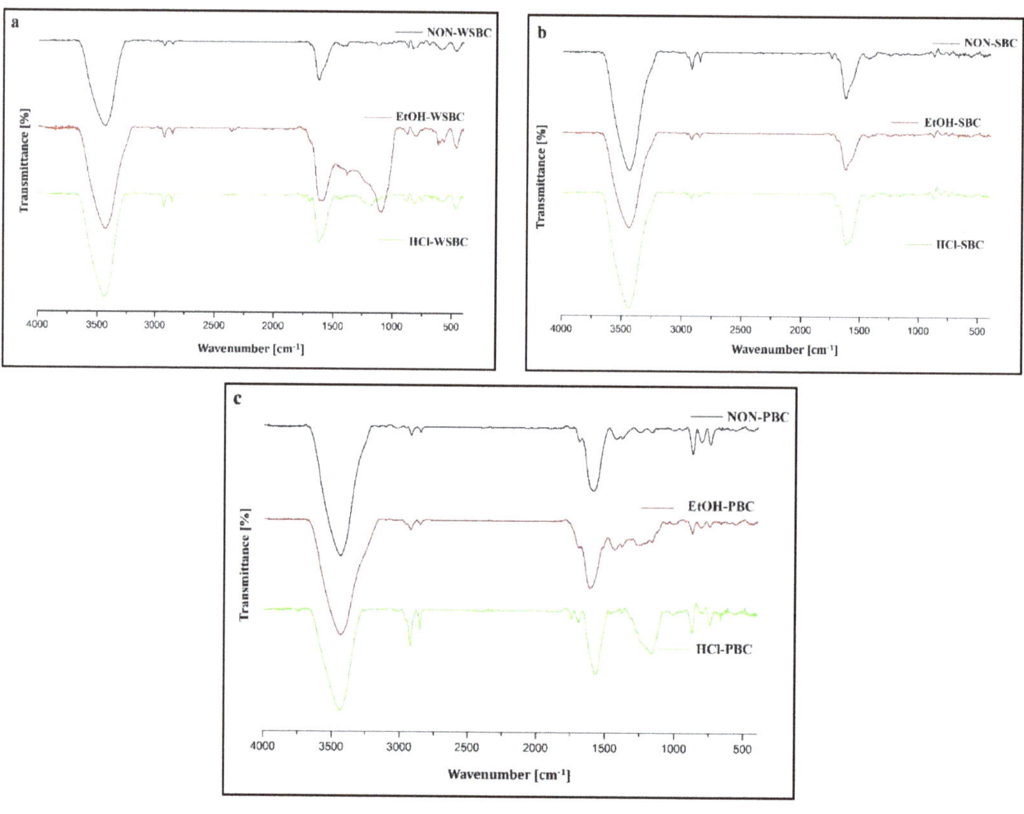


**Figure 1.** FT-IR spectra of investigated biochars: (**a**) pure (non-WSBC) and modified (EtOH-WSBC and HCl-WSBC) wheat straw biochar; (**b**) pure (non-SBC) and modified (EtOH-SBC and HCl-SBC) sunflower husk biochar; (**c**) pure (non-PBC) and modified (EtOH-PBC and HCl-PBC) pinewood biochar.

Calculations of FT-IR integrated areas confirmed the variability of the observation obtained during spectra analysis. For wheat straw biochar, EtOH treatment increased the aromaticity, reducing the number of –C–H and –OH groups on its surface. For SBC, aromaticity increased after HCl treatment, while EtOH increased the number of –OH groups. In PBC, EtOH increased aromaticity, while HCl treatment decreased the number of –OH groups, increasing –C–H (Table 4). Biochars can donate, accept, or transfer electrons in their surrounding environments, which is probably the most important property of the material when concerning environmental applications, such as cation removal [28].

According to the study by Klüpfel et al. [29], different mechanisms can be distinguished between the biochars produced at different temperatures, and spectroscopic analysis provides evidence that the pool of redox-active moieties is dominated by electron-donating, phenolic moieties in the low-temperature biochars, by newly formed electron-accepting quinone moieties in intermediate-temperature biochars, and by electron-accepting quinones and possibly condensed aromatics in the high-temperature pyrolysis. Awan et al. [13] described that biochars containing highly structured, aromatic compounds that induce electronegativity in the form of delocalized $\pi$ electrons become more electrochemically active than original feedstock. This electronegativity may effectively sorb soft base cations (e.g., $Na^+$) to a greater extent than hard base cations (e.g., $Ca^{2+}$ and $Mg^{2+}$) due to the weak hydration and relatively large radii of soft base cations.

**Table 4.** FT-IR integrated areas of biochars.

| Biochar | 3188–3720 | 2800–2989 | 1480–1660 |
|---|---|---|---|
| | –OH | C–H Aliphatic<br>% of Area | C–C Aromatic |
| Non-WSBC | 77.72 | 0.71 | 13.90 |
| HCl-WSBC | 72.35 | 1.47 | 15.04 |
| EtOH-WSBC | 71.62 | 0.96 | 31.50 |
| Non-SBC | 66.91 | 2.21 | 13.02 |
| HCl-SBC | 82.00 | 0.56 | 17.70 |
| EtOH-SBC | 85.55 | 0.87 | 13.65 |
| Non-PBC | 73.83 | 0.74 | 15.60 |
| HCl-PBC | 56.16 | 3.93 | 16.65 |
| EtOH-PBC | 63.31 | 0.97 | 17.16 |

EPR analysis of radicals built in the biochar matrices show that there were no structural changes to the investigated biochar during EtOH and HCl treatment (g-parameter did not change). However, a difference was observed in radical concentrations. The quenching of the radicals was especially observed for biochar originated from pinewood after HCl treatment. Non-WSBC was characterized by the lowest radical concentration, but a decrease in radical concentration was also observed in HCl-WSBC (Table 5).

**Table 5.** Radical concentration and g-factor calculations for non-modified and modified biochars obtained during EPR analysis.

| Biochar | Radical Concentration $\times 10^{-19}$ (spin/gram) | g-Parameter |
|---|---|---|
| Non-WSBC | 0.80 | 2.0028 |
| HCl-WSBC | 0.71 | 2.0028 |
| EtOH-WSBC | 0.42 | 2.0028 |
| Non-SBC | 1.85 | 2.0028 |
| HCl-SBC | 1.88 | 2.0027 |
| EtOH-SBC | 1.20 | 2.0029 |
| Non-PBC | 1.58 | 2.0029 |
| HCl-PBC | 1.63 | 2.0030 |
| EtOH-PBC | 0.46 | 2.0030 |

EPR analysis of biochar suggests that certain chars contain radicals of semiquinone-type character and the highest concentration quinone concentrations can be obtained during intermediate to high temperature pyrolysis [28]. However, this phenomenon was not confirmed in our study as the lowest radical concentrations were determined in high-temperature WSBC.

### 3.3. *Sodium Sorption Experiment*

#### 3.3.1. The Mechanism of the Adsorption

The adsorption of $Na^+$ in the analyzed types of activated and non-activated biochars had different patterns, which have been shown in the appendix (Figures A1–A3). The $Na^+$ adsorption on non-activated PBC and SBC surfaces had the L type character according to Giles classification (Figure A1). In the case of WSBC, S-type is visible. In all the cases, the single layer of the BC surface is not fully covered by Na ions, which means that the adsorption type belongs to the first sub-class [30]. According to IUPAC classification, the adsorption curves of all biochars are characteristic of type I, which is characteristic of microporous adsorbents, for which there is a strong affinity adsorbent–adsorbent [31]. The $Na^+$ adsorption on EtOH-WSBC and EtOH-SBC surfaces had clear S-type character according to the Giles classification (Figure A2). In the case of PBC, S-type is also visible. The S class includes isotherms for which, in the low equilibrium concentration range, the isotherm contains an inflection point behind which the curve increase is sharper. Adsorption, in this case, is preferred at higher concentrations. For this situation to occur, three conditions must be met: (i) the adsorptive molecule must interact with the surface of the condensed phase through interactions of only one type, (ii) have a moderate affinity for the adsorbent surface, and (iii) encounter strong competition from solvent molecules. In the case of EtOH-WSBC and EtOH-SBC, it belongs to the fourth sub-class. The third subgroup, identified for EtOH-PBC, contains isotherms for which the formation of the next layer of adsorption is observed, which is possible only in cases of physisorption [30]. According to IUPAC classification, the adsorption curves are characteristic of type II. Reversible Type II isotherms are given by the physisorption on nonporous or macroporous adsorbents. The shape is the result of unrestricted monolayer-multilayer adsorption. If the knee is sharp, Point B—the beginning of the middle, almost linear, section—usually corresponds to the completion of monolayer coverage of EtOH-WSBC and EtOH-SBC. A more gradual curvature (i.e., a less distinctive Point B) is an indication of a significant amount of overlap of monolayer coverage and the onset of multilayer adsorption, such as in the case of EtOH-PBC (Figure A2).

The $Na^+$ adsorption on HCl-activated PBC, WSBC and SBC surface had clear S-type character according to the Giles classification (Figure A3). In the case of all biochars modified by HCl, the isotherms belong to the second sub-class. The second subgroup includes isotherms for which the surface saturation with adsorbate monolayer has been achieved [30]. According to IUPAC classification, the adsorption curves of all types of HCl-biochars are characteristic of type III. Type III is also characteristic of macroporous adsorbents, but in this case, the adsorbent–adsorbent interactions are weak. In the case of a Type III isotherm, there is no Point B and therefore no identifiable monolayer formation; the adsorbent–adsorbate interactions are now relatively weak, and the adsorbed molecules on $Na^+$ are clustered around the most favorable sites on the surface of a nonporous or macroporous solid. In contrast to a Type II isotherm, the amount of $Na^+$ adsorbed remains finite at the given concentration [31].

#### 3.3.2. The Isotherms of the Na Adsorption

For the investigation of adsorption isotherms, two models were tested: Langmuir's and Freundlich's. The patterns of isotherms in non-linear regression modeling are presented in Appendix B (Figures A4–A6). The performed non-linear regression analyses indicated very high fitting of both models to the experimental data, confirmed by high values of determination coefficients (Table 6). In the case of non-activated biochars, $R^2$ is higher in Langmuir's model. The additional parameter AIC indicates lower values than in the case of the Freundlich model, which confirms that Langmuir's model should be preferred in the case of non-activated biochars. It has been shown that the maximum $Na^+$ adsorption capacity was found in the case of WSBC, which is associated with the highest values of cation exchange capacity and specific surface area. Results showed that the biochar from high-temperature pyrolysis should be preferred. These reactions were previously explained by Klüpfel et al. [29], showing that chars produced at intermediate to high heat treatment temperatures (400–700 °C) have higher capacities to accept and donate electrons compared to low-temperature biochars, such as the tested PBC. Tan et al. [5] described mitigation of soil salinity and showed that biochar derived from lignocellulosic biomass produced at high pyrolysis temperatures developed bigger pore volumes and specific surface area, providing higher binding sites for $Na^+$ ions.

**Table 6.** The comparison of Langmuir's and Freundlich's Na adsorption isotherms parameters on non-activated biochars.

| | Langmuir's Model | | | | Freundlich's Model | | | |
|---|---|---|---|---|---|---|---|---|
| Biochar Type | $q_{max}$, Maximum Adsorption Capacity, mg/g | b, the Ratio of Adsorption Constant Rate to Desorption Constant Rate ($k/k'$) | $R^2$—Determination Coefficient | AIC, Akaike Criterion Coefficient | Kf, $mg \cdot g^{-1} \cdot (L \cdot mg^{-1})^{\wedge}(1/n)$ | $(1/n)$, Hetero-genicity Coefficient | $R^2$—Determination Coefficient | AIC, Akaike Criterion Coefficient |
| WSBC | 308.86 | 0.00014 | 0.9909 | 29.09 | 0.368 | 0.678 | 0.9858 | 31.13 |
| SBC | 245.80 | 0.00017 | 0.9724 | 33.60 | 0.384 | 0.657 | 0.9392 | 37.50 |
| PBC | 214.33 | 0.00019 | 0.9932 | 25.46 | 0.517 | 0.614 | 0.9661 | 33.51 |

Executed modeling of $Na^+$ adsorption isotherms of ethanol-treated biochars indicated a significant (10-fold) increase in the maximum $Na^+$ adsorption capacity (Table 7), which could be related to the elution of organic compounds with high oxygen content. The EtOH-PBC had the highest adsorption capacity with the lowest specific surface area, indicating the importance of the ethanol treatments of biochars produced through the torrefaction conditions—elution of low molecular weight products of torrefaction. In the case of both models, the determination coefficients were high and comparable; however, the lower values of AIC were noted in the case of Freundlich's isotherms in all types of biochars, indicating that the adsorption had physical characteristics and that the biochar surfaces were not uniform [32,33].

**Table 7.** The comparison of Langmuir's and Freundlich's Na adsorption isotherms parameters on Ethanol-activated biochars.

| | Langmuir's Model | | | | Freundlich's Model | | | |
|---|---|---|---|---|---|---|---|---|
| Biochar Type | $q_{max}$, Maximum Adsorption Capacity, mg/g | b, the Ratio of Adsorption Constant Rate to Desorption Constant Rate ($k/k'$) | $R^2$—Determination Coefficient | AIC, Akaike Criterion Coefficient | Kf, $mg \cdot g^{-1} \cdot (L \cdot mg^{-1})^{\wedge}(1/n)$ | $(1/n)$, Hetero-genicity Coefficient | $R^2$—Determination Coefficient | AIC, Akaike criterion Coefficient |
| EtOH-WSBC | 2687.32 | 0.00001 | 0.9167 | 42.60 | 0.002164 | 1.281 | 0.9388 | 41.06 |
| EtOH-SBC | 1029.30 | 0.000028 | 0.906 | 42.73 | 0.009061 | 1.112 | 0.925 | 41.58 |
| EtOH-PBC | 3449.54 | 0.000009 | 0.9869 | 33.91 | 0.021207 | 1.037 | 0.9892 | 32.92 |

The non-linear regression modeling of $Na^+$ adsorption isotherms of HCl-activated biochars indicated a moderate (5-fold) increase in the maximum $Na^+$ adsorption capacity in comparison to non-activated biochars (Table 8). The HCl-WSBC had the highest adsorption capacity; and there were also differences between the two other types of HCl-activated biochars. Similarly, as in the case of ethanol activation, the lower values of AIC were noted in the case of Freundlich's isotherms in all types of biochars. Additionally, the determination coefficients were also higher in Freundlich's isotherms, and this model should be preferred.

A comparative analysis indicated that in the case of non-activated biochars, the high-temperature pyrolysis biochar should be applied for $Na^+$ adsorption. Biochars obtained during low-temperature pyrolysis, mainly torrefaction, are usually not considered as good sorbents of metals due to low sorption capacity, not well-developed specific surface area or, in terms of hard-wood derived biochars, low content of basic cations in exchangeable forms able to replace adsorbents in the solution. However, the results of the study show that to increase $Na^+$ adsorption capacity, ethanol activation of a low-temperature biochar can be performed to improve inorganic contaminants sorption capacity. The tests revealed a lower efficiency of HCl treatment in comparison to ethanol. An additional test of ethanol solution composition should be executed, as this aspect of the industrial wastewater production during chemical activation of biochar may create additional problems to be solved with the development of this technology. The presented results concern the initial stage of the technology readiness level development. Additionally, tests aimed at the determination of $Na^+$ adsorption optimum conditions (pH, adsorbent dosage as independent values), the investigation of $Na^+$ desorption, and the determination of biochar bed breakthrough before the scaling up of this material should be performed.

**Table 8.** The comparison of Langmuir's and Freundlich's $Na^+$ adsorption isotherms parameters on HCl-activated biochars.

| Biochar Type | Langmuir's Model | | | | Freundlich's Model | | | |
|---|---|---|---|---|---|---|---|---|
| | $q_{max}$, Maximum Adsorption Capacity, mg/g | b, the Ratio of Adsorption Constant Rate to Desorption Constant Rate ($k/k'$) | $R^2$—Determination Coefficient | AIC, Akaike Criterion Coefficient | Kf, $mg{\cdot}g^{-1}\cdot(L{\cdot}mg^{-1})^{\wedge}(1/n)$ | (1/n), Heterogenicity Coefficient | $R^2$—Determination Coefficient | AIC, Akaike Criterion Coefficient |
| HCl-WSBC | 1628.50 | 0.000059 | 0.7889 | 54.09 | 0.000259 | 1.707 | 0.9719 | 44.04 |
| HCl-SBC | 1432.91 | 0.000068 | 0.7981 | 53.83 | 0.000083 | 1.858 | 0.9911 | 38.21 |
| HCl-PBC | 1539.88 | 0.000063 | 0.7761 | 54.48 | 0.000027 | 1.999 | 0.9893 | 39.31 |

## 4. Conclusions

The problem of water and soil salinization is a global concern, and there is an urgent need for the development of new, highly efficient, inexpensive and environmentally friendly materials for salt removal. The results of the study showed that waste biomass, such as cereal straw and wood chips, can be converted to biochar, offering a solution for minimizing the agricultural and forestry waste volume and producing valuable $Na^+$ adsorbents, reducing the problem of water salinization. Wheat straw biochar, due to its high aromaticity, cation exchange capacity and specific surface area obtained during high-temperature pyrolysis (550 °C), presented the best sorption capacity for $Na^+$ removal amongst studied biochars. The tested methods of biochar pre-treatments with EtOH and HCl showed that sorption capacity for $Na^+$ can be significantly improved when chemical modifications are applied to biochars produced through torrefaction (<300 °C). The pre-treatment with ethanol of pinewood torrefaction-derived biochar increased the $Na^+$ adsorption capacity up to 10-fold compared to non-modified material. The result of the study showed that biochar produced through torrefaction may be utilized for $Na^+$ immobilization with lower energy demand and carbon footprint by ethanol treatment, becoming a promising method of material application for environmental and industrial purposes.

**Author Contributions:** A.M.-J. participated in the conceptualization and method design of this study. A.M.-J., A.B., M.L.Á. and M.J. conducted all experimental assays and played a part in data acquisition. A.M.-J., A.B. and M.J. considerably contributed to the data analysis and results interpretations. A.M.-J., A.B. and M.J. were involved in drafting, critically revising, and editing this manuscript. All authors have read and agreed to the published version of the manuscript.

**Funding:** The publication is financed under the Leading Research Groups support project from the subsidy increased for the period 2020–2025 in the amount of 2% of the subsidy referred to Art. 387 (3) of the Law of 20 July 2018 on Higher Education and Science, obtained in 2019.

**Institutional Review Board Statement:** Not applicable.

**Informed Consent Statement:** Not applicable.

**Data Availability Statement:** Data sharing is not applicable to this article.

**Acknowledgments:** The presented article results were obtained as part of the activity of the leading research team—Waste and Biomass Valorization Group (WBVG).

**Conflicts of Interest:** The authors declare no conflict of interest.

## Appendix A

**Table A1.** Samples prepared for adsorption experiment.

| Biochar | | NaCl (%) 0.1 | 0.2 | 0.5 | 1.0 | 2.0 |
|---|---|---|---|---|---|---|
| Non-activated | Non-WSBC<br>Non-SBC<br>Non-PBC | Non-WSBC + 0.1<br>Non-SBC + 0.1<br>Non-PBC + 0.1 | Non-WSBC + 0.2<br>Non-SBC + 0.2<br>Non-PBC + 0.2 | Non-WSBC + 0.5<br>Non-SBC + 0.5<br>Non-PBC + 0.5 | Non-WSBC + 1.0<br>Non-SBC + 1.0<br>Non-PBC + 1.0 | Non-WSBC + 2.0<br>Non-SBC + 2.0<br>Non-PBC + 2.0 |
| Activated with 0.1 M HCl | HCl-WSBC<br>HCl-SBC<br>HCl-PBC | HCl-WSBC + 0.1<br>HCl-SBC + 0.1<br>HCl-PBC + 0.1 | HCl-WSBC + 0.2<br>HCl-SBC + 0.2<br>HCl-PBC + 0.2 | HCl-WSBC + 0.5<br>HCl-SBC + 0.5<br>HCl-PBC + 0.5 | HCl-WSBC + 1.0<br>HCl-SBC + 1.0<br>HCl-PBC + 1.0 | HCl-WSBC + 2.0<br>HCl-SBC + 2.0<br>HCl-PBC + 2.0 |
| Activated with 0.1 M $C_2H_5OH$ | EtOH-WSBC<br>EtOH-SBC<br>EtOH-PBC | EtOH-WSBC + 0.1<br>EtOH-SBC + 0.1<br>EtOH-PBC + 0.1 | EtOH-WSBC + 0.2<br>EtOH-SBC + 0.2<br>EtOH-PBC + 0.2 | EtOH-WSBC + 0.5<br>EtOH-SBC + 0.5<br>EtOH-PBC + 0.5 | EtOH-WSBC + 1.0<br>EtOH-SBC + 1.0<br>EtOH-PBC + 1.0 | EtOH-WSBC + 2.0<br>EtOH-SBC + 2.0<br>EtOH-PBC + 2.0 |

**Table A2.** Concentration of base cations in NaCl solutions used in the experiment.

| Solution | Na | Ca | Mg | K |
|---|---|---|---|---|
| | | mg/L | | |
| 0.1% NaCl | 594.3 | 0.07 | 0.02 | 0.09 |
| 0.2% NaCl | 1188.6 | 0.12 | 0.03 | 0.18 |
| 0.5% NaCl | 2971.5 | 0.38 | 0.04 | 0.45 |
| 1.0% NaCl | 5943.2 | 0.8 | 0.05 | 0.9 |
| 2.0% NaCl | 11,886.5 | 1.7 | 0.06 | 1.8 |

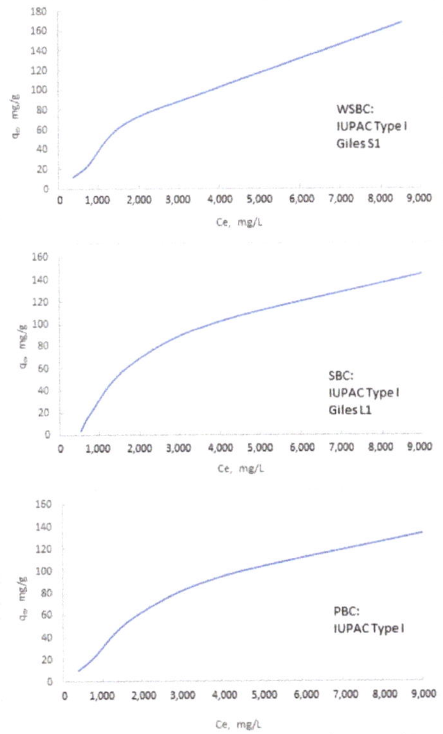


**Figure A1.** The changes of maximum amount of adsorbed Na ions on units of non-activated biochar mass, $q_m$, mg/g depending on $C_e$—equilibrium concentration of Na ions, mg/L. The IUPAC and Giles adsorption types of classification have been given.

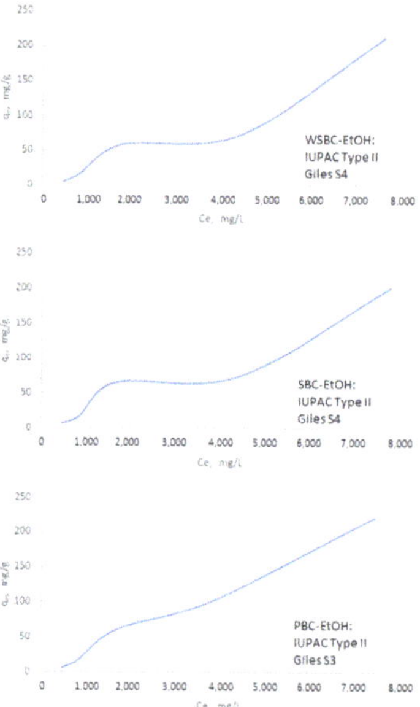


**Figure A2.** The changes of maximum amount of adsorbed Na ions on units of EtOH-activated biochar mass, $q_m$, mg/g depending on $C_e$—equilibrium concentration of Na ions, mg/L. The IUPAC and Giles adsorption types of classification have been given.

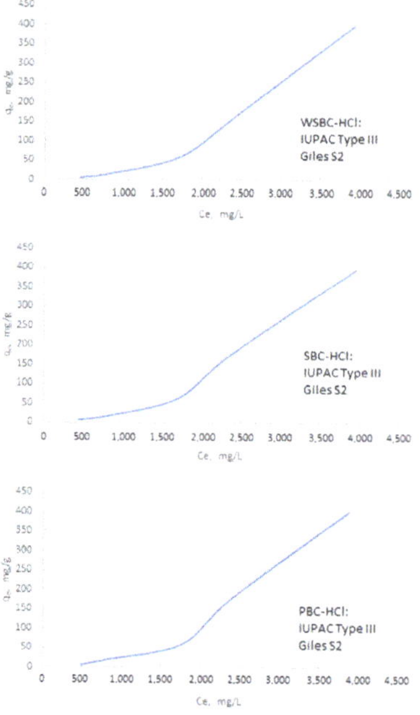


**Figure A3.** The changes of maximum amount of adsorbed Na ions on units of HCl-activated biochar mass, $q_m$, mg/g depending on $C_e$—equilibrium concentration of Na ions, mg/L. The IUPAC and Giles adsorption types of classification have been given.

## Appendix B

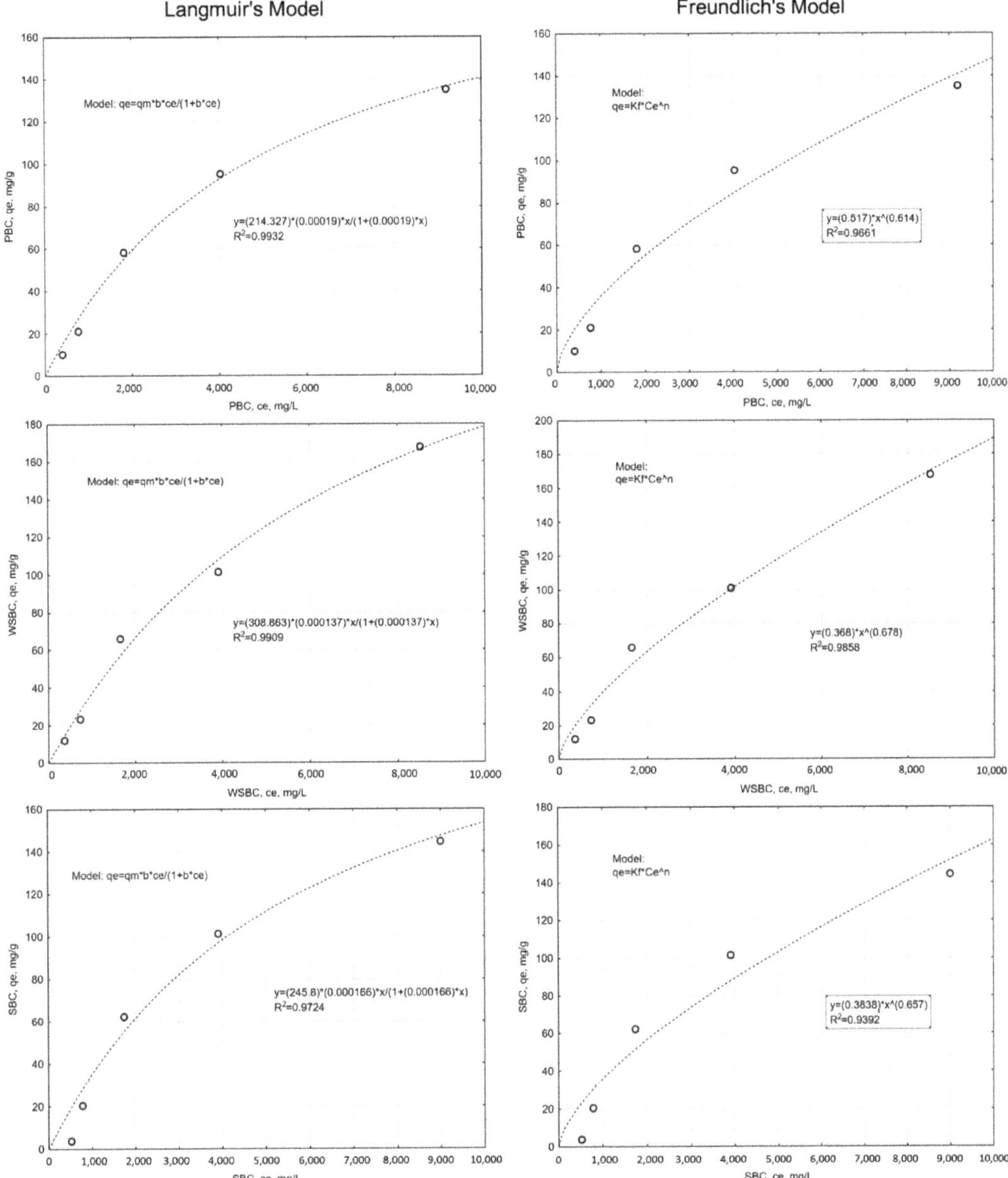


**Figure A4.** The modeling of Na adsorption of Langmuir's and Freundlich's isotherms on non-activated biochar, $q_e$—the equilibrium (instantaneous) amount of adsorbed Na ions on units of biochar mass, mg/g, $C_e$—equilibrium concentration of Na ions, mg/L. The model parameters and $R^2$ values have been given.

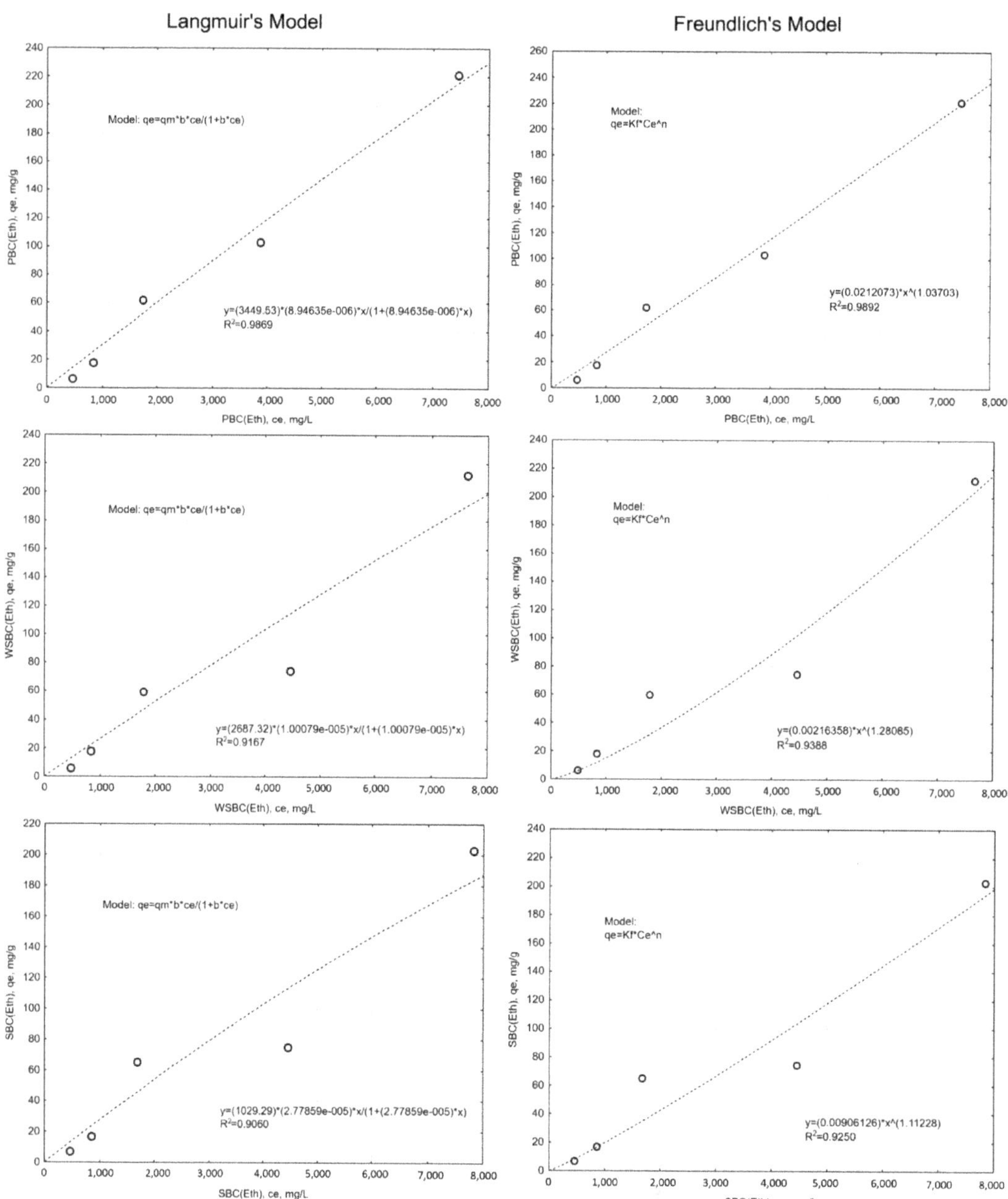


**Figure A5.** The modeling of Na adsorption Langmuir's and Freundlich's isotherms on ethanol-activated biochar, $q_e$—the equilibrium (instantaneous) amount of adsorbed Na ions on units of biochar mass, mg/g, $C_e$—equilibrium concentration of Na ions, mg/L. The model parameters and $R^2$ values have been given.

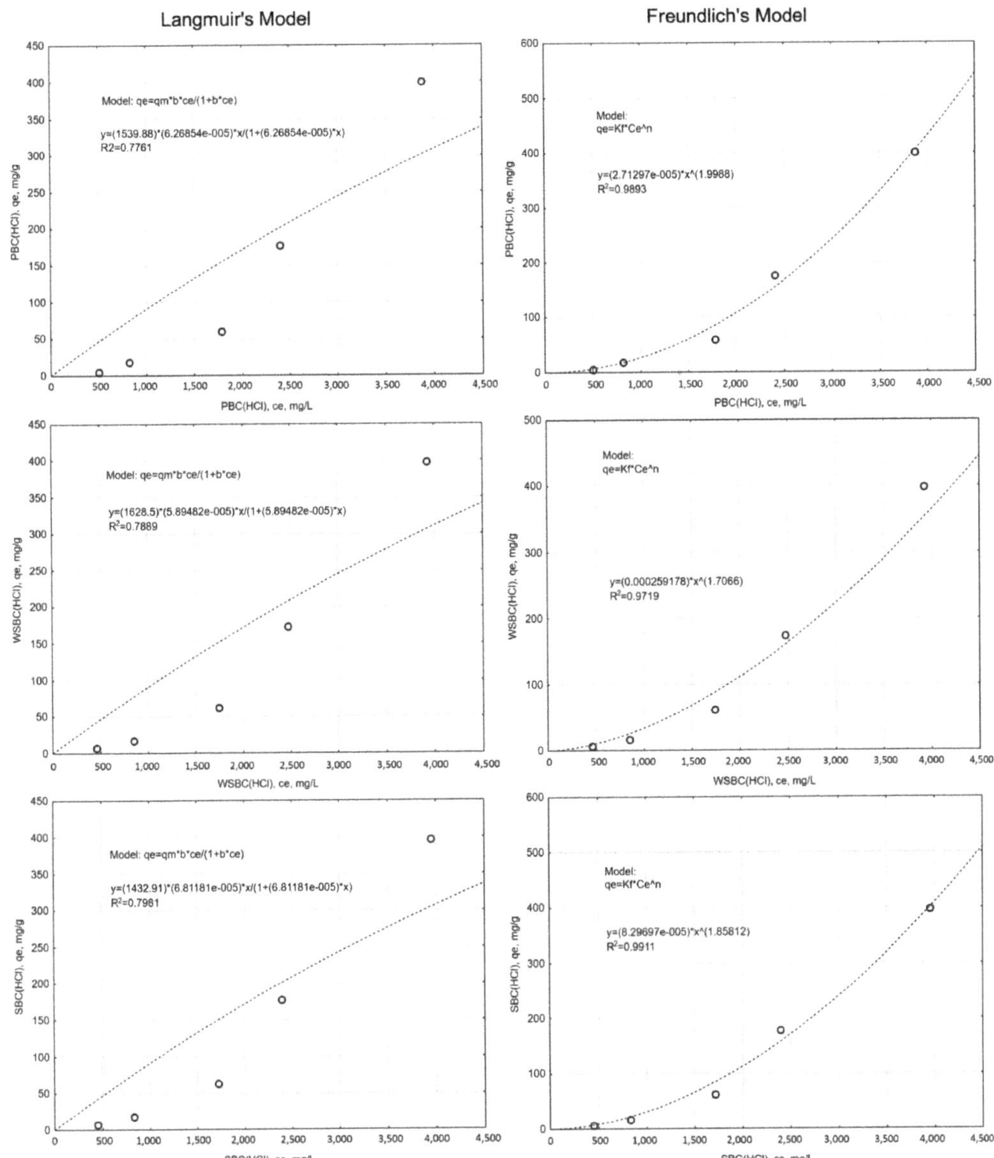


**Figure A6.** The modeling of Na adsorption Langmuir's and Freundlich's isotherms on HCl-activated biochar, $q_e$—the equilibrium (instantaneous) amount of adsorbed Na ions on unit of biochar mass, mg/g, $C_e$—equilibrium concentration of Na ions, mg/L. The model parameters and $R^2$ values have been given.

## References


1. Singh, R.; Mavi, M.S.; Choudhary, O.P. Saline Soils Can Be Ameliorated by Adding Biochar Generated from Rice-Residue Waste. *CLEAN Soil Air Water* **2018**, *47*, 1700656. [CrossRef]
2. Amini, S.; Ghadiri, H.; Chen, C.; Marschner, P. Salt-affected soils, reclamation, carbon dynamics, and biochar: A review. *J. Soils Sediments* **2015**, *16*, 939–953. [CrossRef]
3. Trenouth, W.R.; Gharabaghi, B.; Perera, N. Road salt application planning tool for winter de-icing operations. *J. Hydrol.* **2015**, *524*, 401–410. [CrossRef]

4. Gunarathne, V.; Senadeera, A.; Gunarathne, U.; Biswas, J.K.; Almaroai, Y.A.; Vithanage, M. Potential of biochar and organic amendments for reclamation of coastal acidic-salt affected soil. *Biochar* **2020**, *2*, 107–120. [CrossRef]
5. Tana, H.; Onga, P.Y.; Klemešb, J.J.; Bonga, C.P.C.; Lic, C.; Gaoc, Y.; Leea, C.T. Mitigation of Soil Salinity Using Biochar Derived from Lignocellulosic Biomass. *Chem. Eng. Trans.* **2021**, *83*, 235–240. [CrossRef]
6. Sadegh-Zadeh, F.; Parichehreh, M.; Jalili, B.; Bahmanyar, M.A. Rehabilitation of calcareous saline-sodic soil by means of biochars and acidified biochars. *Land Degrad. Dev.* **2018**, *29*, 3262–3271. [CrossRef]
7. Qadir, M.; Schubert, S.; Ghafoor, A.; Murtaza, G. Amelioration strategies for sodic soils: A review. *Land Degrad. Dev.* **2001**, *12*, 357–386. [CrossRef]
8. Rostamian, R.; Heidarpour, M.; Mousavi, S.F.; Afyuni, M. Characterization and Sodium Sorption Capacity of Biochar and Activated Carbon Prepared from Rice Husk. *J. Agric. Sci. Technol.* **2015**, *17*, 1057–1069.
9. Lee, J.E.; Park, Y.-K. Applications of Modified Biochar-Based Materials for the Removal of Environment Pollutants: A Mini Review. *Sustainability* **2020**, *12*, 6112. [CrossRef]
10. Mohan, D.; Sarswat, A.; Ok, Y.S.; Pittman, C.U. Organic and inorganic contaminants removal from water with biochar, a renewable, low cost and sustainable adsorbent—A critical review. *Bioresour. Technol.* **2014**, *160*, 191–202. [CrossRef]
11. Medyńska-Juraszek, A.; Ćwieląg-Piasecka, I.; Jerzykiewicz, M.; Trynda, J. Wheat Straw Biochar as a Specific Sorbent of Cobalt in Soil. *Materials* **2020**, *13*, 2462. [CrossRef]
12. Karami, N.; Clemente, R.; Moreno-Jiménez, E.; Lepp, N.W.; Beesley, L. Efficiency of green waste compost and biochar soil amendments for reducing lead and copper mobility and uptake to ryegrass. *J. Hazard. Mater.* **2011**, *191*, 41–48. [CrossRef] [PubMed]
13. Awan, S.; Ippolito, J.A.; Ullman, J.L.; Ansari, K.; Cui, L.; Siyal, A.A. Biochars reduce irrigation water sodium adsorption ratio. *Biochar* **2021**, *3*, 77–87. [CrossRef]
14. Akhtar, S.S.; Li, G.; Andersen, M.N.; Liu, F. Biochar enhances yield and quality of tomato under reduced irrigation. *Agric. Water Manag.* **2014**, *138*, 37–44. [CrossRef]
15. Dahlawi, S.; Naeem, A.; Rengel, Z.; Naidu, R. Biochar application for the remediation of salt-affected soils: Challenges and opportunities. *Sci. Total Environ.* **2018**, *625*, 320–335. [CrossRef]
16. Mukherjee, A.; Zimmerman, A.; Harris, W. Surface chemistry variations among a series of laboratory-produced biochars. *Geoderma* **2011**, *163*, 247–255. [CrossRef]
17. Li, J.-H.; Lv, G.-H.; Bai, W.-B.; Liu, Q.; Zhang, Y.-C.; Song, J.-Q. Modification and use of biochar from wheat straw (*Triticum aestivum* L.) for nitrate and phosphate removal from water. *DESALINATION Water Treat.* **2014**, 1–13. [CrossRef]
18. Du, J.; Kim, S.H.; Hassan, M.A.; Irshad, S.; Bao, J. Application of biochar in advanced oxidation processes: Supportive, adsorptive, and catalytic role. *Environ. Sci. Pollut. Res.* **2020**, *27*, 1–27. [CrossRef]
19. Sarpong, K.A.; Amiri, A.; Ellis, S.; Idowu, O.J.; Brewer, C.E. Short-term leachability of salts from Atriplex-derived biochars. *Sci. Total Environ.* **2019**, *688*, 701–707. [CrossRef]
20. Wu, H.; Yip, K.; Kong, Z.; Li, C.-Z.; Liu, D.; Yu, Y.; Gao, X. Removal and Recycling of Inherent Inorganic Nutrient Species in Mallee Biomass and Derived Biochars by Water Leaching. *Ind. Eng. Chem. Res.* **2011**, *50*, 12143–12151. [CrossRef]
21. Dietrich, C.C.; Rahaman, A.; Robles-Aguilar, A.A.; Latif, S.; Intani, K.; Müller, J.; Jablonowski, N.D. Nutrient Loaded Biochar Doubled Biomass Production in Juvenile Maize Plants (*Zea mays* L.). *Agronomy* **2020**, *10*, 567. [CrossRef]
22. Dou, G.; Jiang, Z. Preparation of Sodium Humate-Modified Biochar Absorbents for Water Treatment. *ACS Omega* **2019**, *4*, 16536–16542. [CrossRef]
23. Munera-Echeverri, J.; Martinsen, V.; Strand, L.; Zivanovic, V.; Cornelissen, G.; Mulder, J. Cation exchange capacity of biochar: An urgent method modification. *Sci. Total Environ.* **2018**, *642*, 190–197. [CrossRef]
24. Standard Reference Standard Test Methods for Loss on Ignition (LOI) of Solid Combustion Residues ASTM D7348—13. Available online: https://www.astm.org/Standards/D7348.htm (accessed on 16 August 2021).
25. Ćwieląg-Piasecka, I.; Medyńska-Juraszek, A.; Jerzykiewicz, M.; Debicka, M.; Bekier, J.; Jamroz, E.; Kawałko, D. Humic acid and biochar as specific sorbents of pesticides. *J. Soils Sediments* **2018**, *18*, 2692–2702. [CrossRef]
26. Fathy, M.; Mousa, M.A.; Moghny, T.A.; Awadallah, A.E. Characterization and evaluation of amorphous carbon thin film (ACTF) for sodium ion adsorption. *Appl. Water Sci.* **2017**, *7*, 4427–4435. [CrossRef]
27. Tang, Y.; Alam, S.; Konhauser, K.O.; Alessi, D.; Xu, S.; Tian, W.; Liu, Y. Influence of pyrolysis temperature on production of digested sludge biochar and its application for ammonium removal from municipal wastewater. *J. Clean. Prod.* **2018**, *209*, 927–936. [CrossRef]
28. Heymann, K.; Lehmann, J.; Solomon, D.; Schmidt, M.W.; Regier, T. C 1s K-edge near edge X-ray absorption fine structure (NEXAFS) spectroscopy for characterizing functional group chemistry of black carbon. *Org. Geochem.* **2011**, *42*, 1055–1064. [CrossRef]
29. Klüpfel, L.; Keiluweit, M.; Kleber, M.; Sander, M. Redox Properties of Plant Biomass-Derived Black Carbon (Biochar). *Environ. Sci. Technol.* **2014**, *48*, 5601–5611. [CrossRef] [PubMed]
30. Giles, C.H.; MacEwan, T.H.; Nakhwa, S.N.; Smith, D. Studies in adsorption. Part XI. A system of classification of solution adsorption isotherms, and its use in diagnosis of adsorption mechanisms and in measurement of specific surface areas of solids. *J. Chem. Soc.* **1960**, 3973–3993. [CrossRef]

31. Sing, K.S.W.; Everett, D.H.; Haul, R.A.W.; Moscou, L.; Pierotti, R.A.; Rouquerol, J.; Siemieniewska, T. Reporting physisorption data for gas/solid systems with special reference to the determination of surface area and porosity. *Pure Appl. Chem.* **1985**, *57*, 603–619. [CrossRef]
32. Tomczak, E.; Sulikowski, R. Opis równowagi i kinetyki sorpcji jonów metali ciężkich na klinoptylocie. *Inżynieria Apar. Chem.* **2010**, *49*, 113–114.
33. El-Banna, M.F.; Mosa, A.; Gao, B.; Yin, X.; Ahmad, Z.; Wang, H. Sorption of lead ions onto oxidized bagasse-biochar mitigates Pb-induced oxidative stress on hydroponically grown chicory: Experimental observations and mechanisms. *Chemosphere* **2018**, *208*, 887–898. [CrossRef] [PubMed]

 materials

*Review*

# Clay-Polymer Nanocomposites: Preparations and Utilization for Pollutants Removal

**Abdelfattah Amari [1,2,*], Fatimah Mohammed Alzahrani [3], Khadijah Mohammedsaleh Katubi [3], Norah Salem Alsaiari [3,*], Mohamed A. Tahoon [4,5] and Faouzi Ben Rebah [4,6,*]**

1 Department of Chemical Engineering, College of Engineering, King Khalid University, Abha 61411, Saudi Arabia
2 Department of Chemical Engineering, Research Laboratory: Energy and Environment, National School of Engineers, Gabes University, Gabes 6072, Tunisia
3 Chemistry Department, College of Science, Princess Nourah bint Abdulrahman University, Riyadh 11671, Saudi Arabia; fmalzahrani@pnu.edu.sa (F.M.A.); kmkatubi@pnu.edu.sa (K.M.K.)
4 Department of Chemistry, College of Science, King Khalid University, P.O. Box 9004, Abha 61413, Saudi Arabia; tahooon_87@yahoo.com
5 Chemistry Department, Faculty of Science, Mansoura University, Mansoura 35516, Egypt
6 Higher Institute of Biotechnology of Sfax (ISBS), Sfax University, P.O. Box 263, Sfax 3000, Tunisia
* Correspondence: abdelfattah.amari@enig.rnu.tn (A.A.); nsalsaiari@pnu.edu.sa (N.S.A.); benrebahf@yahoo.fr (F.B.R.)

**Abstract:** Nowadays, people over the world face severe water scarcity despite the presence of several water sources. Adsorption is considered as the most efficient technique for the treatment of water containing biological, organic, and inorganic contaminants. For this purpose, materials from various origins (clay minerals, modified clays, zeolites, activated carbon, polymeric resins, etc.) have been considered as adsorbent for contaminants. Despite their cheapness and valuable properties, the use of clay minerals as adsorbent for wastewater treatment is limited due to many factors (low surface area, regeneration, and recovery limit, etc.). However, clay mineral can be used to enhance the performance of polymeric materials. The combination of clay minerals and polymers produces clay-polymers nanocomposites (CPNs) with advanced properties useful for pollutants removal. CPNs received a lot of attention for their efficient removal rate of various organic and inorganic contaminants via flocculation and adsorption ability. Three main classes of CPNs were developed (exfoliated nanocomposites (NCs), intercalated nanocomposites, and phase-separated microcomposites). The improved materials can be explored as novel and cost-effective adsorbents for the removal of organic and inorganic pollutants from water/wastewater. The literature reported the ability of CPNs to remove various pollutants such as bacteria, metals, phenol, tannic acid, pesticides, dyes, etc. CPNs showed higher adsorption capacity and efficient water treatment compared to the individual components. Moreover, CPNs offered better regeneration than clay materials. The present paper summarizes the different types of clay-polymers nanocomposites and their effective removal of different contaminants from water. Based on various criteria, CPNs future as promising adsorbent for water treatment is discussed.

**Keywords:** water treatment; nanomaterials; adsorbents; clay-polymers nanocomposites

**Citation:** Amari, A.; Mohammed Alzahrani, F.; Mohammedsaleh Katubi, K.; Salem Alsaiari, N.; Tahoon, M.A.; Ben Rebah, F. Clay-Polymer Nanocomposites: Preparations and Utilization for Pollutants Removal. *Materials* **2021**, *14*, 1365. https://doi.org/10.3390/ma14061365

Academic Editor: Andrea Petrella

Received: 11 February 2021
Accepted: 8 March 2021
Published: 11 March 2021



## 1. Introduction

Clean water is a crucial factor of our world and shows a vital role to perform all aspects of life and continuous development [1,2]. Despite the presence of several water sources on earth, clean drinking water resources are limited in many regions of the world. This fact is observed in big cities characterized by large dense populations, and quick industrialization. Related to the quick industrialization, scientists reported the existence of more than 700 carcinogenic and highly toxic inorganic and organic micro-contaminants. They are considered as persistent environmental pollutants non-biotransformable or non-biodegradable. Toxic inorganic metals included for instance chromium, mercury, cadmium, lead, arsenic etc. while organic contaminants included drugs, phenols, plasticizers, polybrominated diphenyl ethers, polychlorinated biphenyls, polynuclear aromatic hydrocarbons, and pesticides.

Research efforts have been conducted to remove these contaminants from water/wastewater using several methods such as adsorption [3–5], electrolysis [6], electrodialysis [7], ion-exchange [8], reverse osmosis [9], conventional coagulation [10], and chemical precipitation [11]. Many of these methods, such as electrodialysis, electrolysis, ion-exchange, and reverse osmosis, are expensive and cannot be applied in developing countries. Conventional coagulation methods and chemical precipitation cause secondary contaminants requiring an additional treatment and increasing the treatment cost. Interestingly, adsorption is the most attractive process in developing countries due to its lower cost and its high efficiency to remove different types of contaminants. Generally, the selection of the useful adsorbent for water treatments is controlled by many factors such as the adsorption capacity of the material toward the target contaminant, cost/efficiency ratio, and the type and concentration of the contaminants present in water. The selected adsorbent must be available, non-expensive, easily regenerable, and have high selectivity toward target contaminants. Interestingly, the adsorption method produces low quantities of sludge, which is largely produced by other expensive methods like chemical precipitation. Efficient adsorbents from biological, organic or mineral origin have been developed for water treatment. The most important used adsorbents are polymeric resins [12], biomass [13], agricultural wastes [14], industrial by-products, silica beads [15], clay minerals, modified clays [16], zeolites [17], and activated carbon [18]. Among the cited materials, polymeric resins have the ability to adsorb a wide variety of organic, high surface area, wide pore structure range, and can be regenerated with no loss in their adsorption capacities. However, they have some disadvantages mainly the sensitivity to particle size, the poor water dissolution, and the pH dependence. Their performances are also affected by the specific surface area and the porosity, the type of the used resin and their non-suitability for all aromatic pollutants. Furthermore, they are not suitable for water treatment in developing countries due to their high cost of application.

Due to their cheapness and availability compared to other materials, clay minerals have been considered as adsorbent for the removal of several organic and inorganic contaminants from water [19]. However, their efficiency remains lower compared to activated carbon and zeolites. Generally, the practical use of clay materials as adsorbents is limited due to many factors such as the smaller surface area, the low adsorption tendency toward organic species, the difficulty in recovering clay particles from aqueous solutions, and their reduction after regeneration. In order to overcome these limitations, researchers have developed clay-polymer nanocomposites (CPNs) combining the advantageous characteristics of both clay minerals (cheapness, availability, eco-friendly, large surface area and stability) and polymers (mainly the high adsorption efficiency, high surface area and better regeneration) in a new adsorbent, which is nano-scaled or micro-scaled according to their modification process [20,21]. Interestingly, CPNs showed advanced properties (mechanical strength, low gas permeability and heat resistance) allowing cost reduction and enhancing their efficiencies in removing various contaminants from water. In addition to that, many of these CPNs can be synthesized from green materials, which are considered sustainable [22]. Consequently, it is very important to collect, analyze, and deepen the CPNs research work in order to develop safe and eco-friendly adsorbents applicable at a large scale for water treatment.

The present paper provides recent data concerning CPNs that have been prepared and used as adsorbents for the removal of pollutants from water/wastewater. All the utilities of CPNs in wastewater treatment will be discussed to clarify the future direction of the research work. This will help promote the development of a strategy for the design of CPNs adsorbents useful and applicable in water treatment process.

## 2. Structures and Types of Clay Minerals

Clay minerals are an assembly of crystalline minerals. This structure consists of thin sheets (phyllosilicates) arranged in layers. Layers are composed of octahedral ($[AlO_3(OH)_3]^{6-}$) and tetrahedral ($[SiO_4]^{4-}$) sheets connected through sharing of apical atoms of oxygen. Clay minerals layers stack via Van der Waals attractions to form interlayer. Clay minerals are classified into seven classes (Table 1): (i) fibrous-layered silicates such as sepiolite and palygorskite, (ii) 1:1 ratio of $[AlO_3(OH)_3]^{6-}$ octahedral and $[SiO_4]^{4-}$ tetrahedral sheets called kaolinite class such as serpentine, halloysite, and kaolinite, (iii) 2:1 non-expanding class in which the ratio between $[SiO_4]^{4-}$ tetrahedral and $[AlO_3(OH)_3]^{6-}$ octahedral sheets is 2:1 such as illite and mica, (iv) 2:1 uncharged class such as talc and pyrophyllite, (v) strongly expanding 2:1 class such as montmorillonite, (vi) limited expanding 2:1 class such as vermiculite, (vii) 2:1:1 class in which there is an extra brucite octahedral sheet sandwiched to form a 2:1 composition [23,24].

**Table 1.** Classification of clay minerals [23,24].

| Classes | Structure | Formulas |
|---|---|---|
| Kaolinite and serpentine | two-sheet phyllosilicates | kaolinite $Al_4[Si_4O_{10}](OH)_8$<br>serpentine $Mg_6[Si_4O_{10}](OH)_8$ |
| Micas | three-sheet phyllosilicates | $_2Al_4[(Si_{>6}Al_{<2})O_{20}](OH)_4{\cdot}nH_2O$ |
| Vermiculite | expanding three sheet phyllosilicates | $(Mg,Fe^{2+}, Fe^{3+})_6[(Si{>}Al)_8O_{20}](OH)_4{\cdot}nH_2O$ |
| Smectites | strongly expanding three-sheet phyllosilicates | montmorillonite:<br>$M^+_{x+y}(Al,Fe^{3+})_{4-y}(Fe^{2+},Mg)_y[Si_{8-x}Al_xO_{20}]{\cdot}(OH)_4{\cdot}nH_2O$,<br>beidellite: $M^+_xAl_4[Si_{8-x}Al_xO_{20}]$ $(OH)_4{\cdot}nH_2O$,<br>nontronite: $M^+_xFe^{3+}[Si_{8-x}Al_xO_{20}]$ $(OH)_4{\cdot}nH_2O$,<br>saponite: $M^+_xMg_6[Si_{8-x}Al_xO_{20}]$ $(OH)_4{\cdot}nH_2O$. |
| Pyrophyllite and talc | nonswelling three-sheet phyllosilicates | pyrophyllite: $Al_4[Si_8O_{20}]$ $(OH)_4$<br>talc: $Mg_6[Si_8O_{20}]$ $(OH)_4$ |
| chlorites | four-sheet silicates | $Al_4[Si_8O_{20}]$ $(OH)_4Al_4(OH)_{12}$ |
| Palygorskite and sepiolite | sheet-fibrous structure | palygorskite: $Mg_5[Si_8O_{20}]$ $(OH)_2(OH_2)_4{\cdot}4H_2O$<br>sepiolite: $Mg_8[Si_{12}O_{30}]$ $(OH)_4(OH_2)_4{\cdot}nH_2O$ |

$M^+$: represents adsorbed alkali cations ($Na^+$).

These clay minerals carry positive and negative charges due to isomorphous substitution of octahedral and tetrahedral sheets. Matrix and support of clay-polymers nanocomposites are provided by 2:1 class silicates like saponite, hectorite, and montmorillonite as well as palygorskite. These types of clay minerals are nano-scaled either in fiber diameter or in sheet thickness. Interestingly, organic compounds can be used for the modification of clay minerals to produce organoclays and this represents the first step of clay-polymer nanocomposites synthesis. Natural clay minerals are characterized by its hydrophilicity that allowed their saturation with lithium and sodium cations. Also, this hydrophilicity makes clay minerals incompatible with non-polar polymers but compatible with hydrophilic polymers like polyvinyl alcohol and polyethylene oxide [23]. Thus, non-polar polymers were made compatible with hydrophilic clay minerals via the synthesis of organoclays as preliminary stage. The conventional method of organoclays preparation is the combination of sulfonium, phosphonium, and ammonium with clay minerals [25]. Ion-exchange reaction is used to replace clay minerals cations by positively charged surfactant like hexadecyl trimethylammonium to make clay hydrophobic and allow its modification with hydrophobic polymers. Polar-nonpolar partition mechanisms provided to natural clay minerals via organoclays are known to improve the adsorption capacity of pollutants. Also, they allow an intercalation of pollutants and an increase of positive charges over clay minerals surfaces, which improve the chelation power [26].

Because of their cheapness, availability, and valuable properties, clay minerals are of large interest for their potential applications in various areas such as wastewater treatment. However, the application of clay minerals is limited due to many factors (low surface area, regeneration, and recovery limit, etc.) [19,22].

## 3. Nanocomposites of Clay-Polymers

Clay minerals can be used to enhance the performance of polymeric materials. The combination of clay minerals and polymers is considered as promising reinforcements to produce clay-polymer nanocomposites with advanced properties useful for pollutants removal. There are three main classes of nanocomposite clay-polymers; (i) exfoliated nanocomposites, (ii) intercalated nanocomposites, and (iii) phase-separated microcomposites [26]. In the exfoliated nanocomposites, the clay layers are fully dispersed within the polymer matrix. They are composed of ~1 nm clay layers separated within the matrix of polymer. In intercalated nanocomposites, the silicate crystalline layers are inserted within the polymeric chains to form nano-scaled clay-polymer composite. The phase-separated microcomposite, resulted from embedded polymer inside silicate layers, have similar properties as conventional microcomposites. This polymer nanocomposites display exceptional properties regarding the biodegradability, self-extinguishing behavior, heat distortion, flexural properties, tensile strength, and high modulus. CPNs may include many polymers in their fabrication like chitosan (CS), polystyrene, polypropylene, polyesters, polyurethanes, epoxies, and polyvinyl chloride (PVC).

### *3.1. Clay-Polymers Magnetic Nanocomposites*

Polymer matrix of CPNs could be filled with magnetic nanoparticles (NPs) to develop their functionality. For instance, magnetically coated nanoparticles were prepared by casting mix of oleic-acid-coated nanoparticles, poly(butyl acrylate) emulsion, and clay (laponite) [27]. However, this nanocomposite with low content of oleic acid exhibits greater saturation magnetization than that of pure oleic acid-coated magnetic nanoparticles component. This magnetic composite was synthesized via three steps; first, clay dispersion and mixed polymer were transformed into gel. Then, oleic acid-magnetic NPs and polymer were separately fixed in clay platelets and finally, all materials were merged in the main composite. Thus, polymeric matrix is used to fix oleic acid magnetic NPs and clay to produce exclusive nanomaterial. In another work, palygorskite–iron-oxide magnetic nanocomposite was synthesized via environmentally safe and economical precipitation synthesis [28].

### *3.2. Clay-Biopolymers Nanocomposites*

Biopolymers (polynucleotides, nucleotides, polypeptides, polysaccharides, and proteins, etc.) are green and biocompatible materials useful in the synthesis of clay NCs with improved properties like functionality and reactivity. The improved properties offered by biopolymers allowed the application of the clay-biopolymers nanocomposites in different fields like packaging materials, tissue engineering, drug delivery, controlled pesticide, and electrochemical sensing. Chitosan is an example of biopolymers extracted from crustacean seafood wastes [29] and composed of units of *N*-acetylglucosamine and glucosamine. This biopolymer was used for the synthesis of clay nanocomposite useful for the removal of pollutants from water [30]. The synthesis of clay-chitosan nanocomposite was described in another work [31] as shown in Figure 1.

### *3.3. Nanoclay of Polyacrylamide-Polymers Nanocomposites*

Bentonite nanoclay polymer composite was synthesized using acrylic acid. First, alkali medium was used for the synthesis of acrylamide from acrylic acid. Then, bentonite was added in nitrogen atmosphere and cross-linking reactions were carried out via *N,N′*-methylenebisacrylamide. After that, the polymerization reaction was carried out at 70 °C in the presence of $(NH_4)S_2O_8$ in order to produce the nanoclay-polymer [32]. The characterization of the obtained composite using various methods (Fourier transform infrared spectroscopy, scanning electron microscopy, and X-ray diffraction) showed a complete exfoliatation and dispersion of the bentonite layers in the composite after the polymerization, leading to an increase of the surface area. The new composite can be exploited in many fields such as agriculture [32].

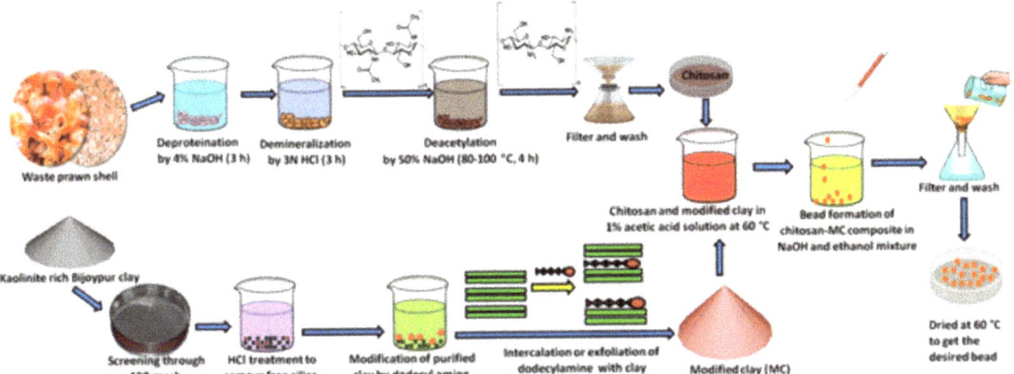


**Figure 1.** The scheme of clay-chitosan nanocomposite synthesis [31].

### 3.4. *Nanocomposites of Clay-Polysiloxane*

Organoclays were usually used as matrix during the synthesis of NCs silicate-polysiloxane in the presence of little amount of $H_2O$. Silanol terminated poly(dimethylsiloxane) was mixed with organo- montmorillonite, which is a mix of dimethyl ditallowammonium alkyl chains of C12, C14, C16, and C18 and cross-linked with tetraethylorthosilicate in the presence of catalyst to form NCs of silicate-polysiloxane [33]. The production of exfoliated NCs needs water and the mixing of the modified clay with silanol-terminated poly(dimethylsiloxane) should be carried out under sonication. Moreover, the nature of the clay modifier and silicon affects considerably the NCs synthesis. Hence, the absence of exfoliation and intercalation when montmorillonite is modified with benzyl dimethyloctadecylammonium cation was reported. Also, the use of silanol-terminated poly(dimethylsiloxane) containing only 14% (mol/mol) of diphenylsiloxane units showed only intercalation. Therefore, the optimization of the layered silicate exfoliation required right fitting between the matrix and organoclay [33].

### 3.5. *Nanocomposites of Clay-Polystyrene*

Polystyrene-clay composites were produced using the addition polymerization technique [34]. A vinyl monomer-montmorillonite intercalate was synthesized by the cation exchange process. Then, free-radical polymerization of styrene in the presence of vinyl monomer-montmorillonite intercalate was achieved leading to the production of polystyrene-montmorillonite materials. The solvation of vinyl monomer-montmorillonite intercalate using acetonitrile enhance the intercalation of styrene between the vinyl monomer- montmorillonite [34]. According to Vaia and Giannelis (1997), the melt intercalation is the best efficient method for the intercalation of polymer into organoclay [35]. The intercalation of polymers depends on the length of the alkyl chain in alkylammonium-exchanged smectites and on the annealing temperature. Under 160 °C, no intercalation was reported for chain having less than 12 carbon atoms. At 180 °C, intercalation occurred independently of the carbon atom number. However, it was reported that the intercalation was not affected by the molecular weight of the polystyrene [35].

### 3.6. Blends of Organoclay-Polymer

Organoclay with polymer blends such as poly(l-lactide), poly(ε-caprolactone)–clay, and poly-ethylene oxide–clay blends were prepared as reported by Ogata et al., (1997) [36]. Distearyl dimethylammonium-modified montmorillonite was mixed and blended with pellets of poly(l-lactide) in warmed chloroform. A 100-mm thick film was formed after evaporation of chloroform. Tactoids of unmodified montmorillonite were aligned on the surface of the film while intercalation of poly(l-lactide) into modified montmorillonite was absent [36]. The modified clay produced an interesting geometric superstructure (tactoids with several silicate monolayers) showing the enhancement of the Young's modulus of the poly(l-lactide)-based composites. Also, storage and loss moduli were enhanced [36].

### 3.7. Hybrid Clay-Polyimide

Clay-polyimide NCs were synthesized by polymerization using pyromellitic dianhydride and 4,4′-diaminodiphenyl ether in dimethylacetamide solution and in the presence of organoclay [37]. The synthesized polyimide–clay nanocomposites with limit fraction of organoclay (2% wt/wt) showed remarkable reduction in the permeability of several gases ($H_2O$, $O_2$, $CO_2$, etc.). Moreover, the increase in ratio of the clay mineral caused a decrease of thermal expansion coefficient and montmorillonite exhibited great exfoliation when NCs were synthesized [37].

### 3.8. Nanocomposites of Clay-Epoxy

To improve the solvent resistance and thermal stability of NCs, highly expanded onium-modified clay minerals were diffused into epoxide [38]. Epoxy support was formed using catalytic curing agents, anhydride, aromatic amines, and aliphatic amines having wide glass-transition temperature. Interestingly, new type of exfoliated structure was observed and the matrix tensile properties were improved by the reinforcement offered by the silicate nanolayers. The tensile properties of the resulted nanocomposites (epoxy−magadiite elastomeric) are comparable to other smectite clays [38].

### 3.9. Nanocomposites of Clay-Polyurethane

The solvation process allowed the intercalation of polyols into onium modified organoclays. Structure and length of onium ions greatly affected the intercalation process. The polymerization process of polyolisocyanate precursor–organoclay allowed the production of NCs with clay phase (50 Å basal spaces) intercalated in the cross-linked polyurethane network. An interesting benefit was observed with montmorillonites exchanged with chain onium ions containing a number of carbon atoms more than 12. The produced NCs showed higher toughness and strength compared to pristine polymer NCs [39].

Here, we reported the techniques used to develop nanocomposites of clay-polymers. Using clay minerals, significant technical advantages were achieved (mechanical, thermal, degradative, rheological properties, etc.) allowing the improvement of the new materials. The improved materials can be explored as novel and cost-effective adsorbents for the removal of organic and inorganic pollutants from water/wastewater.

## 4. Clay-Polymers Nanocomposites for Pollutants Removal

Based on their improved performances, CPNs showed the ability to remove various pollutants (bacteria, metals, phenol, tannic acid, pesticides, dyes, etc.) from water/wastewater. Below, the pollutant removals reported in the literature will be discussed.

### 4.1. Biological Pollutants Removal

Drinking water containing microbes, such as *Entamoeba histolytica*, *Shigella sp.*, and *Salmonella typhi*, is responsible for dangerous water borne infections such as diarrhea, dysentery, and typhoid [40]. The reduction of pathogen population was traditionally ensured by the chlorination technology. However, nowadays, this technology becomes less used due the existence of soluble organic pollutants that allowed the formation of secondary toxic contaminants (chlorophenols, haloacetic acids, trihalomethanes, etc.) [41]. Also, chloramination process showed a higher potential to form nitrosamines than chlorination. Earlier, the use of sand slow filtration was suggested but with limited efficiency due to high cost of filter repair and blocking [42]. Interestingly, composites of clay-polymers were used for antimicrobial activity but the used polymers must be soluble. For instance, antimicrobial efficiency of copper-doped montmorillonite low-density polyethylene (LDPE) nanocomposites toward *E. coli* was examined [43]. The prepared CPNs contain montmorillonite-$Cu^{+2}$ complex allowed high antimicrobial activity against *E. coli* (99%). Cell death was suggested via complexation between microbial nucleic acids and proteins with copper ions [44].

Similarly, other CPNs such as montmorillonite- polydimethyloxane-chlorhexidine acetate, and clay-polydimethyloxane-chitosan-silver showed the ability to retard the growth of different bacterial strains (*E. coli, Candida albicans, Pseudomonas aeruginosa, and Staphylococcus aureus*) [45–48]. *E. coli* was also removed via bentonite clay composite modified with starch-grafted quaternary ammonium ethers. The bacterial removal was achieved due to the presence of cationic monomers on clay surface [47]. The existence of monomer enhanced the composite with positive charges, that electrostatically attracted the negative phospholipids of the bacterial cell membrane causing cell disruption. Thus, CPNs are more efficient than polymers alone for bacterial removals. Also, chitosan-based NCs are known to have an antimicrobial activity. Interestingly, montmorillonite-chitosan NCs was more efficient for the removal of *S. aureus* and *E. coli* than montmorillonite and chitosan alone [49]. Additionally, bentonite silver and zinc oxide NCs were examined for bacterial removal via adhesion and killing mechanism [50]. Furthermore, *E. coli* was removed via NCs of rectorite modified with PVA, chitosan, and sodium-dodecylsulfonate compared to the single components [46,51]. High antimicrobial activity was also reached via chitosan-organoclay and chitosan-montmorillonite modified with $Ag^{+}$ nanoparticles [52,53].

Generally, the microbial removals using CPNs is controlled by various factors such as the properties of the adsorbent (surface charge, composition, etc.), the characteristics of the bacterial strains and the operating conditions (pH, temperature, dosage, etc.). In this context, most chitosan NCs showed antibacterial activity in acidic medium but, *N*-(2-hydroxyl)propyl-3-trimethyl ammonium chitosan Cl displayed antibacterial activity in a large range of pH [46,54,55]. Finally, the reuse of any CPNs is very important due to the cost of application. As reported in the literature, the reuse of CPNs applied for microbial removal was achieved using various reagents such as NaClO, HCl, and steam [47,56]. As a result of their high antimicrobial activity, CPNs can be applied efficiently to disinfect contaminated water.

### 4.2. Organic Pollutants Removal

The presence of dangerous toxic organic pollutants (organic acids, phenols, pesticides, dyes, etc.) may cause environmental problems even at trace amounts. Interestingly, these organic contaminants were removed using CPNs as adsorbent, due to the existence of polymeric hydrophobic parts on the NCs surfaces [57–90]. In this perspective, several CPNs in the form of powder or tablets were prepared using the thin-coated layer methods. These methods provide an enhancement of the adsorbent reusability after adsorption of toxic contaminants [57]. As listed in Table 2, different organic contaminants can be removed from water via adsorption process using CPNs.

**Table 2.** Removal of different organic contaminants from water using clay-polymers nanocomposites (CPNs) adsorbents.

| Adsorbent | Adsorbate | Temp. (°C) | pH | Removal Efficiency (%) or Adsorption Capacity (mg/g) | Isotherm Model | Kinetics Model | Ref. |
|---|---|---|---|---|---|---|---|
| Chitosan-coated attapulgite | Tannic acid | – | 5.5 | 95.2 mg/g | Freundlich | Pseudo-second order | [58] |
| palygorskite/chitosan resin microspheres | Tannic acid | – | 8.0 | 455.0 mg/g | Langmuir | Pseudo-second order | [59] |
| Chitosan/bentonite | Tartrazine | 47 | 2.5 | 294.1 mg/g | Langmuir | Pseudo-second order | [66] |
| Chitosan-g-poly (acrylic acid)/montmorillonite | Methylene blue | – | 6.5 | 1895.0 mg/g | Langmuir | Pseudo-second order | [72] |
| Chitosan/montmorillonite | Congo red | 30 | 4.0 | – | Langmuir | Pseudo-second order | [73] |
| Amino-modified polyacrylamide–bentonite NCs | Malachite green | 30 | 6.0 | 656.5 mg/g | Freundlich | Pseudo-second order | [78] |
| Mixture of bentonite, acrylic polymer, and polyethylene-diamine (Zwitterionic adsorbent) | Acid Red and Brilliant Green | 27 | – | 70.09 and 255.99 mg/g | – | – | [79] |
| AAm-AMPSNa/clay hydrogel nanocomposite and acrylamide (AAm)-2-acrylamide-2-methylpropanesulfonic acid sodium salt (AMPSNa) hydrogel | Brilliant cresyl blue and Safranine-T | 25 | – | 494.2 and 484.2 mg/g | Langmuir | Pseudo-second order | [80] |
| Humic acid-modified bentonite | 2,4-dichlorophenol | 30 | 6.5 | 14.23 mg/g | – | – | [81] |
| montmorillonite/layer double hydroxide | Methyl orange | – | – | 88% | – | – | [82] |
| montmorillonite /layer double hydroxide | Methylene blue | – | – | 74% | – | – | [82] |
| Chitosan/bentonite | Amido Black 10B | 20 | 2.0 | 323.6 mg/g | Langmuir | Pseudo-second order | [83] |
| Hydrogels of Kappa-carrageenan-g-poly(acrylamide)/sepiolite NCs | Crystal violet | Ambient Temp. | 10.0 | 47.0 mg/g | Langmuir | Pseudo-second order | [84] |
| Poly(acrylic acid-co-2-acrylamido-2-methylpropanesulfonic acid)/ montmorillonite | Methylene blue | 25 | 10.0 | 215.0 mg/g | Redlich–Peterson | Pseudo-second order | [85] |
| Alginate–clay quasi-cryogel beads | Methylene blue | 40 | – | 181.8 mg/g | Langmuir | Pseudo-second order | [86] |
| Chitosan/bentonite | Malachite green | 37 | 6.0 | 435.0 mg/g | Langmuir | Pseudo-second order | [87] |
| polyaniline/montmorillonite clay nanocomposites | Green 25 | 20 | 6.0 | 100% | Langmuir | Pseudo-second order | [88] |
| Tetraethoxysilane-functionalized Na-bentonite (2 wt%) incorporated into polysulfone/polyethylenimine membranes | Methylene blue | Room Temp. | – | 98.9% | – | – | [89] |
| carboxy methyl cellulose/ nano-organobentonites | Nine pesticides | – | – | 57–100% | – | – | [89] |

#### 4.2.1. Phenol, Tannic Acid and Pesticides Removal

The direct release of organic pollutants in water greatly affects the environment. Among these organics, polyphenolic tannic acid, which resulted from decomposed organic materials, is highly harmful for organisms in aquatic environment. Commonly, tannic acid compounds exist in surface water and soil. Consequently, it is important to find the proper way for their removal. In this context, attapulgite clay coated with chitosan was used as a new adsorbent for tannic acid removal with an adsorption capacity of 95.3 mg/g. This adsorption resulted from Van der Waals attractions, H-bonding, and electrostatic interactions [58]. Interestingly, this nanocomposite was improved to adsorb more tannic acid (456 mg/g) using solid matrix of protonated palygorskite clay (20% wt/wt) with chitosan resin microspheres [59]. In this study, the reusability of this novel nanocomposite was studied using hydrochloric acid (0.1 M) as an eluent up to three cycles showing a significant decrease of the adsorption capacity (58 mg/g) of tannic acid [59].

Similarly, phenolic materials such as 4-chlorophenol and phenols were removed using alginate polymer support incorporated by clay modified with a surfactant hexadecyl trimethyl ammonium. This adsorbent allowed a removal capacity of 0.119 mg/g and 0.335 mg/g respectively for 4-chlorophenol and phenols [60]. Also, trichlorophenol and trinitrophenol were removed using nano- montmorillonite synthesized from poly-4-vinylpyridine-co-styrene with removal efficiency of 60% and 99.6%, respectively. In this study, it has been suggested that a weak Van der Waals interaction is involved in the adsorption mechanism [61]. In another study, phenolic materials were removed by nanocomposite prepared using cetyltrimethylammonium poly(diallyldimethylammonium) to modify montmorillonite clay. Although the surfactant was not correctly intercalated inside the clay montmorillonite, the adsorption capacity toward phenols was enhanced [62].

Wastewater was also treated to remove pesticides like 2,4-dichlorophenol and atrazine using CPNs as adsorbents. 90% poly-4-vinylpyridine-co-styrene was used to modify montmorillonite allowing the removal of 99% of atrazine within only 40 min [63]. Interestingly, the synthesized NCs decreased the presence of pesticide to 3 μg/L during the use of fixed bed experiment. However, the adsorption capacity of these NCs was decreased in the presence of other organic soluble contaminants due to the competition occurred on the adsorbent surface. In another study, dinitrophenols were removed using membranes filled with NCs of organo- montmorillonite/ polyethersulfone. The NCs were prepared using wet-phase inversion and solution dispersion methods [64]. The addition of 4% of organo-montmorillonite enhanced the nitrophenols removal. Nitrophenols were removed via hydrogen bonding over NCs surfaces at pH 4.6. Furthermore, more than 70% of mecoprop and clopyralid anionic pesticides was removed over hexadimethrine/ montmorillonite NCs due to electrostatic interactions with adsorbent $NH^{4+}$ groups [65].

#### 4.2.2. Dyes Removal

The dyes removal from water/wastewater using clay/polymers nanocomposites was reported by many studies. In this context, many researchers reported the use of chitosan polymer [66–73]. For example, clay-polymer nanocomposite of chitosan/bentonite was synthesized and tested for the removal of azo dye tetrazine, known by its harmful effect on aquatic organisms [66]. According to this study, the pH value is the driving force of the adsorption process. At low pH, the +ve charge was produced allowing the electrostatic attractions between dye and nanocomposite surfaces. However, at higher pH value, the charge became –ve reducing the adsorption process due to the repulsion forces between NCs surfaces and the dye. The cross-linked chitosan in the nanocomposite with epichlorohydrin becomes water insoluble improving the adsorption process. Also, methylene blue (MB) dye was removed using chitosan/clay magnetic beads at pH 3–12 with weight ratio of clay to chitosan of more than 0.5. These magnetic NCs provided faster removal of MB within 13 min (50% removal efficiency and 83 mg/g maximum adsorption capacity) [67]. The adsorption process occurred via a mechanism of electrostatic attractions between clay –ve charge and dye +ve charge for wide pH range. Similarly, chitosan-clay NCs were used for the removal of Rhodamin-6 G with adsorption capacity of 446.43 mg/g [68]. At high pH, the adsorption of this positively charged dye occurred due to the attraction with the negatively charged adsorbent. The pseudo-second-order model was found to explain the adsorption kinetics of Rhodamin-6 G [68]. In the same way, high removal efficiency (99.99%) was obtained for MB using chitin-clay NCs with an adsorption capacity of 152.3 mg/g. The adsorption was greatly linked to the pH and followed the Langmuir isotherm [69]. Chitosan modified nano-organoclay was also applied for the removal of reactive red-141 (adsorption capacity of 440 mg/g) and reactive blue-21 (adsorption capacity of 477 mg/g). The adsorption data of the two dyes followed pseudo-second-order model [70]. For reactive dye (Yellow 3 RS), the adsorption capacity reached 71.39 mg/g over chitosan modified palygorskite clay NCs, which is significantly higher than that obtained with unmodified clay materials. (6.4 mg/g) [71]. At high pH, the adsorption was decreased due to the repulsion forces between similar charges while at low pH the positive charge was formed over chitosan surface due to the existence of amino groups allowing the adsorption via electrostatic attractions [71].

Other polymer types were used to formulate PNCs useful for dyes removal [74–78]. For example, reactive yellow K-4G, and disperse yellow-brown S-2RFL were removed via NCs of poly(epicholorohydrin-dimethylamine) and bentonite with maximum adsorption capacity of 30.9 mg/g and 12.5 mg/g, respectively. The complete dyes removal was reached via the optimization of polymers quantity [74]. Using the same approach, humic acid immobilized-amine-modified bentonite–polyacrylamide nanocomposite was tested for the removal of various dyes (malachite green, MB, and crystal violet) from wastewater [76]. This clay-polymer nanocomposite showed maximum adsorption capacity of malachite green, MB, and crystal violet of 658, 649, and 511 μmol/g, respectively at pH 4.8. In This study, intraparticle diffusion mechanism was suggested [76].

Recently, De Sousa et al. synthesized hybrid clay-polymer nanocomposite using sodium bentonite and ureasil-poly(ethylene oxide) via sol gel method for the removal of MB from water. Interestingly, MB was removed rapidly and efficiently through the adsorption process [77]. In another work, MB was removed from water using hybrid membranes based on kaolin and polystyrene. These membranes were fired at 1000 °C allowing the modification of polystyrene and the formation of cavities inside the clay matrix. The resulted materials have porous structure with expanded surface area. An optimum removal of MB from water was obtained with clay membrane loaded with 20% (wt/wt) polystyrene [78]. The mechanism of MB removal over clay-polymer composites was described in the literature [31] as shown in Figure 2.

**Figure 2.** The mechanism of methylene blue (MB) removal over clay-polymer composite [31].

Generally, dyes removal efficiency offered by the new CPNs varied depending on the polymerization process and the used raw materials. For each dye, the adsorption performance (removal rates and the adsorption capacity) varied from one experiment to another depending on the CPNs nature and operating conditions (dosage, pH, temperature, contact time, dye type, initial dye concentration, etc.).

### 4.3. *Inorganic Pollutants Removal*

#### 4.3.1. Toxic Gases Removal

Clay-polymer nanocomposites can be used for the adsorption of toxic gases like ammonia and metallic gases such as mercury $Hg^0$. For example, chitosan-bentonite composite was used as adsorbent for the elimination of $Hg^0$. In this study, the authors found that the low surface area of composite compared to single bentonite caused less removal of $Hg^0$ [91]. Also, a composite, synthesized from acid-activated bentonite and natural palygorskite via polymerization reaction between *N,N'*-methylenebisacrylamide and acrylic acid, was used for the adsorption of ammonia [92]. To increase the ability to remove ammonia, Lewis acid-base interaction was used to bind copper divalent ions with the composite. Among all acid-treated bentonite/acrylic acid polymer composites, 66% acid-treated bentonite exhibited the highest adsorption capacity. However, 75% palygorskite showed the maximum adsorption capacity among all palygorskite/acrylic acid polymer nanocomposites. The differences in adsorption capacity between acid-treated bentonite/acrylic acid polymer and palygorskite composites are related to the existence of mesopores (diameters between 2 and 50 nm) in the composite (copper-complexed clay/poly-acrylic acid composite), that allowed the interaction between polymers active sites and ammonia [92].

#### 4.3.2. Metalloids and Heavy Metals Removal

Urbanization, fertilizer application, and rapid industrialization are the major sources of water toxic metalloids and heavy metals such as As, U, Se, Sb, Cr, Cd, Ni, Pb, and Cu. Metalloids and heavy metals represent a serious danger for human and living organisms. These pollutants can be accumulated inside different organisms causing harmful effects. Therefore, their removal from water is essential for healthy environment [93].

Efforts have been conducted to remove the toxic heavy metals from water using clay as the support and chitosan as the polymer [94–111]. Divalent copper ions were removed from water using chitosan–silver NPs clay composite with a maximum adsorption capacity (181.6 mg/g) at pH 7 [94]. The high adsorption at pH 7 resulted from the interaction between chitosan functional groups and Cu(II) ions. At high pH, the formed cupric hydroxide decreased the adsorption process of copper ions. Another nanocomposite, poly-methacrylic acid grafted chitosan–bentonite, was tested for the removal of Cd (II), Pb (II), and Hg (II) ions. In this case, removal values of 78%, 89%, and 94%, respectively for Cd (II), Pb (II), and Hg (II) were obtained at high pH [95]. For the same purpose, chitosan/attapulgite composites were synthesized and tested for the removal of Cr(III) and Fe(III) from aqueous solution. The maximum adsorption capacity was 11.65 and 10.41 mg/g for Cr(III) and Fe(III), respectively. These values were obtained with adsorbent dosage of 0.2 g/l. However, the adsorption efficiency was found to be increased by increasing the temperature. Generally, the metal sorption is controlled by the process of electrons sharing between adsorbent and metals or through the covalent bonding [96].

In the same context, Chitosan–Al-pillared montmorillonite nanocomposite was used to remove Cr (VI) with an adsorption capacity of 15.68 mg/g [97]. Lead and copper divalent ions were removed with a similar nanocomposite prepared with clay: chitosan ratio of 1:0.45 to remove Pb (II) and Cu (II) at pH 6.5 with an efficiency of 99.5% and 96%, respectively via chemisorption mechanism. This nanocomposite when treated with nitric acid showed excellent reuse results [98]. Also, clay-polymer nanocomposite was successfully synthesized using chitosan and fibrous clay minerals like palygorskite. Different mass ratios of palygorskite to chitosan (1:1, 2:1, and 1:2) was prepared and applied for the removal of Pb (II) from water. A maximum removal capacity (201.5 mg/g) was obtained using the ratio 1:1 [99]. The availability of pores equally from both clay and polymer is the reason for high removal of Pb (II). In another study, the lead adsorption was increased by many folds from aqueous solution when the clay mass increased by 10% with respect to the polymer polystyrene inside the composite [100]. In a recent research, high amount of lead ions were removed using ion-imprinted polymer/ montmorillonite NCs [101].

For the same purpose, chitosan-grafted polyacrylic acid bentonite was used to remove various metals (Ni (II), Cd (II), Zn (II), and Cu (II)) from water. Maximum removal was obtained at pH of 8, 6, 7, and 6, respectively for Ni (II), Cd (II), Zn (II), and Cu (II) [102]. The maximum adsorption at high pH can be explained by the existence of negative charge over the adsorbent surfaces allowing the chelation of positive heavy metals. Similarly, chemisorption mechanism explained the efficient metal removal (Ni (II), Cu (II), and Pb(II)) from aqueous solution by chitosan immobilized on bentonite using ethylene glycol diglycidyl ether as a cross-linker [103]. This adsorbent offered high adsorption capacity of 15.82, 21.55, and 26.38 mg/g respectively for Ni (II), Cu (II), and Pb(II) [104]. An exothermic adsorption was observed for these metals as concluded from the thermodynamic parameters with the decreasing of the entropy behavior [105–110]. Additionally, adsorption of Cu (II) was spontaneous only at 25 °C, while that of Pb (II) was always spontaneous. Concerning Ni (II), the adsorption was non-spontaneous at 25–55 °C [104]. In the same work, the desorption process was studied showing high recovery of Cu (II) (92%) obtained with HCl solution (pH 1) and under agitation (2 h) [104]. Remarkably, the effectiveness of membrane of chitosan-kaolin-based ceramic synthesized using PVA as the chelating material for the removal of Hg and As from aqueous solution was also demonstrated [111].

The interesting opportunity offered by the implication of chitosan in the adsorption was well performed. The mechanism of heavy metals removal over chitosan-clay nanocomposite was described in the literature [31] as shown in Figure 3.

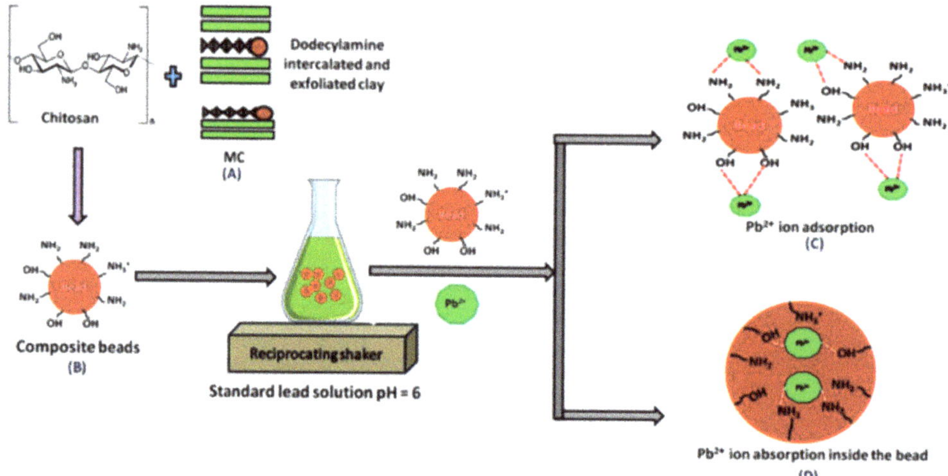


**Figure 3.** The mechanism of $Pb^{+2}$ ions removal over chitosan-clay composite [31].

On the other hand, various clay NCs involving other types of polymers were tested for the removal of metals. For example, montmorillonite modified with *N*-methyl-*D*-glucamine-based monomer through radical polymerization reaction to form water soluble nanocomposite was used with ion-exchange resin for the removal of As [112]. Interestingly, the resin removed 56 mg/g of arsenic via a Langmuir adsorption isotherm and decreased its level below the allowed limits according to the World Health Organization (WHO). Recently, in situ polymerization was used for the preparation of magnetic clay

–polymer nanocomposite by using monomer methyl methacrylate, iron oxide nanoparticles, and bentonite [113]. This magnetic nanocomposite adsorbed 113 mg/g of Cr (VI) from water. Intraparticle diffusion equation was used to approve that the adsorption process was reached via film-diffusion. Another strategy was developed by Ravikumar and Udayakumar et al. 2020 using *M. oleivera* biopolymer and bentonite (mass ration 1:1) to produce green nanocomposite by the solution processing method. The application of this nanocomposite for the removal of Cd (II), Cr (II), and Pb (II) from aqueous solution via coagulo-adsorption process showed an average removal efficiency of heavy metals of 99.99% at pH 6–8, 2–4, and 5–7 for Cd, Cr, and Pb, respectively with a dosage of 5 g/L [114].

The reported data showed various possibilities useful to remove the metal from water as summarized in Table 3. Nevertheless, in some cases, the water can be contaminated by the nanocomposite itself, which should be taken into consideration and studied in more details.

**Table 3.** Water treatment from different inorganic contaminants over clay-polymers nanocomposites (NCs) adsorbents.

| Adsorbent | Adsorbate | Temp. (°C) | pH | Removal Efficiency (%) or Adsorption Capacity (mg/g) | Isotherm Model | Kinetics Model | Ref. |
|---|---|---|---|---|---|---|---|
| Bentonite/humic acid | Cu (II) | 30 | 6.50 | 22.41 mg/g | – | – | [81] |
| Chitosan/attapulgite | Cr (III) | 45 | 5.0 | 65.37 mg/g | Langmuir | Intraparticle diffusion | [96] |
| chitosan grafted poly acrylic acid bentonite composites | Cd (II) | 25 | 6.0 | 51.60 mg/g | Langmuir | – | [96] |
| Chitosan/attapulgite | Fe (III) | 45 | 3.0 | 62.51 mg/g | Langmuir | Intraparticle diffusion | [96] |
| Chitosan–Al-pillared MMT nanocomposite | Cr (VI) | 25 | 6.38 | 15.68 mg/g | Langmuir | Pseudo-second order | [97] |
| Chitosan–Al-pillared montmorillonite NCs | Pb (II) | 25 | 6.5 | 99.6% | Freundlich | Pseudo-second order | [98] |
| Chitosan grafted poly acrylic acid bentonite composites | Ni (II) | 25 | 7.0 | 49 mg/g | Langmuir | – | [102] |
| Chitosan grafted poly acrylic acid bentonite composites | Cu (II) | 25 | 6.0 | 88.60 mg/g | Langmuir | – | [102] |
| Chitosan immobilized on bentonite | Pb (II) | 35 | – | 26.39 mg/g | Freundlich | Pseudo-second order | [104] |
| Polyaniline modified bentonite | U(VI) | 20 | 6.5 | 14.10 mg/g | Langmuir | Pseudo-second order | [115] |
| Bentonite/thiourea-formaldehyde composite | Mn (VII) | – | 4.0 | 14.82 mg/g | Langmuir | Pseudo-second order | [116] |
| Alginate–montmorillonite nanocomposite | Pb (II) | – | 6.0 | 244.7 mg/g | – | – | [117] |
| Alginate–montmorillonite nanocomposite | Mn (II), Fe (III), Ni (II), Zn(II) | – | 6.0 | 100% | – | – | [117] |
| Chitin/bentonite nanocomposite | Cr (VI) | – | 4.0 | 443.72 mg/g | Freundlich | – | [118] |
| Na-montmorillonite /cellulose | Cr (VI) | – | 3.8–5.5 | 22.3 mg/g | Langmuir | Pseudo-second order | [119] |
| Chitosan/PVA/bentonite nanocomposite | Hg (II) | – | – | 360.74 mg/g | – | – | [120] |
| Chitosan and montmorillonite | Se (VI) | – | – | 18.50 mg/g | – | – | [121] |
| Poly(methacrylic acid) grafted chitosan/bentonite grafted chitosan /bentonite | Th (IV) | 30 | 5.0 | 110.60 mg/g | Langmuir | Pseudo-second order | [122] |

**Table 3.** *Cont.*

| Adsorbent | Adsorbate | Temp. (°C) | pH | Removal Efficiency (%) or Adsorption Capacity (mg/g) | Isotherm Model | Kinetics Model | Ref. |
|---|---|---|---|---|---|---|---|
| L-cysteine modified bentonite-cellulose nanocomposite | Cu (II) | 50 | – | 32.37 mg/g | Langmuir | Pseudo-second order | [123] |
| Poly(acrylic acid-co-acrylamide)/attapulgite | Cu (II) | – | 6.0 | 69.76 mg/g | – | – | [124] |
| Cellulose-graft- polyacrylamide/hydroxyapatite | Cu (II) | – | 7.0 | 176.0 mg/g | – | Pseudo-second order | [125] |
| Chitosan/clinoptilolite | Ni (II) | 25 | 5.0 | 247.04 mg/g | Langmuir | Pseudo-second order | [126] |
| Cystene–montmorillonite nanocomposite | Pb (II) | – | – | 0.180 mg/g | – | – | [127] |
| Cloisite–polycaprolactone nanocomposite | Pb (II) | – | – | 88% | – | – | [128] |
| Attapulgite/poly(acrylic acid) | Pb (II) | – | 5.0 | 38.0 mg/g | Freundlich | Pseudo-second order | [129] |

## 5. Factors Making Clay-Polymers Nanocomposites Promising Materials for Wastewater Treatment

This work reviews the most appropriate clay-polymers nanocomposites available for the treatment of water/wastewater containing different types of pollutants. Similar to other fields of fabricated hybrid composites, clay-polymers nanocomposites associate the advantages of their individual parts, like chemical stability, high surface charge, high surface structure, and mechanical stability as well as enhanced adsorption capacities to different pollutants that allow efficient water purification. Also, clay-polymers nanocomposites withstand the fixed bed systems high temperatures and pressures due to the improvement offered by polymers (thermal and the mechanical stability). Therefore, clay-polymers nanocomposites can be used in industrial scale with low exhaustion during long-term use [130]. Additionally, clay-polymers nanocomposites provide great price advantage compared to other famous adsorbents (zeolites, activated carbon, etc.) related to the use of both clay minerals and functional polymers. This cost advantage can be further enhanced with the reuse opportunity of the clay-polymers NCs that reduce the overall cost of treatment [131]. In addition to cost, clay-polymers nanocomposites can be applied at low dose allowing its potential use at industrial scale.

Generally, unmodified clay minerals such as montmorillonite, bentonite, and kaolinite showed poor adsorption capacity for many pollutants as reported by Styszko et al., (2015) [132]. However, several CPNs exhibit higher adsorption capacities than mineral clays. For example, Cu (II) ions were removed from water with adsorption capacity of 44.9 mg/g using natural mineral clay while this value was folded more than twice for CPNs [133]. Similarly, maximum adsorption capacity of dyes was found for clay composites. In this context, raw kaolinite and montmorillonite showed dye adsorption capacity of 29 and 19 mg/g, respectively [134]. However, CPNs showed higher adsorption capacities as reported in Table 3. In addition to that, mineral clays adsorb better in acidic solutions making their application difficult for the treatment of water/wastewater having neutral pH [132].

As reported in Tables 2 and 3, the pollutants removal efficiency allowed by synthesized CPNs varied depending on the polymerization process and the used raw materials. In some cases, the new nanocomposites exhibit an interesting removal capacity. For example, chitosan-g-poly (acrylic acid)/ montmorillonite showed an adsorption capacity of 1895.0 mg/g for methylene blue [72]. However, for poly(acrylic acid-co-2-acrylamido-2- methylpropanesulfonic acid)/montmorillonite, the adsorption capacity was only 215.0 mg/g for the same dye [85]. Result variability was also observed for other pollutants (heavy metals, bacteria, etc.). Therefore, the removal rates and the adsorption capacities varied depending on the CPNs nature and on the operating conditions (dosage, pollutants, pH, temperature, contact time, etc.). Many of these adsorbents have the ability to reduce the level of biological pollutants to the safe limits considered by WHO.

An additional advantage of clay polymer nanocomposites is their ability to be used in batch and fixed bed systems. When comparing both systems (batch and fixed bed), fixed bed system is most efficient for the continuous removal of contaminants and offered larger regeneration capacity of the adsorbent. Therefore, nanocomposites can be easily separated from reaction medium than unmodified mineral clays. This makes clay polymer nanocomposites more cost effective than other conventional adsorbents [135]. Moreover, clay polymer nanocomposites are more stable in water compared to their individual parts.

The advantages of clay polymer nanocomposites can be summarized in the economic cost, the improved performances, and the easy synthesis in large quantities. Therefore, more researches are needed to produce the most promising clay polymer nanocomposites and evaluate the process efficiency at large scale for the removal of different contaminants (biological, organic, and inorganic) from water/wastewater. Furthermore, the future research of clay polymer nanocomposites can include the development of clay polymer NCs membranes. Membranes will provide effective treatment with low- cost embedded adsorbents. Also, study should be addressed to the synthesis of clay polymer NCs hydrogels, which may be considered more attractive for water/wastewater treatment due to their high capacity and ease of regeneration. Hydrogels are usually excluded from water purification due to poor strength and poor mechanical properties that can be overcome via the synthesis of clay polymers NCs hydrogels. Additionally, it will be promising to fabricate multi-functional clay polymer NCs having the ability to adsorb simultaneously different types of cationic and anionic contaminants.

## 6. Conclusions

This review article summarizes the different types of clay polymer nanocomposites and their effective use for the removal of pollutants from water/wastewater. The introduction of CPNs is induced by the advantages offered by clay minerals such as their cheapness, availability, and valuable properties. Therefore, the combination of clay minerals and polymers as promising reinforcements produce nanocomposites of clay-polymers with advanced properties (high capacity, simple synthesis, eco-friendly, cost advantages, etc.) useful for pollutants removal. CPNs may include various polymers in their fabrication such as chitosan, polystyrene, polypropylene, polyesters, polyurethanes, epoxies, and polyvinyl chloride, using appropriate synthesis techniques. Interestingly, the developed CPNs showed significant advantages of established adsorbents for water/wastewater treatment. The current literatures reported the removal of different pollutants (bacteria, metals, phenol, tannic acid, pesticides, dyes, etc.) using clay polymers nanocomposites showing them as promising materials for water/wastewater treatment. However, the majority of results were collected based on research experiments conducted for pollutants in aqueous solutions at laboratory scale. Therefore, further investigations are needed to evaluate the process efficiency using real wastewater at large scale. Moreover, it is very important to introduce sustainable waste materials such as agricultural and industrial wastes in the production of new CPNs-based adsorbents. Also, the competitive applicability of these innovative adsorbents should be evaluated taking into account the various parameters linked to the material proprieties (material degradation, alteration, life cycle, and regeneration) and to the generated wastes including the disposal of the loaded pollutants and the chemicals used for the process of adsorption/desorption.

**Author Contributions:** Writing—original draft preparation, A.A., N.S.A., K.M.K., F.M.A., and M.A.T.; writing—review and editing, A.A., N.S.A., K.M.K., F.M.A., M.A.T., and F.B.R.; visualization; F.B.R. All authors have read and agreed to the published version of the manuscript.

**Funding:** The authors extend their appreciation to the Deanship of Scientific Research at King Khalid University for funding this work through research groups program under grant number R.G.P.2/157/42. Also, this research was funded by the Deanship of Scientific Research at Princess Nourah bint Abdulrahman University through the Fast-track Research Funding Program.

**Institutional Review Board Statement:** Not applicable.

**Informed Consent Statement:** Not applicable.

**Data Availability Statement:** Data sharing is not applicable.

**Conflicts of Interest:** The authors declare no conflict of interest.

## References

1. Jaishankar, M.; Tseten, T.; Anbalagan, N.; Mathew, B.B.; Beeregowda, K.N. Toxicity, mechanism and health effects of some heavy metals. *Interdiscip. Toxicol.* **2014**, *7*, 60–72. [CrossRef]
2. Jonathan, M.; Srinivasalu, S.; Thangadurai, N.; Ayyamperumal, T.; Armstrong-Altrin, J.; Ram-Mohan, V. Contamination of Uppanar River and coastal waters off Cuddalore, Southeast coast of India. *Environ. Geol.* **2008**, *53*, 1391–1404. [CrossRef]
3. Siddeeg, S.M.; Tahoon, M.A.; Ben Rebah, F. Simultaneous Removal of Calconcarboxylic Acid, NH4+ and PO43− from Pharmaceutical Effluent Using Iron Oxide-Biochar Nanocomposite Loaded with Pseudomonas putida. *Processes* **2019**, *7*, 800. [CrossRef]
4. Khulbe, K.C.; Matsuura, T. Removal of heavy metals and pollutants by membrane adsorption techniques. *Appl. Water Sci.* **2018**, *8*, 1–30. [CrossRef]
5. Sahmoune, M.N. Evaluation of thermodynamic parameters for adsorption of heavy metals by green adsorbents. *Environ. Chem. Lett.* **2019**, *17*, 697–704. [CrossRef]
6. Chebotarevaa, R.D.; Remeza, S.V.; Bashtana, S.Y. Water Softening and Disinfection Using an Electrolysis Unit with a Filtering Cartridge. *J. Water Chem. Technol.* **2020**, *42*, 54–59. [CrossRef]
7. Al-Amshawee, S.; Yunus, M.Y.B.M.; Azoddein, A.A.M.; Hassell, D.G.; Dakhil, I.H.; Hasan, H.A. Electrodialysis desalination for water and wastewater: A review. *Chem. Eng. J.* **2020**, *380*, 122231. [CrossRef]
8. Liu, Z.; Lompe, K.M.; Mohseni, M.; Bérubé, P.R.; Sauvé, S.; Barbeau, B. Biological ion exchange as an alternative to biological activated carbon for drinking water treatment. *Water Res.* **2020**, *68*, 115148. [CrossRef]
9. Couto, C.F.; Santos, A.V.; Amaral, M.C.S.; Lange, L.C.; De Andrade, L.H.; Foureaux, A.F.S.; Fernandes, B.S. Assessing potential of nanofiltration, reverse osmosis and membrane distillation drinking water treatment for pharmaceutically active compounds (PhACs) removal. *J. Water Process. Eng.* **2020**, *33*, 101029. [CrossRef]
10. Skaf, D.W.; Punzi, V.L.; Rolle, J.T.; Kleinberg, K.A. Removal of micron-sized microplastic particles from simulated drinking water via alum coagulation. *Chem. Eng. J.* **2020**, *386*, 123807. [CrossRef]
11. Verma, B.; Balomajumder, C. Hexavalent chromium reduction from real electroplating wastewater by chemical precipitation. *Bull. Chem. Soc. Ethiop.* **2020**, *34*, 67–74. [CrossRef]
12. Khan, M.A.; Siddiqui, M.R.; Otero, M.; Alshareef, S.A.; Rafatullah, M. Removal of rhodamine b from water using a solvent impregnated polymeric dowex 5wx8 resin: Statistical optimization and batch adsorption studies. *Polymers* **2020**, *12*, 500. [CrossRef]
13. Coelho, C.M.; De Andrade, J.R.; Da Silva, M.G.C.; Vieira, M.G.A. Removal of propranolol hydrochloride by batch biosorption using remaining biomass of alginate extraction from Sargassum filipendula algae. *Environ. Sci. Pollut. Res.* **2020**, *27*, 16599–16611. [CrossRef] [PubMed]
14. Joseph, J.; Sajeesh, A.K.; Nagashri, K.; Gladis, E.E.; Sharmila, T.M.; Dhanaraj, C.J. Determination of ammonia content in various drinking water sources in Malappuram District, Kerala and its removal by adsorption using agricultural waste materials. *Mater. Today* **2020**. [CrossRef]
15. He, P.; Ding, J.; Qin, Z.; Tang, L.; Haw, K.G.; Zhang, Y.; Valtchev, V. Binder-free preparation of ZSM-5@ silica beads and their use for organic pollutant removal. *Inorg. Chem. Front.* **2020**, *7*, 2080–2088. [CrossRef]
16. Thiebault, T. Raw and modified clays and clay minerals for the removal of pharmaceutical products from aqueous solutions: State of the art and future perspectives. *Crit. Rev. Environ. Sci. Technol.* **2020**, *50*, 1451–1514. [CrossRef]
17. Chen, X.; Yu, L.; Zou, S.; Xiao, L.; Fan, J. Zeolite cotton in tube: A Simple Robust Household Water treatment filter for Heavy Metal Removal. *Sci. Rep.* **2020**, *10*, 1–9. [CrossRef] [PubMed]
18. Nam, S.W.; Choi, D.J.; Kim, S.K.; Her, N.; Zoh, K.D. Adsorption characteristics of selected hydrophilic and hydrophobic micropollutants in water using activated carbon. *J. Hazard. Mater.* **2014**, *270*, 144–152. [CrossRef]
19. Unuabonah, E.I.; Taubert, A. Clay–polymer nanocomposites (CPNs): Adsorbents of the future for water treatment. *Appl. Clay Sci.* **2014**, *99*, 83–92. [CrossRef]
20. LeBaron, P.C.; Wang, Z.; Pinnavaia, T.J. Polymer-layered silicate nanocomposites: An overview. *Appl. Clay Sci.* **1999**, *15*, 11–29. [CrossRef]
21. Pavlidou, S.; Papaspyrides, C.D. A review on polymer-layered silicate nanocomposites. *Prog. Polym. Sci.* **2008**, *32*, 1119–1198. [CrossRef]

22. Bergaya, F.; Detellier, C.; Lambert, J.F.; Lagaly, G. Introduction to Clay Polymer Nanocomposites (CPN). Chapter 13. In *Handbook of Clay Science. Developments in Clay Science*; Bergaya, L., Ed.; Elsevier: Amsterdam, The Netherlands, 2013; Volume 5.
23. Lee, S.M.; Tiwari, D. Organo and inorgano-organo-modifid clays in the remediation of aqueous solutions: An overview. *Appl. Clay Sci.* **2012**, *59–60*, 84–102. [CrossRef]
24. Konta, J. Clay and man: Clay raw materials in the service of man. *Appl. Clay Sci.* **1995**, *10*, 275–335. [CrossRef]
25. Sarkar, B.; Xi, Y.; Megharaj, M.; Krishnamurti, G.S.R.; Bowman, M.; Rose, H.; Naidu, R. Bioreactive organoclay: A new technology for environmental remediation. *Crit. Rev. Environ. Sci. Technol.* **2012**, *42*, 435–488. [CrossRef]
26. Mukhopadhyay, R.; De, N. Nano clay polymer composite: Synthesis, characterization, properties and application in rainfed agriculture. *Glob. J. Biosci. Biotechnol.* **2014**, *3*, 133–138.
27. Liu, Y.; Takafuji, M.; Ihara, H.; Wakiya, T. Saturation Magnetization of inorganic/ polymer Nanocomposites Higher Than That of Their Inorganic Magnetic Component. *arXiv* **2012**, arXiv:1206.2805.
28. Rusmin, R.; Sarkar, B.; Tsuzuki, T.; Kawashima, N.; Naidu, R. Removal of lead from aqueous solution using superparamagnetic palygorskite nanocomposite: Material characterization and regeneration studies. *Chemosphere* **2017**, *186*, 1006–1015. [CrossRef] [PubMed]
29. Santos, V.P.; Marques, N.S.; Maia, P.C.; Lima, M.A.B.D.; Franco, L.D.O.; Campos-Takaki, G.M.D. Seafood Waste as Attractive Source of Chitin and Chitosan Production and Their Applications. *Int. J. Mol. Sci.* **2020**, *21*, 4290. [CrossRef] [PubMed]
30. Gogoi, P.; Thakur, A.J.; Devi, R.R.; Das, B.; Maji, T.K. A comparative study on sorption of arsenate ions from water by crosslinked chitosan and crosslinked chitosan/MMT nanocomposite. *J. Environ. Chem. Eng.* **2016**, *4*, 4248–4257. [CrossRef]
31. Biswas, S.; Rashid, T.U.; Debnath, T.; Haque, P.; Rahman, M.M. Application of chitosan-clay biocomposite beads for removal of heavy metal and dye from industrial effluent. *J. Compos. Sci.* **2020**, *4*, 16. [CrossRef]
32. Saurabh, K.; Kanchikeri, M.M.; Datta, S.C.; Thekkumpurath, A.S.; Kumar, R. Nanoclay polymer composites loaded with urea and nitrification inhibitors for controlling nitrification in soil. *Arch. Agron. Soil. Sci.* **2019**, *65*, 478–491. [CrossRef]
33. Burnside, S.D.; Emmanuel, P.G. Synthesis and properties of new poly (dimethylsiloxane) nanocomposites. *Chem. Mate.* **1995**, *7*, 1597–1600. [CrossRef]
34. Akelah, A.; Moet, A. Polymer-clay nanocomposites: Free-radical grafting of polystyrene on to organophilic montmorillonite interlayers. *J. Mater. Sci.* **1996**, *31*, 3589–3596. [CrossRef]
35. Vaia, R.A.; Giannelis, E.P. Lattice model of polymer melt intercalation in organically-modified layered silicates. *Macromolecules.* **1997**, *30*, 7990–7999. [CrossRef]
36. Ogata, N.; Jimenez, G.; Kawai, H.; Ogihara, T. Structure and thermal/mechanical properties of poly (l-lactide)-clay blend. *J. Polym. Sci. B Polym Phys.* **1997**, *35*, 389–396. [CrossRef]
37. Yano, K.; Usuki, A.; Okada, A.J.; Kurauchi, T.; Kamigaito, O.J. Synthesis and properties of polyimide-clay hybrid films. *J. Polym. Sci. A Polym. Chem.* **1997**, *35*, 2289. [CrossRef]
38. Wang, Z.; Pinnavaia, T.J. Hybrid organic— inorganic nanocomposites: Exfoliation of magadiite nanolayers in an elastomeric epoxy polymer. *Chem. Mater.* **1998**, *10*, 1820–1826. [CrossRef]
39. Wang, Z.; Pinnavaia, T.J. Nanolayer reinforcement of elastomeric polyurethane. *Chem. Mater.* **1998**, *10*, 3769–3771. [CrossRef]
40. Wołejko, E.; Wydro, U.; Butarewicz, A.; Jabłońska-Trypuć, A. Methods Used in Situ for Removal of Waterborne Pathogens. In *Waterborne Pathogens*; Butterworth-Heinemann: Oxford, UK, 2020; pp. 321–337.
41. Kaya, A.U.; Güner, S.; Ryskin, M.; Lameck, A.S.; Benitez, A.R.; Shuali, U.; Nir, S. Effect of Microwave Radiation on Regeneration of a Granulated Micelle–Clay Complex after Adsorption of Bacteria. *Appl. Sci.* **2020**, *10*, 2530. [CrossRef]
42. Farré, M.J.; Insa, S.; Lamb, A.; Cojocariu, C.; Gernjak, W. Occurrence of N-nitrosamines and their precursors in Spanish drinking water treatment plants and distribution systems. *Environ. Sci. Wat. Res.* **2020**, *6*, 210–220. [CrossRef]
43. Bruna, J.E.; Peñaloza, A.; Guarda, A.; Rodríguez, F.; Galotto, M.J. Development of Mt Cu2+ /LDPE nanocomposites with antimicrobial activity for potential use in food packaging. *Appl. Clay Sci.* **2012**, *58*, 79–87. [CrossRef]
44. Lejon, D.P.H.; Pascault, N.; Ranjard, L. Diffrential copper impact on density, diversity and resistance of adapted culturable bacterial populations according to soil organic status. *Eur. J. Soil Biol.* **2010**, *46*, 168–174. [CrossRef]
45. Zhou, N.L.; Liu, Y.; Li, L.; Meng, N.; Huang, Y.X.; Zhang, J.; Wei, S.H.; Shen, J. A new nanocomposite biomedical material of polymer/Clay–Cts–Ag nanocomposites. *Curr. Appl. Phys.* **2007**, *7*, 58–62. [CrossRef]
46. Kang, J.K.; Lee, C.G.; Park, J.A.; Kim, S.B.; Choi, N.C.; Park, S.J. Adhesion of bacteria to pyrophyllite clay in aqueous solution. *Environ. Technol.* **2013**, *34*, 703–710. [CrossRef]
47. Undabeytia, T.; Posada, R.; Nir, S.; Galindo, I.; Laiz, L.; Saiz-Jimenez, C.; Morillo, E. Removal of waterborne microorganisms by fitration using clay–polymer complexes. *J. Hazard. Mater.* **2014**, *279*, 190–196. [CrossRef]
48. Meng, N.; Zhou, N.-L.; Zhang, S.-Q.; Shen, J. Synthesis and antimicrobial activities of polymer/montmorillonite–chlorhexidine acetate nanocomposite fims. *Appl. Clay Sci.* **2009**, *42*, 667–670. [CrossRef]
49. Han, Y.-S.; Lee, S.-H.; Choi, K.H.; Park, I. Preparation and characterization of chitosan–clay nanocomposites with antimicrobial activity. *J. Phys. Chem. Solids* **2010**, *71*, 464–467. [CrossRef]
50. Motshekga, S.C.; Ray, S.S.; Onyango, M.S.; Momba, M.N.B. Preparation and antibacterial activity of chitosan-based nanocomposites containing bentonite-supported silver and zinc oxide nanoparticles for water disinfection. *Appl. Clay Sci.* **2015**, *114*, 330–339. [CrossRef]

51. Deng, H.; Li, X.; Ding, B.; Du, Y.; Li, G.; Yang, J.; Hu, X. Fabrication of polymer/ layered silicate intercalated nanofirous mats and their bacterial inhibition activity. *Carbohydr. Polym.* **2011**, *83*, 973–978. [CrossRef]
52. Shameli, K.; Ahmad, M.B.; Mohsen, Z.; Yunis, W.Z.; Ibrahim, N.A.; Shabanzadeh, P.; Moghaddam, M.G. Synthesis and characterization of silver/montmorillonite/ chitosan bionanocomposites by chemical reduction method and their antibacterial activity. *Int. J. Nanomed.* **2011**, *6*, 271–284. [CrossRef]
53. Rhim, J.W.; Hong, S.I.; Park, H.M.; Ng, P.K.W. Preparation and characterization of chitosan-based nanocomposite fims with antimicrobial activity. *J. Agric. Food Chem.* **2006**, *54*, 5814–5822. [CrossRef] [PubMed]
54. Chi, W.; Qin, C.; Zeng, L.; Li, W.; Wang, W. Microbiocidal activity of chitosan-*N*-2- hydroxypropyl trimethyl ammonium chloride. *J. Appl. Polym. Sci.* **2007**, *103*, 3851–3856. [CrossRef]
55. Unuabonah, E.I.; Ugwuja, C.G.; Omorogie, M.O.; Adewuyi, A.; Oladoja, N.A. Clays for efficient disinfection of bacteria in water. *Appl. Clay Sci.* **2018**, *151*, 211–223. [CrossRef]
56. Unuabonah, E.I.; Kolawole, M.O.; Agunbiade, F.O.; Omorogie, M.O.; Koko, D.T.; Ugwuja, C.G.; Ugege, L.E.; Oyejide, N.E.; Günter, C.; Taubert, A. Novel metal-doped bacteriostatic hybrid clay composites for point-of-use disinfection of water. *J. Environ. Chem. Eng.* **2017**, *5*, 2128–2141. [CrossRef]
57. Momina, M.; Shahadat, S.I. Regeneration performance of clay-based adsorbents for the removal of industrial dyes: A review. *RSC Adv.* **2018**, *8*, 24571–24587. [CrossRef]
58. Deng, Y.; Wang, L.; Hu, X.; Liu, B.; Wei, Z.; Yang, S.; Sun, C. Highly efficient removal of tannic acid from aqueous solution by chitosan-coated attapulgite. *Chem. Eng. J.* **2012**, *181*, 300–306. [CrossRef]
59. Wu, J.; Chen, J. Adsorption characteristics of tannic acid onto the novel protonated palygorskite/chitosan resin microspheres. *J. Appl. Polym. Sci.* **2013**, *127*, 1765–1771. [CrossRef]
60. Hernández-Hernández, K.A.; Illescas, J.; Díaz-Nava, M.D.C.; Martínez-Gallegos, S.; MuroUrista, C.; Ortega-Aguilar, R.E.; Rodríguez-Alba, E.; Rivera, E. Preparation of nanocomposites for the removal of phenolic compounds from aqueous solutions. *Appl. Clay Sci.* **2018**, *157*, 212–217. [CrossRef]
61. Ganigar, R.; Rytwo, G.; Gonen, Y.; Radian, A.; Mishael, Y.G. Polymer–clay nanocomposites for the removal of trichlorophenol and trinitrophenol from water. *Appl. Clay Sci.* **2010**, *49*, 311–316. [CrossRef]
62. Zhu, J.; Wang, T.; Zhu, R.; Ge, F.; Wei, J.; Yuan, P.; He, H. Novel polymer/surfactant modified montmorillonite hybrids and the implications for the treatment of hydrophobic organic compounds in wastewaters. *Appl. Clay Sci.* **2011**, *51*, 317–322. [CrossRef]
63. Zadaka, D.; Nir, S.; Radian, A.; Mishael, Y.G. Atrazine removal from water by polycation–clay composites: Effct of dissolved organic matter and comparison to activated carbon. *Water Res.* **2009**, *43*, 677–683. [CrossRef]
64. Ghaemi, N.; Madaeni, S.S.; Alizadeh, A.; Rajabi, H.; Daraei, P. Preparation, characterization and performance of polyethersulfone/organically modifid montmorillonite nanocomposite membranes in removal of pesticides. *J. Memb. Sci.* **2011**, *382*, 135–147. [CrossRef]
65. Gámiz, B.; Hermosín, M.C.; Cornejo, J.; Celis, R. Hexadimethrine-montmorillonite nanocomposite: Characterization and application as a pesticide adsorbent. *Appl. Surf. Sci.* **2015**, *332*, 606–613. [CrossRef]
66. Wan Ngah, W.S.; Ariff, N.F.M.; Hanafih, M.A.K.M. Preparation, characterization, and environmental application of crosslinked chitosan-coated bentonite for tartrazine adsorption from aqueous solutions. *Water Air Soil Pollut.* **2010**, *206*, 225–236. [CrossRef]
67. Bée, A.; Obeid, L.; Mbolantenaina, R.; Welschbillig, M.; Talbot, D. Magnetic chitosan/clay beads: A magsorbent for the removal of cationic dye from water. *J. Magn. Magn. Mater.* **2017**, *421*, 59–64. [CrossRef]
68. Vanamudan, A.; Pamidimukkala, P. Chitosan, nanoclay and chitosan–nanoclay composite as adsorbents for Rhodamine-6G and the resulting optical properties. *Int. J. Biol. Macromol.* **2015**, *74*, 127–135. [CrossRef]
69. Xu, R.; Mao, J.; Peng, N.; Luo, X.; Chang, C. Chitin/clay microspheres with hierarchical architecture for highly efficient removal of organic dyes. *Carbohydr. Polym.* **2018**, *188*, 143–150. [CrossRef]
70. Vanaamudan, A.; Sudhakar, P.P. Equilibrium, kinetics and thermodynamic study on adsorption of reactive blue-21 and reactive red-141 by chitosan-organically modifid nanoclay (Cloisite 30B) nano-bio composite. *J. Taiwan Inst. Chem. Eng.* **2015**, *55*, 145–151. [CrossRef]
71. Peng, Y.; Chen, D.; Ji, J.; Kong, Y.; Wan, H.; Yao, C. Chitosan-modifid palygorskite: Preparation, characterization and reactive dye removal. *Appl. Clay Sci.* **2013**, *74*, 81–86. [CrossRef]
72. Wang, L.; Zhang, J.; Wang, A. Removal of methylene blue from aqueous solution using chitosan-g-poly(acrylic acid)/montmorillonite superadsorbent nanocomposite. *Colloids Surf. A Physicochem. Eng. Asp.* **2008**, *322*, 47–53. [CrossRef]
73. Wang, L.; Wang, A. Adsorption characteristics of Congo Red onto the chitosan/ montmorillonite nanocomposite. *J. Hazard. Mater.* **2007**, *147*, 979–985. [CrossRef] [PubMed]
74. Li, Q.; Yue, Q.; Su, Y.; Gao, B. Equilibrium and a two-stage batch adsorber design for reactive or disperse dye removal to minimize adsorbent amount. *Bioresour. Technol.* **2011**, *102*, 5290–5296. [CrossRef]
75. Kang, Q.; Zhou, W.; Li, Q.; Gao, B.; Fan, J.; Shen, D. Adsorption of anionic dyes on poly(epicholorohydrin dimethylamine) modifid bentonite in single and mixed dye solutions. *Appl. Clay Sci.* **2009**, *45*, 280–287. [CrossRef]
76. Anirudhan, T.S.; Suchithra, P.S.; Radhakrishnan, P.G. Synthesis and characterization of humic acid immobilized-polymer/bentonite composites and their ability to adsorb basic dyes from aqueous solutions. *Appl. Clay Sci.* **2009**, *43*, 336–342. [CrossRef]

77. De Sousa, E.P.; De Araujo, D.T.; Peixoto, V.G.; Ferreira, B.F.; De Faria, E.H.; Molina, E.F. Effect of sodium bentonite content on structural-properties of ureasil poly (ethylene oxide)-PEO hybrid: A perspective for water treatment. *Appl. Clay Sci.* **2020**, *191*, 105605. [CrossRef]
78. Khalil, A.M.; Sayed, H.K. Hybrid Membranes Based on Clay-Polymer for Removing Methylene Blue from Water. *Acta Chim. Slovenica.* **2020**, *67*, 96–104. [CrossRef]
79. Azha, S.F.; Shamsudin, M.S.; Shahadat, M.; Ismail, S. Low cost zwitterionic adsorbent coating for treatment of anionic and cationic dyes. *J. Ind. Eng. Chem.* **2018**, *67*, 187–198. [CrossRef]
80. Kaşgöz, H.; Durmus, A. Dye removal by a novel hydrogel-clay nanocomposite with enhanced swelling properties. *Polym. Adv. Technol.* **2018**, *19*, 838–845. [CrossRef]
81. Jin, X.; Zheng, M.; Sarkar, B.; Naidu, R.; Chen, Z. Characterization of bentonite modifid with humic acid for the removal of Cu (II) and 2,4-dichlorophenol from aqueous solution. *Appl. Clay Sci.* **2016**, *134*, 89–94. [CrossRef]
82. Zhou, K.; Zhang, Q.; Wang, B.; Liu, J.; Wen, P.; Gui, Z.; Hu, Y. The integrated utilization of typical clays in removal of organic dyes and polymer nanocomposites. *J. Clean. Prod.* **2014**, *81*, 281–289. [CrossRef]
83. Liu, Q.; Yang, B.; Zhang, L.; Huang, R. Adsorption of an anionic azo dye by crosslinked chitosan/bentonite composite. *Int. J. Biol. Macromol.* **2015**, *72*, 1129–1135. [CrossRef] [PubMed]
84. Mahdavinia, G.R.; Asgari, A. Synthesis of kappa-carrageenan-g-poly(acrylamide)/ sepiolite nanocomposite hydrogels and adsorption of cationic dye. *Polym. Bull.* **2013**, *70*, 2451–2470. [CrossRef]
85. Hosseinzadeh, H.; Khoshnood, N. Removal of cationic dyes by poly(AA-co-AMPS)/ montmorillonite nanocomposite hydrogel. *Desalin. Water Treat.* **2016**, *57*, 6372–6383. [CrossRef]
86. Uyar, G.; Kaygusuz, H.; Erim, F.B. Methylene blue removal by alginate–clay quasicryogel beads. *React. Funct. Polym.* **2016**, *106*, 1–7. [CrossRef]
87. Wan Ngah, W.S.; Ariff, N.F.M.; Hashim, A.; Hanafih, M.A.K.M. Malachite green adsorption onto chitosan coated bentonite beads: Isotherms, kinetics and mechanism. *Clean Soil Air Water* **2010**, *38*, 394–400. [CrossRef]
88. Kalotraji, S.; Mehta, R. Synthesis of polyaniline/clay nanocomposites by in situ polymerization and its application for the removal of Acid Green 25 dye from wastewater. *Polym. Bull.* **2020**. [CrossRef]
89. Saki, S.; Senol-Arslan, D.; Uzal, N. Fabrication and characterization of silane-functionalized Na-bentonite polysulfone/polyethylenimine nanocomposite membranes for dye removal. *J. Appl. Polym. Sci.* **2020**, *137*, 49057. [CrossRef]
90. Narayanan, N.; Gupta, S.; Gajbhiye, V.T. Decontamination of pesticide industrial effluent by adsorption–coagulation–flocculation process using biopolymer nanoorganoclay composite. *Int. J. Environ. Sci. Technol.* **2020**, *17*, 4775–4786. [CrossRef]
91. Zhang, A.C.; Sun, L.S.; Xiang, J.; Hu, S.; Fu, P.; Su, S.; Zhou, Y.B. Removal of elemental mercury from coal combustion flue gas by bentonite-chitosan and their modifier. *J. Fuel Chem. Technol.* **2009**, *37*, 489–495. [CrossRef]
92. Liu, E.; Sarkar, B.; Wang, L.; Naidu, R. Copper-complexed clay/poly-acrylic acid composites: Extremely efficient adsorbents of ammonia gas. *Appl. Clay Sci.* **2016**, *121*, 154–161. [CrossRef]
93. Jordao, C.; Pereira, M.; Pereira, J. Metal contamination of river waters and sediments from effluents of kaolin processing in Brazil. *Water Air Soil Pollut.* **2002**, *140*, 119–138. [CrossRef]
94. Azzam, E.M.S.; Eshaq, G.; Rabie, A.M.; Bakr, A.A.; Abd-Elaal, A.A.; El Metwally, A.E.; Tawfik, S.M. Preparation and characterization of chitosan-clay nanocomposites for the removal of Cu(II) from aqueous solution. *Int. J. Biol. Macromol.* **2016**, *89*, 507–517. [CrossRef] [PubMed]
95. Abdel, K.M.A.; Mahmoud, G.A.; El-Kelesh, N.A. Synthesis and characterization of poly-methacrylic acid grafted chitosan-bentonite composite and its application for heavy metals recovery. *Chem. Mater. Res.* **2012**, *2*, 1–12.
96. Zou, X.; Pan, J.; Ou, H.; Wang, X.; Guan, W.; Li, C.; Yan, Y.; Duan, Y. Adsorptive removal of Cr(III) and Fe(III) from aqueous solution by chitosan/attapulgite composites: Equilibrium, thermodynamics and kinetics. *Chem. Eng. J.* **2011**, *167*, 112–121. [CrossRef]
97. Wang, Y.M.; Duan, L.; Sun, Y.; Hu, N.; Gao, J.Y.; Wang, H.; Xie, X.M. Adsorptive removal of Cr(VI) from aqueous solutions with an effective adsorbent: Cross-linked chitosan/montmorillonite nanocomposites in the presence of hydroxy-aluminum oligomeric cations. *Desalin. Water Treat.* **2016**, *57*, 10767–10775. [CrossRef]
98. Duan, L.; Hu, N.; Wang, T.; Wang, H.; Ling, L.; Sun, Y.; Xie, X. Removal of copper and lead from aqueous solution by adsorption onto cross-linked chitosan/montmorillonite nanocomposites in the presence of hydroxyl–aluminum oligomeric cations: Equilibrium, kinetic, and thermodynamic studies. *Chem. Eng. Commun.* **2016**, *203*, 28–36. [CrossRef]
99. Rusmin, R.; Sarkar, B.; Liu, Y.; McClure, S.; Naidu, R. Structural evolution of chitosan–palygorskite composites and removal of aqueous lead by composite beads. *Appl. Surf. Sci.* **2015**, *353*, 363–375. [CrossRef]
100. Alsewailem, F.D.; Aljlil, S.A. Recycled polymer/clay composites for heavy-metals adsorption. *Mater. Technol.* **2013**, *47*, 525–529.
101. Msaadi, R.; Ammar, S.; Chehimi, M.M.; Yagci, Y. Diazonium-based ion-imprinted polymer/clay nanocomposite for the selective extraction of lead (II) ions in aqueous media. *Eur. Polym. J.* **2017**, *89*, 367–380. [CrossRef]
102. Kumararaja, P.; Manjaiah, K.M.; Datta, S.C.; Shabeer, A.T.P.; Sarkar, B. Chitosan-g-poly(acrylic acid)-bentonite composite: A potential immobilizing agent of heavy metals in soil. *Cellulose* **2018**, *25*, 3985–3999. [CrossRef]
103. Grisdanurak, N.; Akewaranugulsiri, S.; Futalan, C.M.; Tsai, W.C.; Kan, C.C.; Hsu, C.W.; Wan, M.W. The study of copper adsorption from aqueous solution using crosslinked chitosan immobilized on bentonite. *J. Appl. Polym. Sci.* **2012**, *125*, E132–E142. [CrossRef]

104. Futalan, C.M.; Kan, C.C.; Dalida, M.L.; Hsien, K.J.; Pascua, C.; Wan, M.W. Comparative and competitive adsorption of copper, lead, and nickel using chitosan immobilized on bentonite. *Carbohydr. Polym.* **2011**, *83*, 528–536. [CrossRef]
105. Nunes, P.; Nagy, N.V.; Alegria, E.C.B.A.; Pombeiro, A.J.L.; Correia, I. The solvation and electrochemical behavior of copper acetylacetonate complexes in ionic liquids. *J. Mol. Struct.* **2014**, *1060*, 142–149. [CrossRef]
106. Yousef, W.M.; Alenezi, K.; Naggar, A.H.; Hassan, T.M.; Bortata, S.Z.; Farghaly, O.A. Potentiometric and Conductometric Studies on Complexes of Folic Acid with Some Metal Ions. *Int. J. Electrochem. Sci.* **2017**, *12*, 1146–1156. [CrossRef]
107. De, S.; Ali, S.M.; Ali, A.; Gaikar, V.G. Micro-solvation of the $Zn^{2+}$ ion—A case study. *Phys. Chem. Chem. Phys.* **2009**, *11*, 8285–8294. [CrossRef] [PubMed]
108. Sarada, B.; Krishna Prasad, M.; Kishore Kumar, K.; Murthy, C.V.R. Biosorption of $Cd^{2+}$ by green plant biomass, Araucaria heterophylla: Characterization, kinetic, isotherm and thermodynamic studies. *Appl. Water Sci.* **2017**, *7*, 3483–3496. [CrossRef]
109. Liu, X.; Lee, D.J. Thermodynamic parameters for adsorption equilibrium of heavy metals and dyes from wastewaters. *Bioresour Technol.* **2014**, *160*, 24–31. [CrossRef]
110. Sahmoune, M.N. The role of biosorbents in the removal of arsenic from water. *Chem. Eng. Technol.* **2016**, *39*, 1617–1628. [CrossRef]
111. Jana, S.; Saikia, A.; Purkait, M.K.; Mohanty, K. Chitosan based ceramic ultrafiltration membrane: Preparation, characterization and application to remove Hg(II) and As(III) using polymer enhanced ultrafiltration. *Chem. Eng. J.* **2011**, *170*, 209–219. [CrossRef]
112. Urbano, B.F.; Rivas, B.L.; Martinez, F.; Alexandratos, S.D. Water-insoluble polymer–clay nanocomposite ion exchange resin based on N-methyl-d-glucamine ligand groups for arsenic removal. *React. Funct. Polym.* **2012**, *72*, 642–649. [CrossRef]
113. Sundaram, E.J.S.; Dharmalingam, P. Synthesis and characterization of magnetized clay polymer nanocomposites and its adsorptive behaviour in removal of Cr(VI) from aqueous phase. *Asian J. Chem.* **2018**, *30*, 667–672. [CrossRef]
114. Ravikumar, K.; Udayakumar, J. Preparation and characterisation of green clay-polymer nanocomposite for heavy metals removal. *Chem Ecol.* **2020**, *36*, 270–291. [CrossRef]
115. Liu, X.; Cheng, C.; Xiao, C.; Shao, D.; Xu, Z.; Wang, J.; Hu, S.; Li, X.; Wang, W. Polyaniline (PANI) modified bentonite by plasma technique for U(VI) removal from aqueous solution. *Appl. Surf. Sci.* **2017**, *411*, 331–337. [CrossRef]
116. El-Korashy, S.A.; Elwakeel, K.Z.; El-Hafeiz, A.A. Fabrication of bentonite/thioureaformaldehyde composite material for Pb(II), Mn(VII) and Cr(VI) sorption: A combined basic study and industrial application. *J. Clean. Prod.* **2016**, *137*, 40–50. [CrossRef]
117. Shawky, H.A. Improvement of water quality using alginate/montmorillonite composite beads. *J. Appl. Polym. Sci.* **2011**, *119*, 2371–2378. [CrossRef]
118. Saravanan, D.; Gomathi, T.; Sudha, P.N. Sorption studies on heavy metal removal using chitin/bentonite biocomposite. *Int. J. Biol. Macromol.* **2013**, *53*, 67–71. [CrossRef] [PubMed]
119. Kumar, A.S.K.; Kalidhasan, S.; Rajesh, V.; Rajesh, N. Application of cellulose-clay composite biosorbent toward the effective adsorption and removal of chromium from industrial wastewater. *Ind. Eng. Chem. Res.* **2012**, *51*, 58–69. [CrossRef]
120. Wang, X.; Yang, L.; Zhang, J.; Wang, C.; Li, Q. Preparation and characterization of chitosan–poly(vinyl alcohol)/bentonite nanocomposites for adsorption of Hg(II) ions. *Chem. Eng. J.* **2014**, *251*, 404–412. [CrossRef]
121. Bleiman, N.; Mishael, Y.G. Selenium removal from drinking water by adsorption to chitosan–clay composites and oxides: Batch and columns tests. *J. Hazard. Mater.* **2010**, *183*, 590–595. [CrossRef]
122. Anirudhan, T.S.; Rijith, S.; Tharun, A.R. Adsorptive removal of thorium (IV) from aqueous solutions using poly(methacrylic acid)-grafted chitosan/bentonite composite matrix: Process design and equilibrium studies. *Colloids Surf. A Physicochem. Eng. Asp.* **2010**, *368*, 13–22. [CrossRef]
123. Ahmad, R.; Hasan, I. L-cystein modified bentonite-cellulose nanocomposite (cellu/cys-bent) for adsorption of $Cu^{2+}$, $Pb^{2+}$, and $Cd^{2+}$ ions from aqueous solution. *Sep. Sci. Technol.* **2016**, *51*, 381–394. [CrossRef]
124. Liu, P.; Jiang, L.; Zhu, L.; Guo, J.; Wang, A. Synthesis of covalently crosslinked attapulgite/poly(acrylic acid-co-acrylamide) nanocomposite hydrogels and their evaluation as adsorbent for heavy metal ions. *J. Ind. Eng. Chem.* **2015**, *23*, 188–193. [CrossRef]
125. Saber-Samandari, S.; Saber-Samandari, S.; Gazi, M. Cellulose-graft-polyacrylamide/hydroxyapatite composite hydrogel with possible application in removal of Cu (II) ions. *React. Funct. Polym.* **2013**, *73*, 1523–1530. [CrossRef]
126. Dinu, M.V.; Dragan, E.S. Evaluation of Cu2+, Co2+ and Ni2+ ions removal from aqueous solution using a novel chitosan/clinoptilolite composite: Kinetics and isotherms. *Chem. Eng. J.* **2010**, *160*, 157–163. [CrossRef]
127. Yadav, V.B.; Gadi, R.; Kalra, S. Clay based nanocomposites for removal of heavy metals from water: A review. *J. Environ. Manage.* **2019**, *232*, 803–817. [CrossRef] [PubMed]
128. Dlamini, D.S.; Mishra, A.K.; Mamba, B.B. ANN modeling in Pb(II) removal from water by clay-polymer composites fabricated via the melt-blending. *J. Appl. Polym. Sci.* **2013**, *130*, 3894–3901. [CrossRef]
129. Liu, P.; Jiang, L.; Zhu, L.; Wang, A. Novel approach for attapulgite/poly(acrylic acid) (ATP/PAA) nanocomposite microgels as selective adsorbent for Pb(II) Ion. *React. Funct. Polym.* **2014**, *74*, 72–80. [CrossRef]
130. Eibagi, H.; Faghihi, K. Preparation of thermally stable magnetic poly (urethane-imide)/nanocomposite containing β-cyclodextrin cavities as new adsorbent for lead and cadmium. *J. Polym. Res.* **2020**, *27*, 1–14. [CrossRef]
131. Sari, A.; Tuzen, M. Cd (II) adsorption from aqueous solution by raw and modified kaolinite. *Appl. Clay Sci.* **2014**, *88*, 63–72. [CrossRef]
132. Styszko, K.; Nosek, K.; Motak, M.; Bester, K. Preliminary selection of clay minerals for the removal of pharmaceuticals, bisphenol A and triclosan in acidic and neutral aqueous solutions. *C. R. Chim.* **2015**, *18*, 1134–1142. [CrossRef]

133. Anirudhan, T.S.; Ramachandran, M. Synthesis and characterization of amidoximated polyacrylonitrile/organobentonite composite for Cu (II), Zn (II), and Cd (II) adsorption from aqueous solutions and industry wastewaters. *Ind. Eng. Chem. Res.* **2008**, *47*, 6175–6184. [CrossRef]
134. Aydin, A.H.; Yavuz, Ö. Removal of acid red 183 from aqueous solution using clay and activated carbon. *Indian J. Chem. Technol.* **2004**, *11*, 201.
135. Unuabonah, E.I.; Adebowale, K.O.; Olu-Owolabi, B.I.; Yang, L.Z. Comparison of sorption of $Pb^{2+}$ and $Cd^{2+}$ on kaolinite clay and polyvinyl alcohol-modified kaolinite clay. *Adsorption* **2008**, *14*, 791–803. [CrossRef]

*Article*

# Application of Granular Biocomposites Based on Homogenised Peat for Absorption of Oil Products

**Kristine Irtiseva [1], Marika Mosina [1], Anastasija Tumilovica [1], Vjaceslavs Lapkovskis [2,3], Viktors Mironovs [2,3], Jurijs Ozolins [1], Valentina Stepanova [1] and Andrei Shishkin [1,*]**

1 Rudolfs Cimdins Riga Biomaterials Innovations and Development Centre of RTU, Institute of General Chemical Engineering, Faculty of Materials Science and Applied Chemistry, Riga Technical University, Pulka 3, LV-1007 Riga, Latvia; kristine.irtiseva@rtu.lv (K.I.); marika.mosina@rtu.lv (M.M.); anastasija.tumilovica@rtu.lv (A.T.); jurijs.ozolins@rtu.lv (J.O.); valentina.stepanova@rtu.lv (V.S.)

2 Scientific Laboratory of Powder Materials, Faculty of Mechanical Engineering, Riga Technical University, 6B Kipsalas Street, Lab 319, LV-1048 Riga, Latvia; vjaceslavs.lapkovskis@rtu.lv (V.L.); viktors.mironovs@rtu.lv (V.M.)

3 Institute of Aeronautics, Riga Technical University, 6B Kipsalas Street, LV-1048 Riga, Latvia

* Correspondence: andrejs.siskins@rtu.lv

**Abstract:** Among the various methods for collecting oil spills and oil products, including from the water surface, one of the most effective is the use of sorbents. In this work, three-component bio-based composite granular adsorbents were produced and studied for oil products' pollution collection. A bio-based binder made of peat, devulcanised crumb rubber from used tyres, and part fly ash as cenospheres were used for absorbent production. The structure, surface morphology, porosity, mechanical properties, and sorption kinetics of the obtained samples were studied. Composite hydrophobicity and sorption capacity to oil products, such as diesel fuel (DF) and motor oil (MO), were determined. The obtained pellets are characterised by a sufficiently pronounced ability to absorb oil products such as DF. As the amount of CR in the granules increases, the diesel absorption capacity increases significantly. The case of 30-70-0 is almost three times higher than the granules from homogenised peat. The increase in q is due to two factors: the pronounced surface hydrophobicity of the samples ($\Theta = 152°$) and a heterogeneous porous granule structure. The presence of the cenosphere in the biocomposite reduces its surface hydrophobicity while increasing the diesel absorption capacity. Relatively rapid realisation of the maximum saturation by the MO was noted. In common, the designed absorbent shows up to 0.7 $g{\cdot}g^{-1}$ sorption capacity for MO and up to 1.55 $g{\cdot}g^{-1}$ sorption capacity for diesel. A possible mechanism of DF absorption and the limiting stages of the process approximated for different kinetic models are discussed. The Weber–Morris diffusion model is used to primarily distinguish the limiting effect of the external and internal diffusion of the adsorbate on the absorption process.

**Keywords:** sorption; kinetics; peat; cenosphere; oil sorption; Weber–Morris; diffusion model; granules; biocomposite

**Citation:** Irtiseva, K.; Mosina, M.; Tumilovica, A.; Lapkovskis, V.; Mironovs, V.; Ozolins, J.; Stepanova, V.; Shishkin, A. Application of Granular Biocomposites Based on Homogenised Peat for Absorption of Oil Products. *Materials* **2022**, *15*, 1306. https://doi.org/10.3390/ma15041306

Academic Editors: Andrea Petrella, Alain Celzard and Michele Notarnicola

Received: 14 December 2021
Accepted: 4 February 2022
Published: 10 February 2022

**Publisher's Note:** MDPI stays neutral with regard to jurisdictional claims in published maps and institutional affiliations.

## 1. Introduction

One of the basic principles of the circular economy is the maximum economy of raw materials and the maximum use as well as processing of secondary raw materials. A large amount of waste is the cenospheres (CS), which are generated from the combustion of coal in coal-fired power plants [1], and waste rubber in the form of crumb rubber (CR) [2,3] from used tyres. By using peat processing products as absorbers from CS and crushed rubber, it is possible to obtain sufficiently porous biocomposites with specific properties, including the ability to absorb oil products on their surface [4,5]. Among the various methods for collecting oil spills and oil products, including from the water surface, one of the most effective is the use of sorbents. Several requirements are imposed on the sorbents used to collect oil products from the water surface: the sorbent should mainly absorb oil products upon contact with the water surface, the maximum saturation of the sorbent with oil should be reached as soon as possible after contact, and afterwards the sorbent should remain on the surface for a long time. Thus, the effectiveness of sorbents for the absorption of oil products from the water surface is assessed by three criteria: the absorption capacity of oil products, water absorption capacity, and buoyancy of the sorbent [4,6].

Commercial adsorbents are effective but expensive. Natural sorbents quickly absorb water and do not selectively absorb the oil product or present a good-enough buoyancy. To minimise water absorption and hydrophilicity, adsorbents need to be modified with hydrophobic agents. An increasing number of oil spill incidents call for an increased necessity for cheaper absorbents made of secondary materials.

Liu et al. [7] reported the hydrophobic plant fibre sponge preparation via a simple treatment by hexamethyldisilazane vapours. The material is characterised by high porosity, high mechanical durability, excellent hydrophobicity with a water contact angle of 145°, high oil absorptive capacity (15–40 $g \cdot g^{-1}$), and good selectivity [7]. Wu and Zhou [8] reported waste tyre rubber (WTR) prepared by the graft copolymerisation-blending method using WTR and 4-tert-butyl styrene as monomers, and divinylbenzene and benzoyl peroxide were employed as a crosslinker and initiator, respectively. Oil absorptive material has a maximum absorption capacity up to 24.0 $g \cdot g^{-1}$. Oil removal from the aqueous state by natural fibrous sorbents has been widely studied [9]. The peat and peat treatment products have also been widely studied for absorbent design [10]. Alameri et al. [11] investigated the use of peat-derived biochar as a bio-sorbent for the sorption and removal of crude oil spills from synthetic seawater, with up to 32.5 $g \cdot g^{-1}$ sorption capacity per crude oil.

For practical application, it is necessary to focus on low-cost (preferable secondary) raw material use, maximal sorption capacity and buoyancy, the ability to collect the absorbent from the water surface, and oil sorption kinetics to collect the spill in a short time.

Table 1 summarises several examples of developed and studied sorbents for crude oil and oil products' spill clean-up and water treatments. As can be seen, sorbents with high sorption capacity require deep chemical and thermal treatments. As a result, they are not eco-friendly, using hazardous and toxic chemicals [7,8,12,13]; are not sustainable, as they produce a lot of wastewater water [8,13] and gases [14]; and have a high cost [8,12] due to the complex production and used equipment.

**Table 1.** Examples of developed and studied sorbents for crude oil and oil products' spill clean-up and water treatments.

| Adsorbent Production Method, Composition | Adsorbate | Adsorption Capacity $g \cdot g^{-1}$ | Form, Shape | Ref. |
|---|---|---|---|---|
| Plant fiber foam hexamethyldiazosilane + $NH_3$ vapours treatment | Olive oil | 38 | Sponge | [7] |
| WTR graft copolymerisation-blending method, 4-tert-butylstyrene + divinylbenzene, and benzoyl peroxide | Crude oil (10%) diluted with toluene | 24 | Spherical beads 0.5–1 mm in diameter | [8] |
| Graphene aerogel | Emulsified diesel oil | 25 | Aerogel sponge | [12] |
| WTR activated through carbonisation and $H_2SO_4$ treatment | Crude oil | 14.7 | Powder | [13] |
| Reduced graphene oxide and natural rubber latex | Crude oil | 12–21 | Aerogel sponge | [15] |

**Table 1.** *Cont.*

| Adsorbent Production Method, Composition | Adsorbate | Adsorption Capacity g·g$^{-1}$ | Form, Shape | Ref. |
|---|---|---|---|---|
| Peat activated through carbonisation at $K_2CO_3$ presence | Crude oil | 7–8 | Powder | [14] |
| Natural peat, no treatment | Motor oil | 2.1–6.5 | Bulk peat fibers, 1–7 mm | [16] |
| WTR powder | Crude oil | 4.7 | Powder | [17] |
| Rice straw, no treatment | Water/Diesel mixture | 1.8–2.3 | Bulk straw | [18] |
| Expanded perlite | Petroleum products | 2–2.5 | Powder | [19] |
| Cellulosicfibre | | 2–3.8 | Bulk fibers | |
| Straw sorbent, no treatment | Water/Oil product mixture | 0.033 | Bulk straw | [20] |
| Peat sorbent, no treatment | | 0.037 | Bulk peat | |
| Mineral sorbent (cemetious residue), no treatment | | 0.00183 | Powder | |

The investigated and described sorbents without deep chemical treatment have significantly lower sorption values. However, the significant disadvantage of the described sorbents (by except [8]) is the hardly suitable practical application in real spill collection: spreading and collecting such sorbents from the water surface due to their form/shape. Additionally, according to the literature, the maximal sorption capacity of the oil products is mainly noted starting from 5 and to a longer time: time can be 5–10 min [16], 20–30 min [17], and 30 min [8].

A previous work [4] performed manufacturing and characterisation of a three-component bio-based composite material. The key components used were a bio-based binder made of peat, devulcanised CR from used tyres, and part of the fly ash a cenospheres. The material was prepared in a block to investigate its mechanical properties and density. Preliminary studies to determine the sorption of water and oil products (in the form of granules) demonstrated promising results. Individual investigation for the granulated material for oil products is necessary because it describes its properties as products–granules but not as a powder-like or sponge material that most of the papers usually describe.

In this work, granular adsorbents from secondary raw materials were produced and studied for oil products' pollution collection. The structure, surface morphology, inner structure of the granules (by mCT), porosity, mechanical properties, and sorption kinetics of the obtained samples were analysed. Their hydrophobicity and sorption capacity were determined in relation to oil products: diesel fuel (DF) and motor oil (MO).

## 2. Materials and Methods

### 2.1. Materials and Composites Used

Using the wet granulation method, nineteen granule compositions were prepared. The composition of the granular composites before drying is shown in Tables 2 and 3. They are designated CR-HP-CS, where CR is a percentage by the mass of devulcanised rubber, HP is a percentage by the mass of homogenised peat (including water), and CS is a percentage by the mass of cenospheres. The composition changes significantly after drying due to the evaporation of water from the absorbent. The composition after drying is also presented in Tables 2 and 3 but for convenience, the initial compositions' (including water from HP) designations are used. The detailed granules production method has been published in a previous work [4]. The production scheme for granular adsorbents and the used drum granulator are represented in Figure 1a,b respectively.

**Table 2.** The composition of block and granules in the raw mixture (wet) and after drying by wt.% (part I) [4].

| | Designation of the Composition | | | | | | | | | | | |
|---|---|---|---|---|---|---|---|---|---|---|---|---|
| | 0-100-0 | 5-95-0 | 10-90-0 | 15-85-0 | 20-80-0 | 30-70-0 | 0-95-5 | 5-90-5 | 10-85-5 | 15-80-5 | 20-75-5 | 30-65-5 |
| | **Wet mixture composition (wt.%)** | | | | | | | | | | | |
| DCR | 0.0 | 5.0 | 10.0 | 15.0 | 20.0 | 30.0 | 0.0 | 5.0 | 10.0 | 15.0 | 20.0 | 30.0 |
| HP | 100 | 95.0 | 90.0 | 85.0 | 80.0 | 70.0 | 95.0 | 90.0 | 85.0 | 80.0 | 75.0 | 65.0 |
| CS | 0.0 | 0.0 | 0.0 | 0.0 | 0.0 | 0.0 | 5.0 | 5.0 | 5.0 | 5.0 | 5.0 | 5.0 |
| | **Dried composite material formulation (wt.%)** | | | | | | | | | | | |
| DCR | 0.0 | 27.3 | 44.2 | 55.8 | 64.1 | 75.4 | 0.0 | 22.1 | 37.2 | 48.1 | 56.3 | 68.0 |
| HP | 100 | 72.7 | 55.8 | 44.2 | 35.9 | 24.6 | 72.7 | 55.8 | 44.2 | 35.9 | 29.6 | 20.6 |
| CS | 0.0 | 0.0 | 0.0 | 0.0 | 0.0 | 0.0 | 27.3 | 22.1 | 18.6 | 16.0 | 14.1 | 11.3 |

**Table 3.** The composition of block and granules in the raw mixture (wet) and after drying by wt.% (part II) [4].

| | Designation of the Composition | | | | | |
|---|---|---|---|---|---|---|
| | 0-90-10 | 5-85-10 | 10-80-10 | 15-75-10 | 20-70-10 | 30-60-10 |
| | **Wet mixture composition (wt.%)** | | | | | |
| DCR | 0.0 | 5.0 | 10.0 | 15.0 | 20.0 | 30.0 |
| HP | 90.0 | 85.0 | 80.0 | 75.0 | 70.0 | 60.0 |
| CS | 10.0 | 10.0 | 10.0 | 10.0 | 10.0 | 10.0 |
| | **Dried composite material formulation (wt.%)** | | | | | |
| DCR | 0.0 | 18.6 | 32.1 | 42.3 | 50.3 | 62.0 |
| HP | 55.8 | 44.2 | 35.9 | 29.6 | 24.6 | 17.4 |
| CS | 44.2 | 37.2 | 32.1 | 28.2 | 25.1 | 20.7 |

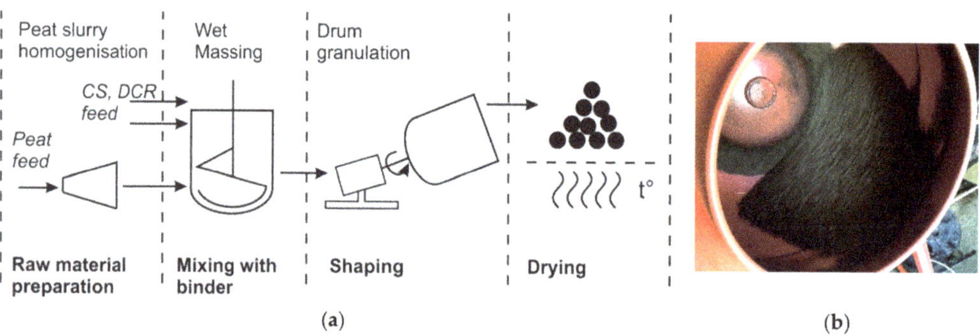


**Figure 1.** Production scheme (**a**) and used drum granulator (**b**) in operation.

### 2.2. Investigations of the Surface Morphology and Structure of Samples

The morphology of the granule surface was studied using a Kyence VHX-1000 digital microscope (Keyence Corp. Osaka, Japan). The Kyence VHX-1000 digital microscope operates in a magnification range from 0 to 200 times and was equipped with VH-Z20R/W lenses. The microscope allows you to study the under-study object's topography from different angles, which will enable you to obtain a three-dimensional picture.

The Tescan Mira/LMU scanning electron microscope (Tescan, Brno, Czech Republic) was used to study the microstructure of the sample surface and visually assess the pore morphology. Before analysis, the samples were coated with a 10–15 nm layer of gold using an Emitech K550X (Emitech, Montigny-le-Bretonneux, France).

The internal structure of the samples was assessed using X-ray micro-computed tomography (microCT). The cycle continued up to 180 degrees. The resulting images were combined to obtain a three-dimensional image during the reconstruction process. MicroCt 50, Scanco Medical (Scanco Medical AG, Wangen-Brüttisellen, Switzerland) was used. Colourless MicroCT scans were made using cabinet cone-beam microCT (μCT50, Scanco Medical AG, Wangen-Bruttisellen, Switzerland) with the settings: 70 kVp energy; 114 μA tube current; 0.5 mm Al filter; 1005 ms integration time; and 5 μm voxel size. Each sample was scanned for 11 h. Reconstruction of 3D datasets from the microCT projection data, including beam-hardening correction, was performed automatically after completing each cone-beam image stack using the SCANCO GPU (Scanco Medical AG, Wangen-Bruttisellen, Switzerland) Accelerated Reconstruction System. The visualisation module performs sophisticated 3D rendering of large datasets using high-quality ray-tracing algorithms.

### 2.3. Mechanical Compressive Strength

The strength of the composite granules was determined using a KAHL mechanical strength tester (Amandus Kahl GmbH & Co. KG, Reinbek, Deutschland). The principle of operation of the device is to fix the force required to compress the granule. The compression of the granules took place at a compression speed of 0.5 mm·min$^{-1}$. For the study of each composition, 30–40 granules with similar geometry and dimensions of 6.0 ± 0.5 mm were tested.

### 2.4. Nitrogen Adsorption Porosimetry

Nitrogen adsorption porosimetry can be used to determine the specific surface area of a sample, total pore volume, and pore size distribution and scatter. The Brunauer–Emmett–Teller (BET) surface analysis method is based on the adsorption and desorption of nitrogen. Samples were degassed in an Autosorb Degasser Model AD-9 (Anton Paar, St Albans, England) before nitrogen adsorption. Degassing is performed to remove excess moisture and impurities from the samples. The samples were degassed for 24 h at 100 °C. The samples were weighed before and after degassing. Higher temperatures speed up the degassing process but can only be used if they do not affect the structure of the sample.

Nitrogen sorption was performed on a QUADRASORB SI (Quantachrome Corporation, Boynton Beach, FL, US) device. The measurement was carried out at a temperature when nitrogen is in a liquid state (−196.15 °C). During this process, nitrogen molecules are physically adsorbed onto the sample surface. When determining the amount of adsorbed gas as a function of pressure under isothermal conditions, adsorption isotherms are constructed, after which the distribution of pores in the material is determined.

### 2.5. Surface Energy and Contact Angle

If a drop of liquid gets on a hard surface, its ability to wet the surface can be characterised by the wetting angle or contact angle. As the surface hydrophobicity increases, the contact angle values increase. In the case of a highly hydrophobic surface, the contact angle is 150° or more. The contact surface and the liquid contact angle of the obtained biocomposites made in blocks were determined using the drop method with the optical tensiometer Theta (Attention, Helsinki, Finland,). Distilled water was used in the experiment, the volume of the liquid drop was 5 μL, and 5 parallel measurements were performed for each sample.

### 2.6. Absorption Capacity of Oil Products

The absorption capacity of oil products was determined for granules of biocomposites by combining various compositions of peat, CS, and CR. For sorption studies, the liquid media-oil products, namely diesel fuel miles PLUS (EN950 standard, by Circle-K Latvia Sia) with a density of 0.815 g·cm$^{-3}$ and motor oil LazerWay 5-W40 (by Circle-K Latvia Sia) with a density of 0.835 g·cm$^{-3}$, were used. The ability of the pellets to absorb DF and MO [21,22] was determined using the gravimetric method and samples weighing 1.0 ± 0.1 g were placed in a stainless-steel mesh and incubated once in one of these media. Measurements were taken after 1, 3, 5, and 10 min. The absorption capacity of oil products was performed by immersion of a single granule (diameter of 5–8 mm) in the metal mesh into a 2 mm thick layer of the studied oil product in a Petri dish, which imitated an oil spill. Each granule was extracted from the Petri dish by the exact time gap for weighting. The exceeding layer or droplets (if any) was wiped by a wet (wetted by the same oil product) paper towel from the metal mesh and from the granules. Since the granules extraction moment sorption time counting is paused until it returns to the Petri dish. The measurements were then repeated every 5 min until equilibrium was reached and the weight of the saturated granules remained unchanged. The mass of the granules was recorded with an accuracy of 0.001 g. For each composition, 6 parallel measurements were performed. The astringency of the oil products was determined by Equation (1) [23,24]:

$$q = (m_t - m_0 - m_s)/m_0 \tag{1}$$

where

$q$—absorption capacity ($g \cdot g^{-1}$);
$m_t$—the mass of the sample and mesh after a certain period of time (g);
$m_s$—mesh mass (g); and
$m_0$—initial mass of the sample (g).

## 3. Results

Biocomposites based on HP slurry processing products were obtained using CS and CR. For the obtained biocomposites in the form of granules, the influence of the composition of CR and CS on their mechanical properties, surface morphology, structure, and hydrophobicity was determined. The ability of the obtained granular biocomposites to absorb oil products to their surface was experimentally determined using DF and MO.

As shown in a previous publication [4], as the content of CR in the biocomposite increases, its compressive strength decreases significantly compared to the compressive strength of a block of dried homogenised peat. An increase in the mass fraction of CR in the original composition from 5% to 30% reduces the mechanical compressive strength from 8 MPa to 2 MPa. In this case, the apparent density of the samples decreased from 1.20 $g \cdot cm^{-3}$ (5 wt.% CR) to 0.95 $g \cdot cm^{-3}$ (30 wt.% CR). Information on the nature of the destruction of samples can be obtained by analysing their SEM images after the loading of the sample. As shown from Figure 2, CR is evenly distributed in the sample volume, which was also demonstrated by digital microscopic images. Cracking and failure of the sample under load occurred at the interface between the aggregate and the absorbent, which can be explained by the insufficient interconnection or adhesion of the CR and HP matrix.

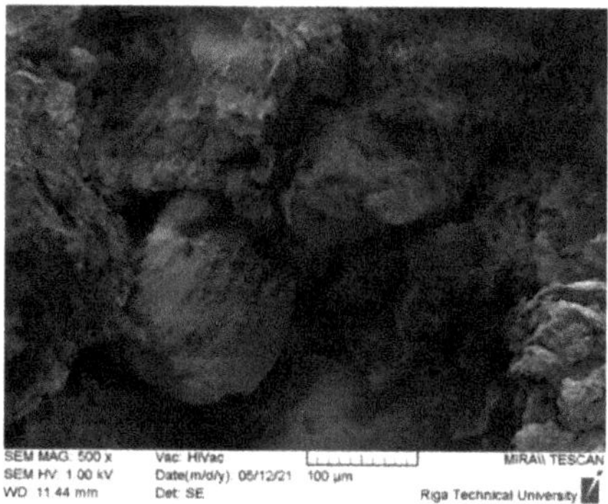


**Figure 2.** Sample images of samples 5-95-0 after loading taken at ×100 and ×500 magnification. The yellow arrows indicate CR particles.

When oil and oil products are collected from the water surface using porous sorbents, the process of their absorption is usually physical. The number of contaminants depends on the hydrophobicity, oliophility, and specific surface area of the material. The contact angle was determined for the test samples by the drop method.

As can be seen from Figure 3, with an increase in the mass fraction of CR in the composite material, the contact angle with water also significantly increases, which indicates that the addition of CR increases the hydrophobicity of the samples. Thus, replacing 5 wt.% peat with CR can show a 42% increase in contact angle. As the amount of CR in the composite increases, the contact angle tends to increase. The most significant contact angle (Θ) 152° was observed for sample 40-60-0. In this case, the sample was barely wetted by water and the drops tended to drain off the surface. The hydrophobicity of the surface of the samples should facilitate better absorption of oil products.

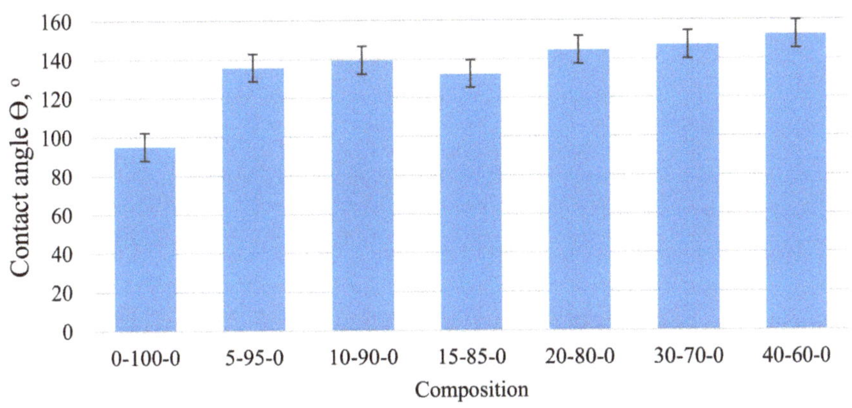


**Figure 3.** Variation of the wetting angle of the biocomposite depending on the amount of CR in the composite.

To increase the buoyancy of the composition, the weight of CS was introduced. Additionally, in this case, the addition of CS, as in the case of CR, reduces the compressive strength of the composite compared to the strength of the HP block and strength of the composite blocks containing CR, as shown in Table 4.

**Table 4.** Influence of CS on the properties of biocomposites in the shape of a block.

| Parameter | CR 0 wt.% | | | CR 5 wt.% | | | CR 10 wt.% | | |
|---|---|---|---|---|---|---|---|---|---|
| | CS 0 wt.% | CS 5 wt.% | CS 10 wt.% | CS 0 wt.% | CS 5 wt.% | CS 10 wt.% | CS 0 wt.% | CS 5 wt.% | CS 10 wt.% |
| σ, MPa | 79.0 | 2.2 | 1.7 | 7.7 | 2.0 | 1.8 | 5.5 | 1.3 | 1.1 |
| ρ, g·cm$^{-3}$ | 1.35 | 0.74 | 0.64 | 1.20 | 0.76 | 0.65 | 1.10 | 0.85 | 0.70 |
| Θ, ° | 95 | 115 | 124 | 135 | 121 | 129 | 139 | 122 | 132 |

The decrease in the mechanical strength of the biocomposite can be explained by the formation of additional voids and structural defects when introduced into the CS system, as evidenced by the decrease in the apparent density ρ (g·cm$^{-3}$) of the samples to 0.65 g·cm$^{-3}$ (5-85-10).

By analysing the SEM images of the samples (Figure 4), it can be concluded that, similarly to the compositions containing CR, in the compositions containing CR and CS, their destruction by compression occurs at the interface between the homogenised peat and the aggregates. In addition, the introduction of CS into the system increases the proportion of structural defects and voids.

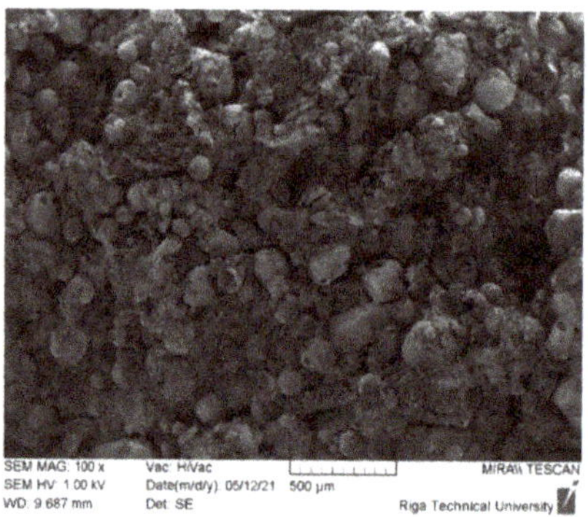


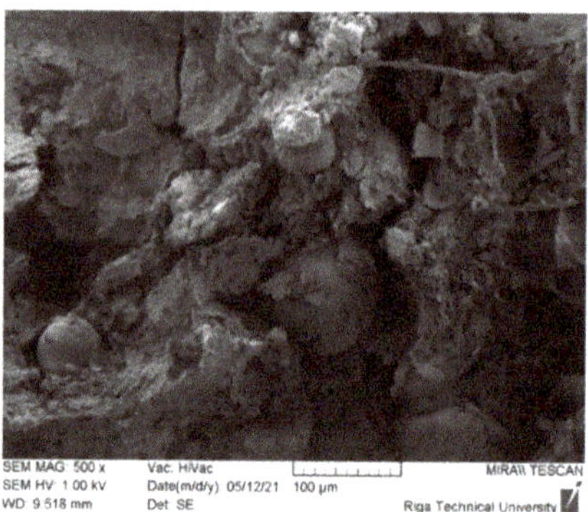


**Figure 4.** Sample images of samples 5-90-5 after destruction taken under ×100 and ×500 magnification.

The incorporation of CS into the composite was expected to reduce the surface hydrophobicity of the samples. As shown from the Table 4 data, this tendency was observed. Still, the experimentally determined values of the wetting angle Θ are sufficiently high, that is, not less than 120°, which is typical for hydrophobic surfaces.

To use the studied biocomposites as sorbents for the absorption of oil products, they were obtained in the form of granules using the granulation technology described in the publication. An essential property of granules for their successful use is mechanical strength. Figure 5 shows the mechanical strength of biocomposite granules when compressed into different granule compositions. As the concentration of CR in the composite increases, the mechanical strength of the granules decreases from 48 N on average if the composition contains 5% CR to 26 N if the composition contains 30% CR (Figure 5). As already noted, a similar trend was observed for biocomposites in the form of a block. In this case, the mechanical strength of the composite decreased, on average, four times with similar increasing CR concentrations. The incorporation of cenospheres into the composition further reduced the mechanical compressive strength of the granules, which may be due to insufficient bonding of the microspheres and homogenised peat, on the one hand, and a high degree of composite filling in the case of dried granules (average 44.2 wt.% 5-90-5 and up to 79.3% by weight in the case of 30-65-5), which of course affects the mechanical strength of the product. The observed changes in the mechanical strength of the granules are also related to the occurrence of structural defects, pores, and voids in the granulation process (Figures 5–7).

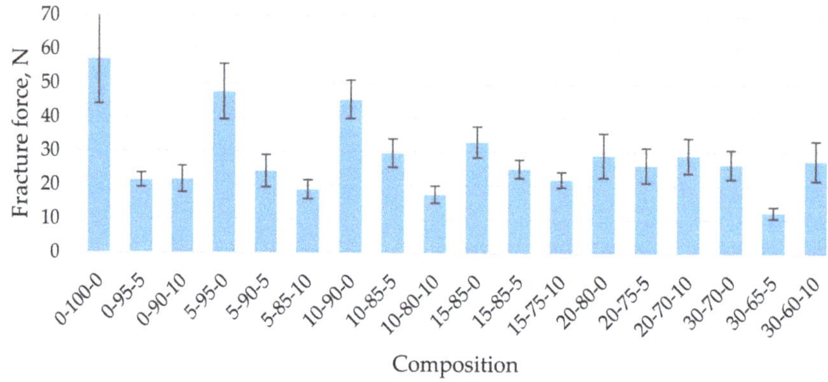


**Figure 5.** Mechanical strength of biocomposite granules depending on their composition.

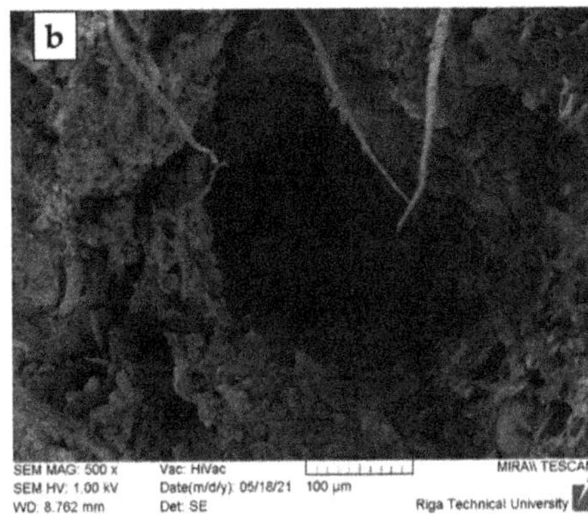


**Figure 6.** Granules with a composition of 5-90-5 SEM images at ×100 (**a**) and ×500 (**b**) magnification.

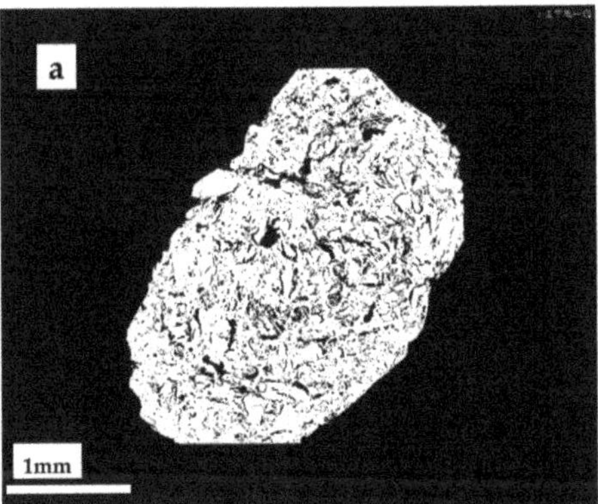


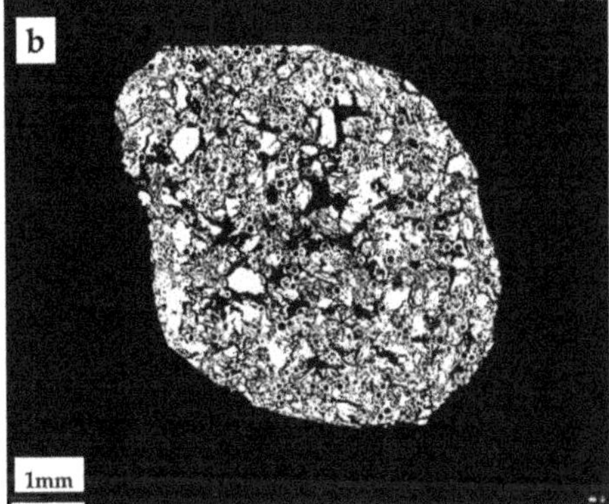


**Figure 7.** Photomicrographs of MCt cross-sections of granules 0-100-0 (**a**) and 5-90-5 (**b**).

The addition of CS to the biocomposite composition promotes and improves the buoyancy of the granules, which is very important when using it for collecting oil products from the water surface. Granules containing 5 to 30 wt.% of the CR are not buoyant. In turn, granules obtained from the initial mixture, which, without CR, additionally contains from 5 to 10 wt.% CS, show good buoyancy. Granules from the initial mixture with different CR content and 10 wt.% CS remain buoyant for more than 10 days, making it possible to collect them from the water surface after the absorption of oil products.

One of the most critical parameters characterising sorbents is their porosity and specific surface area. As shown in Table 5, the porosity of the biocomposite granules was significantly increased when CR and CS were added to the starting mixture.

**Table 5.** Apparent density, porosity, and specific surface area of biocomposite granules of different compositions.

| Granule Composition | Apparent Density $\varrho$, g·cm$^{-3}$ | Porosity, %, BET | Porosity, %, (by MCt) | Specific Surface Area, m$^2$·g$^{-1}$ |
|---|---|---|---|---|
| 0-100-0 | 1.29 | 17.8 | 19.7 | 0.452 |
| 5-95-0 | 1.20 | 32.9 | 34.7 | 0.580 |
| 5-90-5 | 0.71 | 37.6 | 39.5 | 0.702 |

In addition, the data obtained by the BET method are close to the data obtained by the MCt analysis. The porosity of granules with a composition of 5-90-5 increases practically two times compared to the granules obtained from the homogenised peat. At the same time, of course, the apparent density of the samples decreases and the specific surface area of the material increases to 0.702 $m^2 \cdot g^{-1}$.

The SEM microphotographs of samples 5-90-5 (Figure 6) show a distinctly porous structure with characteristic cylindrical voids formed during the granulation process in the rotary granulator. X-ray microtomography (MCt) data provide more objective information on the internal structure of the granular material. Figure 7 shows the cross-sections of the samples. A sufficiently homogeneous structure with small voids due to granulation was observed in the cross-section of sample 0-100-0 (a). In the case of sample 5-90-5 (b), on the other hand, a markedly heterogeneous structure was seen in which CR and CS particles, as well as voids between the particles and the homogenised peat, appear uniform in volume. The heterogeneous structure resulting from the granulation process determines the porosity and specific surface area of the obtained material, which is practically twice as large as in the sample 0-100-0 (Table 5). Increasing the amount of CR and CS in the samples increases the structural heterogeneity and porosity.

The ability of the obtained biocomposite to absorb oil products, namely DF and MO, was studied. Figure 8 shows the experimentally obtained kinetics of DF absorption using the studied biocomposite granules containing CR. As can be seen, with an increasing amount of CR in the composition of the granules, the DF absorption capacity increases significantly and, in the case of 30-70-0, is almost three times higher than the granules from the homogenised peat. The increase in q is due to two factors: pronounced surface hydrophobicity of the samples ($\Theta$ = 152°) and non-uniform porous granule structure. Even though the presence of the cenosphere content in the biocomposite reduces its surface hydrophobicity (Table 4), the DF absorption capacity increases, as shown in Figure 9. This condition could be related to the sufficiently developed porous granule's structure that occurs during their formation in the rotary granulator. Among the studied compositions, granules with a composition of 30-65-5 stand out, in which case the average DF absorption capacity reaches 1.5 $g \cdot g^{-1}$.

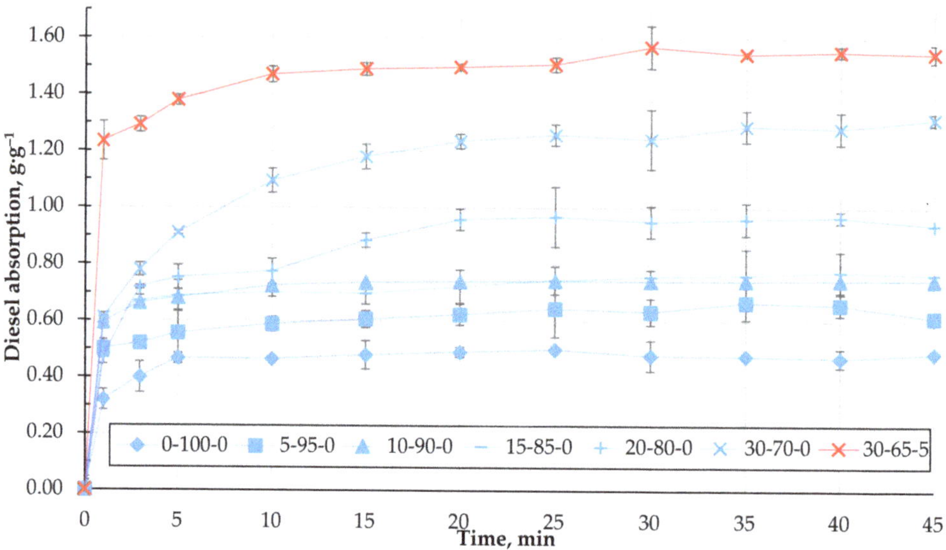


**Figure 8.** Adsorption kinetics of DF with biocomposite pellets containing CR.

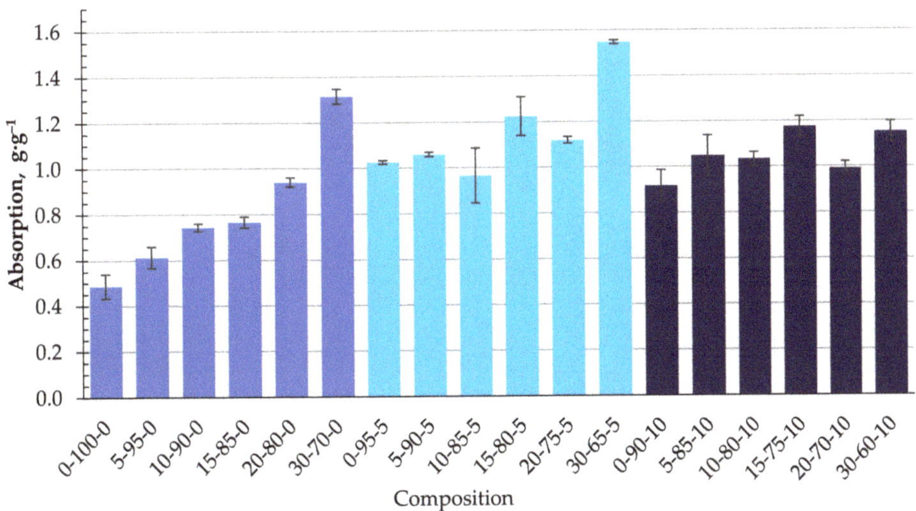


**Figure 9.** Absorption capacity (g·g$^{-1}$) of DF depending on the composition of the composite.

Their ability to bind motor oil, with a viscosity much higher than DF, was also determined for the different granule compositions obtained. For all formulations, the maximum rate of MO retention was achieved in the first minutes of contact and practically did not increase with increasing residence time (Figure 10). This indicates a different mechanism of retention of MO by the biocomposite granules compared to that of diesel fuel. Figure 11 shows the maximum degree of MO retention achieved for all the granule formulations tested. The introduction of CR particles into the composition contributes to motor oil binding insignificantly. At the same time, the use of CS, additionally, as a filler increases the ability to bind motor oil on average by 38% compared with granules containing only CR and reaches the value of 0.65 ± 0.05 g·g$^{-1}$.

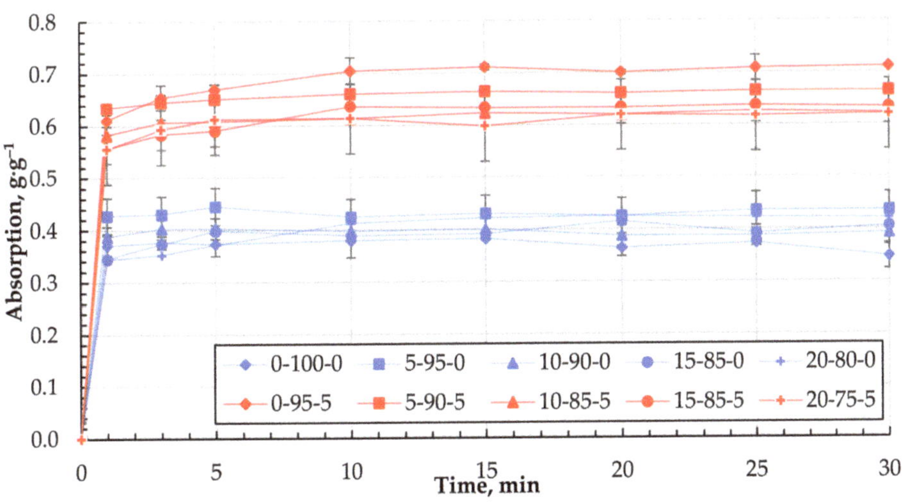


**Figure 10.** Absorption kinetics of MO for granules with 0 and 5 wt.% of CS.

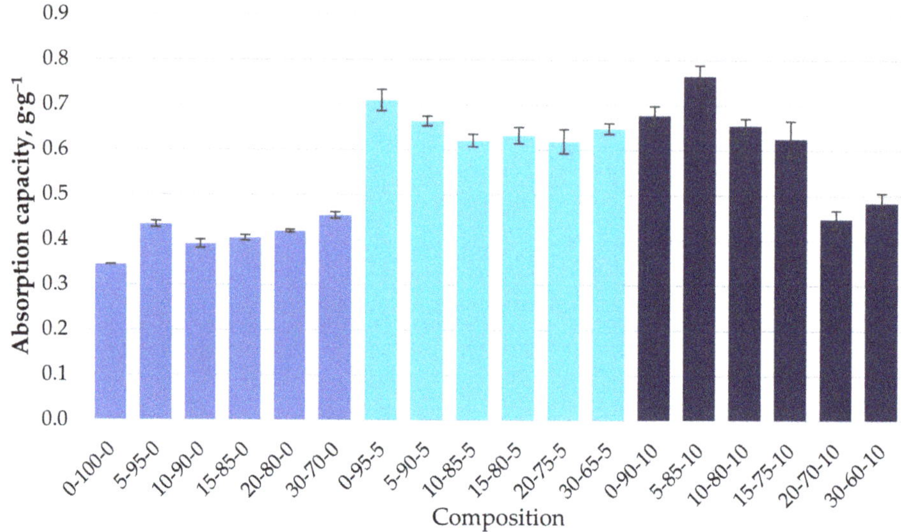


**Figure 11.** Absorption capacity ($g{\cdot}g^{-1}$) of MO depending on the composition of the composite.

## 4. Discussion

To be able to determine the possible mechanism of DF absorption and the limiting stages of the process, the experimental data can be satisfactorily approximated for different kinetic models. To primarily distinguish the limiting effect of external and internal diffusion of the adsorbate on the absorption process, the Weber–Morris diffusion model [25] can be used, the mathematical expression of which is Equation (2):

$$q_t = K_{id}t^{0.5} + C \quad (2)$$

where

$q_t$—amount of adsorbed substance in $g{\cdot}g^{-1}$;

$K_{id}$—Weber–Morris constant or internal diffusion rate constant in pores in $g{\cdot}g^{-1}{\cdot}min^{0.5}$; and

C—a constant parameter related to the boundary layer thickness in $g{\cdot}g^{-1}$.

As shown in Figure 12, the absorption relationships of granules from pure, homogenised peat and DF with a small amount of CR (composition 0-100-0 and 5-95-0) in $q_t - t^{0.5}$ coordinates are practically liberal. It is believed that the diffusion of a substance into the sorbent pores or internal diffusion is a limiting step in the rate of the adsorption process if the relation qt = f ($t^{0.5}$) is linear and crosses the origin of the coordinate (C = 0). In cases where the constant C is not zero, the rate of adsorption is significantly influenced by the absorption of the sorbate to the surface of the granules and the diffusion through the boundary layer [25,26].

As the number of components increases, the granular structure of the biocomposite becomes more porous with pronounced defects and voids. Two stages of sorbate mass transfer can be distinguished in the qt $- t^{0.5}$ coordinates of the DF absorption process (see Figures 12 and 13).

The steepest stage of the relationship $q_t - t^{0.5}$ characterises the mass transfer from the volume of DF medium to the sorbent surface or external diffusion. The other less steep (see Figures 12 and 13) stage can be attributed to the transport and absorption of DF molecules in the pores of the sorbent or the so-called internal diffusion [25,27–30]. The point of transition to the relation $q_t$ = f ($t^{0.5}$), which could correspond to the saturation of the sorbent granule surface, is reached for all compositions, on average, at the same time from 5 to 10 min.

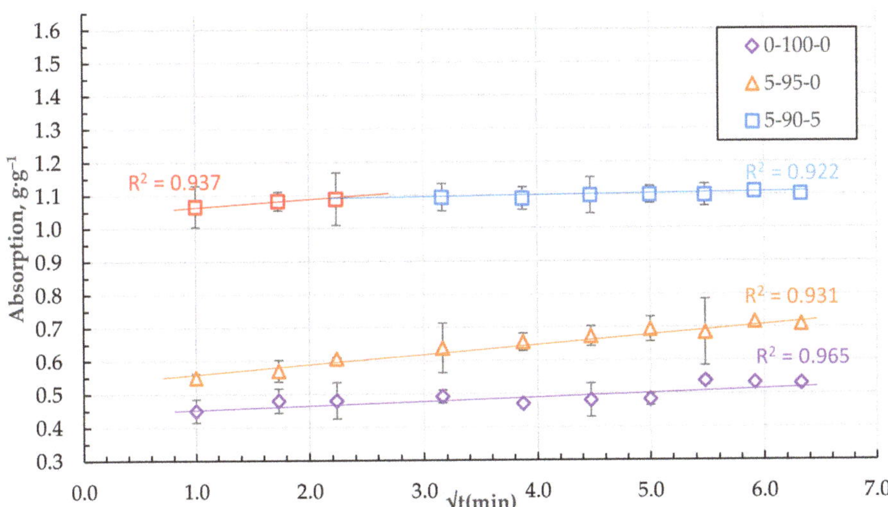


**Figure 12.** Absorption of DF in $q_t - t^{0.5}$ coordinates for the granules with 5 and 0 wt.% of CSand HP only.

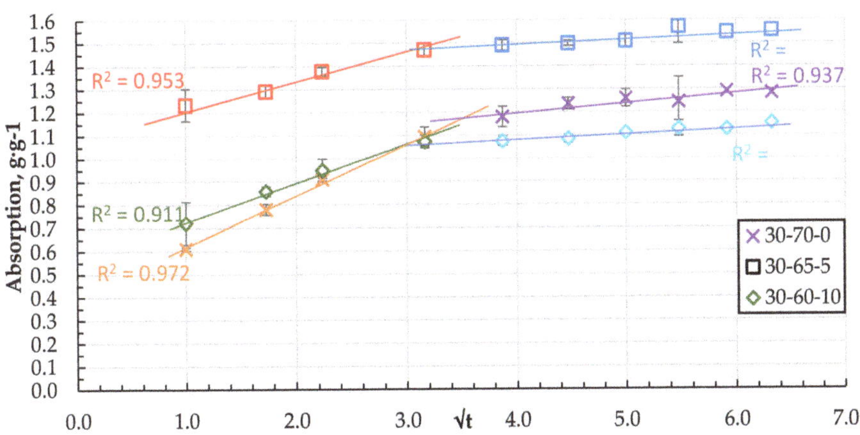


**Figure 13.** Absorption of DF in $q_t - t^{0.5}$ coordinates for the granules with 0, 5, and 10 wt.% of CS.

In the case of MO, as demonstrated above, the saturation of the sorbent is reached in the first minutes (Figure 10) and, most likely, in this case, MO binding mainly occurs only by the surface of the granule. Due to the increased viscosity of MO, its diffusive penetration into the pores of the sorbent practically does not occur. The state of the sorbent surface would determine the ability to bind MO. Sorbents containing particles of CR and cenospheres are characterised by a more developed specific surface and porosity (Table 4 and Figures 6 and 7), which contributes to the increase of MO binding by the granule surface. It is particularly evident for composites containing CS. With good filler content, in our case, the actual mass content of CS in the dried sample 0-95-5 is 27.3% and in sample 0-90-10 is 44.2% (Table 2), while numerous channels, voids, and structural defects can be formed in the near-surface layers and filled with MO. This may lead to a significant increase in the oil absorption of the sorbents (Figure 11).

When using granules to collect oil products from the water surface, they must absorb water to a minimum. As the surface hydrophobicity increased, the ability of the samples to absorb water was expected to decrease. However, as the experiment has shown, pellets with different compositions absorb practically as much water as the g·g$^{-1}$ of DF. A comparative example of water and DF absorption kinetics for granules with a composition of 30-65-5 is shown in Figure 14.

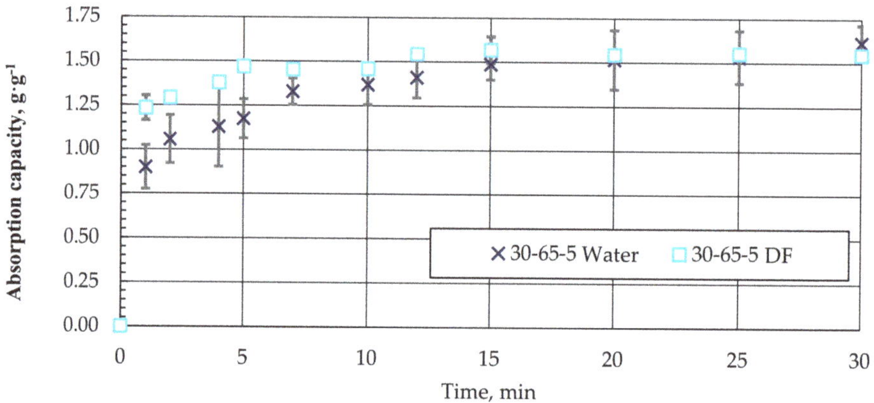


**Figure 14.** Absorption kinetics (approximation) of water and DF for granules with a composition of 30-65-5.

As can be seen, the nature of the absorption of water and DF in the biocomposite pellets is similar. It is only necessary to note that DF absorption and saturation are relatively faster. Obviously, due to the hydrophobicity of the surface, the DF absorption process could be a priority in the case of a water–DF mixture.

When studying the process of MO absorption using the obtained sorbents, it was found that the maximum degree of saturation is already reached in the first minutes of the absorption process (Figure 10). By increasing the time of the absorption process, the achieved degree of saturation for the studied biocomposites barely changes. This indicates that the oil absorption process mainly takes place on the surface of the sorbent. Due to the increased viscosity of the oil, internal diffusion of the products is minimal or does not occur.

Depending on the composition of the biocomposite, there are no apparent differences in the maximum degree of the sorption of MO, which ranges from 0.4 ± 0.05 g·g$^{-1}$ for compositions without CS to 0.65 ± 0.05 g·g$^{-1}$ for compositions with CS 27-11 wt.% in a dry mixture (please see Tables 2 and 3), which can be seen in Figure 11.

## 5. Conclusions

Biocomposites in granules were obtained from industrial waste tyre rubber, fly ash cenospheres, and homogenised peat.

As the amount of CR in the composite increases, its surface hydrophobicity increases, as evidenced by changes in the water wetting angle Θ from 95 to 152°. When applied to the composite CS, the wetting angle tends to decrease, but in no case is it less than 120°, which is typical for hydrophobic surfaces.

As the concentration of CR in the composite increases, the mechanical strength of the granules decreases from 48 N on average if the composition contains 5% CR to 26 N if the composition contains 30% CR. The incorporation of CS into the composition further reduces the mechanical compressive strength of the granules, which may be due to insufficient bonding of the microspheres and the homogenised peat, on the one hand, and a high degree of composite-filling defects and voids, on the other.

The obtained granulated biocomposites based on HP, CR, and CS are characterised by a sufficient capacity to bind petroleum products: diesel fuel and motor oil. With the increase of the CR content in the granules, the absorption capacity of diesel fuel increases significantly on average from 0.5 $g{\cdot}g^{-1}$ in the case of homogenised peat to 1.3 $g{\cdot}g^{-1}$ for granules with the composition 30-70-0. The increase in the binding capacity can be explained by two factors: the clearly expressed hydrophobicity of the surface and the developed porous structure of the granules. External diffusion processes determine diesel binding to the granule surface and the diffusion processes of diesel fuel into the granule's porous structure.

As the amount of CS in the granules increases up to 11.3 wt.% (for a dry composition), the DF absorption capacity increases significantly up to 0.7 $g{\cdot}g^{-1}$ sorption capacity for MO and up to 1.55 $g{\cdot}g^{-1}$ sorption capacity for DF (30-65-5), which is almost three times higher than the granules from the homogenised peat. The increase in q is due to two factors: the pronounced surface hydrophobicity of the samples ($\Theta$ = 152°) and a heterogeneous porous granule structure. The presence of CS in the composition of biocomposites reduces their hydrophobicity. However, due to high porosity, the binding capacity of diesel fuel is not almost reduced, while the buoyancy of the granules is increased.

In the case of motor oil, it was observed to bind mainly only by the surface of the granule and the maximum rate of binding was reached in the first minutes of the experiment. Increasing the CR and CS content in the biocomposite increases the binding capacity of motor oil due to the expanded structure on the surface of the granule.

The obtained pellets are characterised by a sufficiently pronounced ability to absorb MO and DF. Relatively quick (less than 10 min) achievement of the maximal saturation by the MO was noted. The presence of the cenosphere in the biocomposite reduces its surface hydrophobicity while increasing the DF absorption capacity.

**Author Contributions:** Conceptualisation, J.O., A.S., V.M., and V.L.; methodology K.I., J.O., V.S. and A.S.; formal analysis, A.T., J.O., V.S. and A.S.; investigation, K.I., M.M., A.T. and V.S.; resources, V.L., J.O. and A.S.; data curation, A.T., M.M., V.L. and A.S.; writing—original draft preparation, K.I., J.O., A.T. and A.S.; writing—review and editing, A.S., K.I., V.M. and V.L.; visualisation, K.I. and A.S.; supervision, J.O. and A.S.; project administration, V.L. and A.S.; funding acquisition, V.L., V.M. and A.S. All authors have read and agreed to the published version of the manuscript.

**Funding:** This project has been supported by the Latvian Council of Science within the scope of the project "Innovative bio-based composite granules for collecting oil spills from the water surface (InnoGran)" (no. lzp-2020/2-0394). The work is a result within the network collaboration in the frame of COST Actions CA17133 Circular City ("Implementing nature-based solutions for creating a resourceful circular city") and CA18224 GREENERING ("Green Chemical Engineering Network towards upscaling sustainable processes"). COST Actions are funded within the EU Horizon 2020 Programme. The authors are grateful for the support.

**Institutional Review Board Statement:** Not applicable.

**Informed Consent Statement:** Not applicable.

**Conflicts of Interest:** The authors declare no conflict of interest.

## References

1. Ranjbar, N.; Kuenzel, C. Cenospheres: A Review. *Fuel* **2017**, *207*, 1–12. [CrossRef]
2. Lei, Y.; Wei, Z.; Wang, H.; You, Z.; Yang, X.; Chen, Y. Effect of Crumb Rubber Size on the Performance of Rubberized Asphalt with Bio-Oil Pretreatment. *Constr. Build. Mater.* **2021**, *285*, 122864. [CrossRef]
3. Asaro, L.; Gratton, M.; Seghar, S.; Aït Hocine, N. Recycling of Rubber Wastes by Devulcanization. *Resour. Conserv. Recycl.* **2018**, *133*, 250–262. [CrossRef]
4. Irtiseva, K.; Lapkovskis, V.; Mironovs, V.; Ozolins, J.; Thakur, V.K.; Goel, G.; Baronins, J.; Shishkin, A. Towards Next-Generation Sustainable Composites Made of Recycled Rubber, Cenospheres, and Biobinder. *Polymers* **2021**, *13*, 574. [CrossRef] [PubMed]

5. Lapkovskis, V.; Mironovs, V.; Irtiseva, K.; Goljandin, D. Study of Devulcanised Crumb Rubber-Peat Bio-Based Composite for Environmental Applications. *Key Eng. Mater.* **2019**, *799*, 148–152. [CrossRef]
6. Markl, E.; Lackner, M. Devulcanization Technologies for Recycling of Tire-Derived Rubber: A Review. *Materials* **2020**, *13*, 1246. [CrossRef]
7. Liu, Z.; Qin, Z.; Zhao, G.; Aladejana, J.T.; Wang, H.; Huang, A.; Chen, D.; Xie, Y.; Peng, X.; Chen, T. High-Efficiency Oil/Water Absorbent Using Hydrophobic Silane-Modified Plant Fiber Sponges. *Compos. Commun.* **2021**, *25*, 100763. [CrossRef]
8. Wu, B.; Zhou, M.H. Recycling of Waste Tyre Rubber into Oil Absorbent. *Waste Manag.* **2009**, *29*, 355–359. [CrossRef]
9. Wahi, R.; Chuah, L.A.; Choong, T.S.Y.; Ngaini, Z.; Nourouzi, M.M. Oil Removal from Aqueous State by Natural Fibrous Sorbent: An Overview. *Sep. Purif. Technol.* **2013**, *113*, 51–63. [CrossRef]
10. Shishkin, A.; Treijs, J.; Mironovs, V.; Teirumnieks, E.; Ozolinsh, J. Obtaining and Characterization of Peat-Cenosphere Composite Sorbent with Magnetic Properties. In Proceedings of the 23rd International Baltic Conference on Materials Engineering on Kauno Technologijos Universitetas, Kaunas, Lithuania, 23–24 October 2014; p. 25.
11. AlAmeri, K.; Giwa, A.; Yousef, L.; Alraeesi, A.; Taher, H. Sorption and Removal of Crude Oil Spills from Seawater Using Peat-Derived Biochar: An Optimization Study. *J. Environ. Manag.* **2019**, *250*, 109465. [CrossRef]
12. Huang, J.; Yan, Z. Adsorption Mechanism of Oil by Resilient Graphene Aerogels from Oil–Water Emulsion. *Langmuir* **2018**, *34*, 1890–1898. [CrossRef] [PubMed]
13. Aisien, F.A.; Aisien, E.T. Application of Activated Recycled Rubber from Used Tyres in Oil Spill Clean Up. *Turk. J. Eng. Environ. Sci.* **2012**, *36*, 171–177. [CrossRef]
14. Songsaeng, S.; Thamyongkit, P.; Poompradub, S. Natural Rubber/Reduced-Graphene Oxide Composite Materials: Morphological and Oil Adsorption Properties for Treatment of Oil Spills. *J. Adv. Res.* **2019**, *20*, 79–89. [CrossRef] [PubMed]
15. Klavins, M.; Porshnov, D. Development of a New Peat-Based Oil Sorbent Using Peat Pyrolysis. *Environ. Technol.* **2013**, *34*, 1577–1582. [CrossRef] [PubMed]
16. Cojocaru, C.; Macoveanu, M.; Cretescu, I. Peat-Based Sorbents for the Removal of Oil Spills from Water Surface: Application of Artificial Neural Network Modeling. *Colloids Surf. A Physicochem. Eng. Asp.* **2011**, *384*, 675–684. [CrossRef]
17. Odeh, A.O.; Okpaire, L.A. Modelling and optimizing the application of waste TYRE powder (WTP) as oil sorbent, using response surface methodology (RSM). *Afr. J. Health Saf. Environ.* **2020**, *1*, 1–12. [CrossRef]
18. Taufik, S.H.; Ahmad, S.A.; Zakaria, N.N.; Shaharuddin, N.A.; Azmi, A.A.; Khalid, F.E.; Merican, F.; Convey, P.; Zulkharnain, A.; Khalil, K.A. Rice Straw as a Natural Sorbent in a Filter System as an Approach to Bioremediate Diesel Pollution. *Water* **2021**, *13*, 3317. [CrossRef]
19. Teas, C.; Kalligeros, S.; Zanikos, F.; Stournas, S.; Lois, E.; Anastopoulos, G. Investigation of the Effectiveness of Absorbent Materials in Oil Spills Clean Up. *Desalination* **2001**, *140*, 259–264. [CrossRef]
20. Valentukeviciene, M.; Zurauskiene, R. Investigating the Effectiveness of Recycled Agricultural and Cement Manufacturing Waste Materials Used in Oil Sorption. *Materials* **2021**, *15*, 218. [CrossRef]
21. Wu, D.; Fang, L.; Qin, Y.; Wu, W.; Mao, C.; Zhu, H. Oil Sorbents with High Sorption Capacity, Oil/Water Selectivity and Reusability for Oil Spill Cleanup. *Mar. Pollut. Bull.* **2014**, *84*, 263–267. [CrossRef]
22. Vlaev, L.; Petkov, P.; Dimitrov, A.; Genieva, S. Cleanup of Water Polluted with Crude Oil or Diesel Fuel Using Rice Husks Ash. *J. Taiwan Inst. Chem. Eng.* **2011**, *42*, 957–964. [CrossRef]
23. Ali, I.; Asim, M.; Khan, T.A. Low Cost Adsorbents for the Removal of Organic Pollutants from Wastewater. *J. Environ. Manag.* **2012**, *113*, 170–183. [CrossRef] [PubMed]
24. Brovkina, J.; Ozoli, J.; Technical, R. Efektīvu Sorbentu Izstrāde Uz Latvijas Dabīgo Mālu Bāzes. *Sci. J. RTU Mater. Sci. Appl. Chem.* **2012**, *1*, 1–5.
25. Weber, W.; Morris, J.; Sanit, J. Kinetics of Adsorption on Carbon from Solution. *J. Sanit. Eng. Div.* **1963**, *89*, 31–38. [CrossRef]
26. Ahmad, A.L.; Chan, C.Y.; Abd Shukor, S.R.; Mashitah, M.D. Adsorption Kinetics and Thermodynamics of β-Carotene on Silica-Based Adsorbent. *Chem. Eng. J.* **2009**, *148*, 378–384. [CrossRef]
27. Ho, Y.; McKay, G. A Kinetic Study of Dye Sorption by Biosorbent Waste Product Pith. Resources. *Conserv. Recycl.* **1999**, *25*, 171–193. [CrossRef]
28. Escudero, C.; Fiol, N.; Villaescusa, I.; Bollinger, J.C. Arsenic Removal by a Waste Metal (Hydr)Oxide Entrapped into Calcium Alginate Beads. *J. Hazard. Mater.* **2009**, *164*, 533–541. [CrossRef]
29. Qiu, H.; Lv, L.; Pan, B.; Zhang, Q.; Zhang, W.; Zhang, Q. Critical Review in Adsorption Kinetic Models. *J. Zhejiang Univ.-Sci. A* **2009**, *10*, 716–724. [CrossRef]
30. Ho, Y.S.; McKay, G. Pseudo-Second Order Model for Sorption Processes. *Process Biochem.* **1999**, *34*, 451–465. [CrossRef]

*Review*

# Recent Advances on Innovative Materials from Biowaste Recycling for the Removal of Environmental Estrogens from Water and Soil

**Elisabetta Loffredo** 

Dipartimento di Scienze del Suolo, della Pianta e degli Alimenti, Università degli Studi di Bari Aldo Moro, 70126 Bari, Italy; elisabetta.loffredo@uniba.it; Tel.: +39-080-5442282

**Abstract:** New technologies have been developed around the world to tackle current emergencies such as biowaste recycling, renewable energy production and reduction of environmental pollution. The thermochemical and biological conversions of waste biomass for bioenergy production release solid coproducts and byproducts, namely biochar (BC), hydrochar (HC) and digestate (DG), which can have important environmental and agricultural applications. Due to their physicochemical properties, these carbon-rich materials can behave as biosorbents of contaminants and be used for both wastewater treatment and soil remediation, representing a valid alternative to more expensive products and sophisticated strategies. The alkylphenols bisphenol A, octylphenol and nonylphenol possess estrogenic activity comparable to that of the human steroid hormones estrone, 17β-estradiol (and synthetic analog 17α-ethinyl estradiol) and estriol. Their ubiquitous presence in ecosystems poses a serious threat to wildlife and humans. Conventional wastewater treatment plants often fail to remove environmental estrogens (EEs). This review aims to focus attention on the urgent need to limit the presence of EEs in the environment through a modern and sustainable approach based on the use of recycled biowaste. Materials such as BC, HC and DG, the last being examined here for the first time as a biosorbent, appear appropriate for the removal of EEs both for their negligible cost and continuously improving performance and because their production contributes to solving other emergencies, such as virtuous management of organic waste, carbon sequestration, bioenergy production and implementation of the circular economy. Characterization of biosorbents, qualitative and quantitative aspects of the adsorption/desorption process and data modeling are examined.

**Keywords:** biochar; hydrochar; digestate; endocrine disruptor; estrogen; biosorbent; adsorption; water decontamination; soil remediation

**Citation:** Loffredo, E. Recent Advances on Innovative Materials from Biowaste Recycling for the Removal of Environmental Estrogens from Water and Soil. *Materials* **2022**, *15*, 1894. https://doi.org/10.3390/ma15051894

Academic Editors: Andrea Petrella and Michele Notarnicola

Received: 1 February 2022
Accepted: 1 March 2022
Published: 3 March 2022

**Publisher's Note:** MDPI stays neutral with regard to jurisdictional claims in published maps and institutional affiliations.



## 1. Introduction

One of the most important paradigms of the current period is the preservation of the environment. In recent years, in response to the growing global demand for energy and, at the same time, as a solution to the problem of the enormous mass of organic waste produced annually by agricultural, industrial and municipal activities, various innovative technologies have been implemented to convert biowaste in bioenergy. In addition to gaseous and liquid fuels, these processes generate large quantities of solid carbon-rich coproducts and byproducts, such as biochar (BC), hydrochar (HC) and digestate (DG), which are suitable for various agricultural and environmental applications. In agriculture, these materials are primarily used as multifunctional amendments to restore the organic fertility of soil which is increasingly compromised by intensive and superintensive agriculture [1–3]. In soil, these amendments enhance the retention of water and nutrients that are vital for plant growth and production [4]. Another important function of these materials is the immobilization of soil pollutants by adsorption, thus preventing their leaching into groundwater and/or their transport into surface waters. Furthermore, given the recalcitrance of these materials, especially BC and HC, their incorporation into soil allows carbon sequestration for a very long period with a significant reduction in greenhouse gas emissions into the atmosphere [1].

The environmental applications of these materials are based on their excellent ability to behave as adsorbents of a wide variety of both inorganic and organic pollutants [5,6]. This ability is related to their physicochemical properties, such as composition, micromorphology, porosity, functional groups content and degree of aromatization which, in turn, depend on the parameters of the production process, i.e., feedstock, temperature, retention time, pressure and so on. In particular, temperature is a very important parameter of the thermochemical conversion of biomass affecting both the yield of the solid product and its properties and best use. Low temperatures generally prevent a high degree of carbonization and promote the formation of reactive O-containing functional groups that allow the interaction of the material with a wide range of solutes and a consequent more suitable use in agriculture [7]. On the contrary, high temperatures favor a high degree of aromaticity, large specific surface area, low H/C ratio, high C/N ratio and high number of adsorption sites. All these characteristics are indicative of an intense thermal alteration of the raw biomass and suggest a better use of the material for environmental purposes [7]. In any case, in order to select and optimize the applications of these materials, all these properties must be carefully determined [8–10]. Characterization data are also essential to evaluate the chemical structure of the material and the possible mechanisms of interaction with pollutants.

Some aspects of individual BC and HC have been recently reviewed [10,11]. In general, these studies considered a broad context of pollutants, including dyes, pharmaceuticals, polycyclic aromatic hydrocarbons PAHs, heavy metals, antibiotics and single EDCs. Furthermore, DG has been studied so far almost exclusively for agricultural use, although recently it has shown an appreciable ability to retain hydrophobic compounds such as the EDCs bisphenol A (BPA) and 4-tert-octylphenol (OP) [12]. It is reasonable to believe that also DG could be used as a biosorbent, possibly after further studies and improvements. Hence, all three materials examined in this review can be considered as innovative tools for wastewater treatment and soil remediation [13,14]. Furthermore, since these biosorbents are essentially a waste of bioenergy technologies, and consequently have an almost zero cost, their exploitation for sustainable environmental depollution seems very interesting and appropriate. In the last decades, world population growth and industrial development have caused a significant increase in the number and quantity of anthropogenic pollutants released into the environment. Among these compounds, a major attack on wildlife and human health is due to a class of contaminants known as endocrine-disrupting chemicals (EDCs) for their proven ability to alter the normal hormonal functions of animals, especially aquatic animals [15]. Both ascertained and suspected EDCs have been detected in human urine, blood and breast milk [16–18]. Currently, these compounds are ubiquitous in natural waters, wastewaters, soil, sediments, food and consumer products. EDCs have been widely reported in ecosystems surrounding the most urbanized and industrialized areas [19,20] but have also been detected in remote areas of the globe [21]. A long-term monitoring study based on thousands of samples revealed the presence of BPA, which is one of the most representative EDCs, in freshwater, seawater, freshwater sediments and marine sediments at average levels of 29 ng $L^{-1}$, 7 ng $L^{-1}$, 7 ng $g^{-1}$ and <0.03 ng $g^{-1}$, respectively, in Europe, and 5 ng $L^{-1}$, 1 ng $L^{-1}$, 0.7 ng $g^{-1}$ and 1 ng $g^{-1}$, respectively, in North America [22]. In a seasonal monitoring of 14 EDCs, including BPA, in a lagoon located in a coastal area of Southern Italy, single EDCs were detected at concentrations ranging from 132 to 28,000 ng $L^{-1}$ in water and from 0.7 to 155 ng $g^{-1}$ in sediments [23]. In effluents from wastewater treatment plants (WWTPs), BPA has been detected at concentrations up to 370,000 ng $L^{-1}$ [21]. In soil, BPA concentration ranged between <0.01 and 1000 ng $g^{-1}$ in dependence on the amount and type of effluent or waste received [21]. Due to their dangerous effects, EDCs have increasingly attracted the attention of scientists, international organizations, decision-makers and the community [15].

The class of EDCs includes various xenoestrogens, i.e., biologically active synthetic compounds that mimic the activity of the steroid hormone 17β-estradiol. Due to their wide diffusion in the environment, they are also known as environmental estrogens (EEs). Based on their chemical structure, EEs can be grouped as phenolic environmental estrogens (PhEEs) and steroidal environmental estrogens (StEEs). Both groups are often detected in aquatic and terrestrial ecosystems such as rivers, lakes, sea, soil and sediments where they enter through the application, discharge and disposal of urban and industrial effluents, sewage sludge, agro-zootechnical wastes and so on. Wastewater is the main source of EE contamination in the aquatic environment, as the efficiency of EE removal by full-scale WWTPs is quite low [24]. Even in trace amounts, EEs represent a serious risk for aquatic ecosystems and in particular for fish, in which they cause several physiological and reproductive disorders [21].

Contamination of aquatic systems by multiple EEs is frequently detected [19,20], and it has been demonstrated that in such conditions the reproductive disturbance of aquatic fauna can occur even at individual ineffective concentrations [25]. The environmental risks posed by these compounds are also due to their chemical recalcitrance that leads to a long time of persistence in aquatic and terrestrial ecosystems [26]. In the light of all this, it is clear that the environmental consequences of indiscriminate disposal of untreated or not sufficiently decontaminated wastewater and solid waste are of great concern and often go beyond all reasonable expectations.

Various sophisticated and expensive methods are available to remove organic pollutants from aqueous media. Current technologies are mainly based on chemical and physicochemical processes such as flocculation, precipitation–filtration, adsorption on activated carbon, reverse osmosis, advanced chemical and electrochemical oxidation and photocatalytic degradation [27,28], while biological treatments are essentially based on the use of activated sludge. In most cases, these methods fail to completely remove toxic compounds [27,29]. The sorption process plays an important role in wastewater treatment, and well-designed sorption protocols can release high-quality effluent after treatment. The possibility of using coproducts and byproducts of bioenergy technologies to decontaminate wastewater and even soil can certainly be a valid alternative to other complex and less sustainable methods.

For these reasons, the aim of this manuscript is to focus attention on the potential of materials from biowaste recycling to act as low-cost biosorbents in environmental applications according to a modern and sustainable approach that can represent an added value to environmental benefits already achieved with the production of these materials, such as the virtuous recycling of waste in compliance with the principles of the circular economy, the sequestration of carbon, the reduction of climate-altering gas emissions and a significant supply of renewable energy. Recent literature has been included in this review and the majority of references reported are from the last few years.

## 2. Biosorbents from Biowaste Recycling

### 2.1. Biochar

Among the technologies used for bioenergy production there are the thermo-chemical processes of pyrolysis, gasification, flash carbonization, combustion and others [30] (Table 1). Fast pyrolysis, slow pyrolysis and gasification, besides producing gaseous (syngas) and liquid fuels, generate a solid byproduct known as 'biochar' or 'pyrochar' when obtained by pyrolysis [31]. The common operating conditions used to produce BC are low-moisture-containing feedstock, temperatures ranging from 300 to 800 °C, very limited oxygen atmosphere and retention time usually greater than 0.5 h [31]. The raw biomass used to feed the process usually originates from forestry, agriculture, food processing and the organic fraction of municipal solid waste (OFMSW). In Mediterranean countries, the substrate for BC production consists mainly of vineyard residues, olive tree pruning, orchard cuttings, wood chips, OFMSW and sewage sludges [32]. Other biowastes used to produce BC are sawdust, corn cobs, rice husks, coconut shells, residues from coffee preparation and so on [32].

**Table 1.** Operating conditions of biomass conversions that produce biosorbents. From the literature [30,33].

| | Biochar | | | Hydrochar | Digestate |
|---|---|---|---|---|---|
| Type of biomass conversion | Thermochemical | | | Thermochemical | Biochemical |
| Process | Slow pyrolysis | Fast pyrolysis | Gasification | Hydrothermal carbonization | Anaerobic digestion |
| Type of feedstock | Agricultural residues<br>Woody residues | | | Agricultural residues<br>OFMSW | Agricultural residues<br>Livestock wastes<br>Sewage sludge<br>OFMSW |
| Feedstock moisture | Dry | | | Wet | 80–90% |
| Temperature (°C) | 300–650 | 500–650 | 800–900 | 180–260 | Psychrophilic (20–25)<br>Mesophilic (35–37)<br>Thermophilic (>55) |
| Residence time | 1–12 h | <2 s | 10–20 s | 1–12 h | 14–30 days (mesophilic)<br>14–16 (thermophilic) |
| Pressure | - | - | - | Autogenous (2–10 MPa) | - |
| Product yield (%) | | | | | |
| Solid | 25–35 | 12 | <10 | 50–80 | - |
| Liquid | 20–30 | 75 | <5 | 5–20 | - |
| Gases | 25–35 | 13 | >85 | 2–5 | 60–70 (fresh biomass) |

BC is a stable carbonaceous material with a very high carbon content (>60%), and therefore its production significantly contributes to carbon sequestration and, consequently, to climate change mitigation [34]. Due to BC recalcitrance, its residence time in soil can be several centuries [35]. The composition and properties of BC strongly depend on the type of feedstock and on the pyrolysis parameters, primarily temperature [7].

When incorporated into soil, BC adds an important carbon pool that improves the physical, chemical and biological properties of soil, especially in the cases of degraded soil, sandy soils and soils having a very low content of water and nutrients. In addition to improving soil fertility, BC can prevent the movement and leaching of contaminants in soil due to its remarkable retention capacity [10]. Other interesting and innovative applications of BC concern the remediation of soil and water contaminated by inorganic pollutants such as heavy metals and organic pollutants such as polyaromatic hydrocarbons, agrochemicals, dyes and pharmaceuticals [36,37]. Multianalytical characterization of BC is reported in some recent studies [32,34,38,39].

Due to its physicochemical properties, such as porous structure, large specific surface area and numerous reactive functional groups, BC has a remarkable adsorption capacity which makes it a valid and sustainable alternative to the more expensive activated carbon [40]. The pyrolysis temperature notably influences not only the yield of BC but also its properties and utilization. In general, as the pyrolysis temperature increases from 300 to 800 °C, the calorific value of BC decreases and the porosity increases, while the surface area, which does not change markedly up to 700 °C, increases significantly starting from 800 °C [41]. Low-temperature pyrolysis favors the presence of O-containing functional groups on BC surface and allows a better interaction of the material with solutes of a wider range of hydrophilicity and a better use for agricultural application. Differently, high-temperature pyrolysis favors a higher specific surface area and a greater number of adsorption sites on BC, which allow a better use for environmental purposes [7]. An advantage of this material is that by modulating the values of the process parameters it is possible to obtain a production of BC tailored to the needs of use and the local source of waste biomass [32].

A great effort of recent BC-based research is focused on empowering the adsorption efficiency of this material through specific treatments that can make it competitive with other commonly used adsorbents, such as activated carbons, not only in terms of cost but also in terms of performance. The processes of activation, functionalization and engineering of BC are studied and implemented for this purpose [42]. Postproduction treatments are able to increase the specific surface area, porosity and the content of oxygenated and nitrogenated functional groups of BC. Both chemical and physical activations of BC have been carried out, and in some cases, multiple treatments have also been used. Chemical activation aims to alter functional groups and increase the number of active sites on the surface of the material. Common chemical activating agents for BC are acids, such as $H_3PO_4$ [43], HCl [40] and HF; bases, such as NaOH [40] and KOH [44]; and salts. A very recent technique that seems capable of increasing the adsorbing capacity of this material by up to 10 times consists in doping BC with metals such as Fe also in copresence with other chemicals [45]. Very promising results have been reported by codoping BC with Fe and $N_2$ [46]. Another interesting activation technique uses BC and HC for the preparation of magnetic chars by introducing in them metal oxide nanoparticles based on Fe, Ni and Co [47]. Magnetic activation, in addition to improving BC efficiency, has proved effective in facilitating its separation from treated water. Magnetic BC has been used successfully to remove BPA [45] and pharmaceuticals such as 17$\alpha$-ethinyl estradiol [48] from water.

### 2.2. Hydrochar

The hydrothermal carbonization (HTC) technology is a promising thermochemical process for converting an organic feedstock into a carbon-rich product. During the HTC process, biomass dehydration and decarboxylation generate a marked increase in carbon content, compared to the entering raw biomass. Typical conditions adopted for HTC are elevated temperatures (180–250 °C), autogenic saturated pressure (2–10 MPa) and retention times of some hours [33] (Table 1). The solid product of HTC is a lignin-like material named 'hydrochar'.

Although the HTC process has been known for nearly a century, only in the last decade has HC aroused growing interest from researchers of the industrial, environmental and agricultural sectors, having proved to be a valuable material. HC exhibits high C content, low atomic O/C and H/C ratios, O-rich functional groups at the surface and a low aromatic structure [32]. Differently from other thermal processes such as pyrolysis which require dry feedstock, HTC has the advantage of using wet biomass with hydrophilic nature and low calorific value without any pretreatment. Like other thermochemical processes, HTC contributes to carbon sequestration since HC recalcitrance allows the fixing of carbon in soil for a long time, thus limiting the greenhouse effect. Since this technology involves little or no pretreatment of biomass, it represents an economically viable solution for producing fuel from wet organic waste such as agricultural waste, food residues, OFMW and sewage sludge. Compared to the raw biomass, HC features low moisture, lower hydrophilicity, more recalcitrance and therefore easier portability and storage. In addition to the type of feedstock, the operating conditions of HTC such as temperature, residence time and pressure are crucial for the final characteristics of HC and its possible applications [11]. In particular, temperature plays a key role in HTC as it allows the thermo-decomposition of the raw biomass. In general, as the process temperature increases, the yield of the solid components decreases, while the liquid and gaseous fractions increase [49]. High temperatures also promote an increase in pH, possibly due to the breakdown of proteins, EC and ash content, while the surface area of HC usually increases up to about 200 °C and decreases at temperatures >200 °C [49]. The type of feedstock is also important because it influences the final composition of HC. Compared to the raw biomass, HC features lower pH, due to the formation of organic acids from lignocellulose breakdown, and much higher EC and ash content. In any case, the parameter that changes the most during HTC is the specific surface area, which increases by tens of times in HC compared to the raw biomass, which is of paramount importance for the sorption efficiency of HC [49]. Due to its physicochemical characteristics, HC is currently mainly used as a solid fuel in conventional combustion processes [9]. However, the good sorption performance of this material suggests the use of HC for the partial replacement of other less eco-friendly and more expensive synthetic adsorbents [50]. Another less explored use of HC is as a soil amendment [51].

Although HC and BC may have similar applications, their physicochemical properties could be dramatically different. Therefore, an extensive characterization is crucial for a preliminary understanding of the sorption capacity of HC. Jian et al. [5] and Gasco et al. [39] determined several chemical and physical properties, such as pH, EC, cation exchange capacity, available P, porosity, elemental composition, contents of total organic carbon, carbonates, metals, moisture, ash, volatile matter, fixed carbon, functional groups and micromorphology of a BC and an HC obtained from the same feedstock and discussed them comparatively. During the HTC process, the original biomass is subjected to a series of chemical reactions such as dehydration, hydrolysis, decarboxylation, polycondensation and aromatization, which drastically change the physicochemical properties of the biomass. A non-negligible difference between HC and BC is the general higher content of cellulose and hemicellulose of HC and, consequently, its higher content of aliphatic carbon and hydrophilic functional groups, which suggests a greater adsorption efficiency of HC towards polar compounds [32]. HC has been shown to be more appropriate than BC for the adsorption of a wider spectrum of organic contaminants [52,53].

In order to enhance the adsorbing performance of HC, researchers have carried out several specific activation treatments during or after the HTC process [54,55]. Common activating agents adopted for HC are alkali metal hydroxides such as NaOH [56] and KOH [57] and single or binary salts [58].

### 2.3. Digestate

The discharge of livestock waste into soil and natural water represents a danger for the environment as it causes sanitation problems, unpleasant odor emission and the eutrophication phenomenon. On the other hand, manure and livestock slurry have contents of organic matter and plant nutrients that can constitute a low-cost supply for improving soil fertility and reducing the use of synthetic fertilizers. Since the organic matter present in such waste has a low C/N ratio and is not humified, and N and P levels can exceed the environmental safety threshold, zootechnical waste must undergo transformation processes before being incorporated into the soil. An economically and environmentally sustainable strategy that is increasingly being adopted around the world is the recycling of this waste, alone or adequately combined with agro-food waste, to produce bioenergy.

The anaerobic digestion (AD) process consists in the biochemical conversion of biowaste operated by bacterial and archaeal populations [59]. The main product of AD is biogas which is a mixture of methane, $CO_2$ and small quantities of other gases, while the byproduct consists of a semisolid mixture (about 90–95% moisture) which, after a separation process, usually centrifugation, produces a solid phase and a separate clarified liquid [60].

The separated solid fraction with a moisture content usually lower than 20% and a C content of about 50–55% is called digestate (DG) and is easily transportable and storable. As already discussed for BC and HC, the physicochemical properties of this material greatly depend on the type of biomass entering the digestor, the anaerobic technology adopted and the operating parameters (e.g., retention time, temperature, cosubstrates and working volume) adopted. During AD, easily degradable compounds are readily converted into biogas, while the recalcitrant lignocellulosic fraction remains in the byproduct.

Although this technology dates back several decades, the management of byproducts has always been rather problematic, or at least their proper destination has not been satisfactorily explored so far. In fact, both solid and liquid DGs have been considered mainly wastes to be managed carefully due to the still very high N content and the consequent potential danger for the environment. Up to now, moderate doses of DG have been used for the organic amendment of soil [3]. DG has also been directed to bio-oxidative conversion processes such as composting and vermicomposting [60], used in the preparation of biofilters and biobeds in mixtures with other C-rich substrates [61] or converted thermically to produce BC [41]. However, recent studies have shown that some DG properties, such as the presence of surface reactive functional groups, porosity and a fairly large surface area, can make this material a good candidate for the removal of inorganic and organic contaminants by adsorption [12,62].

## 3. Characterization of Biosorbents

Before using biosorbents for soil and water remediation, it is essential to characterize them. Numerous conventional and advanced techniques are used for this purpose. Choosing the proper methods is crucial to gaining a better understanding of the physical, chemical and physicochemical properties of the material and predicting its sorptive potential. A number of studies report broad, multianalytical characterization of these materials, which includes (i) basic and cheap analyses, such as moisture, ash content, volatile matter and fixed carbon (i.e., proximate analyses), pH, EC and CEC; (ii) more expensive elemental analysis (CHNS-O, i.e., ultimate analysis); and (iii) cutting-edge analyses based on advanced analytical techniques aiming at investigating the surface properties of biosorbents. The advanced analytical techniques include total reflection X-ray fluorescence (TXRF) spectroscopy, scanning electron microscopy (SEM), SEM coupled with energy-dispersive X-ray spectroscopy (SEM-EDX), Brunauer–Emmett–Teller (BET) analysis, Fourier transform infrared (FT-IR) spectroscopy, thermogravimetric (TG) and derivative thermogravimetric (DTG) analyses, nuclear magnetic resonance (NMR) spectroscopy and pyrolysis coupled with gas chromatography and mass spectrometry (Py-GC/MS). Table 2 lists the main analytical techniques used in several characterization studies, while Table 3 shows some results of BC, HC and DG characterization reported in the literature.

**Table 2.** Analytical techniques adopted to characterize biosorbents.

| Sample | Feedstock | Process T (°C) | Elemental Analysis | SEM, SEM- EDX | BET | XRF | FTIR | TG, DTG | NMR | Ref. |
|---|---|---|---|---|---|---|---|---|---|---|
| BC | Poultry litter and wheat straw | 400 | • | | • | | • | | • | [52] |
| BC | Pinewood | 500 | • | • | • | • | • | | | [63] |
| BC | Corncob | 700 | • | • | | | | | | [64] |
| BC | Swine solids and poultry litter | 250, 450, 600 | • | • | • | | | | • | [53] |
| BC | Eucalyptus wood | 600 | • | • | • | | | | | [43] |
| BC | Digestate | Various (from 300 to 900) | | • | • | | | | | [41] |
| BC | Dairy manure and sorghum | 600 | • | • | • | | • | | | [37] |
| BC | Algae and sorghum | 500 | • | • | • | | • | | | [38] |
| BC | Red spruce and grapevine wood | 550 | • | • | | • | • | • | | [32] |
| BC | Lotus seedpod | 650 | • | • | • | | • | | | [44] |
| BC | Wheat straw | 700 | | • | • | | • | | | [40] |
| HC | Orange peels | 200 | • | • | • | | • | | | [64] |
| HC | Swine solids and poultry litter | 250 | • | • | • | | | | • | [53] |
| HC | Corncob | 230 | • | • | | | | | | [64] |
| HC | Olive pomace | 180–250 | • | | | | • | • | | [65] |
| HC | Lignin | 240, 300 | • | | | | • | | | [66] |

**Table 2.** *Cont.*

| Sample | Feedstock | Process T (°C) | Elemental Analysis | SEM, SEM- EDX | BET | XRF | FTIR | TG, DTG | NMR | Ref. |
|---|---|---|---|---|---|---|---|---|---|---|
| HC | Rice husk | 200 | | • | • | | • | | | [58] |
| HC | Rice husk | 180 | • | • | • | | • | • | | [57] |
| HC | Urban pruning and OFMW | 180–210 | • | • | | • | • | • | | [32] |
| HC | Sewage sludge | 160, 190, 250 | • | • | • | | • | | | [49] |
| HC | Argan nut shells | 180, 200 | | • | • | • | • | | | [67] |
| HC | Pine fruit shells | 190 | | • | • | | • | | | [56] |
| DG | Swine manure | 34 | • | | | | • | • | • | [68] |
| DG | Food waste | - | • | • | | | • | | | [8] |
| DG | Swine manure | - | • | | | | | • | | [41] |
| DG | Mixed residues and olive pomace | 20–45 | | • | | | • | | | [12] |

BC: biochar; HC: hydrochar; DG: digestate; SEM and SEM-EDS: scanning electron microscopy and SEM coupled to energy-dispersive X-ray spectroscopy; BET: Brunauer–Emmett–Teller analysis; XRF: X-ray fluorescence spectroscopy; FTIR: Fourier transform infrared spectroscopy; TG and DTG: thermogravimetric and derivative thermogravimetric analysis; NMR: nuclear magnetic resonance spectroscopy.

**Table 3.** Some results of biochar (BC), hydrochar (HC) and digestate (DG) characterization.

| | Process | Feedstock | Process T (°C) | Residence Time | Elemental Composition [a] (%) | | | | pH [b] | EC [b] (dS $m^{-1}$) | Ash (%) | Surface Area [c] ($m^2$ $g^{-1}$) | Avg. Pore Size (nm) | Pore Volume ($cm^3$ $g^{-1}$) | Ref. |
|---|---|---|---|---|---|---|---|---|---|---|---|---|---|---|---|
| | | | | | C | H | N | O | | | | | | | |
| BC | Py | Digestate from swine manure | 800 | 1.3 h | – | – | – | – | – | – | – | 101.9 | 3.04 | 0.08 | [41] |
| BC | Py | Algae | 500 | 2 h | 24.6 | 1.3 | 3.2 | 11.4 | 10.2 | 10.70 | 59.7 | 0.5 | 1.88 | 0.16 | [38] |
| BC | Py | Sorghum | 500 | 2 h | 46.7 | 3.0 | 0.0 | 13.0 | 7.4 | 5.95 | 29.4 | 4.1 | 13.22 | 13.27 | |
| BC | Py | Sorghum | 600 | 2 h | 47.4 | 2.3 | 0.0 | 9.8 | 9.6 | 5.92 | 45.1 | 4.1 | 13.29 | 12.99 | |
| BC | Py | Red spruce wood | 550 | 3 h | 84.0 | 1.5 | 0.2 | n.d. | 9.1 | 0.39 | 4.7 | – | – | – | [32] |
| BC | Py | Grapevine pruning | 550 | 3 h | 75.5 | 1.3 | 0.5 | n.d. | 9.9 | 2.23 | 9.9 | – | – | – | |
| BC | Py | Lotus seedpod | 650 | 2 h | 69.9 | 2.1 | 1.1 | 15.6 | – | – | – | 25.2 | 2.55 | 0.03 | [44] |
| *KOH*-BC | Py | Lotus seedpod | 650 | 2 h | 78.8 | 2.4 | 1.3 | 15.1 | – | – | – | 306.2 | 1.90 | 0.13 | |
| BC | Py | Wheat straw | 700 | 2 h | – | – | – | – | – | – | – | 57.2 | 3.70 | 0.05 | [40] |
| *NaOH*-BC | Py | Wheat straw | 700 | 2 h | – | – | – | – | – | – | – | 254.9 | 1.92 | 0.12 | |
| *HCl*-BC | Py | Wheat straw | 700 | 2 h | – | – | – | – | – | – | – | 197.2 | 2.86 | 0.14 | |
| HC | HTC | Pig manure | 200 | 2 h | 33.8 | 4.2 | 2.5 | 15.0 | 8.3 | 19.86 | 44.0 | – | – | 2.10 | [39] |
| HC | HTC | Pig manure | 240 | 2 h | 25.8 | 3.0 | 1.9 | 10.4 | 7.8 | 10.93 | 58.6 | – | – | 2.80 | |
| HC | HTC | Urban pruning | 210 | 8 h | 61.5 | 6.2 | 1.7 | – | 6.6 | 1.03 | 12.5 | – | – | – | [32] |
| HC | HTC | OFMSW | 210 | 8 h | 62.6 | 6.0 | 1.7 | – | 7.7 | 1.09 | 15.7 | – | – | – | |
| HC | HTC | Sewage sludge | 160 | 4 h | 30.8 | 4.9 | 3.2 | 14.0 | 5.1 | 6.10 | 46.5 | 9.5 | – | – | [49] |
| HC | HTC | Sewage sludge | 190 | 4 h | 30.0 | 4.3 | 2.4 | 11.4 | 5.7 | 8.44 | 51.3 | 11.9 | – | – | |
| HC | HTC | Sewage sludge | 250 | 4 h | 31.0 | 4.1 | 2.4 | 8.00 | 6.6 | 16.53 | 53.8 | 2.9 | – | – | |
| DG | AD | Food waste | – | – | 42.1 | 5.2 | 5.8 | 21.3 | – | – | 25.6 | – | – | 0.32 | [8] |
| DG | AD | Mixed residues | – | 30 d | 40.0 [d] | – | 6.5 | – | 8.7 | – | – | 3.10 | – | – | [61] |
| DG | AD | Swine manure | – | – | 37.2 | 5.5 | 4.6 | 31.9 | – | – | 23.0 | – | – | – | [41] |
| DG | AD | Mixed residues | 30 | 40 d | 50.5 [d] | - | - | - | 8.7 | 1.36 | 12.8 | – | – | – | [12] |

Py: pyrolysis; HTC: hydrothermal carbonization; AD: anaerobic digestion; [a] on dry and ash-free basis; [b] 1:10 ($w/v$) in double-distilled water; [c] calculated by the BET method; [d] total organic C; –: not reported.

The CHNS-O analysis provides the elemental composition of the material and the molar ratios of elements, while TXRF analysis allows quantifying trace and polluting elements in the structure and ash composition [32]. Due to the type of production process, BC generally shows a higher C content than HC and DG (Table 3). Sun et al. [52] measured the organic C content of different BC and HC samples and found values between 53.5% and 65.8% for BC and between 40.2% and 47.5% for HC. A high C content of the material is advantageous for both C storage and adsorption of pollutants. The total C content of BC increases and the O and H contents decrease with increasing pyrolysis temperature, which may be attributed to increased dehydration and decarboxylation of biomass occurring during the process [38]. The atomic H/C and O/C ratios are important parameters for evaluating the degree of carbonization of the material, which strongly depends on the process temperature. High H/C ratios are typical of HC obtained at a temperature around 200 °C and suggest the presence of lignin and cellulose (polar fractions), while low H/C and O/C ratios of BC, which is usually obtained at temperatures ranging between 350 and 800–900 °C, are indicative of high degrees of condensation, aromatization and hydrophobicity [32]. A H/C ratio <0.3 denotes the formation of a highly condensed aromatic structure, while a H/C ratio >0.7 is indicative of an uncondensed structure [38]. The lower C content of HC, compared to BC, is attributable to the lower dehydration and decarboxylation reactions occurring during the HTC process. The DG usually features a total C content between 25 and 41%, on a dry matter basis, with a variability depending on the type of raw biomass processed [6].

The surface micromorphology of both chars and DG can be investigated using the SEM technique, while SEM-EDX analysis allows the evaluation of the composition and the distribution of elements on the surface of the material. Both SEM and SEM-EDX analyses are important for the identification of the type of surface, the size and allocation of pores and the mineral elements present [11,64,69]. The BET analysis is commonly used to measure the specific surface area ($m^2$ $g^{-1}$) of the material, and it provides information on the total porosity of the adsorbent, which is very important for the overall sorptive capacity. The surface area of BC has shown positive correlations with the removal of contaminants from soil and water [34]. The BC commonly exhibits diffuse microporosity with pores usually smaller than 1–1.5 nm in diameter, while porosity is less pronounced in HC [32] and much less in DG [12]. The extensive porosity of BC, besides allowing routes for the release of volatile compounds such as $H_2O$, CO, $CO_2$ and $CH_4$ during pyrolysis and providing sites for contaminants, would be an adequate habitat for symbiotic microorganisms once BC is incorporated into soil [10]. The mild temperature used in the HTC process makes HC less or much less porous than BC, which might be attributed to the persistence, during the process, of decomposition products on the surface of HC, which causes pore blockage [52]. Compared to BC, HC has a smaller surface area, although its surface area is sufficiently large to guarantee an appreciable adsorption capacity of molecules with different hydrophobicity [56]. Common features of SEM images of DG are rough surfaces with irregularly shaped ridges, sharp edges, microparticles, channels and cavities mostly smaller than 10 μm. The EDS spectrum of DG usually shows the presence on the surface of elements typical of the raw biomass used in the AD process [12].

The FTIR technique contributes to the understanding of the adsorption mechanisms by providing information on the functional groups of the material and their possible involvement in binding reactions. The surface properties of these materials are dictated by those functional groups that are exposed to the interaction with other surfaces and molecules. FTIR analysis is also used to determine the degree of carbonization and mineralogy of chars. The presence of specific FTIR absorption bands and their relative intensity are indicative of ionizable functional groups (hydroxyl, carboxyl, phenolic, carbonyl, amino and sulfhydryl) capable of interacting with ionizable compounds, as well as the presence of aromatic structure in the material. The FTIR spectra of BC, HC and DG are quite different, denoting different structural properties of these materials. Typical peaks observed in FTIR spectra of BC can be assigned to O-H stretching of inter- and intramolecular hydrogen bonds and N-H stretching (3420–3440 $cm^{-1}$), methyl C-H stretching (~2930 $cm^{-1}$), methylene C-H stretching (~2860 $cm^{-1}$) and aromatic carbonyl/carboxyl C=O (~1700 $cm^{-1}$) and aromatic C=C (~1600 $cm^{-1}$) stretching vibrations [70]. The absorption bands at ~1400 $cm^{-1}$ are assigned to phenolic O-H and C-O bonds which promote the immobilization of pollutants through complexation, while the bands at ~1110 and ~870 $cm^{-1}$ can be assigned to aromatic C-H deformation [71]. In addition to the peak at 3420–3440 $cm^{-1}$ (O-H and N-H stretching), FTIR spectra of HC usually feature peaks at ~2920 and 2850 $cm^{-1}$ (asymmetric and symmetric stretching of aliphatic C-H, respectively); ~1620 $cm^{-1}$ (aromatic C=C stretching); at 1310, 1160, 1110 and 1030 $cm^{-1}$ (stretching vibrations of C-H in hydroxyl, ether or ester and bending vibrations of O-H in cellulose and hemicellulose); 670 (out-of-plane bending vibration of C-H); and 580 $cm^{-1}$ (bending vibration of O-H) [32]. The main features of FTIR spectra of DG are the presence of a wide absorption band at ~3300–3400 $cm^{-1}$ (O-H vibration of carboxylic and alcoholic groups and N-H stretching) and peaks at ~2920 and ~2850 $cm^{-1}$ (aliphatic C-H stretching), ~1630–1600 $cm^{-1}$ (various vibrations, including aromatic C=C stretching), 1385 $cm^{-1}$ (various vibrations, including O-H deformation and C-O stretching of phenolic groups, $COO^-$ asymmetric stretching) and ~1040 $cm^{-1}$ (C-O stretching of polysaccharide-like substances and Si-O vibration of silicate impurities) [12].

The TG and DTG techniques allow the evaluation of the structural stability of the adsorbent and the changes of physical and chemical properties occurring during heating, which can be achieved by monitoring the weight loss pattern caused by the heating rate under controlled atmospheric conditions (air, He or $N_2$). The TG analysis is important for processed materials as it highlights and differentiates their thermal decomposition. The first phase of mass loss is due to the loss of moisture (dehydration) and occurs between 50 and 150 °C, the second one is measurable between 150–200 and 360 °C and is mainly related to the thermal degradation (volatilization and decomposition) of hemicelluloses and cellulose and the third one is attributed to the degradation of lignin and occurs between 360 and 600 °C. Chars from lignocellulosic biomass feature more markedly the three stages of decomposition compared to chars from mixed plant and animal biomass [32]. The TG and DTG analyses were used by Missaoui et al. [65] in studying changes of physical and chemical properties of an HC from olive pomace during heating. The TG analysis of a DG from swine manure clearly showed three distinct stages corresponding to (i) loss of moisture and volatile compounds (up to 200 °C), (ii) decomposition of lignocellulosic residues still present after AD (from 200 to 800 °C) and (iii) decomposition of inorganic minerals such as calcite and calcium phosphates (from 800 to 1000 °C) [41].

NMR is another technique commonly used to investigate the structural composition of biosorbents. This technique provides details on the type of functional groups of the material and the proportion between aliphatic and aromatic fractions. Moreover, it allows the evaluation of the degree of carbonization of chars and their stability. Solid-state $^{13}C$ NMR spectra reveal the main types of functional groups (e.g., aliphatic, aromatic, phenolic, methoxyl) of the material and usually show significant differences between BC and HC. Usually, $^{13}C$ NMR spectra of BC are dominated by a huge peak at ~120–130 ppm that is characteristic of aromatic C [72]. The contribution of aryl C (108–148 ppm) in BC is greater than that in HC, and the contribution of alkyl C (0–45 ppm) in BC is lower than that in HC, denoting a lower degree of carbonization of HC. The NMR spectra of HC are more structured than those of BC and show significant signals of carbohydrates (63–108 ppm) and carboxyl C (165–187 ppm) [52]. Fierro and coworkers [68] used $^{1}H$ NMR to characterize DG obtained from swine manure and found high-intensity peaks in the aliphatic region, especially at 1.22 ppm ascribed to $CH_2$- groups, 1.7 ppm ascribed to unsaturated compounds and 7.2 ppm associated with the coupling effect of methylene protons in benzene rings.

Finally, the Py-GC/MS technique is used to investigate the key marker compounds present within biosorbents. The Py-GC/MS pyrograms allow the identification of a list of compounds, mainly aromatic hydrocarbons and volatile organic compounds, present in the material based on release time and m/z value. The key compounds of the material are identified by the intensity of the peaks present in the pyrogram. Depending on the source materials used, the common components of chars include benzene, ethylbenzene, naphthalene, xylene and styrene [32].

## 4. Environmental Estrogens

### 4.1. Phenolic Estrogens

Some alkylphenols are industrial products and byproducts that have attracted the attention of researchers due to their estrogenic activity and their diffusion in the environment at doses generally higher than those of other EDCs of different chemical nature, such as natural and synthetic estradiols [26]. Due to their toxicity and persistence, PhEEs are of great concern as they have detrimental effects on human health and environmental safety [19]. The role of these compounds has been demonstrated in the context of several pathologies in males and females, such as reproductive system and thyroid dysfunction, immunotoxicity, development of breast and prostate cancer, impaired metabolism and obesity, diabetes and cardiovascular dysfunctions [15,21].

This group of estrogens includes several compounds that are constituents of a wide range of common consumer products and products and byproducts of industrial manufacturing processes, such as surfactants, detergents, pesticides, paints, dyes and pharmaceuticals. They are continuously released into natural water bodies with the discharge of effluents from sewage treatment plants [73]. Primary sources of PhEEs in natural and cultivated soil include the application of sewage sludge, irrigation with wastewater and the discharge of landfill leachate [74].

Among PhEEs, (2,2-bis(4-hydroxyphenyl) propane (BPA), OP and nonylphenol (NP) have similar structural features and behavior in the environment. Some properties of these compounds are shown in Table 4. BPA, OP and NP are widely adopted in industrial processes for the preparation of items manufactured for daily use, such as electrical and electronic parts, medical equipment (e.g., dental prostheses and sealants), flame retardants, adhesives, paints and food and beverage packaging [73].

**Table 4.** Chemical structure and physicochemical properties of phenolic and steroidal estrogens (EEs).

| EE | Chemical Structure | Molecular Weight (g mol$^{-1}$) | Water Solubility (mg L$^{-1}$ at 25 °C) | LogKow | pKa | Vapor Pressure (mm Hg at 25 °C) |
|---|---|---|---|---|---|---|
| BPA | 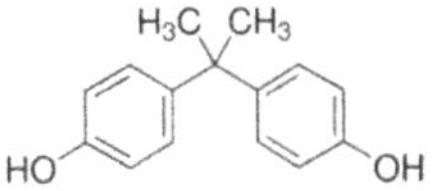 $H_3C$ $CH_3$ HO OH | 228.29 | 300 | 3.32 | 9.6 | $4.00 \times 10^{-8}$ |
| 4-tert-OP | 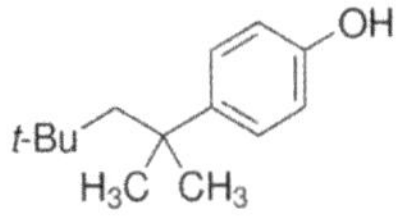 OH t-Bu $H_3C$ $CH_3$ | 206.32 | 5.1 | 5.25 | 10.33 | $4.80 \times 10^{-4}$ |
| NP | 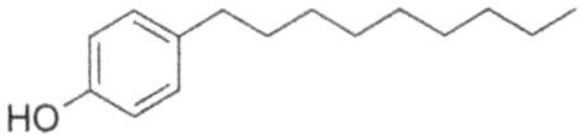 HO | 220.35 | 7.0 | 5.76 | 10.31 | $8.18 \times 10^{-4}$ |
| E1 | 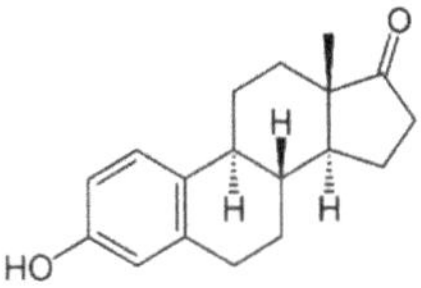 O H H H HO | 270.37 | 30 | 3.13 | 10.40 | $2.49 \times 10^{-10}$ |
| E2 | 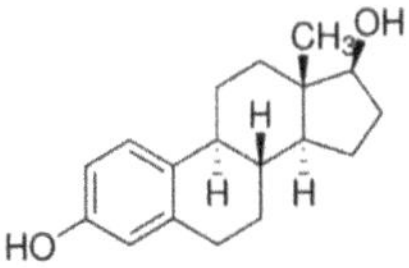 $CH_3$ OH H H H HO | 272.38 | 3.9 | 4.01 | 10.71 | $6.38 \times 10^{-9}$ |
| E3 | 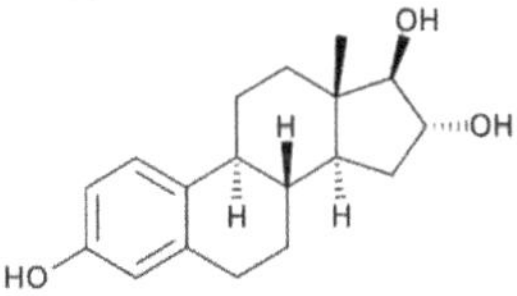 OH OH H H H HO | 288.39 | 27.3 | 2.45 | 10.33 | $9.93 \times 10^{-12}$ |
| EE2 | 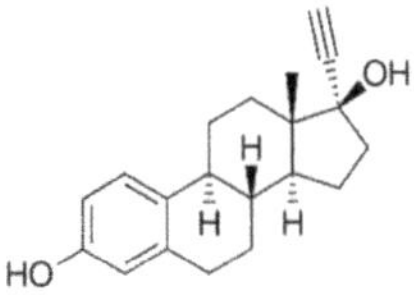 OH H H H HO | 296.41 | 11.3 | 3.67 | 10.33 | $1.95 \times 10^{-9}$ |

Data from PubChem [75].

Currently, BPA is one of the chemicals produced in the highest quantities worldwide, with approximately 8 million tons produced per year [49], and much more is expected to be produced in the future. BPA is the building block for the preparation of epoxy resins and polycarbonate plastics and is used as a stabilizer for plastics such as polyvinyl chloride (PVC). BPA is widely present in everyday items, such as plastic bottles, paints, food cans, plastic toys, electronic equipment, paper and cardboard products, furniture, building materials, water pipes, footwear and leather products [76]. Due to its endocrine-disrupting activity, BPA has been banned in baby bottles in the European Union since June 2011 [25].

BPA reaches the environment from both preconsumer and postconsumer sources. The preconsumer sources derive from the industrial production of BPA-containing plastics and epoxy resins and their discharge via industrial wastewater. Postconsumer sources include effluent discharge from municipal WWTPs, landfill leachates and degradation of plastics in the environment [74]. BPA is a moderately water-soluble compound (Table 4) and is believed to possess moderate bioaccumulation [77]. The low vapor pressure and consequently the low volatility of BPA is the reason for its persistence mainly in the hydrosphere. A long-term BPA monitoring study conducted in North American and European fresh and marine surface waters reports 95th percentile concentrations of 0.30 $\mu g\ L^{-1}$ for both geographical areas and 0.024 $\mu g\ L^{-1}$ and 0.15 $\mu g\ L^{-1}$ in marine water, respectively [22]. In WWTP effluents, BPA concentration ranges from undetectable or a few micrograms per liter to 370 $\mu g\ L^{-1}$ [21]. This molecule is ubiquitous in sewage sludges and biosolids, with concentrations on dry weight ranging from 10 to 100,000 $\mu g\ kg^{-1}$, and can also be much more concentrated in sludge from WWTPs [21]. The BPA concentration in soil varies over several orders of magnitude, i.e., from less than 0.01 to 1000 $\mu g\ kg^{-1}$ depending on the amount and type of effluent or waste received, while more than 20,000 $\mu g\ kg^{-1}$ of BPA was found in downstream sediments of heavily polluted urban areas [21].

Other well-known molecules with estrogenic properties are the two alkylphenols OP and NP which are the precursors of the nonionic surfactants octylphenol polyethoxylates (OPEOs) and nonylphenol polyethoxylates (NPEOs) that are used in the formulation of paints, detergents, personal care products, pesticides, lubricants, emulsifiers and so on. As a result of the microbial degradation of OPEOs and NPEOs, OP and NP are released into the environment where they exert a greater toxicity than their parent compounds and can cause estrogenic effects on aquatic fauna and humans [73]. OP has higher estrogenic activity than NP, and both can bind to estrogen receptors of animals and cause disorders of reproductive functions [78]. OP and NP are widely present in natural waters, especially river water, due to the discharge of domestic and industrial wastewater and rain runoff. The concentrations of OP and NP in natural waters generally range between 23 and 255 $ng\ L^{-1}$ [19,79] but can also reach some micrograms per liter [80]. OP and NP are also present in WWTP effluents and can be released in the environment where they persist for a long time due to their significant recalcitrance to biodegradation [81].

### 4.2. *Steroidal Estrogens*

The StEEs belong to the class of EDCs and include natural compounds involved in the maintenance of female functions, such as estrone (E1), 17$\beta$-estradiol (17$\beta$-E2), 17$\alpha$-estradiol (17$\alpha$-E2) and estriol (E3), as well as the synthetic estrogen 17$\alpha$-ethinyl estradiol (EE2). E1, E2 and E3 are produced by all vertebrates and some insects, while EE2 is widely adopted as a contraceptive pharmaceutical, as a menopause-related drug and in the therapy of human diseases such as breast and prostate cancer and osteoporosis [82]. Some properties of these compounds are reported in Table 4. The estrogenic potency of StEEs, in particular E2 which is the strongest, is significantly greater than that of PhEEs, but they are generally detected less frequently and at lower levels in aquatic environments [82]. The overall estrogenic potential of an EE is usually expressed quantitatively via the estradiol equivalent (EEQ) parameter since the reference for estrogenicity for all EEs is the compound 17$\beta$-estradiol [82].

The StEEs are widespread water contaminants and are of great concern due to their potential hazard to the health of aquatic animals and humans. It is estimated that ~30 tons/year of natural StEEs are released into the environment by the global human population, along with ~0.7 tons/year of synthetic EE2, while in the US and the EU, the overall cattle industry generates ~83 tons/year of estrogens [82]. The StEEs are released as liquid and solid animal and human excreta in municipal and rural waterways, especially in densely populated areas, and accumulate in wastewaters. The direct animal manure application in agricultural land can contribute to the release of steroid estrogens into soil with potential subsequent transfer to aquatic systems. The use of animal manure to produce biogas from AD can determine the presence of these compounds in both solid and liquid digestates which are often applied to soil as organic amendments [83]. Other sources of StEEs are pharmaceutical and hospital wastes. The StEEs are typically excreted into the environment in conjugated form but are rapidly converted to unconjugated molecules [82]. The EE2 molecule containing an ethynyl group is more stable and shows greater resistance to oxidative reactions than other estrogens [82]. Consequently, EE2 has a longer persistence in the environment than natural estrogens and therefore is even more dangerous. Furthermore, EE2 has an estrogenic potency estimated 1.25 times higher than that of E2, which is the most potent natural estrogen [84].

The nonadsorptive and adsorptive methods commonly used for wastewater treatment, such as activated sludge and advanced oxidation processes (AOPs), are quite expensive, generate toxic byproducts during treatments and are unable to completely remove these compounds [28]. The use of not completely depolluted effluents for crop irrigation releases these compounds into soil and natural waters after leaching and runoff. Due to the lack of effective removal treatments, StEEs have been detected in natural waters after the discharge of municipal wastewater effluents and landfill leachate from waste disposal [84]. A survey of StEE levels in effluents from sewage and WWTPs collected at numerous European sites reported that E1, 17α-E2, 17β-E2 and EE2 were present at concentration ranges of 12–197 ng $L^{-1}$, 6–13 ng $L^{-1}$, 6–43 ng $L^{-1}$ and 1–6 ng $L^{-1}$, respectively [85]. An extensive monitoring of the presence of EE2 in the incoming and outgoing effluents of WWTPs of 282 municipalities across 29 countries has detected EE2 concentrations ranging from 0 to 7890 ng $L^{-1}$ in influents and from 0 to 470 ng $L^{-1}$ in effluents originating from the activated sludge processes [84]. In a Chinese river, Wang et al. [79] found levels of E1, E2, EE2 and E3 of up to 56, 24, 31, and 5 ng $L^{-1}$. StEEs have also been detected in wildlife, mostly fish, where they bioaccumulate and reach very high concentrations [82].

Even at ppt levels, these compounds can alter hormonal mechanisms in animals, producing harmful effects such as feminization of aquatic fauna, infertility and cancer [86]. For these reasons, StEEs have increasingly drawn the attention of the community and regulatory authorities. Data from the global monitoring of StEEs are of increasing concern, and therefore more robust environmental assessment and management programs are needed, especially in more urbanized areas.

## 5. Technologies for the Removal of Estrogens

Water scarcity caused by population growth, economic development, current styles of consumption and climate change has become an urgent emergency. The growing demand for water from the industrial sector and, especially, from agriculture, which is the main water user (~70% of the global water demand), requires great attention to water resources [87]. In the next years, water demand will grow, particularly in countries with developing or emerging economies [29]. Pure water is precious for living beings and healthy agricultural productions, and its availability is not unlimited. Another important issue is the expected relevant increase in municipal wastewater due to rapid urbanization. Furthermore, the ever-increasing intensive anthropic activities continuously deteriorate the quality of the environment, and therefore timely and sustainable solutions that rely on the use of low-cost natural resources are required. These problems are emerging more and more widely in both industrialized and developing areas. The elimination of

EEs represents a serious economic and environmental challenge. Finding sustainable solutions for wastewater reclamation and reuse will reduce the risk of environmental contamination and provide an additional supply of water for agricultural, industrial and civil uses (Figure 1). A schematic overview of the current technologies available for the removal of EEs from water and soil is shown in Table 5.

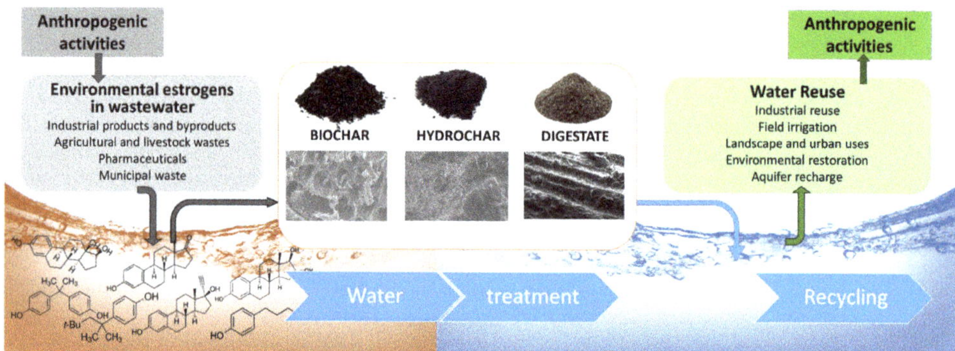


**Figure 1.** Water recycling.

**Table 5.** Methods for the removal of environmental estrogens (EEs) from water and soil.

| Method | Advantages | Limitations | EE | Ref. |
|---|---|---|---|---|
| | | **Water** | | |
| | | Immobilization | | |
| Coagulation/sedimentation | • High efficacy for the elimination of EEs having a logKow > 4 | | BPA, NP, E1, E2, E3 | [26] |
| | | Destruction | | |
| | | Chemical and Photochemical Treatments | | |
| Advanced oxidation processes (AOPs) Photocatalysis using catalysts and nanocatalysts ($TiO_2$, ZnO and others) Electrochemical oxidation | • High efficacy<br>• Low environmental impact<br>• Reduce estrogenic activity | • High costs<br>• Ex situ application<br>• Incomplete elimination<br>• Complexity | BPA, NP, E1, E2, E3 | [26]<br>[27]<br>[88] |
| Ozonation | • Used in tertiary treatments, increases the efficacy of other methods | • May generate byproducts more toxic and persistent than the parent compound | BPA, NP, E1, E2, E3 | [26] |
| Chlorination | • Used in tertiary treatments, increases the efficacy of other methods | • May generate byproducts more toxic and persistent than the parent compound | BPA, NP, E1, E2, E3 | [26] |
| UV photolysis | • High efficacy for StEE removal<br>• Enhancement of the efficacy of other methods | • Ineffective for the removal of other pollutants coexisting with EEs | BPA, NP, E1, E2, E3 | [26] |

**Table 5.** *Cont.*

| Method | Advantages | Limitations | EE | Ref. |
| --- | --- | --- | --- | --- |
| | | Biological Treatments | | |
| Activated sludge treatment | • High efficacy | | BPA, NP, E1, E2, E3 | [26] |
| Membrane bioreactor | • High removal efficacy<br>• Wide diversity of degrading microbial community | • High retention time | BPA, NP, E1, E2, E3 | [26] |
| | | Separation | | |
| Ultrafiltration | • High efficacy when used in combination with other methods and for hydrophobic compounds | | BPA, NP, E1, E2, E3 | [26] |
| | | Adsorption | | |
| Use of activated carbon | • High efficacy | • High cost | | |
| Use of recalcitrant carbon-rich biosorbents (biochar, hydrochar, digestate) | • Innovative approach<br>• Low cost and simplicity<br>• Very effective for hydrophobic compounds | • Lower sorptive performance than other expensive adsorbents | BPA, OP | * |
| | | **Soil** | | |
| | | Containment | | |
| Engineering techniques for the isolation of polluted sites and sources of contamination (containment barriers) | • Reduced leaching/transport into natural waters | • High costs<br>• High environmental impact<br>• High contaminant persistence | | [89] |
| | | Immobilization | | |
| Incorporation of recalcitrant carbon-rich materials, especially carbonaceous adsorbents, as soil amendments (biochar, hydrochar, digestate, compost) | • Reduced leaching/transport<br>• Low cost and simplicity<br>• Very effective for highly hydrophobic compounds<br>• Reduced bioavailability<br>• Increased soil fertility | • Possible desorption for less hydrophobic compounds<br>• Possible degradation of not sufficiently recalcitrant adsorbents | BPA, OP, E1 EE2 | [90] [51] [91] [92] [93] |
| | | Destruction | | |
| | | Chemical remediation | | |
| Advanced oxidation processes (AOPs, Fenton processes, $TiO_2$ photo-catalysis, ozonation, electrochemical oxidation | • Mineralization of contaminants | • High costs<br>• Ex situ application<br>• Incomplete elimination<br>• Complexity | | [89] |

**Table 5.** *Cont.*

| Method | Advantages | Limitations | EE | Ref. |
|---|---|---|---|---|
| | Biological remediation | | | |
| Phytoremediation (plants and algae) Assisted phytoremediation (combination of plants and soil amendments) | • Absorption by the root system and complete elimination by transformation in plant tissues<br>• Low cost and simplicity<br>• Environmentally friendly<br>• Increased soil fertility<br>• Less toxicity for plants using assisted phytoremediation | • Long-lasting process<br>• Use of plants resistant to contaminants<br>• Not suitable for heavily contaminated sites | | [89] [94] |
| Bioremediation (bacteria and fungi) | • Absorption by microbial cells and biodegradation<br>• Low cost and simplicity<br>• Environmentally friendly<br>• Reduced bioavailability | • Selection and inoculation of appropriate consortia of microorganisms or single strains | | [89] |
| | Separation | | | |
| Adsorption on carbon-rich biosorbents without incorporation in soil (biochar, hydrochar, digestate, compost) | • Significant removal<br>• Low cost and simplicity<br>• Environmentally friendly | • Incomplete removal<br>• Needs further study | BPA, OP | [37] [89] |
| Ex situ soil washing (extractants) | • Complete elimination of contaminants | • High costs<br>• High environmental impact<br>• Not suitable for all soil textures | | [89] |

* See references reported in tables of Section 6.2.

Nowadays, the removal of micropollutants such as EEs from the environment is not adequately carried out and perhaps not even sufficiently explored. Therefore, the detection of these compounds in ecosystems continues to be reported [21,22]. The processes conducted in the WWTPs aim mainly to remove the organic load and nutrients from wastewater. Conventional technologies studied and developed in recent years for the treatment of wastewaters and other environmental matrices are essentially based on chemical, physicochemical and microbiological processes. Depending on the nature of the contaminants, the applied processes can be classified into three categories: containment immobilization, separation and destruction [26,89]. Current treatments are based on flocculation, precipitation–filtration, adsorption on activated carbon, membrane filtration (micro-, ultra- and nanofiltration, reverse osmosis), advanced chemical and electrochemical oxidation and photocatalytic degradation [27,29]. Biological treatments are essentially based on the use of activated sludge, which is a mixture of microorganisms and suspended solids. Unfortunately, these techniques are not always resolutive, can generate hazardous byproducts and are extremely expensive.

Despite the numerous studies carried out on this matter, to date, there is no specific process that can be applied to completely remove EEs from wastewater [84], and tertiary-stage treatments usually include chlorine disinfection. Furthermore, multicontamination is the norm in water and wastewaters. The coexistence of many EEs in water is very likely due to intense agricultural and industrial activities and large volumes of wastewater discharged.

Furthermore, the co-occurrence of pollutants exerts highly dangerous synergistic adverse effects on animal health [95]. Gao et al. [28] have recently reviewed scientific information on the available eco-sustainable methods for the removal of EEs in wastewaters with a focus on photocatalysis and biodegradation.

An affordable and ecological solution to avoid, or at least limit, environmental pollution from EEs might be to combine emerging technologies and new biobased materials with existing technologies to improve their performance or overcome limitations such as high costs [13]. Based on this scenario, new sustainability-oriented strategies have been recently proposed to remove EEs from water and solid environmental matrices at affordable costs, including new biobased technologies for full-scale implementation. Biobased technologies appear to be an interesting economic alternative and easily meet people's consensus for their nature-friendly approach. Examples of biobased remediation technologies used for the removal of PhEEs and StEEs from environmental matrices are phytoremediation, which uses algae [96] and higher plants [94] to absorb and transform organic pollutants, and bioremediation, which uses microorganisms [97,98].

Adsorption techniques consist in the retention of organic pollutants present in solution on the surface of a solid adsorbent. Industrial adsorbents, such as activated carbon, carbonaceous resins and modified zeolites, have been widely used for removing single or multiple contaminants. However, the high costs of these synthetic adsorbents limit their use. Innovative solutions explored in recent years for wastewater remediation are based on the use of adsorbents deriving from both raw biowaste [99] and processed biowaste of bioenergy technologies [49,100]. In a medium-term perspective, and especially in low-income countries, the latter category of materials that are effectively waste from waste treatment could be valid competitors of much more expensive synthetic materials and replace them at least in part.

## 6. Sorptive Removal of Estrogens by Biosorbents

### 6.1. Adsorption Kinetics and Equilibrium Isotherms: Type of Interaction and Data Modeling

The characterization of biosorbents, in addition to providing useful information on their structural and functional properties, is useful for understanding the potential binding mechanisms with contaminants. The adsorption process can be generally classified as either physisorption (physical adsorption) or chemisorption (chemical adsorption). Physisorption is generally more common than chemisorption and occurs through weak intermolecular forces, such as van der Waals forces and hydrogen bonding, which do not involve a significant change in the electronic state of interacting species. Chemisorption includes valence forces, such as covalent or ionic bonds, between the binding sites of the adsorbent and solute. The main differences between physisorption and chemisorption relate to the enthalpy of adsorption, reversibility and layered arrangement (one or more layers) of the solute on the adsorbent. Physisorption is a low-enthalpy and reversible process that usually involves multiple layers of solute, while chemisorption is a high-enthalpy and irreversible adsorption that generally involves one monolayer of solute on the adsorbent. Generally, the adsorption of EEs on biosorbents occurs through more than one mechanism and is strongly dependent on the pH of the medium, which influences the electric state of the EE and the adsorbent. Furthermore, the type of prevailing bonds depends on the extent and type of functionalities of the material, which, in turn, depend on the parameters of the production process [100]. In general, low pyrolysis temperatures favor the formation of O-containing groups on BC, while high temperatures favor carbonization and aromatization reactions which make BC more suitable for adsorption of hydrophobic compounds [7]. BC interaction with PhEEs, such as BPA, and StEEs, such as EE2, has been mainly attributed to $\pi$–$\pi$ electron donor–acceptor binding, with the solute acting as $\pi$-donor, along with H bonds between phenols and polar sites of BC [52]. The mild temperature adopted in the HTC process is responsible for the high number of O-containing functional groups and the formation of H bonds with EEs, with both interacting species (HC and EE) generally acting either as acceptors or donors [52]. Other mechanisms involved in the EE–HC interaction are $\pi$–$\pi$

electron acceptor–donor bond, electrostatic interaction and hydrophobic interaction [49,53]. The molecular size of EEs is also an important property for their adsorption on porous chars, and the pore-filling mechanism can play an important role in this process [52]. Very little information on EE–DG interaction is available in the literature. Spectroscopic analyses of DG show the presence of carboxylic, alcoholic and phenolic –OH; amino groups; and aliphatic and aromatic C, suggesting the formation of the same bonds formed by BC and HC [12]. Further studies are needed to investigate the mechanisms of EE–DG interaction.

The adsorptive capacity of a biosorbent for an EE is quantified by studying adsorption kinetics and equilibrium isotherms. Modeling of experimental sorption data is a widely accepted approach to evaluate and compare the adsorption performance of different materials for a given molecule and the affinity of different molecules for the same material.

Kinetics data allow the estimation of the rate of adsorption and the equilibration time and are usually interpreted using various kinetic equations. Based on the type of model that provides the best correlation of experimental data, information can be obtained on the sorption mechanisms, i.e., physisorption or chemisorption. Theoretical kinetic models commonly used to interpret experimental data of EEs on biosorbents are the pseudo-first-order (PFO), the pseudo-second-order (PSO), Elovich's and intraparticle diffusion (IPD) models [43,52,101,102]. The equations and parameters of the four sorption kinetics models are shown in Table 6.

The PFO rate equation of Lagergren [103] is based on the adsorbent capacity and considers that a linear relationship occurs between the speed of occupation of the active sites and the number of sites available on the adsorbent [104]. The PSO kinetic model is based on the sorption at equilibrium and provides information on the type of interaction between the sorbent and the sorbate, such as valence bond and electron exchange [105]. According to the PSO model, the rate-limiting step of adsorption may be the chemisorption of the solute onto the adsorbent through covalent bonds. The Elovich model allows the evaluation of the existence of chemisorption and assumes that the adsorption rate of the solute decreases exponentially as the amount of adsorbed solute increases [106,107]. The IPD model is applied to know the limiting step of adsorption and assumes that the only interaction between solute and adsorbent is the internal diffusion [108]. Numerous studies demonstrated that the PSO kinetic model is the best fit for the adsorption of most EEs on BC [43,46,90] and HC [56–58].

Adsorption isotherms are performed to quantify the distribution of a solute between the adsorbent and the solution at equilibrium. The experimental equilibrium data are generally fitted into a sorption model to calculate the adsorption constants (Table 6). These models can also be useful for understanding the sorption mechanisms occurring at the solid–liquid interface. Two commonly used theoretical sorption isotherm models are the Freundlich and the Langmuir equations. The empirical nonlinear Freundlich model [109] fits well when the solute is adsorbed on heterogeneous substrates, while the Langmuir isotherm [110] is more appropriate for materials having a homogeneous surface and when the solute is adsorbed as a monolayer without interaction between solute molecules. The Freundlich model was the best interpretation of equilibrium sorption data of various EEs on BC [52,53,101], HC [52,58] and DG [12]. Differently, other works demonstrated that the Langmuir model was the most appropriate fit for EEs on BC [43,46,90] and HC [57,102].

**Table 6.** Models of adsorption kinetics and equilibrium isotherm for environmental estrogens (EEs) on biosorbents.

| Ref. | Model | Equation | Parameters |
|---|---|---|---|
| | | | Adsorption Kinetics Models |
| [103] | Pseudo-first order | $q_t = q_e\,(1 - exp^{-k1t})$ | $q_e$ and $q_t$ are the amount of solute adsorbed per mass unit of sorbent (mg $g^{-1}$) at equilibrium and at time $t$, respectively; $k_1$ ($h^{-1}$) and $k_2$ (kg $mg^{-1}$ $h^{-1}$) are the rate constants of sorption |
| [105] | Pseudo-second order | $q_t = \frac{q_e^2 k_2 t}{1+k_2 q_e t}$ | |
| [106] | Elovich | $\frac{dq}{dt} = ae^{-\alpha q}$ | $q$ is the amount of sorbate adsorbed at time $t$, $a$ is the desorption constant and $\alpha$ is the initial adsorption rate |
| [108] | Intraparticle diffusion | $qt = k_{id}\, t^{1/2} + C$ | $q_t$ is the amount of solute adsorbed per mass unit of sorbent (mg $g^{-1}$) at time $t$, $k_{id}$ (mg $g^{-1}$ $min^{-1/2}$) is the particle diffusion rate constant and $C$ (mg $g^{-1}$) is the intercept |
| | | | Adsorption Isotherm Models |
| [109] | Freundlich | $q_e = K_F\, C_e^{1/n}$ | $q_e$ (mg $g^{-1}$) is the amount of solute adsorbed per unit of substrate; $C_e$ (mg $mL^{-1}$) is the equilibrium concentration of the sorbate in solution. $1/n$ indicates the degree of nonlinearity between solution concentration and amount adsorbed, while the reciprocal $n$ is the sorption intensity, $K_F$ is the Freundlich adsorption constant, $b$ (mg $g^{-1}$) is the maximum monolayer adsorption, $K_L$ (mL $mg^{-1}$) is the Langmuir constant which expresses the energy of adsorption and $Kd$ (mL $g^{-1}$) is the distribution coefficient |
| [110] | Langmuir | $q_e = (K_L C_e b)/(1 + K_L C_e)$ | |
| [111] | Henry | $q_e = Kd\, C_e$ | |
| [112] | Temkin | $qe = B\, ln(A_T\, C_e)$ | $q_e$ (mg $g^{-1}$) is the amount of solute adsorbed per unit of substrate, $C_e$ (mg $mL^{-1}$) is the equilibrium concentration of sorbate in solution, $A_T$ is the Temkin isotherm equilibrium binding constant (mL $mg^{-1}$) and $B$ (J $mol^{-1}$) expresses the enthalpy of adsorption. $B = RT/b_T$, where $b_T$ is a constant related to the heat of adsorption, $T$ is the absolute temperature (K) and $R$ is the universal gas constant (8.314 J $mol^{-1}$ $K^{-1}$) |
| [113] | Dubinin–Radushkevich | $qe = (q_s)\, exp\,(-k_{ad}{}^2)$ | $q_s$ (mg $g^{-1}$) is the maximum sorbate adsorption; $k_{ad}$ is the Dubinin–Radushkevich constant ($mol^2/kJ^2$). $k_{ad}$ = A/E, where E (kJ/mol) is the energy associated |

Less common models used to fit equilibrium isotherm data are the Henry, the Temkin and the Dubinin–Radushkevich models (Table 6). The Henry equation assumes a constant proportion between the concentration of the sorbate in solution and that on the adsorbent and allows the calculation of the distribution coefficient, $Kd$ (mL $g^{-1}$), from the slope [111]. The Temkin isotherm is based on the interaction among adsorbate molecules on the adsorbent surface and predicts a logarithmic reduction in adsorptive sites and energy [112]. Finally, the Dubinin–Radushkevich model is widely used to describe the adsorption of molecules in microporous chars and assumes a different energy released in the physical and chemical adsorption [113]. Furthermore, to better compare the capacity of different biosorbents to retain a given EE, the partition coefficient $K_{OC}$, i.e., the quantity of adsorbed compound per unit of organic C of the adsorbent, is generally calculated according to the following: $K_{OC} = (K_d/(\%\ organic\ C)) \times 100$.

An important aspect of biosorption is the capacity of the material to retain/release the compound when external conditions change. Therefore, an adsorption study is usually complemented by a desorption study. The desorption data, i.e., the amounts of a compound that remain adsorbed at each desorption step and the corresponding equilibrium concentrations, are generally interpreted by the Freundlich equation to calculate the desorption parameters $K_{Fdes}$ and $1/n_{des}$. The hysteresis coefficient, H, is calculated from the ratio $H = (1/n_{des})/(1/n_{ads})$.

### 6.2. *Removal from Water and Soil*

#### 6.2.1. Biochar

Searching on the Scopus database for 'biochar' in the article title, abstract and keywords (AT&A&K) resulted in 21187 BC-based documents (articles, conference papers, reviews, book chapters, books and so on) listed since 2000 (queried on 10 January 2022), of which over 78% were published in the last five years. Most of these studies were carried out in China (45%), the United States (15%), India (6%) and Australia (6%), and the remaining ones were carried out in 139 other countries. A large number of these studies focused on the characterization, properties and applications of BC; searching for the combination 'biochar' and 'adsorption' or 'sorption' or 'removal' in the AT&A&K resulted in the identification of 8182 documents in 27 different subject areas, the main one being environmental science (69.2%). Upon limiting this latter search to the seven EEs considered in this review (BPA, OP, NP. E1, E2, E3 and EE2), about 140 documents were identified; as expected, most of them concern BPA (74 documents).

In all works, BC showed an excellent sorptive efficiency which suggests its valid use for water treatment and soil remediation. The addition of BC to soil produces multiple benefits, including the increase in recalcitrant organic matter, the retention of water and plant nutrients and the immobilization of contaminants. The last function is very important because it avoids the entry of contaminants, especially the less hydrophobic ones, into natural waters and into the food chain after plant absorption and accumulation in edible parts [51,114,115].

BC has been used as an adsorbent both unmodified and after a series of chemical and physical postproduction treatments (activation, functionalization, magnetization, engineering). Activation can empower BC performance by increasing its specific surface area, porosity, content of oxygenated functional groups and ability to retain pollutants with a wider range of hydrophobicity.

The mechanisms by which BC interacts with EEs depend on the type and number of chemical functional groups on its surface, which, in turn, vary according to its production conditions, above all pyrolysis or gasification temperature. In general, lower temperatures cause the formation of numerous O-containing groups on BC, while higher temperatures favor carbonization and aromatization reactions which make the material more suitable for retaining hydrophobic compounds [7].

The high aromaticity of BC can explain the strong retention of PhEEs, such as BPA, through the formation of $\pi$–$\pi$ electron donor–acceptor bonds with BPA acting as $\pi$-donor

(–OH substituted aromatic compounds), along with the interaction of phenols with polar aromatic cores of BC [52]. The occurrence of H bonding between BC and polar functional groups of PhEEs is also reported [40]. The smaller molecular size of PhEEs, compared to StEEs, would favor their access to sorption sites of BC. In the StEE–BC interaction, π–π electron donor–acceptor binding is considered to be the dominant mechanism, although H bonding and electrostatic attraction are also common depending on the surface charge of BC and the possible ionization of StEEs [100].

When BC interacts with mixtures of EEs, a positive correlation between the adsorption constants, or organic C-normalized adsorption constants, and the Kow values of EEs is commonly observed [53]. However, this correlation is often not statistically significant, which would suggest that the interaction may be governed by other important factors, such as the molecular size of the EE or the arrangement of functional groups on BC. Furthermore, pore filling is an important mechanism in the adsorption of hydrophobic compounds on BC, which is confirmed by the inverse relationship between the adsorption capacity of the char and the molecular diameter of the EE [52].

A selection of studies on the adsorption of EEs on BC shows that the PSO model is the preferred fit for kinetics data (Table 7), indicating the occurrence of chemisorption, via π–π electron donor–acceptor bonds and electrostatic interactions, between solute and sorbent.

In equilibrium isotherm studies, the experimental data follow different models depending on the type of BC and the EE considered (Table 7). In some studies, the Freundlich equation is the best fit for BPA [52,116,117], OP and NP [118], E1 [53,91], E2 [101] and EE2 [50,119] adsorption on BC. The Freundlich-type model is appropriate for adsorbents featuring heterogeneous surface and when solute molecules interact with each other and form a multilayer coverage on the adsorbent. In other studies, data of BPA [40,43,45,46], OP [120] and any StEE [43] seem better correlated with the Langmuir equation. This model is appropriate when the solute molecules form a monolayer on the homogeneous surface of the adsorbent and do not interact with each other. Furthermore, the activation of BC does not appear to change the preferred isotherm model for BPA [45,46] and E2 [44], compared to the unmodified material.

Freundlich adsorption constants ($K_F$) and Langmuir maximum adsorption capacity ($Q_{max}$) express the amount of solute adsorbed on the sorbent at equilibrium. Loffredo and Taskin [120] found Langmuir-type and Freundlich-type isotherms for sorption of OP and E2, respectively, on BC and PSO kinetic model for both EEs, which indicates chemisorption as the main interaction mechanism. Very high sorption capacity of BC for StEEs is documented in the scientific literature [44,53]. Ahmed et al. [43], investigating the sorption of BPA, E1, E2, E3 and EE2 on functionalized BC, found that the Langmuir equation and the PSO kinetic model were the preferred fit for all compounds and reported H bonds and π–π electron donor–acceptor binding as the main interaction mechanisms.

**Table 7.** Removal of different environmental estrogens (EEs) from water using biochar.

| EE | Feedstock | Process T (°C) | Residence Time (h) | Activation | Kinetics Model | Isotherm Model | Adsorption Capacity (mg $g^{-1}$) * | Ref. |
|---|---|---|---|---|---|---|---|---|
| BPA | Wheat straw | 400 | 2–7 | - | n. e. | F | 9.33 | [52] |
| BPA | Poultry litter | 400 | 2–7 | - | n. e. | F | 57.54 | [52] |
| BPA | Pine chips | 800 | 2 | - | n. e. | F | 9.22 | [116] |
| BPA | Eucalyptus wood | 600 | 2 | o$H_3PO_4$ | PSO | L | 47.65 | [43] |
| BPA | Alfalfa | 650 | 2 | - | PSO | F | 24.30 | [117] |
| BPA | Grapefruit peel | 400 | 2 | - | PSO | L | 123.83 | [45] |
| BPA | Grapefruit peel | 400 | 2 | γ-$Fe_2O_3$ | PSO | L | 342.47 | [45] |
| BPA | Wheat straw | 550, 700 | 0.75 | $CO_2$ | n. e. | L | 17.5, 14.2 | [121] |

Table 7. *Cont.*

| EE | Feedstock | Process T (°C) | Residence Time (h) | Activation | Kinetics Model | Isotherm Model | Adsorption Capacity (mg g$^{-1}$) * | Ref. |
|---|---|---|---|---|---|---|---|---|
| BPA | Wheat straw | 700 | 2 | NaOH | PSO | L | 71.42 | [40] |
| BPA | Sawdust | 800 | 1 | - | PSO | L | 5.08 | [46] |
| BPA | Sawdust | Various (from 500 to 900) | 1 | Fe and $N_2$ | PSO | L | From 0.9 to 54.0 | [46] |
| OP | Red spruce wood pellet | 550 | 3 | - | PSO | L | 1.79 | [120] |
| OP | Sawdust | 450, 650, 850 | 1 | - | n. e. | F | 0.56, 1.05, 0.63 | [118] |
| NP | Sawdust | 450, 650, 850 | 1 | - | n. e. | F | 1.50, 2.07, 1.07 | [118] |
| E1 | Swine solids | Various (from 250 to 600) | 2 | - | n. e. | F | From 1.71 to 665.18 | [53] |
| E1 | Poultry litter | Various (from 250 to 600) | 2 | - | n. e. | F | From 4.77 to 460.47 | [53] |
| E1 | Eucalyptus wood | 600 | 2 | o$H_3PO_4$ | PSO | L | 75.88 | [43] |
| E1 | Lychee fruits | 650 | 2 | - | PSO | F | 0.65 | [91] |
| E2 | Pine chips | 800 | 2 | - | n. e. | F | 30.20 | [116] |
| E2 | Red spruce wood | 550 | 3 | - | PSO | F | 3.385 | [120] |
| E2 | Rice straw | Various (from 400 to 600) | 2 | - | PSO | F | From 6.86 to 13.95 | [101] |
| E2 | Eucalyptus wood | 600 | 2 | o$H_3PO_4$ | PSO | L | 51.26 | [43] |
| E2 | Lotus seedpod | 650 | 2 | - | PSO | L | 135.74 | [44] |
| E2 | Lotus seedpod | 650 | 2 | KOH | PSO | L | 150.10 | [44] |
| E3 | Eucalyptus wood | 400 | 2 | o$H_3PO_4$ | PSO | L | 42.17 | [43] |
| EE2 | Wheat straw | 400 | 2–7 | - | n. e. | F | 29.51 | [52] |
| EE2 | Poultry litter | 400 | 2–7 | - | n. e. | F | 8.32 | [52] |
| EE2 | Peanut shell | 550 | 2 | - | n. e. | F | 80.44 | [119] |
| EE2 | Eucalyptus wood | 400 | 2 | o$H_3PO_4$ | PSO | L | 51.16 | [43] |

* Data refer to $K_F$ or $Q_{max}$ based on preferred Freundlich or Langmuir isotherm, respectively, at temperature of 25 ± 2 °C; $K_F$: Freundlich constant; L: Langmuir isotherm; F: Freundlich isotherm; PFO: pseudo-first order; PSO: pseudo-second order; n. e.: not evaluated; -: no treatment.

Due to its high sorption capacity, BC can also play an important role in soil remediation. The addition of BC to soil significantly increases its ability to retain EEs, thus limiting the danger of their leaching to the deeper soil layers and transport to natural waters. This important BC action has been demonstrated in soils naturally or artificially contaminated with BPA [90,92–94,122], OP [94], E1 and E2 [123] and EE2 [92,93]. Furthermore, the retention of EEs by BC in soil is strongly influenced by the level and type of natural organic matter, in particular the humic fraction, and by dissolved organic matter [122]. Recently, BC and other carbon-rich materials have been used to remove BPA and OP from soil in a new strategy that does not involve the incorporation of BC into soil. This approach consists in a preliminary adsorption of contaminants on BC which is subsequently removed from the soil and exposed to fungal activity for the degradation of contaminants [37]. Using this method, in just 2 days of soil–BC contact, up to 80% and 62% of OP and BPA, respectively, can be permanently removed from soil, and a large part of the contaminants (83% and 75% of OP and BPA, respectively) is successively eliminated through mycodegradation [37].

### 6.2.2. Hydrochar

The HTC process produces a carbonaceous material that exhibits good performance as an adsorbent of organic and inorganic contaminants in aqueous solutions. HTC is a relatively recent and less widespread technique compared to other thermochemical processes such as pyrolysis, gasification and combustion; therefore, in the scientific literature, the number of HC-based studies is much more limited than that of those concerning BC. The number of studies on HC surveyed by Scopus database (AT&A&K) is 1856, and these studies were published in 2009 or later (queried on 8 January 2022). Of all these studies, a very large percentage (82%) was published in the last 5 years, and 50% were published in the last 2 years. The majority of HC-based studies focus on the HTC process and on HC characterization and applications in different sectors. Combining the words 'hydrochar' and 'adsorption' or 'removal' and searching in AT&A&K resulted in 558 articles, of which only about 15 are related to PhEEs and StEEs. Therefore, although HC as an adsorbent has significantly attracted the attention of scientists, its potential for EE retention/removal is still very little explored.

The capacity of HC to retain pollutants strongly depends on the type of raw biomass used and on HTC parameters (temperature, residence time, pressure, raw material/liquid ratio). In general, the sorption efficiency of HC is lower than that of BC, which is due to a smaller specific surface area and lower porosity that provide fewer active sorption sites for pollutants. However, due to the presence of different types of functional groups on HC, this material is effective for the removal of both polar and nonpolar contaminants [52,53,102].

The mechanisms by which EEs are bound by HC are different and depend on the EE considered and the pH of the medium. They can be listed as π–π electron acceptor–donor bond, H bond, electrostatic interaction and hydrophobic interaction [49,53]. Due to its high content of oxygenated functional groups, HC can form H bonds by acting either as an acceptor or a donor. In addition, EEs such as BPA and EE2 can act either as acceptors or donors in H bond formation, which is supported by the inverse relationship between the Kow values of the compounds and their adsorption constants [52].

Although the adsorption efficiency of HC is more limited than that of other adsorbents, such as activated carbon or BC, its production and use are simple, eco-compatible and inexpensive, making HC attractive for wastewater treatment or immobilization of contaminants in soil.

Table 8 shows the main results of some studies on EE adsorption on HC, including the most appropriate kinetics and isotherm models and adsorption capacities. Most studies concern the adsorption of BPA on HC [49,52,56], which is undoubtedly due to the wide diffusion of BPA in both terrestrial and aquatic environments, although a significant number of studies concern StEEs. Some works investigated the sorption activity of unmodified HC [49,52,102], while others included a preliminary activation of HC [56].

**Table 8.** Removal of different environmental estrogens (EEs) from water using hydrochar and digestate.

| EE | Feedstock | Process T (°C) | Residence Time (h) | Activation | Kinetics Model | Isotherm Model | Adsorption Capacity ($mg\ g^{-1}$) * | Ref. |
|---|---|---|---|---|---|---|---|---|
| | | | | Hydrochar | | | | |
| BPA | Poultry litter | 250 | 20 | - | n. e. | F | 77.62 | [52] |
| BPA | Swine solids | 250 | 20 | - | n. e. | F | 33.11 | [52] |
| BPA | Magonia pubescens wood | 180 | 6 | - | PFO | L | 21.26 | [102] |
| BPA | Sewage sludge | 160, 190, 250 | 4 | - | n. e. | L | From 10.86 to 18.37 | [49] |
| BPA | Argan nut shells | 200 | 6 | - | PSO | L | 1162.79 | [67] |
| BPA | Pine fruit shells | 190 | 24, 48, 72 | NaOH | PSO | L | From 332.52 to 378.77 | [56] |
| E1 | Swine solids | 250 | 8 | - | n. e. | F | 100.48 | [53] |
| E1 | Poultry litter | 250 | 8 | - | n. e. | F | 51.88 | [53] |
| E2 | Rice husk | 200 | 6 | - | n. e. | F | 15.87 | [58] |
| E2 | Rice husk | 200 | 6 | $KMnO_4 + FeCl_2$ | PSO | F | 22.31 | [58] |
| E2 | Montmorillonite and rice husk | 180 | 16 | KOH | PSO | L | 138.00 | [57] |
| EE2 | Swine solids | 250 | 20 | - | n. e. | F | 18.62 | [52] |
| EE2 | Poultry litter | 250 | 20 | - | n. e. | F | 181.97 | [52] |
| EE2 | Montmorillonite and rice husk | 180 | 16 | KOH | PSO | L | 69.00 | [57] |

**Table 8.** *Cont.*

| EE | Feedstock | Process T (°C) | Residence Time (h) | Activation | Kinetics Model | Isotherm Model | Adsorption Capacity (mg $g^{-1}$) * | Ref. |
|---|---|---|---|---|---|---|---|---|
| | | | | Digestate | | | | |
| BPA | Mixed residues and olive pomace | 20–45 | 28 days | - | PFO/PSO | F | 0.13 | [12] |
| OP | Mixed residues and olive pomace | 20–45 | 28 days | - | PFO | F | 1.07 | [12] |

* Data refer to $K_F$ or $Q_{max}$ based on preferred Freundlich or Langmuir isotherm, respectively, at temperature of 25 ± 2 °C; $K_F$: Freundlich constant; L: Langmuir isotherm; F: Freundlich isotherm; PFO: pseudo-first order; PSO: pseudo-second order; n. e.: not evaluated.

In general, kinetics data of EEs on HC show a better correlation with the PSO model (Table 8), which suggests that chemisorption is the main adsorption mechanism involving prevalently π–π electron acceptor–donor bonds and electrostatic interactions. Differently, equilibrium isotherm data of EEs follow almost equally the Freundlich and Langmuir equations (Table 8), which indicates the importance of the type of HC considered. The Freundlich model yields the best correlation of adsorption data when the adsorbent has a heterogeneous surface and the solute forms a multilayer on the sorbent with multiple interactions among the solute molecules. The Langmuir isotherm is the most appropriate when the adsorbent has a homogeneous surface and the solute molecules do not interact with each other and form a monolayer on the adsorbent. Similar to what was observed in the studies on BC, the activation of HC does not appear to change the preferred isotherm model, compared to unmodified HC [58].

The sorption capacity of HC is quite variable as it greatly depends on the specific surface area, porosity and the number and type of functional groups present on HC. In a recent work, de Lima et al. [56] found a very high sorption capacity of BPA on a NaOH-activated HC. Likewise, a KOH-activated HC was very efficient in the removal of E2 and EE2 [57].

Like BC, HC, besides being applied for water treatment, can be used for soil remediation. In particular, HC added to soil can immobilize organic pollutants, especially those with high hydrophobicity, thus regulating their bioavailability for plants and microorganisms and limiting the leaching of the less hydrophobic ones. Recently, a pioneering study by Loffredo and Parlavecchia [37] tested HC, BC and spent coffee grounds to permanently remove some EDCs, including BPA and OP, from a loamy soil. Although in that study HC showed a lower removal capacity than BC, it clearly favored the subsequent mycodegradation of the contaminants more than BC [37].

The regeneration and reusability of adsorbents is a key criterion for practical applications in water treatment. This aspect was investigated by various researchers who tested different methodologies to separate the contaminants from the adsorbents. Common protocols adopted for the desorption of contaminants are based on the use of chemical agents, such as acids, bases, salts and organic solvents such as methanol and ethanol [56,67] or acetone [58]. Physical treatments, for example, a new HTC cycle [49], are also adopted for HC regeneration.

### 6.2.3. Digestate

In parallel with the development of the AD technology for biowaste recycling and bioenergy production, there has been an increase in scientific efforts to study and valorize the solid and liquid byproducts of AD in compliance with environmental and economic sustainability. According to the Scopus database and searching in AT&A&K, of all DG-based studies of the past ~70 years (3644 documents), 93% were published in the last decade (queried on 26 January 2022). Italy ranks first in the world for scientific production concerning DG, followed by China, the United States and Germany. Most of the entire scientific production on DG concerns the AD process and its optimization, biogas production, the use of DG as soil fertilizer and new research trends in DG valorization [124]. A significant number of works in recent years have focused on the thermochemical conversion of DG to produce BC or HC for environmental applications [36] or the bio-oxidative conversion of DG to obtain compost and vermicompost for agricultural applications [125,126].

Although agriculture is currently the end-user of DG, some aspects of this material deserve to be investigated and valorized with a view to potential environmental benefits. In particular, when DG is incorporated into soil, besides improving soil fertility and contributing to carbon sequestration, it plays the important role of interacting with both natural and xenobiotic organic compounds present in soil, thus modulating their bioavailability and reducing the risk of leaching/transport of contaminants to natural waters. The presence of pollutants such as EEs in soil depends on agricultural practices such as irrigation with wastewater or addition of contaminated biomass. Studies published so far on the interaction of DG with organic contaminants are limited, and those on DG–EE interaction are almost lacking. However, this is an important topic given the increasing extent of contaminated agricultural land and the ability shown by DG to significantly retain pollutants of different hydrophobicity [12].

Compared to other materials originating from biowaste such as BC and HC, DG shows a lower but not negligible adsorption capacity. The adsorption efficiency of DG is strictly related to the raw biomass used and the AD parameters adopted (temperature, retention time, microbial consortia and so on), which, in turn, determine the composition, microstructure, content and type of functional groups of DG. Exploring and valorizing DG as an innovative biosorbent is economically and ecologically interesting if one considers that DG has long been considered a hazardous waste to be disposed of. The first work (on Scopus database) on the use of DG for decontamination purposes was carried out by Mukherjee et al. [61], who used soil/DG biomixtures to obtain biobeds and biofilters to immobilize pesticides with contrasting physicochemical properties. Yao et al. [62] used an unmodified DG from OFMW for the removal of dyes from textile wastewater and reported the satisfactory performance of this material, thus demonstrating that DG could be a promising renewable and cost-effective decontaminant. A very recent study on adsorption/desorption of EEs on/from DG demonstrated that this material has a remarkable efficiency in removing BPA and, especially, OP from water and that the occurrence of a positive hysteresis guarantees a long-lasting retention of the pollutants [12].

## 7. Conclusions

Global concerns about the entry of dangerous organic pollutants, such as phenolic and steroidal EEs, into surface and groundwater and the urgent need to reclaim and reuse wastewaters have prompted the search for new sustainable strategies. A promising approach would be the use of biosorbents originating from the thermochemical and biological conversion of biowaste. In particular, coproducts and byproducts of bioenergy production, such as BC, HC and DG, show a relevant ability to remove organic pollutants, including EEs, from water and contribute significantly to soil remediation. The efficiency of these materials is related to their physicochemical properties and the type of contaminant, in particular its hydrophobicity. This review discusses the main aspects of these biosorbents, EEs and the related adsorption process both as it regards the interaction mechanisms and the modeling of the sorption data which allows the quantitative evaluation of the process. The main analytical methods used for the characterization of biosorbents and the evaluation of EE–biosorbent adsorption are also described. The remediation approach presented in this review could in perspective integrate with the other currently available strategies as it represents a valid and sustainable alternative to more sophisticated and expensive methods. However, to realize large-scale applications of these biosorbents, further studies are needed to make them competitive in terms of performance. The most modern and advanced characterization techniques and further improvement of the activation treatments of the materials will be extremely useful to achieve this goal, with great benefits for the economy and the environment. Finally, considering that these materials are 'waste from waste' at negligible costs and their performance as biosorbents is constantly improving, their valorization for sustainable use in water treatment and soil remediation is certainly convenient. Furthermore, it appears useful to highlight that the production and use of these materials are in line with the modern approach to sustainable resource management, carbon sequestration, implementation of the circular economy and environmental remediation.

**Funding:** This research was funded by the University of Bari Aldo Moro, Italy.

**Conflicts of Interest:** The author declares no conflict of interest.

## References

1. Schimmelpfennig, S.; Müller, C.; Grünhage, L.; Koch, C.; Kammann, C. Biochar, hydrochar and uncarbonized feedstock application to permanent grassland—Effects on greenhouse gas emissions and plant growth. *Agric. Ecosyst. Environ.* **2014**, *191*, 39–52. [CrossRef]
2. Igalavithana, A.D.; Ok, Y.S.; Usman, A.R.A.; Al-Wabel, M.I.; Oleszczuk, P.; Lee, S.S. The effects of biochar amendment on soil fertility. In *Agricultural and Environmental Applications of Biochar: Advances and Barriers*; Guo, M., He, Z., Uchimiya, S.M., Eds.; Soil Science Social America: Madison, WI, USA, 2016; pp. 123–144. [CrossRef]
3. Peng, W.; Pivato, A. Sustainable Management of Digestate from the Organic Fraction of Municipal Solid Waste and Food Waste Under the Concepts of Back to Earth Alternatives and Circular Economy. *Waste Biomass. Valoriz.* **2017**, *10*, 465–481. [CrossRef]
4. Rombolà, A.G.; Torri, C.; Vassura, I.; Venturini, E.; Reggiani, R.; Fabbri, D. Effect of biochar amendment on organic matter and dissolved organic matter composition of agricultural soils from a two-year field experiment. *Sci. Total Environ.* **2021**, *812*, 151422. [CrossRef] [PubMed]
5. Jian, X.; Zhuang, X.; Li, B.; Xu, X.; Wei, Z.; Song, Y.; Jiang, E. Comparison of characterization and adsorption of biochars produced from hydrothermal carbonization and pyrolysis. *Environ. Technol. Innov.* **2018**, *10*, 27–35. [CrossRef]
6. Cesaro, A. The valorization of the anaerobic digestate from the organic fractions of municipal solid waste: Challenges and perspectives. *J. Environ. Manag.* **2021**, *280*, 111742. [CrossRef]
7. Wang, Z.; Han, L.; Sun, K.; Jin, J.; Ro, K.; Libra, J.; Liu, X.; Xing, B. Sorption of four hydrophobic organic contaminants by biochars derived from maize straw, wood dust and swine manure at different pyrolytic temperatures. *Chemosphere* **2016**, *144*, 285–291. [CrossRef]
8. Opatokun, S.A.; Kan, T.; Al Shoaibi, A.; Srinivasakannan, C.; Strezov, V. Characterization of Food Waste and Its Digestate as Feedstock for Thermochemical Processing. *Energy Fuels* **2015**, *30*, 1589–1597. [CrossRef]
9. Sharma, H.B.; Sarmah, A.K.; Dubey, B. Hydrothermal carbonization of renewable waste biomass for solid biofuel production: A discussion on process mechanism, the influence of process parameters, environmental performance and fuel properties of hydrochar. *Renew. Sustain. Energy Rev.* **2020**, *123*, 109761. [CrossRef]
10. Zheng, H.; Zhang, C.; Liu, B.; Liu, G.; Zhao, M.; Xu, G.; Luo, X.; Li, F.; Xing, B. Biochar for water and soil remediation: Production, characterization, and application. In *A New Paradigm for Environmental Chemistry and Toxicology*; Jiang, G.B., Li, X.D., Eds.; Springer: Singapore, 2020; pp. 153–196. [CrossRef]
11. Fang, J.; Zhan, L.; Ok, Y.S.; Gao, B. Minireview of potential applications of hydrochar derived from hydrothermal carbonization of biomass. *J. Ind. Eng. Chem.* **2018**, *57*, 15–21. [CrossRef]
12. Loffredo, E.; Carnimeo, C.; Silletti, R.; Summo, C. Use of the Solid By-Product of Anaerobic Digestion of Biomass to Remove Anthropogenic Organic Pollutants with Endocrine Disruptive Activity. *Processes* **2021**, *9*, 2018. [CrossRef]
13. Wu, S.; Wu, H. Incorporating Biochar into Wastewater Eco-treatment Systems: Popularity, Reality, and Complexity. *Environ. Sci. Technol.* **2019**, *53*, 3345–3346. [CrossRef] [PubMed]
14. Zhang, Z.; Zhu, Z.; Shen, B.; Liu, L. Insights into biochar and hydrochar production and applications: A review. *Energy* **2019**, *171*, 581–598. [CrossRef]
15. European Commission (EC). Endocrine Disruptors. 2020. Available online: https://ec.europa.eu/environment/chemicals/endocrine/index_en.htm (accessed on 10 January 2022).
16. Cho, S.H.; Choi, Y.; Kim, S.H.; Kim, S.J.; Chang, J. Urinary bisphenol A versus serum bisphenol A concentration and ovarian reproductive outcomes among IVF patients: Which is a better biomarker of BPA exposure? *Mol. Cell. Toxicol.* **2017**, *13*, 351–359. [CrossRef]
17. Gao, Q.; Niu, Y.; Wang, B.; Liu, J.; Zhao, Y.; Zhang, J.; Wang, Y.; Shao, B. Estimation of lactating mothers' daily intakes of bisphenol A using breast milk. *Environ. Pollut.* **2021**, *286*, 117545. [CrossRef]
18. Ougier, E.; Zeman, F.; Antignac, J.-P.; Rousselle, C.; Lange, R.; Kolossa-Gehring, M.; Apel, P. Human biomonitoring initiative (HBM4EU): Human biomonitoring guidance values (HBM-GVs) derived for bisphenol A. *Environ. Int.* **2021**, *154*, 106563. [CrossRef]
19. Lei, K.; Pan, H.-Y.; Zhu, Y.; Chen, W.; Lin, C.-Y. Pollution characteristics and mixture risk prediction of phenolic environmental estrogens in rivers of the Beijing–Tianjin–Hebei urban agglomeration, China. *Sci. Total Environ.* **2021**, *787*, 147646. [CrossRef]

20. Zhao, J.-L.; Huang, Z.; Zhang, Q.-Q.; Ying-He, L.; Wang, T.-T.; Yang, Y.-Y.; Ying, G.-G. Distribution and mass loads of xenoestrogens bisphenol a, 4-nonylphenol, and 4-tert-octylphenol in rainfall runoff from highly urbanized regions: A comparison with point sources of wastewater. *J. Hazard. Mater.* **2020**, *401*, 123747. [CrossRef]
21. Corrales, J.; Kristofco, L.A.; Steele, W.B.; Yates, B.S.; Breed, C.S.; Williams, E.S.; Brooks, B.W. Global Assessment of Bisphenol A in the Environment: Review and Analysis of Its Occurrence and Bioaccumulation. *Dose-Response* **2015**, *13*, 1559325815598308. [CrossRef]
22. Staples, C.; van der Hoeven, N.; Clark, K.; Mihaich, E.; Woelz, J.; Hentges, S. Distributions of concentrations of bisphenol A in North American and European surface waters and sediments determined from 19 years of monitoring data. *Chemosphere* **2018**, *201*, 448–458. [CrossRef]
23. Russo, G.; Laneri, S.; Di Lorenzo, R.; Ferrara, L.; Grumetto, L. The occurrence of selected endocrine-disrupting chemicals in water and sediments from an urban lagoon in Southern Italy. *Water Environ. Res.* **2021**, *93*, 1944–1958. [CrossRef]
24. Wan, Y.-P.; Chai, B.-W.; Wei, Q.; Hayat, W.; Dang, Z.; Liu, Z.-H. 17α-ethynylestradiol and its two main conjugates in seven municipal wastewater treatment plants: Analytical method, their occurrence, removal and risk evaluation. *Sci. Total Environ.* **2021**, *812*, 152489. [CrossRef]
25. Qiu, W.; Liu, S.; Chen, H.; Luo, S.; Xiong, Y.; Wang, X.; Xu, B.; Zheng, C.; Wang, K.-J. The comparative toxicities of BPA, BPB, BPS, BPF, and BPAF on the reproductive neuroendocrine system of zebrafish embryos and its mechanisms. *J. Hazard. Mater.* **2021**, *406*, 124303. [CrossRef] [PubMed]
26. Ben, W.; Zhu, B.; Yuan, X.; Zhang, Y.; Yang, M.; Qiang, Z. Occurrence, removal and risk of organic micropollutants in wastewater treatment plants across China: Comparison of wastewater treatment processes. *Water Res.* **2018**, *130*, 38–46. [CrossRef] [PubMed]
27. De Luna, Y.; Bensalah, N. Review on the electrochemical oxidation of endocrine-disrupting chemicals using BDD anodes. *Curr. Opin. Electrochem.* **2021**, *32*, 100900. [CrossRef]
28. Gao, X.; Kang, S.; Xiong, R.; Chen, M. Environment-Friendly Removal Methods for Endocrine Disrupting Chemicals. *Sustainability* **2020**, *12*, 7615. [CrossRef]
29. Sousa, J.C.; Ribeiro, A.R.; Barbosa, M.O.; Pereira, M.F.; Silva, A. A review on environmental monitoring of water organic pollutants identified by EU guidelines. *J. Hazard. Mater.* **2018**, *344*, 146–162. [CrossRef]
30. Tursi, A. A review on biomass: Importance, chemistry, classification, and conversion. *Biofuel Res. J.* **2019**, *6*, 962–979. [CrossRef]
31. International Biochar Initiative (IBI). Standardized Product Definition and Product Testing Guidelines for Biochar that Is Used in Soil-Version 2.1. Published Online by International Biochar Initiative. 2015. Available online: https://www.biochar-international.org/wp-content/uploads/2018/04/IBI_Biochar_Standards_V2.1_Final.pdf (accessed on 5 December 2021).
32. Taskin, E.; Bueno, C.D.C.; Allegretta, I.; Terzano, R.; Rosa, A.H.; Loffredo, E. Multianalytical characterization of biochar and hydrochar produced from waste biomasses for environmental and agricultural applications. *Chemosphere* **2019**, *233*, 422–430. [CrossRef]
33. Singh, L.; Kalia, V.C. *Waste Biomass Management—A Holistic Approach*; Springer Science and Business Media LLC.: Berlin/Heidelberg, Germany, 2017.
34. Igalavithana, A.D.; Mandal, S.; Niazi, N.K.; Vithanage, M.; Parikh, S.J.; Mukome, F.N.D.; Rizwan, M.; Oleszczuk, P.; Al-Wabel, M.; Bolan, N.; et al. Advances and future directions of biochar characterization methods and applications. *Crit. Rev. Environ. Sci. Technol.* **2017**, *47*, 2275–2330. [CrossRef]
35. Kuzyakov, Y.; Bogomolova, I.; Glaser, B. Biochar stability in soil: Decomposition during eight years and transformation as assessed by compound-specific 14C analysis. *Soil Biol. Biochem.* **2014**, *70*, 229–236. [CrossRef]
36. Jiang, B.; Lin, Y.; Mbog, J.C. Biochar derived from swine manure digestate and applied on the removals of heavy metals and antibiotics. *Bioresour. Technol.* **2018**, *270*, 603–611. [CrossRef]
37. Loffredo, E.; Parlavecchia, M. Use of plant-based sorbents and mycodegradation for the elimination of endocrine disrupting chemicals from soil: A novel facile and low-cost method. *Environ. Technol. Innov.* **2021**, *21*, 101358. [CrossRef]
38. Yang, X.; Igalavithana, A.D.; Oh, S.-E.; Nam, H.; Zhang, M.; Wang, C.-H.; Kwon, E.E.; Tsang, D.; Ok, Y.S. Characterization of bioenergy biochar and its utilization for metal/metalloid immobilization in contaminated soil. *Sci. Total Environ.* **2018**, *640–641*, 704–713. [CrossRef] [PubMed]
39. Gascó, G.; Paz-Ferreiro, J.; Álvarez, M.; Saa, A.; Méndez, A. Biochars and hydrochars prepared by pyrolysis and hydrothermal carbonisation of pig manure. *Waste Manag.* **2018**, *79*, 395–403. [CrossRef] [PubMed]
40. Tang, Y.; Li, Y.; Zhan, L.; Wu, D.; Zhang, S.; Pang, R.; Xie, B. Removal of emerging contaminants (bisphenol A and antibiotics) from kitchen wastewater by alkali-modified biochar. *Sci. Total Environ.* **2021**, *805*, 150158. [CrossRef]
41. Hung, C.-Y.; Tsai, W.-T.; Chen, J.-W.; Lin, Y.-Q.; Chang, Y.-M. Characterization of biochar prepared from biogas digestate. *Waste Manag.* **2017**, *66*, 53–60. [CrossRef]
42. Qian, F.; Zhu, X.; Liu, Y.; Hao, S.; Ren, Z.J.; Gao, B.; Zong, R.; Zhang, S.; Chen, J. Synthesis, characterization and adsorption capacity of magnetic carbon composites activated by $CO_2$: Implication for the catalytic mechanisms of iron salts. *J. Mater. Chem. A* **2016**, *4*, 18942–18951. [CrossRef]
43. Ahmed, M.B.; Zhou, J.L.; Ngo, H.H.; Johir, A.H.; Sornalingam, K. Sorptive removal of phenolic endocrine disruptors by functionalized biochar: Competitive interaction mechanism, removal efficacy and application in wastewater. *Chem. Eng. J.* **2018**, *335*, 801–811. [CrossRef]

44. Liu, N.; Liu, Y.; Zeng, G.; Gong, J.; Tan, X.; Wen, J.; Liu, S.; Jiang, L.; Li, M.; Yin, Z. Adsorption of 17β-estradiol from aqueous solution by raw and direct/pre/post-KOH treated lotus seedpod biochar. *J. Environ. Sci.* **2020**, *87*, 10–23. [CrossRef]
45. Wang, J.; Zhang, M. Adsorption Characteristics and Mechanism of Bisphenol A by Magnetic Biochar. *Int. J. Environ. Res. Public Health* **2020**, *17*, 1075. [CrossRef]
46. Xu, L.; Wu, C.; Chai, C.; Cao, S.; Bai, X.; Ma, K.; Jin, X.; Shi, X.; Jin, P. Adsorption of micropollutants from wastewater using iron and nitrogen co-doped biochar: Performance, kinetics and mechanism studies. *J. Hazard. Mater.* **2021**, *424*, 127606. [CrossRef] [PubMed]
47. Rocha, L.S.; Pereira, D.; Sousa, Érika; Otero, M.; Esteves, V.I.; Calisto, V. Recent advances on the development and application of magnetic activated carbon and char for the removal of pharmaceutical compounds from waters: A review. *Sci. Total Environ.* **2020**, *718*, 137272. [CrossRef] [PubMed]
48. Reguyal, F.; Sarmah, A.K. Adsorption of sulfamethoxazole by magnetic biochar: Effects of pH, ionic strength, natural organic matter and 17α-ethinylestradiol. *Sci. Total Environ.* **2018**, *628–629*, 722–730. [CrossRef] [PubMed]
49. Chen, L.; Li, D.; Huang, Y.; Zhu, W.; Ding, Y.; Guo, C. Preparation of sludge-based hydrochar at different temperatures and adsorption of BPA. *Water Sci. Technol.* **2020**, *82*, 255–265. [CrossRef]
50. Yu, J.; Zhu, Z.; Zhang, H.; Di, G.; Qiu, Y.; Yin, D.; Wang, S. Hydrochars from pinewood for adsorption and nonradical catalysis of bisphenols. *J. Hazard. Mater.* **2020**, *385*, 121548. [CrossRef]
51. Parlavecchia, M.; Carnimeo, C.; Loffredo, E. Soil Amendment with Biochar, Hydrochar and Compost Mitigates the Accumulation of Emerging Pollutants in Rocket Salad Plants. *Water Air Soil Pollut.* **2020**, *231*, 1–12. [CrossRef]
52. Sun, K.; Ro, K.; Guo, M.; Novak, J.; Mashayekhi, H.; Xing, B. Sorption of bisphenol A, 17α-ethinyl estradiol and phenanthrene on thermally and hydrothermally produced biochars. *Bioresour. Technol.* **2011**, *102*, 5757–5763. [CrossRef]
53. Han, L.; Ro, K.S.; Sun, K.; Sun, H.; Wang, Z.; Libra, J.A.; Xing, B. New Evidence for High Sorption Capacity of Hydrochar for Hydrophobic Organic Pollutants. *Environ. Sci. Technol.* **2016**, *50*, 13274–13282. [CrossRef]
54. Fang, J.; Gao, B.; Zimmerman, A.R.; Ro, K.S.; Chen, J. Physically ($CO_2$) activated hydrochars from hickory and peanut hull: Preparation, characterization, and sorption of methylene blue, lead, copper, and cadmium. *RSC Adv.* **2016**, *6*, 24906–24911. [CrossRef]
55. Fang, J.; Gao, B.; Mosa, A.; Zhan, L. Chemical activation of hickory and peanut hull hydrochars for removal of lead and methylene blue from aqueous solutions. *Chem. Speciat. Bioavailab.* **2017**, *29*, 197–204. [CrossRef]
56. de Lima, H.H.C.; Llop, M.E.G.; dos Santos Maniezzo, R.; Moisés, M.P.; Janeiro, V.; Arroyo, P.A.; Guilherme, M.R.; Rinaldi, A.W. Enhanced removal of bisphenol A using pine-fruit shell-derived hydrochars: Adsorption mechanisms and reusability. *J. Hazard. Mater.* **2021**, *416*, 126167. [CrossRef]
57. Tian, S.-R.; Liu, Y.-G.; Liu, S.-B.; Zeng, G.-M.; Jiang, L.-H.; Tan, X.-F.; Huang, X.-X.; Yin, Z.-H.; Liu, N.; Li, J. Hydrothermal synthesis of montmorillonite/hydrochar nanocomposites and application for 17β-estradiol and 17α-ethynylestradiol removal. *RSC Adv.* **2018**, *8*, 4273–4283. [CrossRef]
58. Ning, Q.; Liu, Y.; Liu, S.; Jiang, L.; Zeng, G.; Zeng, Z.; Wang, X.; Li, J.; Kare, Z. Fabrication of hydrochar functionalized Fe–Mn binary oxide nanocomposites: Characterization and 17β-estradiol removal. *RSC Adv.* **2017**, *7*, 37122–37129. [CrossRef]
59. Braguglia, C.M.; Gallipoli, A.; Gianico, A.; Pagliaccia, P. Anaerobic bioconversion of food waste into energy: A critical review. *Bioresour. Technol.* **2018**, *248*, 37–56. [CrossRef]
60. Wang, W.; Lee, D.-J. Valorization of anaerobic digestion digestate: A prospect review. *Bioresour. Technol.* **2021**, *323*, 124626. [CrossRef] [PubMed]
61. Mukherjee, S.; Weihermüller, L.; Tappe, W.; Hofmann, D.; Köppchen, S.; Laabs, V.; Vereecken, H.; Burauel, P. Sorption–desorption behaviour of bentazone, boscalid and pyrimethanil in biochar and digestate based soil mixtures for biopurification systems. *Sci. Total Environ.* **2016**, *559*, 63–73. [CrossRef] [PubMed]
62. Yao, S.; Fabbricino, M.; Race, M.; Ferraro, A.; Pontoni, L.; Aimone, O.; Chen, Y. Study of the Digestate as an Innovative and Low-Cost Adsorbent for the Removal of Dyes in Wastewater. *Processes* **2020**, *8*, 852. [CrossRef]
63. Abdel-Fattah, T.M.; Mahmoud, M.E.; Ahmed, S.B.; Huff, M.D.; Lee, J.W.; Kumar, S. Biochar from woody biomass for removing metal contaminants and carbon sequestration. *J. Ind. Eng. Chem.* **2015**, *22*, 103–109. [CrossRef]
64. Ma, X.; Zhou, B.; Budai, A.; Jeng, A.; Hao, X.; Wei, D.; Zhang, Y.; Rasse, D. Study of Biochar Properties by Scanning Electron Microscope—Energy Dispersive X-Ray Spectroscopy (SEM-EDX). *Commun. Soil Sci. Plant Anal.* **2016**, *47*, 593–601. [CrossRef]
65. Missaoui, A.; Bostyn, S.; Belandria, V.; Cagnon, B.; Sarh, B.; Gökalp, I. Hydrothermal carbonization of dried olive pomace: Energy potential and process performances. *J. Anal. Appl. Pyrolysis* **2017**, *128*, 281–290. [CrossRef]
66. Ruan, X.; Liu, Y.; Wang, G.; Frost, R.L.; Qian, G.; Tsang, D.C. Transformation of functional groups and environmentally persistent free radicals in hydrothermal carbonisation of lignin. *Bioresour. Technol.* **2018**, *270*, 223–229. [CrossRef] [PubMed]
67. Zbair, M.; Bottlinger, M.; Ainassaari, K.; Ojala, S.; Stein, O.; Keiski, R.L.; Bensitel, M.; Brahmi, R. Hydrothermal Carbonization of Argan Nut Shell: Functional Mesoporous Carbon with Excellent Performance in the Adsorption of Bisphenol A and Diuron. *Waste Biomass Valoriz.* **2020**, *11*, 1565–1584. [CrossRef]
68. Fierro, J.; Martinez, E.J.; Rosas, J.G.; Fernández, R.A.; López, R.; Gomez, X. Co-Digestion of Swine Manure and Crude Glycerine: Increasing Glycerine Ratio Results in Preferential Degradation of Labile Compounds. *Water Air Soil Pollut.* **2016**, *227*, 1–13. [CrossRef]

69. Suman, S.; Panwar, D.S.; Gautam, S. Surface morphology properties of biochars obtained from different biomass waste. *Energy Sources Part A Recover. Util. Environ. Eff.* **2017**, *37*, 1007–1012. [CrossRef]
70. Senesi, N.; Loffredo, E. The chemistry of soil organic matter. In *Soil Physical Chemistry*, 2nd ed.; Sparks, D.L., Ed.; CRC Press: Boca Raton, FL, USA, 2018; pp. 239–370. [CrossRef]
71. Senesi, N.; Loffredo, E.; D'Orazio, V.; Brunetti, G.; Miano, T.M.; La Cava, P. Adsorption of pesticides by humic acids from organic amendments and soils. In *Humic Substances and Chemical Contaminants*; Clapp, C.E., Hayes, M.H.B., Senesi, N., Bloom, P.R., Jardine, P.M., Eds.; Soil Science Social America: Madison, WI, USA, 2015; pp. 129–153. [CrossRef]
72. Marra, R.; Vinale, F.; Cesarano, G.; Lombardi, N.; D'Errico, G.; Crasto, A.; Mazzei, P.; Piccolo, A.; Incerti, G.; Woo, S.L.; et al. Biochars from olive mill waste have contrasting effects on plants, fungi and phytoparasitic nematodes. *PLoS ONE* **2018**, *13*, e0198728. [CrossRef]
73. Metcalfe, C.; Bayen, S.; Desrosiers, M.; Muñoz, G.; Sauvé, S.; Yargeau, V. An introduction to the sources, fate, occurrence and effects of endocrine disrupting chemicals released into the environment. *Environ. Res.* **2022**, *207*, 112658. [CrossRef]
74. Boonnorat, J.; Treesubsuntorn, C.; Phattarapattamawong, S.; Cherdchoosilapa, N.; Seemuang-On, S.; Somjit, M.; Huadprom, C.; Rojviroon, T.; Jutakanoke, R.; Prachanurak, P. Effect of leachate effluent water reuse on the phytotoxicity and micropollutants accumulation in agricultural crops. *J. Environ. Chem. Eng.* **2021**, *9*, 106639. [CrossRef]
75. U.S. National Library of Medicine. PubChem Open Chemistry Database at the National Institutes of Health (NIH). Available online: https://pubchem.ncbi.nlm.nih.gov (accessed on 10 January 2022).
76. Huang, R.-P.; Liu, Z.-H.; Yuan, S.-F.; Yin, H.; Dang, Z.; Wu, P.-X. Worldwide human daily intakes of bisphenol A (BPA) estimated from global urinary concentration data (2000–2016) and its risk analysis. *Environ. Pollut.* **2017**, *230*, 143–152. [CrossRef]
77. Diao, P.; Chen, Q.; Wang, R.; Sun, D.; Cai, Z.; Wu, H.; Duan, S. Phenolic endocrine-disrupting compounds in the Pearl River Estuary: Occurrence, bioaccumulation and risk assessment. *Sci. Total Environ.* **2017**, *584–585*, 1100–1107. [CrossRef]
78. Li, Z.; Zhang, W.; Shan, B. The effects of urbanization and rainfall on the distribution of, and risks from, phenolic environmental estrogens in river sediment. *Environ. Pollut.* **2019**, *250*, 1010–1018. [CrossRef]
79. Wang, S.; Zhu, Z.; He, J.; Yue, X.; Pan, J.; Wang, Z. Steroidal and phenolic endocrine disrupting chemicals (EDCs) in surface water of Bahe River, China: Distribution, bioaccumulation, risk assessment and estrogenic effect on Hemiculter leucisculus. *Environ. Pollut.* **2018**, *243*, 103–114. [CrossRef] [PubMed]
80. Gu, Y.; Yu, J.; Hu, X.; Yin, D. Characteristics of the alkylphenol and bisphenol A distributions in marine organisms and implications for human health: A case study of the East China Sea. *Sci. Total Environ.* **2016**, *539*, 460–469. [CrossRef] [PubMed]
81. Olaniyan, L.W.B.; Okoh, O.O.; Mkwetshana, N.T.; Okoh, A.I. Environmental Water Pollution, Endocrine Interference and Ecotoxicity of 4-tert-Octylphenol: A Review. *Rev. Environ. Contam. Toxicol.* **2018**, *248*, 81–109. [CrossRef]
82. Adeel, M.; Song, X.; Wang, Y.; Francis, D.; Yang, Y. Environmental impact of estrogens on human, animal and plant life: A critical review. *Environ. Int.* **2017**, *99*, 107–119. [CrossRef] [PubMed]
83. Rodriguez-Navas, C.; Björklund, E.; Halling-Sørensen, B.; Hansen, M. Biogas final digestive byproduct applied to croplands as fertilizer contains high levels of steroid hormones. *Environ. Pollut.* **2013**, *180*, 368–371. [CrossRef]
84. Tang, Z.; Liu, Z.-H.; Wang, H.; Dang, Z.; Liu, Y. Occurrence and removal of 17α-ethynylestradiol (EE2) in municipal wastewater treatment plants: Current status and challenges. *Chemosphere* **2021**, *271*, 129551. [CrossRef] [PubMed]
85. Jarošová, B.; Erseková, A.; Hilscherová, K.; Loos, R.; Gawlik, B.M.; Giesy, J.P.; Bláha, L. Europe-wide survey of estrogenicity in wastewater treatment plant effluents: The need for the effect-based monitoring. *Environ. Sci. Pollut. Res.* **2014**, *21*, 10970–10982. [CrossRef]
86. Yu, W.; Du, B.; Yang, L.; Zhang, Z.; Yang, C.; Yuan, S.; Zhang, M. Occurrence, sorption, and transformation of free and conjugated natural steroid estrogens in the environment. *Environ. Sci. Pollut. Res.* **2019**, *26*, 9443–9468. [CrossRef]
87. UNESCO. The United Nations World Water Development Report 2018: Nature-Based Solutions for Water. 2018. Available online: https://www.unwater.org/publications/world-water-development-report-2018/ (accessed on 11 January 2022).
88. Javaid, R.; Qazi, U.Y.; Ikhlaq, A.; Zahid, M.; Alazmi, A. Subcritical and supercritical water oxidation for dye decomposition. *J. Environ. Manag.* **2021**, *290*, 112605. [CrossRef]
89. Morillo, E.; Villaverde, J. Advanced technologies for the remediation of pesticide-contaminated soils. *Sci. Total Environ.* **2017**, *586*, 576–597. [CrossRef]
90. Liu, J.; Liu, L.; Shu, Y.; Jiang, S.; Huang, R.; Jia, Z.; Wei, D. Effect of ageing process on bisphenol A sorption and retention in agricultural soils amended with biochar. *Environ. Sci. Pollut. Res.* **2020**, *27*, 17401–17411. [CrossRef] [PubMed]
91. Tao, H.-Y.; Ge, H.; Shi, J.; Liu, X.; Guo, W.; Zhang, M.; Meng, Y.; Li, X.-Y. The characteristics of oestrone mobility in water and soil by the addition of Ca-biochar and Fe–Mn-biochar derived from Litchi chinensis Sonn. *Environ. Geochem. Health* **2020**, *42*, 1601–1615. [CrossRef] [PubMed]
92. Xu, N.; Zhang, B.; Tan, G.; Li, J.; Wang, H. Influence of biochar on sorption, leaching and dissipation of bisphenol A and 17α-ethynylestradiol in soil. *Environ. Sci. Process. Impacts* **2015**, *17*, 1722–1730. [CrossRef] [PubMed]
93. Zhang, B.; Li, J.; Tan, G.; Xu, N. Influence of biochar on the sorption of bisphenol A and 17α-ethynylestradiol in soil. *Chin. J. Environ. Eng.* **2016**, *10*, 5255–5261. [CrossRef]
94. Loffredo, E.; Picca, G.; Parlavecchia, M. Single and combined use of Cannabis sativa L. and carbon-rich materials for the removal of pesticides and endocrine-disrupting chemicals from water and soil. *Environ. Sci. Pollut. Res.* **2021**, *28*, 3601–3616. [CrossRef]

95. Altenburger, R.; Scholze, M.; Busch, W.; Escher, B.I.; Jakobs, G.; Krauss, M.; Krüger, J.; Neale, P.; Ait-Aissa, S.; Almeida, A.; et al. Mixture effects in samples of multiple contaminants—An inter-laboratory study with manifold bioassays. *Environ. Int.* **2018**, *114*, 95–106. [CrossRef]
96. Hom-Diaz, A.; Llorca, M.; Rodriguez-Mozaz, S.; Vicent, T.; Barceló, D.; Blánquez, P. Microalgae cultivation on wastewater digestate: β-estradiol and 17α-ethynylestradiol degradation and transformation products identification. *J. Environ. Manag.* **2015**, *155*, 106–113. [CrossRef]
97. Castellana, G.; Loffredo, E. Simultaneous Removal of Endocrine Disruptors from a Wastewater Using White Rot Fungi and Various Adsorbents. *Water Air Soil Pollut.* **2014**, *225*, 1872. [CrossRef]
98. Madadi, R.; Bester, K. Fungi and biochar applications in bioremediation of organic micropollutants from aquatic media. *Mar. Pollut. Bull.* **2021**, *166*, 112247. [CrossRef]
99. Alves, T.C.; Mota, J.A.X.; Pinheiro, A. Biosorption of organic micropollutants onto lignocellulosic-based material. *Water Sci. Technol.* **2020**, *82*, 427–439. [CrossRef]
100. Peiris, C.; Nawalage, S.; Wewalwela, J.J.; Gunatilake, S.R.; Vithanage, M. Biochar based sorptive remediation of steroidal estrogen contaminated aqueous systems: A critical review. *Environ. Res.* **2020**, *191*, 110183. [CrossRef]
101. Wang, X.; Liu, N.; Liu, Y.; Jiang, L.; Zeng, G.; Tan, X.; Liu, S.; Yin, Z.; Tian, S.; Li, J. Adsorption Removal of 17β-Estradiol from Water by Rice Straw-Derived Biochar with Special Attention to Pyrolysis Temperature and Background Chemistry. *Int. J. Environ. Res. Public Health* **2017**, *14*, 1213. [CrossRef] [PubMed]
102. Antero, R.V.P.; Alves, A.C.F.; Sales, P.D.T.F.; De Oliveira, S.B.; Ojala, S.A.; Brum, S.S. A new approach to obtain mesoporous-activated carbon via hydrothermal carbonization of Brazilian Cerrado biomass combined with physical activation for bisphenol-A removal. *Chem. Eng. Commun.* **2019**, *206*, 1498–1514. [CrossRef]
103. Lagergren, S. Zur theorie der sogenannten adsorption gelöster stoffe, Kungliga Svenska Vetenskapsakademiens. *Handlingar* **1898**, *24*, 1–39.
104. Kumar, K.V. Linear and non-linear regression analysis for the sorption kinetics of methylene blue onto activated carbon. *J. Hazard. Mater.* **2006**, *137*, 1538–1544. [CrossRef]
105. Ho, Y.-S. Second-order kinetic model for the sorption of cadmium onto tree fern: A comparison of linear and non-linear methods. *Water Res.* **2006**, *40*, 119–125. [CrossRef]
106. McLintock, I.S. The Elovich Equation in Chemisorption Kinetics. *Nature* **1967**, *216*, 1204–1205. [CrossRef]
107. Kajjumba, G.W.; Emik, S.; Öngen, A.H.; Özcan, K.; Aydın, S. Modelling of adsorption kinetic processes—Errors, theory and application. In *Advanced Sorption Process Applications*; Edebali, S., Ed.; IntechOpen: London, UK, 2018; Available online: https://www.intechopen.com/chapters/63161 (accessed on 10 January 2022). [CrossRef]
108. Weber, W.J., Jr.; Morris, J.C. Kinetics of Adsorption on Carbon from Solution. *J. Sanit. Eng. Div.* **1963**, *89*, 31–59. [CrossRef]
109. Freundlich, H.M.F. Over the adsorption in solution. *J. Phys. Chem.* **1906**, *57*, 385–470.
110. Langmuir, I. The adsorption of gases on plane surfaces of glass, mica and platinum. *J. Am. Chem. Soc.* **1918**, *40*, 1361–1403. [CrossRef]
111. Barrer, R.M.; Rees, L.V.C. Henry's law adsorption constants. *Trans. Faraday Soc.* **1961**, *57*, 999–1007. [CrossRef]
112. Alothman, Z.A.; Badjah, A.Y.; Ali, I. Facile synthesis and characterization of multi walled carbon nanotubes for fast and effective removal of 4-tert-octylphenol endocrine disruptor in water. *J. Mol. Liq.* **2019**, *275*, 41–48. [CrossRef]
113. Nguyen, C.; Do, D. The Dubinin–Radushkevich equation and the underlying microscopic adsorption description. *Carbon* **2001**, *39*, 1327–1336. [CrossRef]
114. Eibisch, N.; Schroll, R.; Fuß, R.; Mikutta, R.; Helfrich, M.; Flessa, H. Pyrochars and hydrochars differently alter the sorption of the herbicide isoproturon in an agricultural soil. *Chemosphere* **2015**, *119*, 155–162. [CrossRef] [PubMed]
115. Parlavecchia, M.; D'Orazio, V.; Loffredo, E. Wood biochars and vermicomposts from digestate modulate the extent of adsorption-desorption of the fungicide metalaxyl-m in a silty soil. *Environ. Sci. Pollut. Res.* **2019**, *26*, 35924–35934. [CrossRef]
116. Kim, E.; Jung, C.; Han, J.; Her, N.; Park, C.M.; Jang, M.; Son, A.; Yoon, Y. Sorptive removal of selected emerging contaminants using biochar in aqueous solution. *J. Ind. Eng. Chem.* **2016**, *36*, 364–371. [CrossRef]
117. Choi, Y.-K.; Kan, E. Effects of pyrolysis temperature on the physicochemical properties of alfalfa-derived biochar for the adsorption of bisphenol A and sulfamethoxazole in water. *Chemosphere* **2019**, *218*, 741–748. [CrossRef]
118. Del Bubba, M.; Anichini, B.; Bakari, Z.; Bruzzoniti, M.C.; Camisa, R.; Caprini, C.; Checchini, L.; Fibbi, D.; El Ghadraoui, A.; Liguori, F.; et al. Physicochemical properties and sorption capacities of sawdust-based biochars and commercial activated carbons towards ethoxylated alkylphenols and their phenolic metabolites in effluent wastewater from a textile district. *Sci. Total Environ.* **2020**, *708*, 135217. [CrossRef]
119. Zhou, J.; Chen, H.; Huang, W.; Arocena, J.M.; Ge, S. Sorption of Atrazine, 17α-Estradiol, and Phenanthrene on Wheat Straw and Peanut Shell Biochars. *Water Air Soil Pollut.* **2016**, *227*, 1–13. [CrossRef]
120. Loffredo, E.; Taskin, E. Adsorptive removal of ascertained and suspected endocrine disruptors from aqueous solution using plant-derived materials. *Environ. Sci. Pollut. Res.* **2017**, *24*, 19159–19166. [CrossRef]
121. Kozyatnyk, I.; Oesterle, P.; Wurzer, C.; Mašek, O.; Jansson, S. Removal of contaminants of emerging concern from multicomponent systems using carbon dioxide activated biochar from lignocellulosic feedstocks. *Bioresour. Technol.* **2021**, *340*, 125561. [CrossRef] [PubMed]

122. Deng, Y.; Yan, C.; Nie, M.; Ding, M. Bisphenol A adsorption behavior on soil and biochar: Impact of dissolved organic matter. *Environ. Sci. Pollut. Res.* **2021**, *28*, 32434–32445. [CrossRef] [PubMed]
123. Mann, S.; Qi, Z.; Prasher, S.O.; Li, L.; Gui, D.; Jiang, Q. Effect of biochar amendment on soil's retention capacity for estro-genic hormones from poultry manure treatment. *Front. Agric. Sci. Eng.* **2017**, *4*, 208–219. [CrossRef]
124. Selvaraj, P.S.; Periasamy, K.; Suganya, K.; Ramadass, K.; Muthusamy, S.; Ramesh, P.; Bush, R.; Vincent, S.G.T.; Palanisami, T. Novel resources recovery from anaerobic digestates: Current trends and future perspectives. *Crit. Rev. Environ. Sci. Technol.* **2020**, *in press*. [CrossRef]
125. Wang, N.; Huang, D.; Shao, M.; Sun, R.; Xu, Q. Use of activated carbon to reduce ammonia emissions and accelerate humification in composting digestate from food waste. *Bioresour. Technol.* **2022**, *347*, 126701. [CrossRef] [PubMed]
126. Le Pera, A.; Sellaro, M.; Bencivenni, E.; D'Amico, F. Environmental sustainability of an integrate anaerobic digestion-composting treatment of food waste: Analysis of an Italian plant in the circular bioeconomy strategy. *Waste Manag.* **2022**, *139*, 341–351. [CrossRef]

MDPI
St. Alban-Anlage 66
4052 Basel
Switzerland
Tel. +41 61 683 77 34
Fax +41 61 302 89 18
www.mdpi.com

*Materials* Editorial Office
E-mail: materials@mdpi.com
www.mdpi.com/journal/materials

www.ingramcontent.com/pod-product-compliance
Lightning Source LLC
LaVergne TN
LVHW071550170726
843515LV00008B/2257

* 9 7 8 3 0 3 6 5 5 2 3 4 7 *